Precalculus Mathematics

THOMAS W. HUNGERFORD

RICHARD MERCER

SYBIL R. BARRIER

UNIVERSITY OF WASHINGTON

1980 SAUNDERS COLLEGE

PHILADELPHIA

Library of Congress Cataloging in Publication Data
Hungerford, Thomas W.
 Precalculus Mathematics: Algebra, Trigonometry and
 Elementary Functions
 Includes index.
 1. Mathematics—1961– I. Mercer, Richard,
joint author. II. Barrier, Sybil R., joint author.
III. Title.
QA39.2.H857   512′.1   79-23510
ISBN 0-03-020346-5
Printed in the United States of America
0 1 2 3 032 9 8 7 6 5 4 3 2 1

TO OUR PARENTS

Preface

This book is intended for students who have had two to three years of high school mathematics. It is designed to provide the essential mathematical background needed in calculus and other college level courses involving mathematics or quantitative methods. It is suitable for use in courses such as college algebra, trigonometry, algebra and trigonometry, elementary functions, and precalculus mathematics.

Although there are many books on the market that treat algebra and trigonometry, we were unable to find one that satisfactorily met the needs of our students. Our classroom experience has led to two inescapable conclusions:

1. It is not sufficient merely to present the necessary technical tools, without explaining both how and why they work *in language that the student understands.*
2. Many students taking a course such as this at the college level simply do *not* understand the terse "definition-example-theorem-proof" style of many mathematics texts. Such an approach does not appear to them as reasonable and logical but rather as an arbitrary, unreasonable, and often incomprehensible foreign language.

We have attempted to write a book that deals effectively with these facts, one that helps students discover that mathematics can be both comprehensible and reasonable. As a first step toward this goal, the text is designed to be understood by an average student with a minimal amount of outside assistance. This has been accomplished without any sacrifice of rigor. But excessive and unnecessary formalism has been omitted in both format and content. We have done our best to present sound mathematics in an informal manner that stresses

detailed explanations of basic ideas, techniques and results;

extensive use of pictures and diagrams;

the fundamental reasons *why* a given technique works, as well as numerous examples showing *how* it is used;

the origins of many concepts (especially functions) in the real world and the ways that the common features of such real life situations are abstracted to obtain a mathematical definition.

There is, alas, no way to do this in a short space. A given topic may well occupy more space here than in some other texts, but this length is misleading. We have found that

> **because of the detailed explanations and numerous examples, a typical student can usually read the longer discussion here more easily and with greater understanding than a terse, compact presentation elsewhere.**

So in the long run the extra length should cause no serious scheduling problems. Most important, it provides the added benefit of greater understanding.

Another reason for the length of the book is that it has been designed for the instructor's convenience: it is extremely flexible and adaptable to a wide variety of courses. In many cases entire chapters or sections within chapters may be omitted, shortened, or covered in several different orders *without* impairing the book's readability by students.

A complete discussion of the possible ways of using this text, including:

the interdependence of the various chapters,
the interdependence of the sections within each chapter,
sample course outlines,
section by section pedagogical comments,
numerous exam questions, and answers to exercises

is given in the Instructor's Manual. It is available on request from the publisher.

Since calculators are now a fact of mathematical life, the book discusses both the uses and limitations of calculators in the study of functions. A student does not need a calculator to use the text. Exercises involving calculators are clearly labeled. But those students who do have one (the majority in our experience) will benefit from seeing that calculators are not a substitute for learning the underlying theory but can make that theory much easier to deal with in practical problems.

There are an unusually large number of exercises of widely varying types. The exercises labeled A are routine drills designed primarily to develop algebraic and manipulative skills. The B exercises are somewhat less mechanical and may occasionally require some careful thought. But any student who has read the text carefully should be able to do the great majority of the B exercises. *The C label is used for exercises that are unusual* for one reason or another. A few of the C exercises are difficult mathematically, but most of them should be well within reach of most students.

Finally, there are scattered throughout the text sections labeled DO IT YOURSELF! Some of these include topics that are not absolutely essential but which some instructors may want to include as a regular part of the course. Others provide interesting mathematical background or useful applications of the topics discussed in the text. Still others are needed by some students but not by others. Although they vary in level of difficulty, all of them can be read by students on their own.

T. W. H.
R. E. M.
S. R. B.

Seattle, Washington
September, 1979

Acknowledgments

The first version of this book was the joint work of all three authors, with some crucial assistance from our colleague Caspar Curjel. The manuscript was written during the summer of 1975 with financial support from the Innovative Fund and the Department of Mathematics of the University of Washington. It would never have appeared on time had it not been for the fast and efficient typing of Geri Button, Sonja Ogle, Judy Rieben, and Laurie Walton Elmer. Special mention must be given to Alison Ogle, then seven months old, whose cheerful smile brightened many a hectic day during that summer.

Based on classroom experience during the next three years, the book was completely revised and rewritten, section by section. This rewriting was done primarily by Thomas W. Hungerford, with part-time assistance from Richard Mercer. Due to other commitments, Sybil Barrier left the project at the end of 1975 and did not participate in the rewriting. The successive revisions were ably typed by Diane Hamm, Cristina Ignacio, Debbie Kolb, Judy Rieben, and Anita Tabares.

Many of our colleagues and students, who have used the various preliminary editions of *Precalculus Mathematics*, have offered helpful comments and advice. Vivian Klein provided some useful suggestions about our treatment of slopes. Steve Monk and Caspar Curjel wrote many of the more innovative graph reading problems. We thank them all. But very special thanks are due to Caspar Curjel. He suggested the basic approach to functions and functional notation and even wrote the original version of this material. The book has benefited greatly from his pedagogical insight.

Finally, Thomas W. Hungerford wants to acknowledge the continual interest and support of his wife Mary and their children Anne and Tom. He is deeply grateful to them.

Contents

To the Student

If you want to succeed in this course, remember that *mathematics is not a spectator sport*. You can't learn math simply by listening to your instructor lecture or work problems, by looking at examples in the text, by reading the answers in the back of the book, or by borrowing your neighbor's homework. You have to take an active role, using pencil and paper and working out many problems yourself. And you *can* do this—even if you haven't taken math for a while or if you're a bit afraid of math—by making wise use of your chief resources: your instructor and this book.

When it comes to math textbooks, many students make a serious mistake, they use their books only for finding out what the homework problems are. If they get stuck on a problem, they page back through the text until they find a similar example. If the example doesn't clarify things, they may try reading part of the text (as little as possible). On a really bad day they may end up reading most of the section—piecemeal, from back to front. Rarely, if ever, do such students read through an entire section (or subsection) from beginning to end.

If this description fits you, don't feel guilty. Some mathematics texts *are* unreadable. But don't use your bad past experiences as an excuse for not reading this book. It has been classroom tested for years by students like yourself. Some parts were rewritten several times to improve their clarity. Consequently we can assure you that this book is readable and understandable by an average student, with a minimal amount of outside assistance. So if you want to get the most out of this course, we strongly suggest that you follow these guidelines:

1. Read the pages assigned by your instructor from beginning to end *before* starting the homework problems.
2. You can't read a math book as you would read a newspaper or a novel, so go slowly and carefully. If you find calculations you don't understand, take pencil and paper and try to work them out. If you don't understand a particular statement, reread the preceding material to see if you missed something.
3. Don't get bogged down on the first reading. If you have spent a reasonable amount of time trying to figure something out, mark the place with a question mark and continue reading.
4. When you've read through the assigned material once, go back and reread the parts you marked with question marks. You will often find that you can now understand many of them. Plan to ask your instructor about the rest.
5. *Now* do the homework problems. You should be able to do all, or almost all, of the assigned problems. After you've worked at the homework for a reasonable amount of time and answered as many problems as you can, mark the exercises that are still causing trouble. Plan to ask your instructor about them.

If you follow these five guidelines, you will get the most out of this book. But no book can answer all your questions. That's why your instructor is there. Unfortunately, many students are afraid to ask questions in class for fear that the questions will seem "dumb." Such students should remember this:

> **If you have honestly followed the five guidelines above and still have unanswered questions, then there are *at least* six other students in your class who have the same questions.**

So it's not a dumb question. Furthermore, your instructor will welcome questions that arise from a serious effort on your part. In any case, your instructor is being paid (with your tuition money) to answer questions. So do yourself a favor and get your money's worth—*ask questions.*

Preliminaries

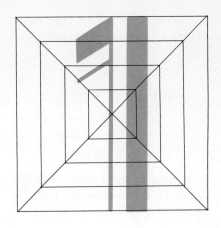

In grade school and high school mathematics, you were introduced to the system of real numbers. These are the numbers used in everyday life. The first part of this chapter summarizes the basic facts about the real numbers that will be needed in this course and in calculus.

1. THE REAL NUMBER SYSTEM

The most familiar real numbers are the **integers** (or whole numbers), that is, the numbers $0, 1, -1, 2, -2, 3, -3, \ldots$, etc. The numbers $1, 2, 3, 4, \ldots$ are called the natural numbers or the positive integers.

The real number system also includes all **rational numbers,** that is, all fractions r/s, where r and s are integers and $s \neq 0$. Each of the following is a rational number:

$$\tfrac{1}{2}, \qquad -\tfrac{4}{3}, \qquad \tfrac{99642}{716}, \qquad -\tfrac{1400}{2} = -\tfrac{700}{1} = 700, \qquad 3\tfrac{5}{16} = \tfrac{53}{16} = 3.3125, \qquad -.07 = -\tfrac{7}{100}$$

The word "rational" in this context has no psychological implications. It refers to the ratio or quotient of two integers.

A crucial fact is that

Some real numbers are *not* rational numbers.

For example, consider a right triangle° with two equal sides of length 1 as shown in Figure 1-1.

Figure 1-1

According to the Pythagorean Theorem,° the length of the hypotenuse° is a real number c that satisfies the equation $c^2 = 1^2 + 1^2 = 2$. (The number c is called the square root of 2 and is denoted $\sqrt{2}$.) We claim that

$$c = \sqrt{2} \text{ is } not \text{ a rational number.}$$

In other words, it is not possible to find integers a and b such that $(a/b)^2 = 2$.

A proof of this claim is given in Exercise C.1 on page 200. For now you can convince yourself of the *plausibility* of this claim by *trying* to find a rational number whose square is 2. (Of course, you won't succeed.) For example, if we square $\frac{1414}{1000}$ ($= 1.414$), we obtain 1.999396, which is *close* to 2, but not exactly *equal* to 2. It doesn't take much of this to convince most people that no matter what rational number they square, the answer will never be *exactly* 2.°° In other words, $\sqrt{2}$ cannot be a rational number.

A real number that is *not* a rational number (such as $\sqrt{2}$) is called an **irrational number** (ir-rational = not rational = not a ratio). Another well-known irrational number is the number pi (π) used to calculate the area of a circle. In grade school you probably used $\frac{22}{7}$ or 3.1416 as the number π. But these rational numbers are just *approximations* of π (close to, but not quite *equal* to, π).

Although we have only mentioned two irrational numbers, there are in fact infinitely many of them. Rationals and irrationals are further discussed in the DO IT YOUR-SELF! segment at the end of this section.

THE NUMBER LINE

The real number system can be pictured geometrically via the real **number line** (or **coordinate line**). Take a straight line and choose a point on it; label this point 0 and call it the **origin**. Now choose some unit of measurement and label the point that is one unit to the *right* of 0 by the number 1. Using this unit length over and over, label the point one unit to the right of 1 as 2, the point one unit to the right of 2 as 3, and so on, as shown in Figure 1-2. Then do the same thing toward the left. The point one unit to

° The terms right triangle, Pythagorean Theorem, and hypotenuse are explained in the Geometry Review on page 581.
°° A calculator may display $\sqrt{2}$ as a decimal (rational number) such as 1.414213562. It may also display the square of this number as 2. But this is due to the calculator's rounding off long decimals. If you perform the squaring by hand, you will find that the answer is *not* 2.

the left of 0 is labeled -1, the point one unit to the left of -1 is labeled -2, and so on. By now the scheme should be clear: the point $1\frac{1}{2}$ units to the *right* of 0 is labeled $1\frac{1}{2} = \frac{3}{2}$, the point that is 5.78 units to the *left* of 0 is labeled -5.78, and so on (see Figure 1-2).

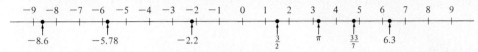

Figure 1-2

In the construction of the number line, one thing is certainly clear:

Every real number is the label of a unique point on the line.

We shall also assume the converse of this statememt as a basic axiom:

Every point on the line has a unique real number label.

Strictly speaking, the number 3.6 and the point on the line labeled 3.6 are two different things. Nevertheless, we shall frequently use such phrases as "the point 3.6" or refer to a "number on the line." In context this usage will not cause any confusion. Indeed, the mental identification of real numbers with points on the line is often extremely helpful in understanding and solving various problems.

ORDER

The statement "a is less than b" (written $a < b$) and the statement "b is greater than a" (written $b > a$) mean exactly the same thing, namely, a lies to the *left* of b on the number line (or equivalently, b lies to the *right* of a on the number line).

EXAMPLE $-50 < 10$ and $10 < 30$ since on the number line -50 lies to the left of 10 and 10 lies to the left of 30 (Figure 1-3).

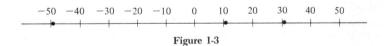

Figure 1-3

Often we shall write $a \leq b$ (or $b \geq a$), which means "a is less than or equal to b" (or "b is greater than or equal to a"). The statement $a \leq b \leq c$ means

$$a \leq b \text{ and } b \leq c$$

Geometrically, the statement $a \leq b \leq c$ means that a lies to the left of c and b lies between a and c on the number line (and may possibly be equal to one or both of them). The preceding example shows that $-50 \leq 10 \leq 30$. The statement $a \leq b < c$ means "$a \leq b$ and $b < c$," and so on.

The following facts are used frequently. For any real numbers a, b, c,

> a $\leq$ a;
> *if* a $\leq$ b *and* b $\leq$ a, *then* a = b;
> *if* a $\leq$ b *and* b $\leq$ c, *then* a $\leq$ c.

These properties are easily verified by looking at the number line. The last property above, as well as all four of those below, is also true if $\leq$ is replaced by $<$.

> *Let* a, b, c *be real numbers.*
> (*i*)   *If* a $\leq$ b, *then* a + c $\leq$ b + c.
> (*ii*) *If* a $\leq$ b, *and* c $>$ 0, *then* ac $\leq$ bc.
> (*iii*) *If* a $\leq$ b *and* c $<$ 0, *then* ac $\geq$ bc
>   (*note direction of last inequality*).
> (*iv*) *If* 0 $<$ a $\leq$ b, *then* $\dfrac{1}{a} \geq \dfrac{1}{b}$
>   (*note direction of last inequality*).

EXAMPLES (i), (ii) Suppose $a = -6$, $b = 2$, and $c = 5$. Then we have:

$$
\begin{array}{ccc}
a \leq b & a + c \leq b + c & ac \leq bc \\
-6 \leq 2 & -6 + 5 \leq 2 + 5 & (-6)5 \leq (2)5 \\
 & -1 \leq 7 & -30 \leq 10
\end{array}
$$

EXAMPLES (iii), (iv) Suppose $a = 3$, $b = 7$, and $c = -4$. Then we have:

$$
\begin{array}{ccc}
a \leq b & ac \geq bc & \dfrac{1}{a} \geq \dfrac{1}{b} \\[2ex]
3 \leq 7 & 3(-4) \geq 7(-4) & \dfrac{1}{3} \geq \dfrac{1}{7}
\end{array}
$$

$$-12 \geq -28$$

The numbers to the right of 0 on the number line, that is,

$$\text{all numbers } a \text{ with } a > 0,$$

are called **positive** numbers. The numbers to the left of 0, that is,

$$\text{all numbers } b \text{ with } b < 0,$$

are called **negative** numbers. Finally, the **nonnegative** numbers are

$$\text{all numbers } a \text{ with } a \geq 0.$$

The various rules that you have learned about the products of positive and negative numbers are just special cases of the properties listed in the last box. For example, the fact that $(-3)(-4) = +12$ reflects the fact that multiplying both sides of the inequality $-3 < 0$ by -4 reverses the direction of the inequality: $(-3)(-4) > 0$.

NEGATIVE NUMBERS AND NEGATIVES OF NUMBERS

Unfortunately for students, the word "negative" has *two* different meanings in mathematics. We have already seen one of these:

A negative number is a number less than 0.

Here is the other meaning of the word "negative":

The negative of a number *is the number obtained by changing the sign of the original number.*

For example, the negative of 3 is the number -3 and the negative of -7 is the number 7. It is customary to consider 0 to be its own negative since $0 = -0$. Naturally, the use of the same word to describe two different situations can be confusing. Unfortunately, however, the usage is so widespread that we are stuck with it.

Observe that you can *always* change the sign of a given number by inserting a minus sign in front of it. For instance, to change the sign of 3 we insert a minus sign and get -3. To change the sign of the number -7 we insert a minus sign in front and obtain $-(-7) = 7$. Consequently, the following statement is always true:

> *The negative of the number c is the number* $-c$.

When you see a statement such as the one in the box just given, it is important to remember that the number c may be either positive or negative or zero. As we have just seen, the negative of the positive number 3 is the negative number -3 and the negative of the negative number -7 is the positive number 7. Thus

> *If c is a positive number, then* $-c$ *is a negative number.*
> *If c is a negative number, then* $-c$ *is a positive number.*
> *In every case, the sum of a number and its negative is 0.*

SQUARE ROOTS

Since $(-5)^2 = (-5)(-5) = 25$ and $5^2 = 5 \cdot 5 = 25$, we see that there are two real numbers whose square is 25. But only *one* of them is nonnegative. More generally, the real number system has this important property:

> *If* d *is a nonnegative real number, then there is one and only one nonnegative real number* c *such that* $c^2 = d$. *This number* c *is called the square root of* d *and is denoted* $\sqrt{d}$.°

Thus we write $\sqrt{25} = 5$. Similarly, $\sqrt{9/4} = \frac{3}{2}$ since $\frac{3}{2}$ is the one and only nonnegative number whose square is $\frac{9}{4}$. The square root of a *rational* number need not be a rational number. For example, we saw above that the real number $\sqrt{2}$ is not a rational number.

Warning   In high school you may have said that both 5 and -5 were square roots of 25 and written $\sqrt{25} = \pm 5$. But here, as in most advanced texts, the term *"square root"* and the symbol $\sqrt{}$ *always denote a nonnegative number*. If we want to express -5 in terms of square roots, we write $-5 = -\sqrt{25}$. Finally, remember that symbols such as $\sqrt{-9}$ (with the minus sign *inside* the radical) have no meaning in the real number system, since the square of a real number could not possibly be the negative number -9. Indeed, the rules for multiplying positive and negative numbers show that c^2 is always nonnegative, no matter what c itself is. Thus

No negative real number has a square root in the real number system.°°

Here is a summary of the basic properties of square roots.

> $\sqrt{a}$ *is defined only when* $a \geq 0$;
> $\sqrt{a} \geq 0$ *for all* $a \geq 0$;
> $\sqrt{a} = 0$ *only when* $a = 0$;
> $\sqrt{a}\,\sqrt{a} = a$ *for all* $a \geq 0$;
> $\sqrt{a}\,\sqrt{b} = \sqrt{ab}$ *provided* $a \geq 0$ *and* $b \geq 0$;
> $\sqrt{\dfrac{a}{b}} = \dfrac{\sqrt{a}}{\sqrt{b}}$ *provided* $a \geq 0$ *and* $b > 0$.

EXAMPLE $\sqrt{48} = \sqrt{16 \cdot 3} = \sqrt{16}\,\sqrt{3} = 4\sqrt{3}$; similarly, $\sqrt{2025} = \sqrt{25 \cdot 81} = \sqrt{25}\,\sqrt{81} = 5 \cdot 9 = 45$.

Beware of symbols such as $\sqrt{a + b}$: in general,

$$\sqrt{a + b} \text{ is } not \text{ the same number as } \sqrt{a} + \sqrt{b}.$$

EXAMPLE $\sqrt{9 + 16} = \sqrt{25} = 5$,   but $\sqrt{9} + \sqrt{16} = 3 + 4 = 7$.   Thus $\sqrt{9 + 16} \neq \sqrt{9} + \sqrt{16}$.

° The symbol $\sqrt{}$ is called a **radical**.
°° If you are familiar with complex numbers, then you know that every real number (including negative ones) is the square of some *complex* number. But our discussion here is confined only to *real* numbers. Complex numbers will be discussed in Chapter 8.

CUBE ROOTS

If c and d are real numbers such that

$$d = c^3 = c \cdot c \cdot c$$

then c is called the **cube root** of d and is denoted $\sqrt[3]{d}$. There are no exceptions or possible ambiguities with cube roots as there were with square roots:

Every real number has a unique real cube root.

EXAMPLE $\sqrt[3]{27} = 3$ since $27 = 3 \cdot 3 \cdot 3$; and $\sqrt[3]{-1} = -1$ since $-1 = (-1)(-1)(-1)$.

Cube roots have the same multiplicative property as square roots, namely,

$$\sqrt[3]{ab} = \sqrt[3]{a} \cdot \sqrt[3]{b} \text{ for all real numbers } a, b$$

In general, $\sqrt[3]{a + b}$ is *not equal* to $\sqrt[3]{a} + \sqrt[3]{b}$.

EXERCISES

A.1.   Draw a number line and mark the location of each of the following numbers:

$$0, \ -7, \ \tfrac{4}{3}, \ 10, \ -1, \ \tfrac{1}{2}, \ -3, \ 2$$

A.2.   Express the following statements in symbols:
(a)   7 is greater than 5
(b) -4 is greater than -8
(c) x is nonnegative
(d) y is positive
(e) z lies strictly between -3 and -2
(f) x is positive but not more than 7
(g) c is less than 4 and d is at least 4

A.3.   Insert *one* of the symbols, $<$, $>$, or $=$ between each of the following pairs of numbers, so as to form a true statement. (When in doubt, locate the numbers on the number line.)
Example 7   10: we have $7 < 10$. Similarly, $\sqrt{5}$ 3: we have $\sqrt{5} < 3$.
(a) -1000 $.01$
(b) $\sqrt{756}$ 657
(c)   77.77   77.777
(d) $\sqrt{2}$ 2
(e) $\sqrt{19}$ 4
(f)   1.4 $\sqrt{2}$
(g) π 3.1415
(h)   2 -2
(i) $-\tfrac{1}{10}$ -6

A.4.   What happens when you multiply each of the following inequalities by 3? by -3?
(a) $7 < 10$
(b) $-7 < 10$
(c) $-7 > -10$

A.5.   State the negative of each of the following numbers:
$0; \quad 1; \quad -\pi; \quad -67.43; \quad [4 + (-6)]; \quad [-(37 - 2)]; \quad (\sqrt{3} - \sqrt{2})$

B.1.   (a)   Use one of the irrational numbers mentioned in the text to show that the product of an irrational number with itself may turn out to be a rational number.

(b) Is the product of two rational numbers always a rational number? (This wasn't discussed in the text, but you should be able to figure out the answer.)

B.2. (a) Find a rational number which is strictly *between* the numbers $\frac{2}{7}$ and $\frac{5}{9}$ on the number line.

(b) Let r/s and a/b be rational numbers with $r/s \neq a/b$. Find a rational number which is strictly between r/s and a/b on the number line.

B.3. Let h be a nonzero real number. Illustrate the truth of each of the following statements by three examples (that is, choose three different numbers that suggest the truth of the statement).

(a) If h is a large positive number, then $1/h$ is a small positive number.

(b) If h is a very small positive number, then $1/h$ is a very large positive number.

(c) If h is a very small negative number, then $1/h$ is very close to zero.

(d) If h is a negative number which is very close to 0, then $1/h$ is a very small negative number.

(*Hint:* remember that as you move to the *right* on the number line, the numbers get larger and larger; but as you move to the *left*, the numbers get smaller and smaller.)

B.4. In each of the following statements, insert either $\leq$ or $\geq$ in the blank so as to make a *true* statement *and then* give an example of the statement, using specific numbers. For instance, in the statement

$$\text{if } 0 < a \leq b, \text{ then } \frac{1}{a} \underline{\quad} \frac{1}{b}$$

insert $\geq$; example: $0 < 2 \leq 3$ and $1/2 \geq 1/3$.

(a) if $a \leq b < 0$, then $1/a \underline{\quad} 1/b$ (d) if $a \leq b \leq 0$, then $a^2 \underline{\quad} b^2$

(b) if $a < 0 < b$, then $1/a \underline{\quad} 1/b$ (e) if $a \leq 0 \leq b$, then $a^3 \underline{\quad} b^3$

(c) if $0 \leq a \leq b$, then $a^2 \underline{\quad} b^2$ (f) (sneaky) if $a \leq 0 \leq b$, then $a^2 \underline{\quad} b^2$

B.5. Fill the blanks in the following table, which shows the equivalence of arithmetic statements about real numbers and geometric statements about the corresponding points on the number line.

Arithmetic Statement	Geometric Statement
(a) a is negative	The point a lies to the left of the point 0
(b) $a \geq b$	_____
(c) _____	a lies c units to the right of b
(d) _____	a lies between b and c
(e) $a - b > 0$	_____

(f) *a* is positive _____

(g) _____ *a* lies strictly to the left of *b*

B.6. Find the square roots of the following numbers, without using a calculator or tables.

(a) 2304 [*Hint:* $2304 = 6(384) = 6 \cdot 4 \cdot 96 = 6 \cdot 4 \cdot 6 \cdot 16 = 6^2 \cdot 2^2 \cdot 4^2$]

(b) 2116

(c) .1764

(d) 10.89

(e) (1092 + 64)

(f) (2000 − 844)

C.1. If *c* is any real number, what is $\sqrt{c^2}$? (Think before you answer—what if $c < 0$?)

DO IT YOURSELF!

DECIMAL REPRESENTATION

We take it for granted that you are familiar with decimal notation for rational numbers whose denominators can be expressed as a power of ten, such as:

$$\tfrac{1}{10} = .1, \qquad -\tfrac{4}{1000} = -.004, \qquad 2\tfrac{3}{4} = 2.75 = \tfrac{275}{100}, \qquad -\tfrac{745}{100} = -7.45$$

Terminating decimals, such as these, are usually easier to compute with than are the corresponding fractions. Of course, some rational numbers cannot be expressed as terminating decimals. Nevertheless, such rational numbers can be expressed as repeating decimals, as the following example illustrates.

EXAMPLE In order to express $\tfrac{4359}{925}$ as a decimal, we divide the numerator (top) by the denominator (bottom) and obtain:

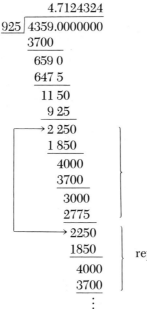

$$
\begin{array}{r}
4.7124324 \\
925\,\overline{\smash{)}\,4359.0000000} \\
\underline{3700} \\
659\ 0 \\
\underline{647\ 5} \\
11\ 50 \\
\underline{9\ 25} \\
2\ 250 \\
\underline{1\ 850} \\
4000 \\
\underline{3700} \\
3000 \\
\underline{2775} \\
2250 \\
\underline{1850} \\
4000 \\
\underline{3700} \\
\vdots
\end{array}
$$

repeats as above

Since the remainder at the third step (namely, 225) occurs again at the sixth step, it is clear that the division process goes on forever, with the three-digit block "243" repeating over and over in the quotient: $\frac{4359}{925} = 4.71243243243\cdots$. Similarly, division shows that $\frac{1}{12} = .083333\cdots$, with the 3's continually repeating.

It is easy to see that the method used in the preceding example (namely, divide the top by the bottom) can be used to express any rational number (fraction) as a decimal. If at some stage of the long division process, you obtain a remainder of 0, then the process stops and the result is a terminating decimal. Otherwise, some nonzero remainder *necessarily* repeats during the division process° and you obtain a repeating decimal.

When convenient we shall consider a terminating decimal as a repeating decimal ending in zeros (for instance, $.75 = .75000\cdots$). There is also a simple method for converting a repeating decimal into a rational number (see Exercise C.1.). Consequently, we conclude that

> Every rational number can be expressed as a terminating or repeating decimal and every such decimal represents a rational number.

Although we shall not prove it here, irrational numbers can also be expressed as decimals. Of course, a decimal that represents an *irrational* number is necessarily nonterminating and *nonrepeating* (that is, no block of digits repeats forever). If you learned a method for finding square roots in grade school, you have already seen one example of this. That method (whose details need not concern us here) shows that $\sqrt{2} = 1.414213562373095\cdots$. As usual,. the three dots indicate that the decimal goes on forever. But this time there is no block of digits that repeats forever. The only way to find out what digit comes next is to carry out the computation that far. Another irrational number is π, whose decimal expansion begins $3.14159265358979\cdots$; this expansion has been carried out to over one million decimal places by computer.

Conversely, every nonrepeating decimal represents an irrational real number (no proof to be given here). Consequently, we can summarize the preceding discussion as follows.

(*i*)   *Every real number can be expressed as a decimal.*
(*ii*) *Every such decimal represents a real number.*
(*iii*) *The repeating decimals are precisely the rational numbers.*
(*iv*) *The nonrepeating decimals are precisely the irrational numbers.*

EXERCISES

A.1. Express each of the following rational numbers as a repeating decimal.
    (a) $\frac{7}{9}$;   **(b)** $\frac{2}{13}$;   **(c)** $\frac{23}{14}$;   **(d)** $\frac{19}{88}$;   **(e)** $\frac{1}{19}$ (long);   **(f)** $\frac{9}{11}$.

° For instance, if you divide a number by 23, the only possible remainders at each step are the numbers 0, 1, 2, 3, ..., 22. Hence after *at most* 23 steps, some remainder must occur for a second time.

B.1. If two real numbers have the same decimal expansion through three decimal places, how far apart can they be on the number line?

C.1. Here is a method for expressing a repeating decimal as a rational number (fraction). For example, let $d = 52.31272727\cdots$. Assuming that the usual laws of arithmetic hold, we see that

$$10{,}000d = 523127.272727\cdots \quad \text{and} \quad 100d = 5231.272727\cdots$$

Now subtract $100d$ from $10{,}000d$:

$$
\begin{aligned}
10{,}000d &= 523127.272727\cdots \\
-\ 100d &= -5231.272727\cdots \\
\hline
9{,}900d &= 517896.
\end{aligned}
$$

Dividing both sides of this last equation by 9900 yields $d = \frac{517896}{9900}$. The idea here is to subtract two *suitable* multiples of d, so as to eliminate the repeating part of the decimal. Here is a second example. If $c = .272727\cdots$, then verify that $100c - c = 27$. Hence $99c = 27$, so that $c = \frac{27}{99} = \frac{3}{11}$. Express each of the following repeating decimals as rational numbers.
   (a) $.373737\cdots$;   (b) $.929292\cdots$;   (c) $76.63424242\cdots$;
   (d) $13.513513\cdots$.

C.2. Use the methods in Exercise C.1 to show that both $.75000\cdots$ and $.749999\cdots$ are decimal expansions of $\frac{3}{4}$. [In general it is true that every terminating decimal (that is, a repeating decimal ending in zeros) can also be expressed as a decimal ending in repeated 9's. It can be proved that these are the only real numbers with more than one decimal expansion.]

2. SETS AND INTERVALS

The *language* of set theory is frequently convenient, so we introduce a small amount of it here. Indeed, all this section amounts to is a short list of names for various concepts with which you are already familiar.

Roughly speaking, a **set** is any collection of objects. In this book the objects will usually be numbers or other mathematical entities. For example, we have the set of integers, the set of positive real numbers, and the set of points on the number line that lie to the left of 3.

The objects in a set are called **elements** or **members** of the set. For instance, -6 is an element of the set of integers. But -7 is *not* an element of the set of all positive real numbers. The number $\sqrt{2}$ is a member of the set of irrational numbers.

Suppose a and b are real numbers with $a < b$. The word **interval** is used to describe the set of all real numbers between a and b. Actually, there are four such sets, depending on whether one, both, or neither of a and b are included:

$[a, b]$ denotes the set of all real numbers x such that $a \leq x \leq b$;
(a, b) denotes the set of all real numbers x such that $a < x < b$;
$[a, b)$ denotes the set of all real numbers x such that $a \leq x < b$;
$(a, b]$ denotes the set of all real numbers x such that $a < x \leq b$.

All four of these sets are called **intervals** from a to b. The numbers a and b are the **endpoints** of the interval. $[a, b]$ is called the **closed interval** from a to b (both endpoints included) and (a, b) is called the **open interval** from a to b (neither endpoint included).

If we think of the real numbers as the points on the number line, then an interval from a to b is just the line segment from a to b, which may or may not include these endpoints. Figure 1-4 shows some specific examples.

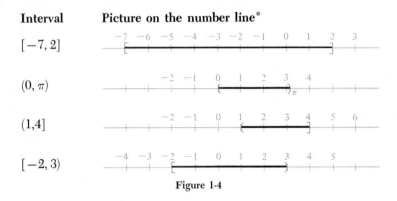

Interval   **Picture on the number line°**

$[-7, 2]$

$(0, \pi)$

$(1,4]$

$[-2, 3)$

Figure 1-4

There are other sets of real numbers which are also called intervals. Let b be a real number. Then

$(-\infty, b)$ denotes the set of all real numbers less than b;
$(-\infty, b]$ denotes the set of all real numbers less than or equal to b;
(b, ∞) denotes the set of all real numbers greater than b;
$[b, \infty)$ denotes the set of all real numbers greater than or equal to b.

If we think of the real numbers as the points on the number line, then these four intervals are simply the half-lines extending to the left or right of b, either including or not including b, as shown in Figure 1-5.

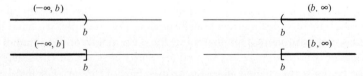

Figure 1-5

Note   The symbol ∞ is sometimes read "infinity" and one sometimes says, for example, that $[b, \infty)$ is the "interval from b to infinity." Don't be misled by this casual use of language. We have simply used the symbol ∞ and the word "infinity" as part of a convenient notation for certain *sets of real numbers*. In particular,

---

° In Figures 1-4 and 1-5, a round bracket, such as) or (, at the endpoint of an interval indicates that this endpoint is *not* included. A square bracket, such as] or [, indicates that the endpoint *is* included.

The concept of "infinity", whatever it may be, has *not* been defined here;
There is no real number called "infinity" or "minus infinity";
The symbols ∞ and $-\infty$ are *not* real numbers.

EXERCISES

A.1. Draw a picture on the number line of each of the following intervals:
(a) $(0, 8]$; (b) $(0, \infty)$; (c) $[-2, 1]$; (d) $(-1, 1)$; (e) $(-\infty, 0]$.

B.1. Let **Z** denote the set of all integers, **Q** the set of all rational numbers, A the interval $(-7, 12]$, and B the set of all rational numbers in the interval $[0, 4]$. The number 2, for instance, is an element of all four of these sets. Of which sets are the following numbers elements?
(a) $-\frac{3}{2}$ (c) -2 (e) π (g) $\pi - 2$
(b) 4.001 (d) -7 (f) -12 (h) $\sqrt{2}$

3. ABSOLUTE VALUE

On an *informal* level, many students think of absolute value as follows:

The absolute value of a nonnegative number is the number itself;
The absolute value of a negative number is obtained by "erasing the minus sign."

If $|c|$ denotes the absolute value of the number c, then, for example:

$$|7| = 7, \qquad |-3| = 3, \qquad \left|\frac{\pi}{7}\right| = \frac{\pi}{7}, \qquad \left|\frac{-37}{5}\right| = \frac{37}{5}, \qquad |0| = 0$$

But when asked to find the absolute value of the negative number $\pi - 6$, such students are often puzzled: what does "erase the minus sign" mean here?

Is $|\pi - 6| = \pi$ 6?   Is $|\pi - 6| = \pi + 6$?   or what?

This quandary suggests that a more formal and precise algebraic definition of absolute value is needed. The earlier examples, where things were clear, can guide us here.
Since the absolute value of a nonnegative number is just the number itself, we have:

$$|c| = c \text{ whenever } c \geq 0$$

When c is a negative number, the informal idea of "erasing the minus sign" means, for example, that

$$|-4| = 4, \qquad |-3| = 3, \qquad |-\tfrac{37}{5}| = \tfrac{37}{5}$$

Observe that

$$|-4| = 4 = -(-4), \qquad |-3| = 3 = -(-3), \qquad |-\tfrac{37}{5}| = \tfrac{37}{5} = -(-\tfrac{37}{5})$$

These examples suggest that the absolute value of a *negative* number c is the *positive* number $-c$. In other words,

$$\text{If } c < 0, \text{ then } |c| = -c.$$

For instance,

$$\text{If } c = -7, \text{ then } |-7| = |c| = -c = -(-7) = 7;$$
$$\text{If } c = -12, \text{ then } |-12| = |c| = -c = -(-12) = 12.$$

The preceding discussion leads to this formal definition of absolute value:

> The **absolute value** of a real number c is denoted $|c|$ and defined as follows:
>
> $$\text{if } c \geq 0, \text{ then } |c| = c$$
> $$\text{if } c < 0, \text{ then } |c| = -c$$

EXAMPLE   In order to find $|\pi - 6|$, we observe that $\pi - 6 < 0$, so that

$$|\pi - 6| = -(\pi - 6) = -\pi + 6 = 6 - \pi$$

EXAMPLE   Since $\pi - 3 > 0$, we have $|\pi - 3| = \pi - 3$.

BASIC PROPERTIES OF ABSOLUTE VALUES

We list below some frequently used facts about absolute values. You should be familiar with all of them and should be able to give numerical examples of each.

> **FACT** $|c| \geq 0$ *and* $|c| > 0$ *when* $c \neq 0$

EXAMPLES $|3| = 3 > 0$; $|-7| = 7 > 0$; $|0| = 0$; $\left|-\frac{1}{3}\right| = \frac{1}{3} > 0$.

> **FACT** $|c| = |-c|$

EXAMPLES $|4| = 4 = |-4|$; $|\pi| = \pi = |-\pi|$; $|0| = 0 = |-0|$; $\left|\frac{7}{2}\right| = \frac{7}{2} = \left|-\frac{7}{2}\right|$.

> **FACT** $|c|^2 = c^2 = |c^2|$

EXAMPLE Let $c = -3$; then

$$c^2 = (-3)^2 = 9$$
$$|c^2| = |(-3)^2| = |9| = 9$$
$$|c|^2 = |-3|^2 = |-3||-3| = 3 \cdot 3 = 9$$

FACT $\sqrt{c^2} = |c|$

EXAMPLES Let $c = -3$. Then

$$\sqrt{c^2} = \sqrt{(-3)^2} = \sqrt{9} = 3 = |-3| = |c|$$

Let $c = 7$; then

$$\sqrt{c^2} = \sqrt{7^2} = \sqrt{49} = 7 = |7| = |c|$$

FACT $|cd| = |c||d|$ *and if* $d \neq 0$, $\left|\dfrac{c}{d}\right| = \dfrac{|c|}{|d|}.$

EXAMPLES Let $c = 6$ and $d = -2$; then

$$|cd| = |6(-2)| = |-12| = 12 \quad \text{and} \quad |c||d| = |6||-2| = 6 \cdot 2 = 12$$

Let $c = -7$ and $d = -\frac{1}{4}$; then

$$|cd| = |(-7)(-\tfrac{1}{4})| = |\tfrac{7}{4}| = \tfrac{7}{4} \quad \text{and} \quad |c||d| = |-7||-\tfrac{1}{4}| = 7(\tfrac{1}{4}) = \tfrac{7}{4}$$

Let $c = -\sqrt{3}$ and $d = \frac{7}{3}$; then

$$\left|\frac{c}{d}\right| = \left|\frac{-\sqrt{3}}{\frac{7}{3}}\right| = \left|\frac{-3\sqrt{3}}{7}\right| = \frac{3\sqrt{3}}{7} \quad \text{and} \quad \frac{|c|}{|d|} = \frac{|-\sqrt{3}|}{|\frac{7}{3}|} = \frac{\sqrt{3}}{\frac{7}{3}} = \frac{3\sqrt{3}}{7}$$

DISTANCE ON THE NUMBER LINE

We would like to have a formula for calculating the distance between any two numbers on the number line. We shall now see that such a formula can be obtained by using absolute values.

EXAMPLE In Figure 1-6 the distance between -6 and 0 on the number line is easily seen to be 6 units.

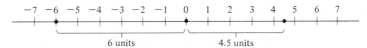

Figure 1-6

Observe that $|-6| = 6 =$ distance between -6 and 0. Figure 1-6 also shows that the distance between 4.5 and 0 on the number line is 4.5 units. Note that $|4.5| =$ distance between 4.5 and 0.

The preceding example is an illustration of this general fact:

> $|c|$ *is the distance between* c *and* 0 *on the number line.*

This geometric interpretation of absolute value as a distance leads to another useful property of absolute values.

EXAMPLE How many real numbers have absolute value 3? Since absolute value can be interpreted as distance to 0, this question is the same as: how many real numbers are 3 units from 0 on the number line? For the answer, look at Figure 1-7.

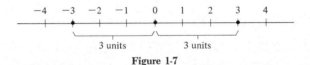

Figure 1-7

There are just two numbers whose distance to 0 is 3 units, namely, 3 and -3.

The argument used in the preceding example works in the general case as well, and proves this fact:

> *If* k *is a positive number, then the only numbers with absolute value* k *are* k *and* $-$k.

In order to determine the distance between two (possibly) nonzero numbers on the line, we consider another example.

EXAMPLE The distance between -5 and 3 on the number line is 8 units, as shown in Figure 1-8.

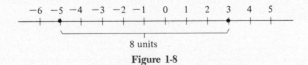

Figure 1-8

Note that $|-5 - 3| = 8 =$ distance between -5 and 3. It is also true that $|3 - (-5)| = |3 + 5| = 8 =$ distance between -5 and 3.

The preceding examples suggest that the distance between two numbers on the number line is always given by the absolute value of their difference. In other words,

> *The distance between the numbers* c *and* d *on the number line is the number* $|c - d|$.

As the examples illustrate, it doesn't matter in which order you take the difference—the same distance results in either order. The reason is that for any two numbers c and d,

$$-(c - d) = -c + d = d - c$$

Since any number and its negative have the same absolute value, we see that

$$|c - d| = |-(c - d)| = |d - c|$$

This is just an algebraic statement of the geometric fact that the distance from c to d is the same as the distance from d to c.

ALGEBRA AND GEOMETRY

We have already seen several examples of how absolute values may be interpreted either algebraically or geometrically. This close relationship between algebra and geometry often suggests two different ways of thinking about the same problem, which usually increases the chances of solving the problem.

EXAMPLE The geometric statement "the distance between t and 7 is $\frac{3}{2}$" can be expressed algebraically as $|t - 7| = \frac{3}{2}$.

EXAMPLE The inequality $|x - 3| < |z + 4|$ can be rewritten as $|x - 3| < |z - (-4)|$. This is just an algebraic way of saying,

The distance from x to 3 is less than the distance from z to -4,

or equivalently, x is closer to 3 than z is to -4.

EXAMPLE The solution of the equation $|x + 5| = 3$ can easily be found geometrically. If we rewrite the equation as $|x - (-5)| = 3$, it states that

The distance from x to -5 is 3 units.

On the number line in Figure 1-9, it's easy to see that there are only two numbers that are exactly 3 units from -5, namely, -8 and -2.

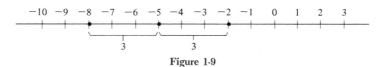

Figure 1-9

Thus $x = -8$ and $x = -2$ are the solutions of $|x + 5| = 3$.

EXAMPLE The solutions of the inequality $|x - 7| < 2.5$ are all numbers x such that

The distance from x to 7 is less than 2.5.

If we look at the number line in Figure 1-10,

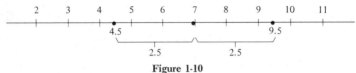

Figure 1-10

we see that any number x between 4.5 and 9.5 is *at most* 2.5 units from 7 and all other numbers are *more than* 2.5 units from 7. Thus the set of all solutions of $|x - 7| < 2.5$ is the interval (4.5, 9.5).

LINEAR ABSOLUTE VALUE EQUATIONS

For slightly more complicated equations, such as $|3x + 5| = 7$ or $\left|\dfrac{5x - 4}{3}\right| = 2$, the following algebraic method is usually easier than the geometric method illustrated above.

EXAMPLE In order to solve $\left|\dfrac{5x - 4}{3}\right| = 2$, we recall that only two numbers have absolute value 2, namely, 2 and -2. Therefore $\dfrac{5x - 4}{3}$ must be either 2 or -2. Each possibility leads to a solution of the original equation.

$$\frac{5x - 4}{3} = 2 \qquad \text{or} \qquad \frac{5x - 4}{3} = -2$$

$$5x - 4 = 6 \qquad\qquad\qquad 5x - 4 = -6$$

$$5x = 10 \qquad\qquad\qquad\quad 5x = -2$$

$$x = 2 \qquad\qquad\qquad\quad x = -\frac{2}{5}$$

You should verify that both these solutions check.

EXERCISES

A.1. Compute:
  (a) $-|-7|$
  (b) $|3 - 14|$
  (c) $|(-13)^2|$

(d) $||-7|-|-4||$ **(f)** $|\sqrt{2}-2|$ **(h)** $|(-2)(3)|$

(e) $\dfrac{|-5|}{-5}$ **(g)** $|\pi-\sqrt{2}|$

A.2. Fill the blank with one of $<$, $=$, or $>$ so that the resulting statement is true.

(a) $|-2|$ _____ $|-5|$; **(d)** $|-3|$ _____ 0; **(f)** $-|-4|$ _____ 0;
(b) 5 _____ $|-2|$; **(e)** -7 _____ $|-1|$; **(g)** $|-4+|-4||$ _____ 0;
(c) $|3|$ _____ $-|4|$.

A.3. In each of the following, find two pairs of numbers that make the statement true *and* two pairs that make it false. (For instance $|x|<|y|$ is true for 1,2 and $-1,7$ but is false for 3,1 and $-3,-2$.)

(a) $|x|+|y|=1$ **(c)** $|x+y|=|x|+|y|$ **(e)** $|x-y|=y-x$
(b) $|y|-|x|<0$ **(d)** $|x+y|<|x|+|y|$ **(f)** $|x-y|=|x|-|y|$

A.4. In each part, find the distance between the two given numbers.

(a) -7 and $\frac{15}{2}$ **(c)** 7 and 107 **(e)** π and -3
(b) $-\frac{3}{4}$ and -10 **(d)** π and 3 **(f)** $\sqrt{2}$ and $\sqrt{3}$

B.1. **(a)** Explain why the statement $|a|+|b|+|c|>0$ is algebraic shorthand for "at least one of the numbers a,b,c is different from zero."
(b) Find an algebraic shorthand version of the statement "the numbers a,b,c are all different from zero."

B.2. Explain why each of the following is a true statement, no matter what the numbers c and d may be. (*Hint:* look back at the Facts about absolute values in the text.)
(a) $|c-d|=|d-c|$
(b) $|(c-d)^2|=c^2-2cd+d^2$
(c) $\sqrt{9c^2-18cd+9d^2}=3|c-d|$

B.3. Fill the blanks in the following table.

Algebraic Statement	Equivalent Geometric Statement
(a) $|x-c|>6$	x is more than 6 units from c.
(b) $|x-3|<2$	_____
(c) _____	c is closer to 0 than b is
(d) $|b|<|c-3|$	_____
(e) $|x+7|\le 3$	_____
(f) _____	x is 5 units from c.
(g) _____	x is at most 17 units from -4.
(h) $|c|\ge 12$	_____

B.4. Explain geometrically why each of the following statements is *always false*.
(a) $|c - 1| < 2$ and $|c - 12| < 3$.   (b) $|d + 1| > 3$ and $|d - 1| < 1$.

B.5. For what values of x is each of the following statements true? For example, the statement $|x| = x$ is true for all nonnegative numbers x and the statement $|x| \geq 0$ is true for all real numbers x.
(a) $x \leq |x|$ (b) $|x| \leq x$ (c) $|x| \leq -x$ (d) $-|x| \leq x$

B.6. Use the geometric approach explained in the text to solve these equations.
(a) $|x| = 1$ (c) $|x - 2| = 1$ (e) $|x + \pi| = 4$
(b) $|x| = \frac{3}{2}$ (d) $|x + 3| = 2$ (f) $|x - \frac{3}{2}| = 5$

B.7. Solve the following equations. For example, to solve $|x^2 - (x + 2)(x - 3)| = 1$, note that $|x^2 - (x + 2)(x - 3)| = |x^2 - (x^2 - x - 6)| = |x^2 - x^2 + x + 6| = |x + 6|$ so that the original equation becomes $|x + 6| = 1$. This can be solved as in Exercise B.6.
(a) $|x^2 - (x + 1)(x - 2)| = 2$ (c) $|(y + 4)(y - 3) - y^2| = 1$
(b) $|(z + 1)^2 - z(1 + z)| = 3$ (d) $|(x + 5)(x - 5) - (x + 6)(x - 7)| = 1$

B.8. Use the algebraic approach to solve the following equations.
(a) $|2x + 3| = 4$ (c) $|3 - 4x| = 7$ (e) $|3x - \frac{1}{2}| = 2$
(b) $|3x - 5| = 2$ (d) $|2 - 3x| = \frac{4}{3}$ (f) $|-5x + \frac{1}{3}| = 4$

B.9. Suppose $|x - 2| = 3$. In this case, $|x - 1| = ?$ $|x + 1| = ?$ $|x| = ?$ $|x - 3| = ?$

B.10. Suppose $|x + 1| = 4$. In this case, $|x| = ?$ $|x - 2| = ?$ $|x - 4| = ?$ $|x + 4| = ?$

B.11. Use the geometric approach explained in the text to solve these inequalities
(a) $|x| < 7$ (c) $|x + 3| < 1$ (e) $|x| \geq 5$
(b) $|x - 5| < 2$ (d) $|x + \frac{1}{2}| < 2$ (f) $|x - 6| > 2$

B.12. Suppose $|x - 2| < 1$. In this case, what can you say about $|x|$ and $|x - 3|$?

4. LINEAR INEQUALITIES

Methods for solving inequalities involving only the first power of x are discussed here. Inequalities involving higher powers of x are considered in Section 5 of Chapter 4. We begin with inequalities that don't involve absolute values, such as

$$5x + 3 \leq 6 + 7x, \qquad \frac{3x + 5}{3} + 2x < \frac{2x - 1}{3}, \qquad -5 < 7 - 2x \leq 9$$

With one exception, noted below, the strategy for solving such inequalities is the same as that used to solve equations. It depends on these

---

**BASIC PRINCIPLES FOR
SOLVING INEQUALITIES**

Performing any one of the following operations on an inequality produces an inequality with the same solutions as the original inequality.

1. *Add or subtract the same quantity from both sides of the inequality.*
2. *Multiply or divide both sides of the inequality by a* positive *quantity.*
3. *Multiply or divide both sides of the inequality by a* negative *quantity and* reverse the direction of the inequality.

You should note Principle 3 carefully: it is the major difference from the situation with equations. For example, $-3 < 5$, but multiplying both sides by the negative number -2 produces the inequality $+6 > -10$ (note direction of inequality sign).

The fundamental idea in solving inequalities is to use these basic principles to transform a given inequality into a simpler inequality with the same solutions; then solve the simpler inequality

EXAMPLE Following is a step-by-step solution of the inequality $5x + 3 \leq 6 + 7x$. The reason for each step is given at the right.

$$5x + 3 \leq 6 + 7x$$
$$-2x + 3 \leq 6 \qquad \text{(subtract } 7x \text{ from both sides)}$$
$$-2x \leq 3 \qquad \text{(subtract 3 from both sides)}$$
$$\qquad \text{(divide both sides by } -2 \text{ and } reverse \text{ the direction}$$
$$x \geq -\tfrac{3}{2} \qquad \text{of the inequality)}$$

The solutions of this last inequality are obvious: every real number that is greater than or equal to $-\frac{3}{2}$ is a solution. In other words, the solutions are all numbers in the interval $[-\frac{3}{2}, \infty)$, as shown in Figure 1-11.

Figure 1-11

EXAMPLE   The first step in solving the inequality $\dfrac{3x+5}{2} + 2x < \dfrac{2x-1}{3}$ is to eliminate the fractions. Since 6 is a common denominator of the two fractions, we multiply both sides of the inequality by 6 and obtain:

$$6\left(\frac{3x-5}{2}\right) + 6 \cdot 2x < 6\left(\frac{2x+1}{3}\right)$$

$$3(3x - 5) + 12x < 2(2x + 1)$$

Now multiply the various terms and proceed as before to collect all x terms on one side and constants on the other:

$$
\begin{array}{ll}
3(3x - 5) + 12x < 2(2x + 1) & \\
9x - 15 + 12x < 4x + 2 & \text{(multiply out both sides)} \\
17x - 15 < 2 & \text{(subtract } 4x \text{ from both sides)} \\
17x < 17 & \text{(add 15 to both sides)} \\
x < 1 & \text{(divide both sides by 17)}
\end{array}
$$

The solutions of this last inequality, and hence the solutions of the original one, are all numbers in the interval $(-\infty, 1)$, as shown in Figure 1-12.

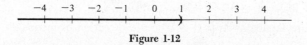

Figure 1-12

EXAMPLE   A solution of the two-part inequality $-5 < 7 - 2x \le 9$ is any number that is a solution of *both* of these inequalities:

$$-5 < 7 - 2x \quad \text{and} \quad 7 - 2x \le 9$$

Each of these last inequalities can be solved by the methods just shown. For the first one, we have:

$$
\begin{array}{ll}
-5 < 7 - 2x & \\
-12 < -2x & \text{(subtract 7 from both sides)} \\
6 > x, \text{ or equivalently, } x < 6 & \text{(divide both sides by } -2 \text{ and} \\
& \text{reverse the direction} \\
& \text{of the inequality)}
\end{array}
$$

The second inequality is solved similarly:

$$
\begin{array}{ll}
7 - 2x \le 9 & \\
-2x \le 2 & \text{(subtract 7 from both sides)} \\
x \ge -1, \text{ or equivalently, } -1 \le x. & \text{(divide both sides by } -2 \text{ and} \\
& \text{reverse the direction} \\
& \text{of the inequality)}
\end{array}
$$

The solutions of the original inequality are the numbers x that satisfy *both* $-1 \leq x$ *and* $x < 6$. Therefore the solutions are precisely the numbers in the interval $[-1, 6)$, as shown in Figure 1-13.

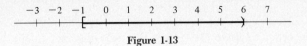

Figure 1-13

Alternate Shorthand Method You probably noticed that in the solution process above we followed exactly the same steps for both inequalities. Consequently, it is usually easier and faster to handle double inequalities such as this one by solving both halves at the same time:

$$-5 < 7 - 2x \leq 9$$
$$-12 < -2x \leq 2 \qquad \text{(subtract 7 from each part)}$$
$$6 > x \geq -1 \qquad \text{(divide each part by } -2 \text{ and reverse the direction of the inequalities)}$$

Reading this last inequality from right to left, we see once again that the solutions are precisely the numbers in the interval $[-1, 6)$.

LINEAR ABSOLUTE VALUE INEQUALITIES

The solution of inequalities such as

$$|3x - 7| \leq 5, \qquad |5x + 2| > 3, \qquad |4x - 1.5| < 10, \qquad |8 - \tfrac{3}{4}x| \geq \tfrac{5}{3}$$

depends on the following basic facts about absolute value (which are also true with $<$ and $>$ in place of $\leq$ and $\geq$).

Let k *be a fixed positive number. Then for any real number* r:

$$|r| \leq k \ exactly \ when \ -k \leq r \leq k.$$

$$|r| \geq k \ exactly \ when \ r \leq -k \ or \ r \geq k.$$

This statement says, for example, that $|r| \leq 5$ exactly when r is a number satisfying $-5 \leq r \leq 5$, and that $|r| \geq 5$ exactly when r is a number satisfying either $r \leq -5$ or $r \geq 5$. This is easily seen to be true on the number line. Since $|r|$ is the distance from r to 0, the condition $|r| \leq 5$ means that r lies *at most* 5 units from 0, as shown in Figure 1-14.

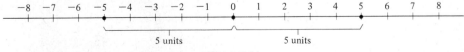

Figure 1-14

But this means that r lies between -5 and 5, that is, $-5 \leq r \leq 5$. Similarly, the condition $|r| \geq 5$ means that r lies 5 *or more* units from 0. And this occurs only when r lies on or to the left of -5 (that is, $r \leq -5$) or when r lies on or to the right of 5 (that is, $r \geq 5$). This same argument works in the general case (with k in place of 5).

EXAMPLE In order to solve $|3x - 7| \leq 5$, apply the fact in the box above with $3x - 7$ in place of r and 5 in place of k. It states that

$$|3x - 7| \leq 5 \qquad \text{exactly when} \qquad -5 \leq 3x - 7 \leq 5$$

Consequently, we need only solve the inequality on the right. This can be done by the methods discussed earlier:

$$-5 \leq 3x - 7 \leq 5$$
$$2 \leq 3x \leq 12 \qquad \text{(add 7 to each part)}$$
$$\tfrac{2}{3} \leq x \leq 4 \qquad \text{(divide each part by 3)}$$

The solutions of this last inequality, and hence of the original one, are all of the numbers in the interval $[\tfrac{2}{3}, 4]$, as shown in Figure 1-15.

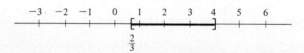

Figure 1-15

EXAMPLE In order to solve $|5x + 2| > 3$, we apply the fact in the box above with $5x + 2$ in place of r and 3 in place of k, and $>$ in place of $\geq$:

$$|5x + 2| > 3 \qquad \text{exactly when} \qquad 5x + 2 < -3 \qquad \text{or} \qquad 5x + 2 > 3$$

Now we need only solve the two inequalities at the right:

$$5x + 2 < -3 \qquad 5x + 2 > 3$$
$$5x < -5 \qquad 5x > 1$$
$$x < -1 \qquad x > \tfrac{1}{5}$$

The solutions of the original inequality are the numbers that satisfy *either* of these last two inequalities. Thus the solutions are all numbers that are in either of the intervals $(-\infty, -1)$, or $(\tfrac{1}{5}, \infty)$, as shown in Figure 1-16.

Figure 1-16

EXERCISES

A.1. Solve these inequalities.
  (a) $6x + 3 \leq x - 5$
  (b) $3 + 5x \leq 2x + 7$
  (c) $5 - 7x < 2x - 4$
  (d) $3x + 5 > 7x - 3$
  (e) $2x + 7(3x - 2) < 2(x - 1)$
  (f) $x + 3(x - 5) \geq 2x + 2(x + 1)$

B.1. Solve these inequalities.
  (a) $\dfrac{x + 1}{2} - 3x \leq \dfrac{x + 5}{3}$
  (b) $\dfrac{x - 1}{4} + 2x \geq \dfrac{2x + 1}{3} + 2$
  (c) $\dfrac{2x - 5}{3} + 1 < 2 - \dfrac{x + 7}{6}$
  (d) $\dfrac{3(x - 1)}{2} + 4x > \dfrac{2(x + 1)}{3} + 4$

B.2. Solve these inequalities.
  (a) $\dfrac{1}{2}x + \dfrac{3}{4} < 3 - \dfrac{1}{4}x$
  (b) $\dfrac{3}{2} - \dfrac{5x}{2} \geq 2x - \dfrac{1}{4}$
  (c) $2.3 - 1.1x \leq 2.9x - 1.7$
  (d) $4.2x + 5.3 > 2.7x + 2.3$

B.3. Solve these inequalities.
  (a) $(x - 1)(x + 2) \leq (x + 5)(x - 2)$
  (b) $(x - 3)^2 + 4 > (x - 1)(x + 1) + 3$
  (c) $\dfrac{(x - 2)(x + 5)}{2} < (x + 1)^2 + 4 - \dfrac{x^2}{2}$
  (d) $\dfrac{(x - 1)(x + 3)}{3} \geq \dfrac{(x + 1)(x - 2)}{2} + 1 - \dfrac{x^2}{6}$

B.4. Solve these inequalities.
  (a) $2 < 3x - 4 < 8$
  (b) $1 < 5x + 6 < 9$
  (c) $-3 \leq 4x + 5 \leq 2$
  (d) $-5 \leq -3x + 1 < 10$
  (e) $2x + 3 \leq 5x + 6 < -3x + 7$
  (f) $4x - 2 < x + 8 < 9x + 1$

B.5. Solve these inequalities.
  (a) $-4 \leq \dfrac{4 - 2x}{3} \leq 4$
  (b) $-1 \leq \dfrac{2 - 3x}{2} < 1$
  (c) $\dfrac{2x + 1}{3} < \dfrac{5x - 2}{2} \leq \dfrac{x - 1}{6}$
  (d) $1 + 2x \geq \dfrac{x - 5}{2} \geq \dfrac{3x - 1}{2} + 3$

B.6. Solve these inequalities.
  (a) $|3x + 2| \leq 2$
  (b) $|5x - 1| < 3$
  (c) $|3 - 2x| < \frac{2}{3}$
  (d) $|4 - 5x| \leq 4$
  (e) $|2x + 3| - 2 < 0$
  (f) $3 + |1 - 2x| < 7$

B.7. Solve these inequalities.
  (a) $|2x + 3| > 1$
  (b) $|3x - 1| \geq 2$
  (c) $|5x + 2| \geq \frac{3}{4}$
  (d) $|2 - 3x| > 4$
  (e) $|4 - 7x| > 1$
  (f) $|2x - 3| \geq \frac{1}{2}$

B.8. Solve these inequalities.

(a) $|\frac{12}{5} + 2x| > \frac{1}{4}$ (c) $\left|\dfrac{3 - 4x}{2}\right| > 5$ (e) $\left|\dfrac{2 - 3x}{4}\right| \leq \frac{3}{2}$

(b) $|\frac{5}{6} + 3x| < \frac{7}{6}$ (d) $\left|\dfrac{2 - 5x}{3}\right| \geq 2$ (f) $\left|\dfrac{5 - 2x}{3}\right| < \frac{3}{4}$

B.9. Use a calculator to solve these inequalities.
(a) $7.35x - 6.42 > 5.37 - 12.24x$
(b) $8.21 - 6.75x \leq 3.59x + 2.74$
(c) $8.53 (2.11x + 5.32) < 2.65 (3.21 - 6.42x)$
(d) $(2.57 - 3.26x) 6.25 \geq 1.73 (2.71x + 4.32)$

B.10. Solve these inequalities and explain your answers.
(a) $|2x + 1| \geq -1$ (c) $|5x - 1| < 0$ (e) $|x - 3| \leq -2$
(b) $|3x + 2| < 0$ (d) $|7x + 3| \geq 0$

C.1. Let a and b be fixed real numbers with $a < b$. Show that the solutions of

$$\left|x - \dfrac{a + b}{2}\right| < \dfrac{b - a}{2}$$

are all x with $a < x < b$.

C.2. Let E be a fixed real number. Show that every solution of $|x - 3| < E/5$ is also a solution of $|(5x - 4) - 11| < E$.

Introduction to Analytic Geometry

One of the great discoveries of seventeenth-century mathematics was that the techniques of algebra could be applied to geometry. The resulting subject was called analytic geometry. This chapter contains the essential parts of analytic geometry needed in calculus. The discussion of straight lines in Sections 3 and 4 is especially important since tangent lines play a crucial role in the development of calculus.

1. THE COORDINATE PLANE

In Chapter 1, we associated numbers with the points on the line. Now we are going to associate pairs of real numbers with the points in the plane. This will allow us to interpret many algebraic ideas geometrically, so that our geometric intuition can aid us. At the same time we can use algebraic methods to help solve geometric problems.

Let's use this page to represent a plane. Draw a horizontal and a vertical line. The point where they intersect is called the **origin.** Taking the origin as zero, make the horizontal line into a number line with positive numbers to the right and negative numbers to the left. Make the vertical line into a number line with positive numbers going up and negative numbers going down (Figure 2-1).

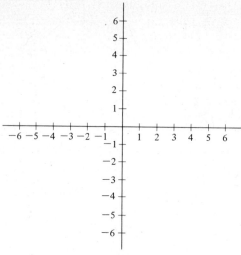

Figure 2-1

We call the horizontal line the **horizontal axis** and the vertical line the **vertical axis.** Together these two axes are called the **coordinate axes.** A plane equipped with coordinate axes is called a **coordinate plane.**

We often label the axes by letters of the alphabet. It is traditional to label the horizontal axis by x and the vertical axis by y. Then the horizontal axis is called the **x-axis,** the vertical axis the **y-axis,** and the coordinate plane the **xy-plane.** However, there is nothing sacred about the letters x and y. Whenever convenient we shall use other letters.

THE COORDINATES OF A POINT

Let P be any point in the plane. Draw two straight lines through P, one vertical, the other horizontal. These lines intersect the horizontal axis and the vertical axis at some numbers c and d, as illustrated in Figure 2-2 for two typical situations.

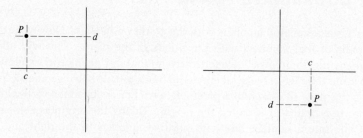

Figure 2-2

The point P is now associated with an ordered pair of numbers (c, d).° We say (ungrammatically) that the **coordinates** of P are the ordered pair (c, d), or more briefly, that P has coordinates (c, d). This means that c is the number directly above or below P on the horizontal axis, and d is the number diretly to the left or right of P on the vertical axis. The number c is called the **first coordinate** of P and d is called the **second coordinate** of P. If the horizontal and vertical axes are labeled by letters—say, x and y—and the point P has coordinates (c, d), then we say that P has **x-coordinate** c and **y-coordinate** d.

EXAMPLE Convince yourself that the points in Figure 2-3 are labeled with the correct coordinates.

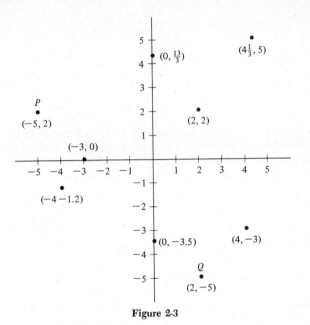

Figure 2-3

It is important to realize what the word "ordered" means in the term "ordered pair." In Figure 2-3, the point P with coordinates $(-5, 2)$ is quite different from the point Q with coordinates $(2, -5)$. The same numbers -5 and 2 occur both times, but in *different order*. The first coordinate of P is -5, but the first coordinate of Q is 2. Likewise, the second coordinates of P and of Q differ.

We have begun with a point P and labeled it by an ordered pair of numbers, the coordinates of P. Conversely, given an ordered pair of numbers (r, s), it is easy to find the point with coordinates (r, s). Because of this correspondence between points and

° The notation (c, d) is also used for an open interval on the number line (see p. 11). This won't cause any difficulty since the contest will always make it clear whether (c, d) denotes an interval or the coordinates of a point.

ordered pairs, we often "identify" a point with its coordinates. For example, we usually say "the point $(3.9, -5)$" instead of "the point with coordinates $(3.9, -5)$."

QUADRANTS

The coordinate axes divide the plane into four regions which are numbered counterclockwise starting from the upper right, as shown in Figure 2-4.

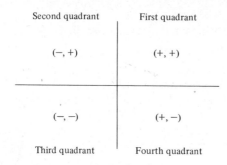

Second quadrant   First quadrant

$(-, +)$ $(+, +)$

$(-, -)$ $(+, -)$

Third quadrant   Fourth quadrant

Figure 2-4

We can easily tell which quadrant a point is in by noticing which of its coordinates are positive and which are negative. Points in the first quadrant have both coordinates positive, points in the second quadrant have negative first coordinates and positive second coordinates, and so on, as shown in Figure 2-4. Note that points on the axes do not lie in any of the quadrants.

STANDARD NOTATION FOR COORDINATES

Let P and Q be two points in the plane. We could use letters such as (c, d) for the coordinates of P and (r, s) for those of Q as we did above. But whenever two points, P and Q, are to be discussed simultaneously, it is more convenient to denote the coordinates of P by (x_1, y_1) and those of Q by (x_2, y_2). The symbol "x_1" (read "x-one") is a single symbol denoting the first coordinate of the first point, just as the letter c did above. Similarly, y_1 is a single symbol, as are x_2 and y_2.

The chief advantage of using (x_1, y_1) and (x_2, y_2) is that it makes it easy to keep everything straight; for instance,

All first coordinates are x's: x_1, x_2.
All second coordinates are y's: y_1, y_2.
The coordinates of the first point have subscript 1: (x_1, y_1).
The coordinates of the second point have subscript 2: (x_2, y_2).

THE DISTANCE FORMULA

A natural question to ask about two points in the plane is, "How far apart are they?" Here is a simple formula for answering this question:

THE DISTANCE FORMULA

The distance between the point (x_1, y_1) *and the point* (x_2, y_2) *is the number*

$$\sqrt{(x_1 - x_2)^2 + (y_1 - y_2)^2}$$

Before we prove that the distance formula is correct, let's see how it works in some examples.

EXAMPLE   Find the distance from $(-8, -5)$ to $(1, -2)$. Let $(-8, -5)$ play the role of (x_1, y_1) and $(1, -2)$ that of (x_2, y_2) in the distance formula. Then the distance between them is

$$\sqrt{(x_1 - x_2)^2 + (y_1 - y_2)^2} = \sqrt{(-8 - 1)^2 + (-5 - (-2))^2}$$
$$= \sqrt{(-9)^2 + (-5 + 2)^2} = \sqrt{81 + (-3)^2}$$
$$= \sqrt{81 + 9} = \sqrt{90} = \sqrt{9}\sqrt{10} = 3\sqrt{10}$$

The order in which the two points are listed doesn't make any difference. If we use $(1, -2)$ for (x_1, y_1) and $(-8, -5)$ for (x_2, y_2), we obtain the same answer:

$$\sqrt{(x_1 - x_2)^2 + (y_1 - y_2)^2} = \sqrt{(1 - (-8))^2 + (-2 - (-5))^2}$$
$$= \sqrt{9^2 + 3^2} = \sqrt{81 + 9} = \sqrt{90} = 3\sqrt{10}$$

EXAMPLE   Let a and b be fixed real numbers. What is the distance from the point $(a, -b)$ to the point $(2a, -b)$? Don't let the letters confuse you. We can apply the distance formula just as before. In this case (a, b) is the first point [corresponding to (x_1, y_1) in the distance formula] and $(2a, -b)$ is the second point [corresponding to (x_2, y_2) in the formula]. Consequently, in the distance formula we substitute a for x_1, b for y_1, $2a$ for x_2, and $-b$ for y_2:

$$\sqrt{(x_1 - x_2)^2 + (y_1 - y_2)^2} = \sqrt{(a - 2a)^2 + (b - (-b))^2}$$
$$= \sqrt{(-a)^2 + (b + b)^2} = \sqrt{a^2 + (2b)^2} = \sqrt{a^2 + 4b^2}$$

Since $a^2 + 4b^2 \geq 0$, no matter what a and b are, the distance $\sqrt{a^2 + 4b^2}$ is a well-defined real number.

DERIVATION OF THE DISTANCE FORMULA

Let P with coordinates (x_1, y_1) and Q with coordinates (x_2, y_2) be two points in the plane that do not lie on the same vertical or on the same horizontal line. We want to find the distance from P to Q. We shall illustrate the argument with pictures of one possible case, but the validity of the argument does *not* depend on the pictures.

First, draw a *vertical* line through P and a *horizontal* line through Q. Denote the point where these two lines intersect by R, as shown in Figure 2-5.

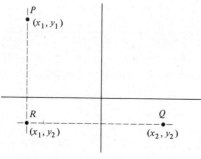

Figure 2-5

Since R is on the same vertical line as P, it has the same first coordinate as P, namely, x_1. Since R is on the same horizontal line as Q, it has the same second coordinate as Q, namely, y_2.

We now find the length of the line segments from P to R and Q to R. The line segment RQ is parallel to the horizontal axis, as shown in Figure 2-6.

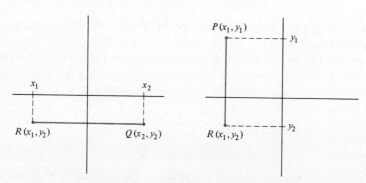

Figure 2-6

Consequently, the length of RQ is the same as the distance from x_1 to x_2 on the axis, namely, $|x_1 - x_2|$. Similarly, the segment PR is parallel to the vertical axis (see Figure 2-6.) Thus its length is the distance from y_1 to y_2 on the axis, namely, $|y_1 - y_2|$.

We now know the lengths of two sides of the right triangle PRQ (Figure 2-7).

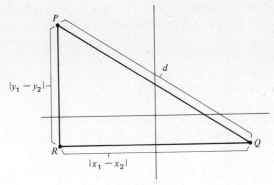

Figure 2-7

The length d of the third side of the triangle is precisely the distance from P to Q. According to the Pythagorean Theorem,[*]

$$d^2 = |x_1 - x_2|^2 + |y_1 - y_2|^2$$

Since $|c|^2 = c^2$ for any real number c (see page 14), this equation becomes

$$d^2 = (x_1 - x_2)^2 + (y_1 - y_2)^2$$

Thus d is a number whose square is the number $(x_1 - x_2)^2 + (y_1 - y_2)^2$. Since d is a length, we know that $d \geq 0$. Therefore we must have

$$d = \sqrt{(x_1 - x_2)^2 + (y_1 - y_2)^2}$$

Thus we have proved the distance formula for any two points that don't lie on the same vertical or horizontal line.

Suppose now that (x_1, y_1) and (x_2, y_2) *do* lie on the same vertical line. Then they must have the same first coordinate; that is, $x_1 = x_2$, as shown in Figure 2-8.

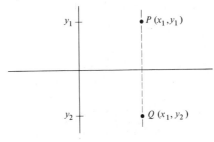

Figure 2-8

[*] If you don't remember this, see the Geometry Review on page 581.

The distance from P to Q is the same as the distance from y_1 to y_2 on the number line, namely, $|y_1 - y_2|$. Since $|y_1 - y_2| \geq 0$ and since $|c|^2 = c^2$ for any number c, we have:

$$|y_1 - y_2| = \sqrt{|y_1 - y_2|^2} = \sqrt{(y_1 - y_2)^2} = \sqrt{0 + (y_1 - y_2)^2}$$

Since $x_1 = x_2$ here, $x_1 - x_2 = 0$, so that

$$|y_1 - y_2| = \sqrt{0 + (y_1 - y_2)^2} = \sqrt{(x_1 - x_2)^2 + (y_1 - y_2)^2}$$

Therefore the distance formula is correct in this case too. A similar argument takes care of the case when (x_1, y_1) and (x_2, y_2) lie on the same horizontal line; that is, when $y_1 = y_2$.

EXERCISES

A.1. Find the coordinates of each of the points in Figure 2-9.

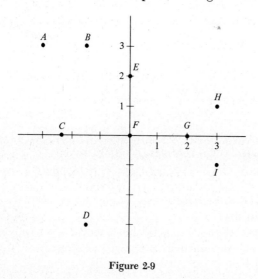

Figure 2-9

A.2. We often call the process of drawing coordinate axes and locating a point on the plane **plotting a point.** Plot the points $(0, 0), (-3, 2.1), (2.1, -3), (-\frac{4}{3}, 1), (5, \pi),$ $(2, \sqrt{2}), (-3, \pi), (4, 6), (-\sqrt{3}, \sqrt{3}), (\sqrt{3}, -\sqrt{3}), (\frac{5}{2}, \frac{17}{3})$.

A.3. Various geometric arguments, using similar triangles, show that the **midpoint** of the line segment joining points (x_1, y_1) and (x_2, y_2) is the point with the coordinates

$$\left(\frac{x_1 + x_2}{2}, \frac{y_1 + y_2}{2} \right)$$

An easy way to remember this is that the first coordinate of the midpoint is the average of the two first coordinates x_1 and x_2; and similarly for the second coordinate. Plot each of the following pairs of points on a coordinate plane. Find the midpoint of the line segment joining them and plot it.

(a) $(1, 2), (-4, 6)$ (c) $(-2, 3), (7, 9)$ (e) $(\frac{5}{2}, 1), (4, -\frac{4}{3})$

(b) $(0, 1), (5, -2)$ (d) $(0, 0), (-6, -3)$ (f) $(-3, \sqrt{2}), (-8, -2)$

A.4. Find the distance between each pair of points.

(a) $(-3, 5), (2, -7)$ (c) $(\frac{1}{2}, 3), (\frac{9}{2}, 6)$ (e) $(a, b), (b, a)$

(b) $(1, -5), (2, -1)$ (d) $(\sqrt{2}, 1), (\sqrt{3}, 2)$ (f) $(s, t), (0, 0)$

B.1. (a) Plot the following points on one coordinate plane: $(-4, 5), (-4, 1),$ $(-4, -7.7), (-4, 0), (-4, -\frac{9}{2}), (-4, \frac{8}{13})$.

(b) Describe the geometric figure formed by the set of all points whose first coordinate is -4.

B.2. Fill the blanks: if the first and second coordinates of a point P are equal, then either P is $(0, 0)$ or P is in the _____ quadrant, or the _____ quadrant.

B.3. One diagonal of a square has endpoints $(-3, 1)$ and $(2, -4)$. What are the coordinates of the endpoints of the other diagonal? (*Hint:* draw a sketch.)

B.4. (a) Find the vertices of all possible squares with this property: two of the vertices are $(2, 1)$ and $(2, 5)$. (*Hint:* there are three such squares.)

(b) Do part (a) with (c, d) in place of $(2, 1)$ and (c, k) in place of $(2, 5)$.

B.5. What are the coordinates of the point which lies on the line from $(-8, -3)$ to $(6, 9)$, one-fourth of the way from $(-8, -3)$? (Exercise A.3 may be helpful.)

B.6. Give a geometric description of the set of *all* points (x, y) in the plane that satisfy the given condition. (When in doubt plot some points that satisfy the given condition.) *Example:* Condition: $y = 5$; geometric description: all points (x, y) with $y = 5$ form a horizontal straight line 5 units above the x-axis.

(a) $x = 3$ (c) $x = y$ (e) $xy > 0$ (g) $\dfrac{y}{x} > 0$

(b) $y = -7$ (d) $xy = 0$ (f) $xy < 0$ (h) $x + 2 > 0$

B.7. Plot all points (x, y) whose coordinates satisfy the given conditions. *Example:* if the condition is $|x| \leq 1$, then the points are those whose first coordinate satisfies $|x| \leq 1$, that is, $-1 \leq x \leq 1$. The set of all such points is shaded in Figure 2-10.

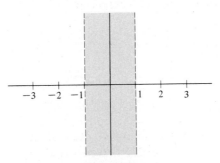

Figure 2-10

(a) $|x| \leq 1$ and $|y| \leq 1$

(b) x and y are positive

(c) $1 \leq x \leq 3$ and $|y| \leq 2$

(d) $|x| \geq 1$

(e) $|x| \geq 1$ and $|y| \leq 1$

(f) $|x| \geq 2$ and $|y| \geq 1$

B.8.   Show that each triple of points given below forms a right triangle. In each case, give the length of the hypotenuse and sketch the triangle. (*Hint:* you may assume that a triangle with sides of length a, b, and c is a right triangle with hypotenuse c, *provided* $c^2 = a^2 + b^2$.)

(a) $(0, 0)$, $(1, 1)$, $(2, -2)$

(c) $(3, -2)$, $(0, 4)$, $(-2, 3)$

(b) $\left(\frac{\sqrt{2}}{2}, 0\right)$, $\left(\frac{\sqrt{2}}{2}, \frac{\sqrt{2}}{2}\right)$, $(0, 0)$

B.9.   What is the perimeter of the triangle with vertices $(1, 1)$, $(5, 4)$, and $(-2, 5)$?

B.10.   (a)   Find all points P on the x-axis that are 5 units from the point $(3, 4)$. (*Hint: P* must have coordinates $(x, 0)$ for some number x and the distance from P to $(3, 4)$ is 5.)

(b)   Find all points on the y-axis that are 8 units from $(-2, 4)$. (*Hint:* reduce the equation you set up to the form $(y - 4)^2 = $ constant and solve as in the example on page 572.)

(c)   Find all points with first coordinate 3 that are 6 units from $(-2, -5)$.

(d)   Find all points with second coordinate -1 that are 4 units from $(2, 3)$.

B.11.   Let S be the triangle with vertices $(\frac{1}{2}, 4)$, $(-3, \frac{1}{2})$, $(1, -\frac{1}{2})$, and T the triangle with vertices $(4, 5)$, $(3, 1)$, $(7, 2)$. Are S and T congruent? (Remember two triangles are congruent if the corresponding sides have the same length.)

C.1.   Suppose every point on the coordinate plane gets moved 5 units straight up.

(a)   To what points do each of these points go: $(0, -5)$, $(2, 2)$, $(5, 0)$, $(5, 5)$, $(4, 1)$?

(b)   Which points go to each of the points in part (a)?

(c)   To what point does (a, b) go?

(d)   To what point does $(a, b - 5)$ go?

(e)   What point goes to $(-4a, b)$?

(f)   What points go to themselves?

2. GRAPHS AND GRAPH READING

In its most general usage the word **graph** means any set of points in the plane. The graphs that are of the most interest and usefulness are those that provide a geometric visualization of various algebraic or verbal statements. To deal effectively with such graphs you must have certain translation skills. You must be able to

> Translate verbal and algebraic statements into equivalent graphical pictures;
> Translate graphical information into equivalent verbal or algebraic statements.

The first of these translation skills will be considered at various times throughout the rest of the book. In this section, we concentrate on the second skill, which is usually referred to as "graph reading," Since there are no hard and fast rules to handle all situations, we shall present several varied examples.

EXAMPLE The graph in Figure 2-11 shows the distance traveled by a car during an hour's time.

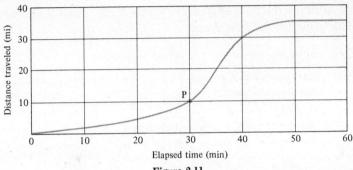

Elapsed time (min)

Figure 2-11

For example, the point (30, 10) is on the graph. This indicates that after 30 minutes (min), the car has traveled 10 miles (mi). The question,

How many miles does the car travel in the first 40 min?

can be answered by examining the point on the graph with first coordinate 40 (representing the elapsed time of 40 min). This point is (40, 30). Therefore the car traveled 30 mi in the first 40 min. The question,

How long does it take the car to travel the first 5 mi?

requires some careful measurement to answer. If you examine the point on the graph with *second* coordinate 5 (representing 5 mi traveled) you will see that it is (approximately) the point (20,5). Therefore it took 20 min to travel the first 5 mi. The graph also provides the answer to this question:

Does the car ever stop during the hour? If so, for how long?

Observe that the point (50, 35) is on the graph, meaning that after 50 min the car has traveled 35 mi. But the graph is *horizontal* to the right of this point, so that every point on the graph to the right of (50, 35) has the same second coordinate, 35. This means that at every time from 50 min to 60 min, the total distance traveled remained 35 mi. In other words, the car did not move during the 10-min period from 50 to 60 min. Finally,

During what 10-min period was the car traveling at the fastest speed?

In the 10-min period from 30 to 40 min the graph rises more steeply than in any other 10-min period. This means that a greater distance was traveled during this 10-min period than during any other one (why?). But a greater distance can be traveled in the same length of time only by going at a faster rate. Therefore the speed was fastest from 30 to 40 min. In fact, it was too fast for safety: the graph shows that the car has traveled 10 mi after 30 min and 30 mi after 40 min. Thus it traveled 20 mi during that 10 min period. But 20 mi per 10 min is equivalent to 120 miles per hour.

EXAMPLE The weather bureau has a device which records the temperature over the 24 hr of a given day in the form of a graph (Figure 2-12). Time is measured in hours after midnight along the horizontal axis.

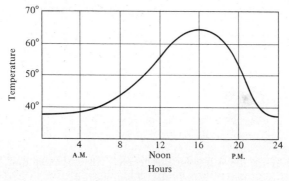

Figure 2-12

This graph can be used to answer several questions, including:

At what times during the day was the temperature 40° or more?

Any time at which temperature is 40° or more is represented on the graph by a point whose *second* coordinate is ≥ 40. Thus the time period when the temperature was 40° or more is the period during which the graph lies on or above the horizontal line through 40°. An examination of the graph shows that it lies on or above this line (approximately) from 5:24 A.M. ($= 5\frac{2}{5}$ hr after midnight) to 10:08 P.M. ($= 22\frac{2}{15}$ hr after midnight).°

During what period did the temperature rise most slowly?

We can't answer this question exactly, but we can give a reasonable estimate. The graph is almost horizontal from midnight to 4:00 A.M., but rises very slightly during this time interval. This means that at 4:00 A.M. the temperature was only a tiny bit higher than at midnight. No other portion of the graph is both rising and so close to horizontal. So this is the period during which the temperature rose most slowly.

° The measurements here and below were made on the original hand-drawn graphs. Consequently they may occasionally differ slightly from measurements made on the printed graph that appears here.

EXAMPLE At a city reservoir water is let in at one end and out at the other. At each of these two gates there is a meter measuring the *rate* at which the water is flowing in or out (in gallons per hr). The two graphs in Figure 2-13 record the incoming and outgoing rate of flow at any time from noon to 5:00 P.M.

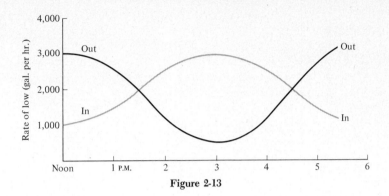

Figure 2-13

First question:

> *If you watch the reservoir from noon to 1:00 P.M.,*
> *what change do you see in the water level?*

Note that the water level will rise if more water is coming in than going out and it will fall if more water is going out than in. Figure 2-13 shows that between noon and 1:00 the outflow graph lies above the inflow graph. This means that at any time from noon to 1:00, water was flowing out at a greater rate than it was flowing in. Therefore the water level in the reservoir was *falling* during this hour.

> *At what time between 2:00 and 4:00 P.M. is the water level*
> *in the reservoir rising at the fastest rate?*

At any given time the difference between the inflow rate and the outflow rate is the rate at which the amount of water in the reservoir is increasing. When this difference is greatest, the water level will be rising at the fastest rate. Between 2:00 and 4:00 P.M. the inflow graph is always above the outflow graph, so the water level is rising. The vertical distance between the two graphs is greatest at approximately 3:00 P.M. This means that at 3:00 P.M. the rate at which the water in the reservoir is increasing, and hence the rate at which the water level is rising, is greatest.

EXERCISES

C.1. A certain mathematics class recently turned in a six-problem homework assignment. Each problem was graded either right or wrong (no partial credit!), so the number of correct problems a student did was one of these numbers: 0, 1, 2, 3,

4, 5, or 6. The graph in Figure 2-14 shows how many students got 0, 1, 2, 3, 4, 5, or 6 correct problems. For example, the point $(1, 4)$ indicates that 4 students got only 1 problem correct.

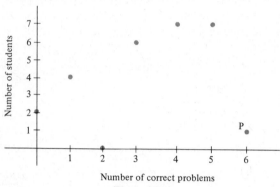

Figure 2-14

(a) How many people got exactly four correct problems?
(b) Find a number x so that exactly two students have x correct problems.
(c) How many students got less than three problems correct?
(d) If exactly one student got t correct answers, then what is t?
(e) How many students got more than four problems correct?
(f) How many students handed in this assignment?

C.2. The Pink Chopper Company produces motorbikes especially designed to please those people between 18 and 21 years old. In order to estimate the demand for motorbikes in the years to come, the company uses the graph in Figure 2-15. It shows the expected population of the 18–21 age group in future years.

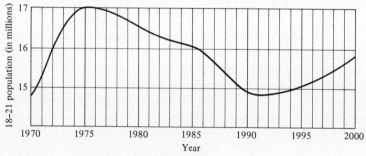

Figure 2-15

(a) How many 18–21 year olds are there in 1980?
(b) How many 18–21 year olds will there be in 1984? in 1988? in 1992?
(c) Find a year in which the population of 18–21 year olds is 16 million.
(d) Find all years in which the population of 18–21 year olds is *less* than 15 million.
(e) If from one year to the next, the number of 18–21 year olds increases, the

demand for motorbikes will probably also increase and the Pink Chopper Company will have to make plans ahead of time to increase production. Similarly, if from one year to the next the number of 18–21 year olds decreases, the Pink Chopper Company must make plans to decrease production. With this in mind, indicate in the following table whether production has to be increased or decreased in the next year.

Year	Increased	Decreased
1982		
1988		
1990		
1995		

(f)   It takes three years to ge a new plant ready for production of motorbikes. In which years would it be a definite mistake to start construction of a new plant? Explain.

C.3.   In Figure 2-16 are the graphs showing the price per share of four stocks during the period from March 1 to June 1.

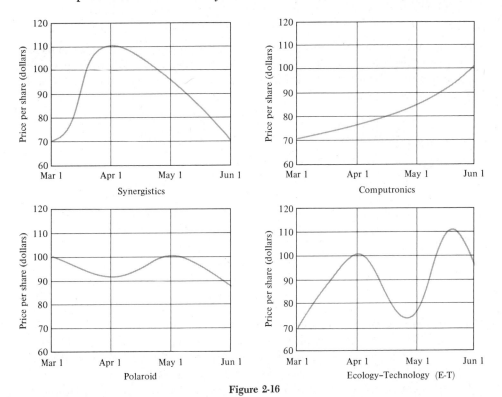

Figure 2-16

In the questions below, when we speak of "holding" a given stock over some time period we mean to buy the stock at the beginning of the period and sell it at the end.

(a)   For maximum profit, would Computronics or E-T have been a better holding over the period March 1 to June 1?

(b)   The increase in the price per share for Computronics from March 1 to June 1 is the same as the increase in price per share of E-T from March 1 to April 1. Is there any basis for arguing that holding Computronics for the three months is better or worse than holding E-T for one? Explain.

(c)   Synergistics did not do well over the three-month period while Computronics did very well. On the other hand, is there some 10-day period when a holding of Synergistics would have brought more profit than any 10-day period of Computronics? Explain.

(d)   The graphs are not sufficiently detailed to calculate profits or losses for holding periods of three days in length. However, is there something about the shapes of these graphs to justify holding one stock for three days rather than another stock? Explain.

C.4.   The graph in Figure 2-17 gives the cost of producing a certain number of items. The items bear serial numbers in the order they are produced, the first one bearing the number 1. At 0 the curve shows $1000, meaning that the fixed costs for rent, equipment, and so on, are $1000 even if no items are produced.

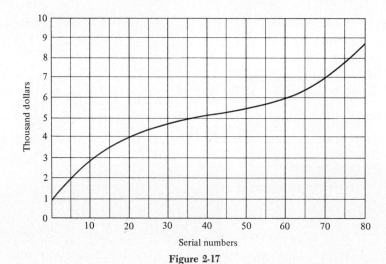

Figure 2-17

(a)   How many items can you produce if you want to spend $5000?
(b)   What is the cost of producing the items numbered 35 through 80?
(c)   What is the average cost per item of the first 25 items? the first 45 items?
(d)   What is the average cost of items 25 through 45?

(e) Which consecutively numbered 20 items cost the most to produce? Which cost the least? How can you decide this question with your ruler?

(f) Suppose you want to stop production when the average cost per item is least. At which serial number would you have to stop?

C.5. Figure 2-18 is a graph showing the distance traveled by each of two cars in a drag race.

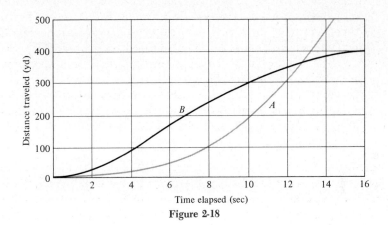

Figure 2-18

(a) Estimate the distance traveled by each car in the time intervals indicated in the following table:

Interval	Car A	Car B
From 0 to 5 sec		
From 5 to 8 sec		
From 5 to 6 sec		
From 10 to 12 sec		

(b) In the time interval from 10 sec to 11 sec car A has gone about 50 yards (yd) while car B has gone about 25 yd. Over this time period, which car was going faster? Over this time period which car was leading the race? Explain both answers.

(c) If you were going to describe the race from one second to the next, you would presumably want to say which car was in the lead and which car was going faster at each second. Give such a description for the period from 6 sec to 14 sec.

(d) If you wanted to describe drag races from graphs similar to the one above, what would you look for to find out who is ahead at a given time? What would you look for to find out which car is going faster over a particular period?

C.6. Two cars start a trip at the same time, but from opposite ends, of a 40-mi-long road. The distance traveled after time t is graphed for each car in Figure 2-19.

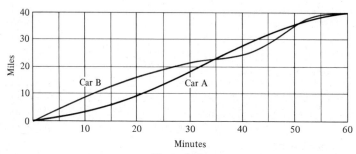

Figure 2-19

(a) How far did car A travel in the first 25 min?
(b) How long did it take car A to travel the first 25 mi?
(c) How far did car B travel between 15 and 30 min after start?
(d) Over which 10-min interval did car A cover the most distance? and car B?
(e) What was the average speed (in miles per minute) for car A over the entire trip? for car B?
(f) What was the average speed for car A over the time interval from 10 to 35 min after start? And for car B?
(g) Over which 10-min interval did car A have the greatest average speed? And car B? How can you decide this question quickly with a ruler?
(h) How far had car A traveled in the first 10 min? And car B? How far were they apart after the first 10 min of travel? How far apart were they after the first 15 min?

C.7. A gauge is attached to the city water reservoir which measures the depth of the water. City water officials can then figure out how much water is in the reservoir. Water is being pumped both in and out of the reservoir so that the water level may rise if the water is being pumped in faster than it is being pumped out, or the water level may fall if the water is being pumped out faster than it is being pumped in. The graph in Figure 2-20 shows the recorded depths over a 12-hr period.

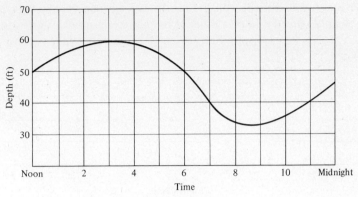

Figure 2-20

(a) Give the change in the depth for the following time periods:
    (i) noon to 3:00. [Solution: the depth at noon is 50 feet (ft) and the depth at 3:00 is 60 ft. The *change* in the depth would be (60 ft − 50 ft) or 10 ft.]
    (ii) noon to 2:00.   (iii) noon to 6:00.   (iv) 7:00 to midnight.
(b) Name the 1-hr period (that is, from 1:00 to 2:00, or 4:00 to 5:00, and so forth) in which the amount of change was the greatest. Name the 1-hr period in which the amount of change was the least.
(c) The actual change in depth between noon and 3:00 was 10 ft. The actual change in depth between 7:30 and midnight was also 10 ft. Is it possible to compare the *rates* of change over these two time periods? In which of these two time periods would you say that the rate of change was the greatest, or in other words, over which time period did the water level rise the fastest? Explain.
(d) The actual change in depth between noon and 2:00 was 8 ft, and the change between 10:00 and 11:00 was 4 ft. Can you compare the rates of change in these two time periods? Does one rate of change seem bigger than the other, or are they about the same? Explain.
(e) The change in depth was 26 ft between 3:00 and 8:00. Can you find some 1-hr period between 3:00 and 8:00 where the *rate* of change over the 1-hr period is the same as the *rate* of change between 3:00 and 8:00? If so, which period? Can you find a 1-hr period where the rate of change over that hour is less than the rate of change from 3:00 to 8:00? If so, which period?

3. SLOPES OF LINES

Throughout this section and the next, the word "line" is understood to mean "straight line." The question of how steeply a line rises is easy to understand *geometrically*. In

Figure 2-21 line *L* does not rise at all. Line *M* rises, but not very steeply, and line *N* rises quite steeply. (It may help to think of lines as roads on which you are walking.)

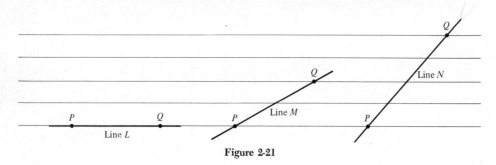

Figure 2-21

We would like to find a *numerical* way to describe the steepness of these lines. Since line *L* doesn't rise at all, it is reasonable to use the number 0 to describe its steepness. Line *M* rises 2 units as you go from *P* to *Q*. But the steeper line *N* rises 4 units as you go from *P* to *Q*. So steepness seems to be related to the *vertical rise* of the line.

On the other hand, consider the three lines in Figure 2-22.

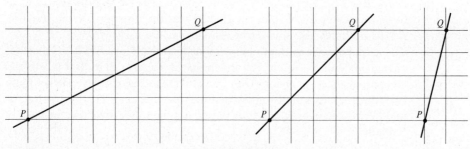

Figure 2-22

All of them have a vertical rise of 4 units as you go from *P* to *Q*. But the lines obviously don't have the same steepness. The difference is this: on the left-hand line, as you go from *P* to *Q*, you run 8 units horizontally while you rise 4 units vertically. But on the middle line you run only 4 units horizontally while rising 4 units vertically. And on the right-hand line you run just 1 unit horizontally while rising 4 units vertically.

The preceding examples suggest that the steepness of a straight line between points *P* and *Q* on the line is determined by *comparing* two factors:

The **vertical rise** from *P* to *Q*.
The **horizontal run** from *P* to *Q*.

The three lines above show how these two factors should be compared. In the left-hand line (the least steep) the rise (4 units) is only *half* the run (8 units). In the steeper middle line, the rise (4 units) is the *same* as the run (4 units). In the right-hand line (the steepest one) the rise (4 units) is *four times* larger than the run (1 unit). We can express these facts in terms of fractions by looking at the fraction $\dfrac{\text{rise}}{\text{run}}$, as illustrated in Figure 2-23.

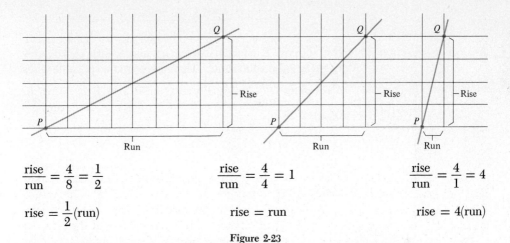

$$\frac{\text{rise}}{\text{run}} = \frac{4}{8} = \frac{1}{2} \qquad\qquad \frac{\text{rise}}{\text{run}} = \frac{4}{4} = 1 \qquad\qquad \frac{\text{rise}}{\text{run}} = \frac{4}{1} = 4$$

$$\text{rise} = \frac{1}{2}(\text{run}) \qquad\qquad \text{rise} = \text{run} \qquad\qquad \text{rise} = 4(\text{run})$$

Figure 2-23

The numbers $\frac{1}{2}$, 1, and 4 are increasing just as the steepness of the corresponding lines increases. Thus the fraction $\dfrac{\text{rise}}{\text{run}}$ seems to provide a numerical means of describing the steepness of a line. But there are still some unanswered questions:

Question (i)   Is there some easy way to compute $\dfrac{\text{rise}}{\text{run}}$ without using a ruler or graph paper?

Question (ii)   If we choose two points on the line other than P and Q to compute $\dfrac{\text{rise}}{\text{run}}$, will we get the same answer?

Answer to Question (i)   Suppose P and Q are points on the line L. If P has coordinates (x_1, y_1) and Q has coordinates (x_2, y_2), then Figure 2-24

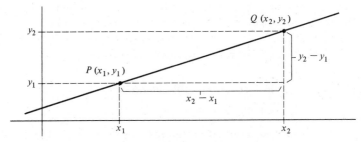

Figure 2-24

shows clearly that the vertical rise as you move from P to Q is the number $y_2 - y_1$ and the horizontal run is the number $x_2 - x_1$. Therefore

$$\frac{\text{rise}}{\text{run}} = \frac{y_2 - y_1}{x_2 - x_1}$$

Answer to Question (ii) Now suppose R coordinates (x_3, y_3) and S with coordinates (x_4, y_4) are also points on L. Then there are two possible ways to compute $\dfrac{\text{rise}}{\text{run}}$, using P and Q or R and S, as shown in Figure 2-25.

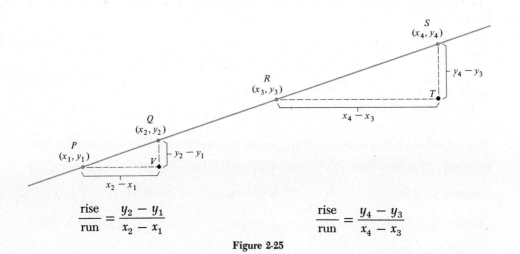

$$\frac{\text{rise}}{\text{run}} = \frac{y_2 - y_1}{x_2 - x_1} \qquad\qquad \frac{\text{rise}}{\text{run}} = \frac{y_4 - y_3}{x_4 - x_3}$$

Figure 2-25

We *claim* that the end result in each case is the same, namely,

---

For any distinct points (x_1, y_1), (x_2, y_2), (x_3, y_3), (x_4, y_4) *on* L,

$$\frac{y_2 - y_1}{x_2 - x_1} = \frac{y_4 - y_3}{x_4 - x_3}$$

In other words, the answer to Question (ii) is yes: we can use *any two points* on L to compute $\dfrac{\text{rise}}{\text{run}}$ and we'll get the *same answer* every time.

Here is a short proof of the claim made in the box above. Consider the triangles PQV and RST (as in Figure 2-25). Since angle V and angle T are both right angles, they are equal. Basic facts about parallel lines show that

$$\text{angle } P = \text{angle } R \qquad \text{angle } Q = \text{angle } S$$

Therefore triangles PQV and RST are similar.° Consequently, the ratios of the corresponding sides of the triangles are equal. In particular,

$$\frac{\text{length } QV}{\text{length } PV} = \frac{\text{length } ST}{\text{length } RT}$$

But we can express these lengths in terms of the coordinates of the points, as shown in Figure 2-25:

$$\frac{y_2 - y_1}{x_2 - x_1} = \frac{\text{length } QV}{\text{length } PV} = \frac{\text{length } ST}{\text{length } RT} = \frac{y_4 - y_3}{x_4 - x_3}$$

This proves the claim made in the preceding box.

We can now make an important definition:

> *Let* L *be a nonvertical straight line. The **slope** of* L *is the number*
> $$\frac{y_2 - y_1}{x_2 - x_1}$$
> *where* (x_1, y_1) *and* (x_2, y_2) *are any two distinct points on* L. *In short,*
> $$slope = \frac{rise}{run}.$$

As the preceding discussion suggests,

The slope of a line is a number which measures the steepness of the line.

The slope is independent of the points on the line used to compute it [see the answer to Question (ii) above]. Finally, since

$$\frac{y_1 - y_2}{x_1 - x_2} = \frac{-(y_2 - y_1)}{-(x_2 - x_1)} = \frac{y_2 - y_1}{x_2 - x_1}$$

it doesn't matter in what order we use the points: we can take (x_2, y_2) as the first point and (x_1, y_1) as the second, or the other way around.

° The basic facts about similar triangles are discussed in the Geometry Review on page 584.

EXAMPLE The lines L_1, L_2, L_3, L_4, and L_5 shown in Figure 2-26 are determined by these points:

L_1: $(-2, -1)$ and $(-1, 6)$ L_2: $(-2, -1)$ and $(1, 8)$ L_3: $(-2, -1)$ and $(2, 3)$
L_4: $(-2, -1)$ and $(1, 0)$ L_5: $(-2, -1)$ and $(5, -1)$

We can compute the slopes of these lines as shown in Figure 2-26.

$$\text{slope } L_1 = \frac{6 - (-1)}{-1 - (-2)} = \frac{7}{1} = 7$$

$$\text{slope } L_2 = \frac{8 - (-1)}{1 - (-2)} = \frac{9}{3} = 3$$

$$\text{slope } L_3 = \frac{3 - (-1)}{2 - (-2)} = \frac{4}{4} = 1$$

$$\text{slope } L_4 = \frac{0 - (-1)}{1 - (-2)} = \frac{1}{3}$$

$$\text{slope } L_5 = \frac{-1 - (-1)}{5 - (-2)} = \frac{0}{7} = 0$$

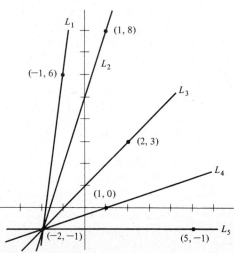

Figure 2-26

Observe how the slopes correspond to the steepness of the lines: horizontal lines have slope 0 and *the larger the slope, the more steeply the line rises* from left to right.

As you may have noticed, we have not yet considered any lines which *fall* from left to right, such as those shown in Figure 2-27.

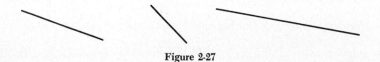

Figure 2-27

But the slope, as defined above, can be computed for *any* line. So let's see what the slopes of falling lines look like.

EXAMPLE The lines L_6, L_7, L_8, and L_9 shown below are determined by these points:

L_6: $(-3, 2)$ and $(4, 1)$ L_7: $(-3, 2)$ and $(2, -4)$
L_8: $(-3, 2)$ and $(0, -6)$ L_9: $(-3, 2)$ and $(-2, -3)$

We can compute the slopes of these lines as shown in Figure 2-28.

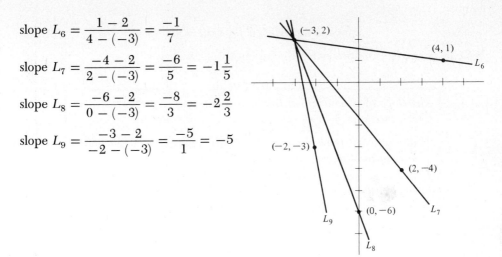

$$\text{slope } L_6 = \frac{1-2}{4-(-3)} = \frac{-1}{7}$$

$$\text{slope } L_7 = \frac{-4-2}{2-(-3)} = \frac{-6}{5} = -1\frac{1}{5}$$

$$\text{slope } L_8 = \frac{-6-2}{0-(-3)} = \frac{-8}{3} = -2\frac{2}{3}$$

$$\text{slope } L_9 = \frac{-3-2}{-2-(-3)} = \frac{-5}{1} = -5$$

Figure 2-28

Each of these lines moves *downward* from left to right and each slope is a *negative number*. In this case too, the slopes correspond to the steepness with which the lines fall: *the larger the slope in absolute value,° the more steeply the line falls* from left to right.

Thus our definition of slope is a good numerical measurement of how steeply *any* nonvertical line rises or falls. Note that the slope formula doesn't work for vertical lines. The reason is that any two points on the same vertical line have the same first coordinate, say $x = c$ (see Figure 2-29).

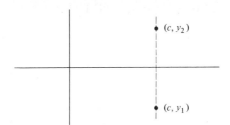

Figure 2-29

° Remember that as negative numbers get farther and farther from 0 (such as $-\frac{1}{7}$, $-\frac{6}{5}$, $-\frac{8}{3}$, -5, -10, -50, -1000), their absolute values get *larger*. Thus a negative number of large absolute value is a number, such as -1000, which is far from 0. Such numbers are sometimes inaccurately but succinctly called "large negative numbers." Lines whose slopes are "large negative numbers" fall very steeply from left to right.

If we try to apply the slope formula, we get $\dfrac{y_2 - y_1}{c - c} = \dfrac{y_2 - y_1}{0}$. But this is *not* a real number. Therefore,

Slope is not defined for vertical lines.

We can summarize the preceding discussion as follows:

> *The slope of a nonvertical line is a number* m
> *which measures how steeply the line rises or falls:*
>
> *If* m *is positive, the line* rises *to the right; the larger* |m| *is, the more
>   steeply the line rises.*
> *If* m *is negative, the line* falls *to the right; the larger* |m| *is, the more
>   steeply the line falls.*
> *If* m = 0, *the line is horizontal.*
>
> *Slope is not defined for vertical lines.*

Here are some examples that show how the slope of a line can be used to find out other information about the line.

EXAMPLE   If L is a line with slope $\dfrac{1}{3}$ and $(2, -4)$ is a point on L, find three more points on L. *Solution:* to say that the slope of L is $\dfrac{\text{rise}}{\text{run}} = \dfrac{1}{3}$ means that if you start at a point on L, run 3 units to the *right*, then *rise* 1 unit, you will end up at another point on L. For instance, see Figure 2-30.

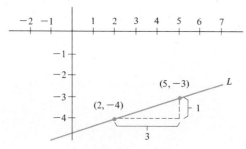

Figure 2-30

Therefore, $(5, -3)$ is a point on L. But $\dfrac{\text{rise}}{\text{run}} = \dfrac{1}{3} = \dfrac{2}{6}$, so we can also say that if you start at a point on L, run 6 units to the *right*, then *rise* 2 units, you will end up at a point on L. If you do this, beginning at $(5, -3)$, you end up at the point $(11, -1)$ on L, as in Figure 2-31. Finally, it is also true that $\dfrac{\text{rise}}{\text{run}} = \dfrac{1}{3} = \dfrac{-1}{-3}$. This means that if you start at a point on L, say, $(11, -1)$, and run 3 units to the *left* (negative horizontal direction), then *fall* 1 unit (negative vertical direction), you will end up at a point on L; namely, $(8, -2)$, as shown in Figure 2-31.

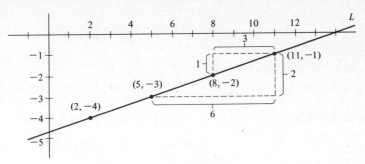

Figure 2-31

The reason the techniques in the preceding example work is that the slope is a *ratio*. Each of the fractions $\dfrac{1}{3}, \dfrac{2}{6}, \dfrac{-1}{-3}$, and so on, expresses the same basic fact: the rise is one-third of the run.

EXAMPLE   If L is the line with slope -3, passing through the point $(1, 2)$, find the point P where L intersects the vertical line through $(5, 0)$. Since L has negative slope, it falls to the right and we can sketch a *rough* picture of the situation (Figure 2-32).

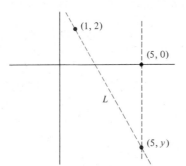

Figure 2-32

Since P lies on the same vertical line as $(5, 0)$, P has coordinates $(5, y)$ for some number y. We must find the number y. To do this we shall find an equation involving only y and various constants, as follows. Use the points $(1, 2)$ and $(5, y)$ to compute the slope of L:

$$\text{slope of } L = \frac{y - 2}{5 - 1} = \frac{y - 2}{4}$$

But L has slope -3. Therefore we have an equation that can be solved for y:

$$\frac{y-2}{4} = \text{slope} = -3$$

$$4\left(\frac{y-2}{4}\right) = 4(-3)$$

$$y - 2 = -12$$

$$y = -10$$

Thus the point P is the point $(5, -10)$.

PARALLEL LINES

The slope of a line measures how steeply it rises or falls. Since parallel lines rise or fall *equally steeply*, it seems plausible that:

> *Two nonvertical straight lines are parallel exactly when they have the same slope.*

Here is a partial proof of this statement. Suppose L and M are parallel lines, with M lying b vertical units above L, as shown in Figure 2-33. Let (x_1, y_1) and (x_2, y_2) be two points on L.

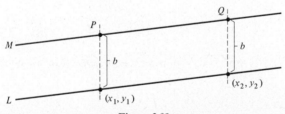

Figure 2-33

Then the slope of L is $(y_2 - y_1)/(x_2 - x_1)$. Consider the point P on M which lies directly above (x_1, y_1). Since it lies on the same vertical line as (x_1, y_1), the first coordinate of P must be x_1. Since the vertical distance from L to M is b, the second coordinate of P must be $y_1 + b$. Thus P is the point $(x_1, y_1 + b)$. A similar argument shows that the point Q which lies on M directly above (x_2, y_2) has coordinates $(x_2, y_2 + b)$. We can now use P and Q to compute the slope of M:

$$\text{slope } M = \frac{(y_2 + b) - (y_1 + b)}{x_2 - x_1} = \frac{y_2 + b - y_1 - b}{x_2 - x_1} = \frac{y_2 - y_1}{x_2 - x_1} = \text{slope } L$$

Therefore parallel lines have the same slope. The rest of the proof (namely, that two lines with the same slope are parallel) is considered in Exercise C.3 in Section 4.

EXAMPLE Let L be the line through $(0, 2)$ and $(1, 5)$. Let M be the line through $(2, 1)$ and $(3, 4)$. Then

$$\text{slope } L = \frac{5 - 2}{1 - 0} = \frac{3}{1} = 3 \quad \text{and} \quad \text{slope } M = \frac{4 - 1}{3 - 2} = \frac{3}{1} = 3$$

Therefore L and M are parallel. You can check this visually by drawing the lines on the coordinate plane.

PERPENDICULAR LINES

Two lines that meet at a right angle (90-deg angle) are said to be perpendicular. As you might suspect, there is a close relationship between the slopes of two perpendicular lines.

> *Let* L *be a line with slope* k *and* M *a line with slope* m. *Then* L *and* M *are perpendicular exactly when* $km = -1$.

We shall illustrate this fact with an example. If you're interested in a proof of it, see Exercise C.4 in Section 4.

EXAMPLE Let L be the line through $(0, 2)$ and $(1, 5)$. Let M be the line through $(-6, -1)$ and $(3, -4)$. Then

$$\text{slope } L = \frac{5 - 2}{1 - 0} = 3 \quad \text{and} \quad \text{slope } M = \frac{-4 - (-1)}{3 - (-6)} = \frac{-3}{9} = -\frac{1}{3}$$

Since $3(-\frac{1}{3}) = -1$, the lines L and M are perpendicular. You can check this visually by drawing the lines on the coordinate plane.

EXERCISES

A.1. Find the slope of the line through the two given points.
   (a) $(1, 2), (3, 7)$ (d) $(-7, -7), (-5, -5)$ (g) $(3, \sqrt{2}), (-4, -\sqrt{5})$
   (b) $(-1, -2), (2, -1)$ (e) $(3, -2), (-4, 6)$ (h) $(\sqrt{2}, -1), (2, -9)$
   (c) $(\frac{1}{4}, 0), (\frac{3}{4}, 2)$ (f) $(\frac{1}{3}, 0), (0, \frac{1}{3})$ (i) $(\pi, 1), (-1, \pi)$

A.2. On one graph, sketch five lines satisfying these conditions: (i) one line has slope 0, two lines have positive slope, and two lines have negative slope; (ii) all five lines meet at a single point.

A.3. (a) On one graph sketch five lines, not all meeting at a single point, whose slopes are five different positive numbers. Do this in such a way that the left-hand line has the largest slope, the second line from the left the next largest slope, and so on.

(b) Do part (a) with nine lines, four of which have positive slope, one of which has slope 0, and four of which have negative slope.

A.4. In each case determine whether the line through P and Q is parallel or perpendicular to the line through R and S, or neither. Then draw the two lines.

(a) $P = (1, 0)$ $\qquad Q = (2, -3)$ $\qquad R = (-1, -6)$ $\qquad S = (2, -15)$
(b) $P = (1, -4)$ $\qquad Q = (2, -3)$ $\qquad R = (-1, 3)$ $\qquad S = (2, 0)$
(c) $P = (2, 5)$ $\qquad Q = (-1, -1)$ $\qquad R = (4, 2)$ $\qquad S = (6, 1)$
(d) $P = (0, \frac{3}{2})$ $\qquad Q = (1, 1)$ $\qquad R = (2, 7)$ $\qquad S = (3, 9)$
(e) $P = (-3, \frac{1}{3})$ $\qquad Q = (1, -1)$ $\qquad R = (2, 0)$ $\qquad S = (4, -\frac{2}{3})$
(f) $P = (3, 3)$ $\qquad Q = (-3, -1)$ $\qquad R = (2, -2)$ $\qquad S = (4, -5)$

B.1. Find five points on the given line, in addition to the given points.
(a) line with slope $\frac{3}{2}$ through $(3, 4)$
(b) line with slope 5 through $(1, -2)$ (Remember $5 = \frac{5}{1}$)
(c) line with slope $-\frac{7}{4}$ through $(0, -2)$

B.2. Graph the following lines:
(a) through $(1, 2)$ with slope 1 $\qquad$ **(d)** through $(-2, 3)$ with no slope
(b) through $(-3, 4)$ with slope -1 $\qquad$ **(e)** through $(0, 0)$ with slope 3
(c) through $(6, 1)$ with slope 0 $\qquad$ **(f)** through $(0, 1)$ with slope 3

B.3. The door of a campus building is 5 ft above ground level. To allow wheelchair access the steps in front of the door are to be replaced by a straight ramp with constant slope $\frac{1}{12}$, as shown in Figure 2-34. How long must the ramp be?

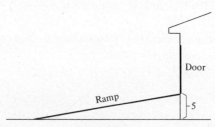

Door

Ramp

5

Figure 2-34

B.4. Use slopes to show that the points $(-4, 6)$, $(-1, 12)$, and $(-7, 0)$ all lie on the same straight line.

B.5. (a) Use slopes to show that the points $(-5, -2)$, $(-3, 1)$, $(3, 0)$, and $(5, 3)$ are the vertices of a parallelogram.
(b) Are the points $(-10, 9)$, $(-\frac{16}{3}, \frac{13}{3})$, $(-3, -2)$, and $(4, -9)$ the vertices of a parallelogram?

B.6. Use slopes to determine if the three given points are the vertices of a right triangle.

(a) $(9, 6), (-1, 2), (1, -3)$ (b) $(1, 6), (-5, -8), (5, -4)$

B.7. Let L be the line through $(-4, 5)$ which is perpendicular to the line through $(1, 3)$ and $(-4, 2)$. Find three points on L.

B.8. Find a number t such that the line passing through the two given points has slope -2.

(a) $(0, t), (9, 4)$ (c) $(-2, t), (-1, -7)$ (e) $\left(\dfrac{t}{3}, -2\right), \left(4, \dfrac{t}{4}\right)$

(b) $(1, t), (-3, 5)$ (d) $(t, t), (5, 9)$ (f) $(t + 1, 5), (6, -3t + 7)$

C.1. (a) Let L be a nonvertical straight line through the origin. L intersects the vertical line through $(1, 0)$ at a point P. Show that the second coordinate of P is the slope of L.

(b) Use part (a) to sketch quickly eleven lines through the origin having these slopes: 6, 4, 3, 2, 1, 0, -1, -2, -3, -5, -7.

(c) Let b be a real number and L a nonvertical line through $(0, b)$. Let P be the point where L intersects the vertical line through $(1, 0)$. Show that the second coordinate of P is (slope of L) $+ b$.

4. EQUATIONS OF LINES

We now begin a systematic program of relating simple algebraic equations and inequalities with familiar geometric shapes. Using our knowledge of slopes, for example, we shall show that every straight line is the graph of a first-degree equation. We begin by defining some of these terms and introducing several concepts that will play an important role in the discussion.

GRAPHS OF EQUATIONS AND INEQUALITIES

Throughout the rest of this chapter, the words "equation" and "inequality" always mean an equation or inequality in *two* variables. Consequently, we shall interpret equations such as $x = 7$ or $6y + 3 = 0$ as equations in two variables, one of which has coefficient zero: $x + 0y = 7$ or $0x + 6y + 3 = 0$; and similarly for inequalities.

A **solution of an equation or inequality** in two variables x and y, such as

$$7y = 4x - 2 \quad \text{or} \quad (x - 2)^2 + (y + 1)^2 = 16 \quad \text{or} \quad \frac{x^2}{9} + \frac{y^2}{4} \leq 1$$

is a pair of numbers such that the substitution of the first number for x and the second for y in the equation yields a true statement. In this case, we say that the given pair of numbers **satisfies the equation.**

EXAMPLE   The pair $(2, \frac{6}{7})$ is a solution of $7y = 4x - 2$ since substituting $x = 2$ and $y = \frac{6}{7}$ yields the true statement $7\left(\dfrac{6}{\underline{7}}\right) = 4 \cdot \underline{2} - 2$. But $(3, 4)$ is *not* a solution since the substitutions $x = 3$ and $y = 4$ produce a *false* statement $7 \cdot \underline{4} = 4 \cdot \underline{3} - 2$.

The **graph of an equation or inequality** in two variables is the set of points in the plane whose coordinates are solutions of the given equation or inequality. Thus the graph is a *geometric picture of the solutions*.

EXAMPLE   The point $(4, 2)$ is on the graph of the equation $7y = 4x - 2$ since $7(2) = 4(4) - 2$. The point $(-3, -2)$ is also on the graph since $7(-2) = 4(-3) - 2$. Later in this section we shall see that the entire graph consists of the points on the straight line joining $(4, 2)$ and $(-3, -2)$, as shown in Figure 2-35. In the next section we shall see that the graph of the inequality $(x - 2)^2 + (y + 1)^2 \le 16$ consists of the points on the circular disc shown in Figure 2-35.

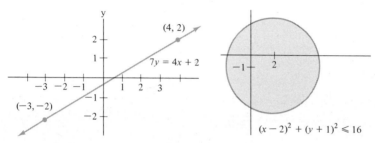

Figure 2-35

INTERCEPTS

The points (if any) where a graph crosses the x- or y-axis often play an important role. If the graph crosses the x-axis at the point $(a, 0)$, then the number a is called an **x-intercept** of the graph. Similarly if the graph crosses the y-axis at the point $(0, b)$, then the number b is called a **y-intercept** of the graph. For example, see Figure 2-36.

x-intercept -2 x-intercepts 3 and -3
y-intercept 1 y-intercepts 2 and -2

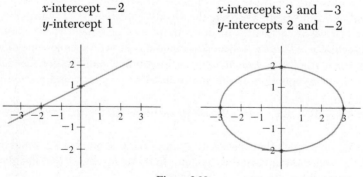

Figure 2-36

The intercepts of the graph of an equation can be determined *algebraically*, without drawing the graph. For instance, the y-intercepts are the numbers b such that $(0, b)$ satisfies the given equation. They may be found by setting $x = 0$ in the equation and solving for y. The x-intercepts may be determined in a similar fashion by setting $y = 0$ and solving for x.

EXAMPLE To find the y-intercepts of the equation $3x - 5y - 12 = 0$, we set $x = 0$ and solve for y:

$$3 \cdot 0 - 5y - 12 = 0$$
$$-5y = 12$$
$$y = -\tfrac{12}{5}$$

Therefore the only y-intercept is $-\tfrac{12}{5}$ and the graph crosses the y-axis at $(0, -\tfrac{12}{5})$. To find the x-intercepts, we set $y = 0$ and solve for x:

$$3x - 5 \cdot 0 = 12$$
$$x = 4$$

Consequently, the only x-intercept is 4 and the graph crosses the x-axis at $(4, 0)$.

VERTICAL LINES

Vertical lines are just lines parallel to the y-axis. Suppose a vertical line L has x-intercept c [that is, L crosses the x-axis at the point $(c, 0)$] as shown in Figure 2-37.

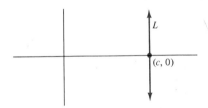

Figure 2-37

Then every point on L has the *same* x-coordinate—namely, c—and every point with x-coordinate c lies on L. Therefore

> *The vertical line through* (c, 0) *is the graph of the equation* x = c.

We often say **"the equation of the line"** instead of "the equation whose graph is the line."

EXAMPLE The vertical line L through $(14, 79)$ is the same as the vertical line through $(14, 0)$. Therefore the equation of L is $x = 14$.

NONVERTICAL LINES

If (x_1, y_1) is a point and m is a real number, it is geometrically clear that there is exactly one line through (x_1, y_1) with slope m. Here is an algebraic description of this line.

POINT-SLOPE FORM OF THE EQUATION OF A LINE

The line L *with slope* m *through the point* (x_1, y_1) *is the graph of the equation*

$$y - y_1 = m(x - x_1)$$

In order to *prove* that the line L coincides with the graph of $y - y_1 = m(x - x_1)$ we must show *two* things: (1) every point on L satisfies the equation *and* (2) every point that satisfies the equation is actually on L. First, suppose (c, d) is any point on L other than (x_1, y_1). Using these two points to compute the slope of L yields

$$\frac{d - y_1}{c - x_1} = \text{slope } L = m, \quad \text{or equivalently,} \quad d - y_1 = m(c - x_1)$$

Therefore (c, d) satisfies the equation $y - y_1 = m(x - x_1)$. Since (x_1, y_1) obviously satisfies this equation, we have proved part 1.

Now for part 2: we know (x_1, y_1) is on L. If (r, s) is any point other than (x_1, y_1) which satisfies the equation, then

$$s - y_1 = m(r - x_1), \quad \text{or equivalently,} \quad \frac{s - y_1}{r - x_1} = m$$

The last equation says that the line M joining (r, s) and (x_1, y_1) has slope m, the same slope as the line L. In Section 2 we saw that two *distinct* lines with the same slope are parallel. Since (x_1, y_1) lies on both L and M, the only possibility here is that L and M are actually the *same line*. Thus (r, s) does lie on L and the proof is complete.

What we have just proved is that if you know *one point* and the *slope* of a line, then you know the equation of the line.

EXAMPLE The line L with slope 5 through the point $(3, -4)$ is the graph of the equation $y - y_1 = m(x - x_1)$, with $m = 5$ and $(x_1, y_1) = (3, -4)$; that is,

$$y - (-4) = 5(x - 3)$$
$$y - 4 = 5x - 15$$
$$y = 5x - 19$$

Observe that the slope of the line L is 5 and the y-intercept of L (obtained by setting $x = 0$ in the equation) is precisely the number -19.

The equation of a line can also be found when you know only *two points* on the line.

EXAMPLE Let M be the line through $(1, 2)$ and $(-5, 6)$. Then the slope of M is $\dfrac{6-2}{-5-1} = \dfrac{4}{-6} = -\dfrac{2}{3}$. By using $m = -\dfrac{2}{3}$ and $(x_1, y_1) = (1, 2)$, we see that the equation of M is

$$y - y_1 = m(x - x_1)$$
$$y - 2 = -\tfrac{2}{3}(x - 1)$$

We can simplify this equation as follows:

$$y - 2 = -\tfrac{2}{3}x + \tfrac{2}{3}$$
$$y = -\tfrac{2}{3}x + (\tfrac{2}{3} + 2) = -\tfrac{2}{3}x + \tfrac{8}{3}$$

You might be wondering what would happen if we used the point $(-5, 6)$ for (x_1, y_1) instead of $(1, 2)$:

$$y - y_1 = m(x - x_1)$$
$$y - 6 = -\tfrac{2}{3}(x - (-5))$$

Now this doesn't look much like the equation obtained above, but some routine algebra shows that it is actually an equivalent form of the same equation:

$$y - 6 = -\tfrac{2}{3}(x - (-5)) = -\tfrac{2}{3}x - \tfrac{10}{3}$$
$$y = -\tfrac{2}{3}x - \tfrac{10}{3} + 6$$
$$y = -\tfrac{2}{3}x + \tfrac{8}{3}$$

Note that the slope of the line M is $-\tfrac{2}{3}$ and the y-intercept is $\tfrac{8}{3}$ [since $(0, \tfrac{8}{3})$ satisfies the equation].

The final form of the equations in the preceding examples suggests that you can find the equation of a line, provided you know its *slope* and *y-intercept*:

SLOPE-INTERCEPT FORM OF THE EQUATION OF A LINE

The line with slope m *and y-intercept* b *is the graph of the equation* y = mx + b.

This fact is easily proved as follows. A line with y-intercept b goes through the point $(0, b)$. But as we saw above, the line through $(0, b)$ with slope m is just the graph of the equation

$$y - b = m(x - 0) = mx$$
$$y = mx + b$$

As the earlier examples showed, the equation of any nonvertical line can be obtained in several ways. But in every case, the equation can always be written in the slope-intercept form $y = mx + b$. This is probably the most useful form of the equation of a line, since you can *immediately* read off the slope and the y-intecept of the line.

EXAMPLE In order to find the graph of the equation $2x - 3y + 14 = 6x + 5y$, we first rewrite the equation in an equivalent form:

$$2x - 3y + 14 = 6x + 5y$$
$$-3y - 5y = 6x - 2x - 14$$
$$-8y = 4x - 14$$
$$y = -\tfrac{4}{8}x - (-\tfrac{14}{8}) = -\tfrac{1}{2}x + \tfrac{7}{4}$$

Now that the equation is in the form $y = mx + b$, we can just *read off the basic facts:* the graph is a line with slope $m = -\tfrac{1}{2}$ and y-intercept $b = \tfrac{7}{4}$. The y-intercept $\tfrac{7}{4}$ tells us that the point $(0, \tfrac{7}{4})$ is on the graph. To obtain another point on the graph, we can substitute any nonzero number for x—say, $x = 2$—in the equation $y = -\tfrac{1}{2}x + \tfrac{7}{4}$ and obtain $y = -\tfrac{1}{2}(2) + \tfrac{7}{4} = \tfrac{3}{4}$. Therefore $(2, \tfrac{3}{4})$ is on the graph and the graph looks like the one in Figure 2-38.

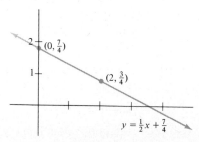

Figure 2-38

EXAMPLE The equation $y = 3$ can be written as $y = 0x + 3$. Therefore its graph is a straight line with slope 0 and y-intercept 3, that is, a horizontal line through $(0, 3)$, as shown in Figure 2-39.

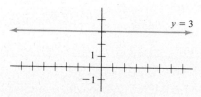

Figure 2-39

EXAMPLE Find the equation of the line L through $(2, -1)$, which is parallel to the line M whose equation is $3x - 2y + 6 = 0$. The equation $3x - 2y + 6 = 0$ may be rewritten

$$-2y = -3x - 6$$
$$y = \tfrac{3}{2}x + 3$$

The slope-intercept form of the equation of M tells us that M has slope $\frac{3}{2}$. Therefore the parallel line L has the *same* slope $\frac{3}{2}$. Using the point $(2, -1)$ on L and the slope $\frac{3}{2}$, we see that the equation of L is

$$y - (-1) = \tfrac{3}{2}(x - 2) = \tfrac{3}{2}x - 3$$
$$y = \tfrac{3}{2}x - 4$$

FIRST-DEGREE EQUATIONS

The preceding examples show that the equation of any straight line is a **first-degree equation,** that is, an equation which can be written in the form

$$Ax + By + C = 0,$$

where A, B, C are real numbers with at least one of A or B nonzero. For instance,

$y = 5x - 19$	can be written	$5x - y - 19 = 0$
$x = -6$	can be written	$x + 0y + 6 = 0$
$y - 2 = -3(x - 4)$	can be written	$3x + y - 14 = 0$

Conversely, consider any first-degree equation $Ax + By + C = 0$. Either $B = 0$ or $B \neq 0$. If $B = 0$, then $A \neq 0$ and the equation becomes

$$Ax + C = 0$$
$$Ax = -C$$
$$x = \frac{-C}{A}$$

Its graph is a vertical line. On the other hand, if $B \neq 0$, then the equation becomes

$$Ax + By + C = 0$$
$$By = -Ax - C$$
$$y = \frac{-A}{B}x - \frac{C}{B}$$

Its graph is a line with slope $-A/B$ and y-intercept $-C/B$. We have proved:

The graph of every first-degree equation $Ax + By + C = 0$ *is a straight line and every straight line is the graph of a first-degree equation.*

For obvious reasons, first-degree equations are often called **linear equations.** We frequently refer to "the line $Ax + By + C = 0$" instead of using the proper phrase "the line that is the graph of $Ax + By + C = 0$."

LINEAR INEQUALITIES

The techniques for graphing linear equations can also be used to graph linear inequalities.

EXAMPLE The graph of $y \leq 2x + 1$ consists of all points (x, y) in the plane whose coordinates satisfy this inequality. Since all points on the line $y = 2x + 1$ obviously satisfy the inequality, this line is *part* of the graph of $y \leq 2x + 1$. Any point (x, y) that is *not* on the line $y = 2x + 1$ must lie either above or below the line (see Figure 2-40).

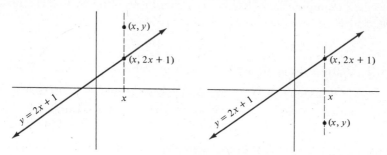

Figure 2-40

As Figure 2-40 shows, (x, y) lies directly above or directly below the point $(x, 2x + 1)$ on the line. Now the relative position of two points on the same vertical line is determined by their second coordinates. So if (x, y) lies above $(x, 2x + 1)$, we must have $y > 2x + 1$. Thus no point (x, y) above the line satisfies the inequality $y \leq 2x + 1$. On the other hand, if (x, y) lies below $(x, 2x + 1)$, we must have $y < 2x + 1$, so that (x, y) *does* satisfy the inequality $y \leq 2x + 1$. Therefore the graph of $y \leq 2x + 1$ is the half-plane consisting of *all points on or below the line* $y = 2x + 1$, as shown in Figure 2-41.

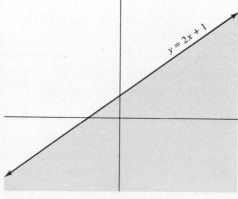

Figure 2-41

The argument in the preceding example also shows that the graph of the inequality $y \geq 2x + 1$ consists of the points *on or above* the line $y = 2x + 1$. A similar argument works for all linear inequalities:

The graph of $y \leq mx + b$ *is the half-plane consisting of all points on or below the line* $y = mx + b$.

The graph of $y \geq mx + b$ *is the half-plane consisting of all points on or above the line* $y = mx + b$.

EXERCISES

A.1   Is the given point on the graph of the given equation? Why or why not?

(a) $(1, -2)$; $3x - y - 5 = 0$

(b) $(2, -1)$; $x^2 + y^2 - 6x + 8y = -15$

(c) $(6, 2)$; $3y + x = 12$

(d) $(1, -2)$; $3y + x = 12$

(e) $(3, 4)$; $(x - 2)^2 + (y + 5)^2 = 4$

(f) $(1, -1)$; $\dfrac{x^2}{2} + \dfrac{y^2}{3} = 1$

A.2.   Find the equation of the line with given slope m and passing through the given point and graph the line.

(a) $m = 1$; $(3, 5)$

(b) $m = 2$; $(-2, 1)$

(c) $m = -1$; $(6, 2)$

(d) $m = 0$; $(-4, -5)$

(e) $m = \frac{1}{2}$; $(-3, 1)$

(f) $m = -3$; $(2, \sqrt{3})$

A.3.   Find the equation of the line through the two given points.

(a) $(0, -5)$ and $(-3, -2)$

(b) $(4, 3)$ and $(2, -1)$

(c) $(\frac{4}{3}, \frac{2}{3})$ and $(\frac{1}{3}, 3)$

(d) $(-10, 11)$ and $(12, 1)$

(e) $(8.7, 1)$ and $(0, 0)$

(f) $(6, 7)$ and $(6, 15)$

(g) $(\sqrt{8}, \sqrt{2})$ and $(-\sqrt{2}, \sqrt{6})$

(h) $(\frac{1}{2}, 4)$ and $(\frac{5}{2}, \frac{6}{5})$

A.4.   Find the equation of the line with given slope m and given y-intercept b; and graph the line.

(a) $m = 1$; $b = 2$

(b) $m = 2$; $b = 5$

(c) $m = -4$; $b = 2$

(d) $m = -5$; $b = -2$

(e) $m = \frac{1}{2}$; $b = 3$

(f) $m = \frac{4}{3}$; $b = -\frac{5}{3}$

A.5.   Find the slope and y-intercept of each line whose equation is given below and graph the line. (*Hint:* put the equation in slope-intercept form.)

(a) $2x - y + 5 = 0$

(b) $3x + 4y = 7$

(c) $7 + 2y = -8x$

(d) $2(x + y + 1) = 5x - 3$

(e) $3(x - 2) + y = 7 - 6(y + 4)$

(f) $2(y - 3) + (x - 6) = 4(x + 1) - 2$

B.1. Find an equation for the line satisfying the given conditions and graph the line.
  (a) through $(-2, 1)$ with slope 3
  (b) y-intercept -7 and slope 1
  (c) through $(2, 3)$ and parallel to $3x - 2y = 5$
  (d) through $(1, -2)$ and perpendicular to $y = 2x - 3$
  (e) x-intercept 5 and y-intercept -5
  (f) through $(-5, 2)$ and parallel to the line through $(1, 2)$ and $(4, 3)$
  (g) through $(-1, 3)$ and perpendicular to the line through $(0, 1)$ and $(2, 3)$
  (h) through $(-5, 6)$ and perpendicular to the x-axis
  (i) through $(-5, 2)$ and parallel to the x-axis
  (j) y-intercept 3 and perpendicular to $2x - y + 6 = 0$
  (k) y-intercept 0 and parallel to $x - 3y + 7 = 0$
  (l) parallel to $3x - 3y + 5 = 7$, with the same y-intercept as $2x - 5y + 7 = 4$

B.2. Determine whether the following pairs of lines are parallel, perpendicular, or neither. (*Hint:* put the equations in slope-intercept form and use well-known facts about slopes.)
  (a) $2x + y - 2 = 0$ and $4x + 2y + 18 = 0$
  (b) $3x + y - 3 = 0$ and $6x + 2y + 17 = 0$
  (c) $x + 2y - 3 = 0$ and $2x - y + 3 = 0$
  (d) $x - 3y + 7 = 0$ and $y - 3x + 3 = 0$
  (e) $x + y - 3 = 0$ and $y - x + 2 = 0$
  (f) $y - ax + 5 = 0$ and $ay + x - 7 = 0$ (where a is a constant)

B.3. (a) Find a real number k such that $(3, -2)$ is on the line $kx - 2y + 7 = 0$.
  (b) Find a real number k such that the line $3x - ky + 2 = 0$ has y-intercept -3.

B.4. Let A, B, C, D be nonzero real numbers. Show that the lines $Ax + By + C = 0$ and $Ax + By + D = 0$ are parallel.

B.5. Suppose A and B are real numbers, at least one of which is nonzero. Prove that the graph of $Ax + By = 0$ is a straight line through the origin.

B.6. Let L be a line which is neither vertical nor horizontal and which does *not* pass through the origin. Show that L is the graph of $\dfrac{x}{a} + \dfrac{y}{b} = 1$, where a is the x-intercept and b the y-intercept of L.

B.7. Graph the equation $|x| + |y| = 1$. [*Hint:* just look at one quadrant at a time. For instance, all points (x, y) in the second quadrant have $x < 0$ and $y > 0$, so that $|x| = -x$ and $|y| = y$. Thus in the second quadrant the graph of $|x| + |y| = 1$ is the same as the graph of $-x + y = 1$. Deal with the other quadrants similarly.]

B.8. Graph these inequalities.

(a) $y \leq -3x + 2$

(b) $y \geq x + 4$

(c) $y < 2x - 1$

(Hint: follow the same procedure above; but the line $y = 2x - 1$ is *not* part of the graph.)

(d) $y > 3x + 1$

(e) $y > -3x + 1$

(f) $4x + 2y - 1 \leq 0$

(Hint: first solve for y.)

(g) $-3x + 4y + 8 > 0$

(h) $-5x + 3y - 6 < 0$

B.9. Graph these systems of inequalities (that is, plot the points that are on the graphs of *all* the inequalities in the given system).

(a) $y \leq x + 1$
$y \leq -2x + 3$

(b) $y \leq 2x - 1$
$y \geq -x$

(c) $3x - 2y - 2 \leq 0$
$y \leq 1$

(d) $y \geq x + 1$
$y \leq -2x + 1$
$x \geq -2$

(e) $2x - y \leq -1$
$4x + y \leq 3$
$x + y \geq -1$

C.1. Let (x_1, y_1), (x_2, y_2), (x_3, y_3) be any three points on the line $y = mx + b$, where $x_1 < x_2 < x_3$. Show that the distance from (x_1, y_1) to (x_3, y_3) is the sum of the distance from (x_1, y_1) to (x_2, y_2) and the distance from (x_2, y_2) to (x_3, y_3). (Hint: $y_1 = mx_1 + b$.)

C.2. A road is being built along the floor of a valley. At the end of the valley it must pass over the hill, as shown in Figure 2-42.

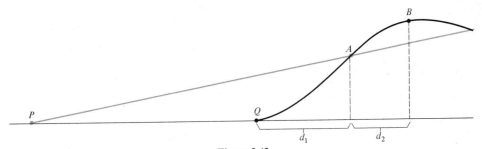

Figure 2-42

Because of soil conditions and other factors, the road is to lie along a straight line of slope $\frac{1}{25}$ which passes through point A (as indicated by the colored line above). The top of the hill will be cut away, and the bottom of the hill will be filled in to road level. The point A lies 400 ft above the floor of the valley. The peak of the hill (point B) lies 500 ft above the valley floor. The horizontal distance d_1 is 600 ft; distance d_2 is 300 ft.

(a) Find the point where the road should meet the valley floor by determining the distance from P to Q. (Hint: sketch the picture on a coordinate system, with the valley floor as the x-axis and the vertical line through A as the y-axis.)

   (b) Find out how much of the hilltop must be cut off by determining the vertical distance from point B to the road.

C.3. This exercise completes the proof of the statement in the box on page 54. Show that two nonvertical lines with the same slope are parallel. (*Hint:* write the equations of the lines in slope-intercept form and show that the vertical distance between the two lines is the same for any value of x.)

C.4. This exercise provides a proof of the statement in the box on page 55. Let L be a line with slope k and M a line with slope m and assume *both* L and M pass through the origin.

   (a) Show that L passes through $(1, k)$ and M passes through $(1, m)$.

   (b) Compute the length of each side of the triangle with vertices $(0, 0)$, $(1, k)$ and $(1, m)$.

   (c) Suppose L and M are perpendicular. Then the triangle of part (b) has a right angle at $(0, 0)$. Use part (b) and the Pythagorean Theorem to find an equation involving k, m, and various constants. Simplify this equation to show that $km = -1$.

   (d) Suppose instead that $km = -1$ and prove that L and M are perpendicular. [*Hint:* you may assume that a triangle whose sides a, b, c satisfy $a^2 + b^2 = c^2$ is a right triangle with hypotenuse c. Use this fact and $km = -1$ to "reverse" the argument in part (c).]

   (e) Finally, assume L and M are any two nonvertical lines (which don't necessarily go through the origin), with slope $L = k$ and slope $M = m$. Use the preceding material to prove that L is perpendicular to M exactly when $km = -1$. (*Hint:* every line is parallel to a line through the origin and parallel lines have the same slope.)

5. CONIC SECTIONS

In this section we discuss algebraic descriptions (in terms of second-degree equations) of several frequently seen geometric curves: circles, ellipses, and hyperbolas. These curves, together with parabolas (which will be discussed in Chapter 3), are called **conic sections.**

   Warning All the graphs in this section are obtained with a minimal amount of plotting of points. Usually geometric or algebraic information is used to produce the entire graph very quickly. In one sense, this illustrates the power of analytic geometry. In another sense, however, it is quite misleading. The techniques presented here work very well for uncomplicated second-degree equations. But determining the graph of an arbitrary equation almost always involves plotting points, as well as some educated guesswork, as we shall see in Chapter 3.

CIRCLES

If P is a point in the plane and r is a positive real number, then the **circle with center P and radius r** is the set of all points in the plane whose distance from P is precisely r units, as shown below. Suppose the point P has coordinates (a, b). Then the point (x, y) is on the circle with center P and radius r exactly when the distance from (x, y) to (a, b) is r, as shown in Figure 2-43.

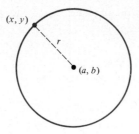

Figure 2-43

Using the distance formula on the points (x, y) and (a, b), we can rewrite this statement as

$$\sqrt{(x - a)^2 + (y - b)^2} = r$$

Squaring both sides of this equation yields

$$(x - a)^2 + (y - b)^2 = r^2$$

Therefore we have proved that

> *The circle with center* (a, b) *and radius* r *is the graph of the equation*
> $$(x - a)^2 + (y - b)^2 = r^2$$

We sometimes say that $(x - a)^2 + (y - b)^2 = r^2$ is the **equation of the circle** with center (a, b) and radius r. If the center of the circle is at the origin (so that $a = 0$ and $b = 0$ in the discussion above), then the equation of the circle takes a simple form:

> *The circle with center* $(0, 0)$ *and radius* r *is the graph of the equation*
> $$x^2 + y^2 = r^2.$$

EXAMPLE   The circle with radius 1 and center $(0,0)$ is the graph of the equation $x^2 + y^2 = 1$, as shown in Figure 2-44. This circle is usually called the **unit circle.**

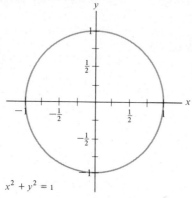

Figure 2-44

EXAMPLE   The circle with center $(-3, 2)$ and radius 2 is the graph of the equation

$$(x - (-3))^2 + (y - 2)^2 = 2^2$$

which can also be written as

$$(x + 3)^2 + (y - 2)^2 = 4$$

The graph is shown in Figure 2-45.

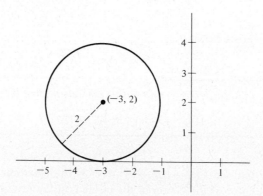

Figure 2-45

EXAMPLE   What is the equation whose graph is the circle with center $(2, -1)$ which passes through the origin? In order to write the equation of this circle, we must first find its radius. The radius is the distance from any point on the circle to the center, $(2, -1)$. Since $(0, 0)$ is on the circle, the radius is the distance from $(0, 0)$ to $(2, -1)$, namely,

$$\sqrt{(0 - 2)^2 + (0 - (-1))^2} = \sqrt{4 + 1} = \sqrt{5}$$

Finally, the equation of the circle with radius $\sqrt{5}$ and center $(2, -1)$ is

$$(x - 2)^2 + [y - (-1)]^2 = (\sqrt{5})^2$$

which can be written as:

$$(x - 2)^2 + (y + 1)^2 = 5$$
$$(x^2 - 4x + 4) + (y^2 + 2y + 1) = 5$$
$$x^2 + y^2 - 4x + 2y = 0$$

EXAMPLE If (x, y) satisfies the inequality

$$(x - 1)^2 + y^2 \geq 4$$

then (x, y) necessarily satisfies

$$\sqrt{(x - 1)^2 + (y - 0)^2} \geq 2$$

But this is just an algebraic way of stating the geometric fact that

The distance from (x, y) to $(1, 0)$ is greater than or equal to 2.

Therefore the points (x, y) which satisfy $(x - 1)^2 + y^2 \geq 4$ lie *outside or on* the circle with center $(1, 0)$ and radius 2. In other words, the graph of the inequality $(x - 1)^2 + y^2 \geq 4$ is the shaded area in Figure 2-46.

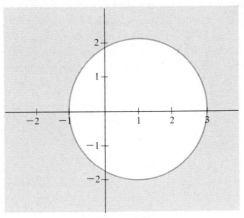

Figure 2-46

Similarly, the points (x, y) that satisfy the inequality

$$(x - 1)^2 + y^2 < 4$$

also satisfy

$$\sqrt{(x - 1)^2 + (y - 0)^2} < 2$$

This last inequality states that the distance from (x, y) to $(1, 0)$ is less than 2, so that (x, y) lies *inside* the circle with center $(1, 0)$ and radius 2. Therefore the graph of $(x - 1)^2 + y^2 < 4$ is the white circular disc in Figure 2-46.

ELLIPSES

Let P and Q be two points in the plane and r a positive real number. The set of all points X such that

$$(\text{distance from } P \text{ to } X) + (\text{distance from } Q \text{ to } X) = r$$

is called an **ellipse.** Points P and Q are called the **foci** of the ellipse ("foci" is the plural of "focus"). The midpoint of the line segment from P to Q is called the **center** of the ellipse. The number r is called the **distance sum.**

One way to see what an ellipse looks like is to take a string of length r and pin its ends on the points P and Q. Then put your pencil against the string so that the string is taut (as indicated by the dotted line in Figure 2-47). If you move the pencil point X, keeping the string taut, you will trace out an ellipse, as shown in Figure 2-47.

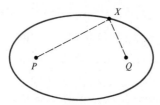

Figure 2-47

Ellipses arise frequently in nature and science. The orbit of a planet around the sun is an ellipse with the sun as one focus. Satellites travel in elliptical orbits around the earth. Elliptical gears and cams are used in various kinds of machinery.

Ellipses can be described as graphs of equations, in much the same way we did for circles above. However, we shall present only the case of ellipses whose center is at the origin.

Let a *and* b *be positive real numbers. The graph of the equation*

$$\frac{x^2}{a^2} + \frac{y^2}{b^2} = 1$$

is an ellipse centered at the origin with x-intercepts $\pm$a *and y-intercepts* $\pm$b. *Every ellipse centered at the origin, with foci on one of the coordinate axes, is the graph of an equation of this form.*

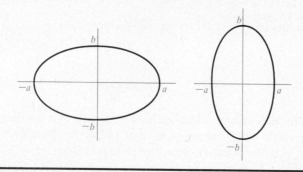

The proof of this claim is basically an exercise in the use of the distance formula. Those interested should consult Exercise C.3.

EXAMPLE In order to graph the equation $\frac{x^2}{9} + \frac{y^2}{4} = 1$, we first rewrite the equation

as $\frac{x^2}{3^2} + \frac{y^2}{2^2} = 1$. Now the equation has the form $\frac{x^2}{a^2} + \frac{y^2}{b^2} = 1$ with $a = 3$ and $b = 2$.

Therefore the graph is an ellipse with x-intercepts ± 3 and y-intercepts ± 2, as shown in Figure 2-48.

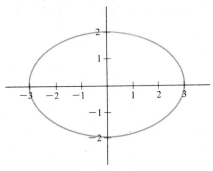

Figure 2-48

EXAMPLE In order to graph the equation $3x^2 + y^2 = 36$, we divide both sides by 36 and rewrite the equation:

$$\frac{3x^2}{36} + \frac{y^2}{36} = \frac{36}{36}$$

$$\frac{x^2}{12} + \frac{y^2}{36} = 1$$

$$\frac{x^2}{(\sqrt{12})^2} + \frac{y^2}{6^2} = 1$$

We now see that the graph is an ellipse with x-intercepts $\pm\sqrt{12} = \pm 2\sqrt{3}$ and y-intercepts ± 6 (Figure 2-49).

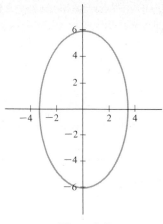

Figure 2-49

HYPERBOLAS

Let P and Q be two points in the plane and r a positive real number. The set of all points X such that

$$|(\text{distance from } P \text{ to } X) - (\text{distance from } Q \text{ to } X)| = r$$

is called a **hyperbola** with **foci** P and Q. Unlike the situation with circles and ellipses, there is no easy mechanical way to sketch a hyperbola. However, it can be shown that every hyperbola has the general shape indicated by the solid lines in Figure 2-50.

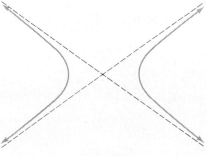

Figure 2-50

The hyperbola gets closer and closer to the dotted straight lines, but never touches them. These two lines are called the **asymptotes** of the hyperbola. The point where the asymptotes intersect is called the **center** of the hyperbola.

Hyperbolas centered at the origin can be described by means of equations quite similar to (but *not* the same as) the equations of ellipses.

Let a *and* b *be positive real numbers. The graph of the equation*

$$\frac{x^2}{a^2} - \frac{y^2}{b^2} = 1$$

is a hyperbola with x-intercepts $\pm$a *and asymptotes the lines*

$$y = \frac{b}{a}x \ and \ y = -\frac{b}{a}x,$$

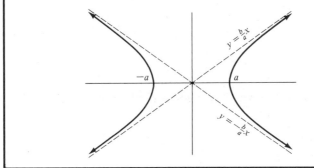

The proof of the statement in the box is another complicated exercise in the use of the distance formula and will be omitted here.

EXAMPLE   To find the graph of $9x^2 - 4y^2 = 36$ we first rewrite the equation

$$9x^2 - 4y^2 = 36$$

$$\frac{9x^2}{36} - \frac{4y^2}{36} = \frac{36}{36}$$

$$\frac{x^2}{4} - \frac{y^2}{9} = 1$$

$$\frac{x^2}{2^2} - \frac{y^2}{3^2} = 1$$

The equation now has the form $\dfrac{x^2}{a^2} - \dfrac{y^2}{b^2} = 1$ with $a = 2$ and $b = 3$. Therefore the graph is a hyperbola with x-intercepts ± 2 and asymptotes $y = (b/a)x = (3/2)x$ and $y = -(b/a)x = -(3/2)x$. We first draw the asymptotes and plot the points $(-2, 0)$

and $(2, 0)$. Then the graph of the hyperbola can be sketched easily, as shown in Figure 2-51.

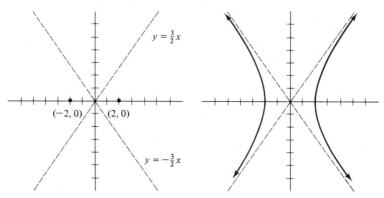

Figure 2-51

Let a *and* b *be positive real numbers. The graph of the equation*

$$\frac{y^2}{b^2} - \frac{x^2}{a^2} = 1$$

is a hyperbola with y-intercepts $\pm b$ *and asymptotes the lines* $y = \dfrac{b}{a}x$ *and* $y = -\dfrac{b}{a}x$,

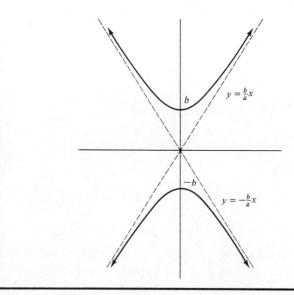

EXAMPLE To find the graph of the equation $75x^2 - 4y^2 + 100 = 0$, we first rewrite the equation

$$75x^2 - 4y^2 + 100 = 0$$

$$75x^2 - 4y^2 = -100$$

$$\frac{75x^2}{-100} - \frac{4y^2}{-100} = \frac{-100}{-100}$$

$$-\frac{3}{4}x^2 + \frac{y^2}{25} = 1$$

$$\frac{y^2}{25} - \frac{x^2}{(4/3)} = 1$$

$$\frac{y^2}{5^2} - \frac{x^2}{(2/\sqrt{3})^2} = 1$$

The equation now has the form $\dfrac{y^2}{b^2} - \dfrac{x^2}{a^2} = 1$ with $b = 5$ and $a = \dfrac{2}{\sqrt{3}}$. Therefore the graph is a hyperbola with y-intercepts ± 5 and asymptotes the lines

$$y = \frac{b}{a}x = \frac{5}{2/\sqrt{3}}x = \frac{5\sqrt{3}}{2}x \qquad \text{and} \qquad y = -\frac{b}{a}x = \frac{-5}{2/\sqrt{3}}x = \frac{-5\sqrt{3}}{2}x$$

as shown in Figure 2-52.

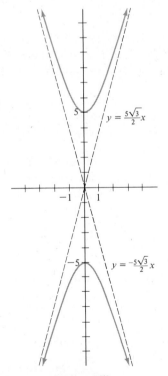

$$y = \frac{5\sqrt{3}}{2}x$$

$$y = \frac{-5\sqrt{3}}{2}x$$

Figure 2-52

Note The two hyperbola equations presented above are easy to confuse with one another. Here is a "memory hook" for distinguishing the graphs of the equations

$$\frac{x^2}{a^2} - \frac{y^2}{b^2} = 1 \quad \text{and} \quad \frac{y^2}{b^2} - \frac{x^2}{a^2} = 1$$

In the first equation, the coefficient of x^2 (namely, $1/a^2$) is *positive* and the coefficient of y^2 (namely, $-1/b^2$) is *negative:* the hyperbola crosses the x-axis but *not* the y-axis. In the second equation, the coefficient of y^2 is positive and that of x^2 negative: the hyperbola crosses the y-axis but not the x-axis.

SECOND-DEGREE EQUATIONS

A **second-degree equation** in x and y is any equation of the form

$$Ax^2 + By^2 + Cxy + Dx + Ey + F = 0$$

where A, B, C, D, E, F are fixed real numbers and at least one of A, B, or C is nonzero.

EXAMPLE The circle with center $(1, 2)$ and radius 5 is the graph of the equation $(x - 1)^2 + (y - 2)^2 = 5^2$. This equation can easily be seen to be a second-degree equation.

$$(x - 1)^2 + (y - 2)^2 = 5^2$$
$$(x^2 - 2x + 1) + (y^2 - 4y + 4) = 25$$
$$x^2 + y^2 - 2x - 4y - 20 = 0$$

This last version of the equation is of the form $Ax^2 + By^2 + Cxy + Dx + Ey + F = 0$ with $A = 1$, $B = 1$, $C = 0$, $D = -2$, $E = -4$, and $F = -20$.

Although we have not discussed all the possibilities, it is in fact true that

> *Every conic section is the graph of a second-degree equation.*

The converse of this statement (namely, the graph of every second-degree equation is a conic section) is *essentially* true, provided you allow for a few so-called degenerate cases. These exceptions and the graphs of various second-degree equations are considered in the DO IT YOURSELF! segment at the end of this section.

EXERCISES

A.1. Find the equation of the circle with the given center and given radius r.

(a) $(-3, 4)$; $r = 2$

(b) $(-2, -1)$; $r = 3$

(c) $(0, 0)$; $r = \sqrt{2}$

(d) $(5, -2)$; $r = 1$

(e) $(4, 7)$; $r = \frac{1}{2}$

(f) $(\frac{1}{2}, \frac{5}{2})$; $r = \frac{7}{3}$

(g) $(3, \frac{8}{3})$; $r = \sqrt{3}$

(h) $(100, \sqrt{50})$; $r = \dfrac{4\sqrt{2}}{3}$

A.2. Sketch the graph of each equation:

(a) $(x - 2)^2 + (y - 4)^2 = 1$

(f) $\dfrac{x^2}{25} + \dfrac{y^2}{4} = 1$

(b) $(x + 1)^2 + (y - 3)^2 = 9$

(g) $\dfrac{x^2}{6} + \dfrac{y^2}{16} = 1$

(c) $(x - 5)^2 + (y + 2)^2 = 5$

(h) $\dfrac{x^2}{10} - 1 = \dfrac{y^2}{36}$

(d) $(x + 6)^2 + y^2 = 4$

(i) $\dfrac{y^2}{49} + \dfrac{x^2}{81} = 1$

(e) $x^2 + y^2 = 16$

A.3. Find an equation whose graph is an ellipse with the given intercepts.
(a) x-intercepts ± 7 and y-intercepts ± 2
(b) x-intercepts ± 1 and y-intercepts ± 8
(c) x-intercepts ± 9 and y-intercepts ± 10.

A.4. The area of the ellipse $\dfrac{x^2}{a^2} + \dfrac{y^2}{b^2} = 1$ is πab. Find the area of these ellipses:

(a) $\dfrac{x^2}{16} + \dfrac{y^2}{4} = 1$ 　　(c) $3x^2 + 4y^2 = 12$ 　　(e) $6x^2 + 2y^2 = 14$

(b) $\dfrac{x^2}{9} + \dfrac{y^2}{5} = 1$ 　　(d) $7x^2 + 5y^2 = 35$ 　　(f) $5x^2 + y^2 = 5$

A.5. Find an equation whose graph is a hyperbola
(a) with asymptotes $y = \frac{3}{2}x$ and $y = -\frac{3}{2}x$ and x-intercepts ± 2
(b) with asymptotes $y = \frac{3}{4}x$ and $y = -\frac{3}{4}x$ and x-intercepts ± 4.

B.1. Find the equation of each of the circles described below.
(a) center $(2, 2)$; passes through the origin
(b) center $(-1, -3)$; passes through $(-4, -2)$
(c) center $(1, 2)$; x-intercepts -1 and 3
(d) center $(3, 1)$; diameter 2
(e) center $(4, 3)$; area 81π (*Hint:* the area of a circle of radius r is πr^2)
(f) center $(0, 0)$; circumference 2π (*Hint:* the circumference of a circle of radius r is $2\pi r$)
(g) center $(-5, 4)$; tangent (touching at one point) to the x-axis
(h) center $(2, -6)$; tangent to the y-axis.

B.2. Without drawing any graphs, determine whether the given point lies inside, outside, or on the given circle.
(a) $(1.5, -1.5)$; circle of radius 4.5 centered at the origin
(b) $(2, 1)$; circle with center $(-1, 4)$ and radius 4
(c) $(-2, 0)$; circle with center $(-1, -2)$ and radius 2.1
(d) $(-2, 3)$; circle with center $(3, 4)$ and radius 5
(e) $(\sqrt{3}, \sqrt{3})$; circle with center $(2, -5)$ and radius $\sqrt{43}$.

B.3. Sketch the graph of these inequalities (see the example on p. 71).

(a) $x^2 + y^2 \leq 1$ (d) $(x + 3)^2 + (y - 5)^2 \leq 9$

(b) $x^2 + y^2 \geq 1$ (e) $(x - 2)^2 + y^2 \leq 4$

(c) $x^2 + y^2 > 4$ (f) $(x - 1)^2 + (y + 2)^2 < 5$

B.4. Graph these systems of inequalities (that is, plot the points that satisfy *all* the inequalities in the given system).

(a) $x^2 + y^2 \leq 25$
$y \leq 2x$

(b) $(x - 4)^2 + (y - 1)^2 \leq 36$
$y > 3x - 1$

(c) $x^2 + y^2 \leq 16$
$x - 2y + 4 \geq 0$
$3x + 4y + 12 \geq 0$

(d) $x^2 + y^2 \geq 9$
$x + 1 \geq 0$
$x + y - 7 \leq 0$

(e) $x^2 + y^2 \leq 16$
$x^2 + y^2 \geq 9$

(f) $x^2 + y^2 \geq 1$
$x^2 + y^2 \leq 4$

(g) $(x + 3)^2 + (y - 1)^2 \geq 1$
$(x + 3)^2 + (y - 1)^2 \leq 9$

(h) $(x - 2)^2 + (y + 2)^2 \leq 36$
$(x - 2)^2 + (y + 2)^2 \geq 16$

B.5. For each real number k, the graph of $(x - k)^2 + y^2 = k^2$ is a circle. Describe all possible such circles.

B.6. Sketch the graph of each equation. If the graph is a hyperbola, also sketch the asymptotes.

(a) $4x^2 - y^2 = 16$ (i) $3x^2 + 2y^2 = 6$

(b) $4x^2 + 3y^2 = 12$ (j) $2x^2 - y^2 = 4$

(c) $\dfrac{x^2}{6} - \dfrac{y^2}{16} = 1$ (k) $18y^2 - 8x^2 - 2 = 0$

(d) $3y^2 - 5x^2 = 15$ (l) $x^2 - 2y^2 = -1$

(e) $\dfrac{x^2}{4} + \dfrac{y^2}{9} = 2$ (m) $(y - 3)^2 - 10 = -x^2 + 2x - 1$

(f) $4x^2 + 4y^2 = 1$ (n) $(x + y)(x - y) + x^2 = 4$

(g) $x^2 + 4y^2 = 1$ (o) $(2x - y)(x + 4y) - 7xy = 8$

(h) $x^2 - 4y^2 = 1$ (p) $15y^2 + 12x^2 - 30 = 0$

B.7. If $a > b > 0$, then the number $e = (\sqrt{a^2 - b^2})/a$ is called the **eccentricity** of the ellipse $\dfrac{x^2}{a^2} + \dfrac{y^2}{b^2} = 1$.

(a) Show that $0 < e < 1$.

(b) Determine the eccentricity of each of the following ellipses:

$$\frac{x^2}{100} + \frac{y^2}{99} = 1, \qquad \frac{x^2}{25} + \frac{y^2}{18} = 1, \qquad \frac{x^2}{16} + \frac{y^2}{2} = 1$$

(c)   Describe the general shape of an ellipse when the eccentricity is very close to 1; when the eccentricity is very close to 0. [*Hint:* part (b) may be helpful.]

B.8.   Find a number k such that $(-2, 1)$ lies on the graph of $3x^2 + ky^2 = 4$. Then graph the equation.

C.1.   The punchbowl and a table holding the punch cups are placed 50 ft apart at a yard party. A portable fence is then set up so that any guest inside the fence can walk straight to the table, then to the punchbowl, and then return to his or her starting point without traveling more than 150 ft. Describe the longest possible such fence.

C.2.   An arched footbridge over a 100-ft wide river is shaped like half an ellipse. The maximum height of the bridge over the river is 20 ft. Find the height of the bridge over a point in the river, exactly 25 ft from the center of the river.

C.3.   Here is a partial proof that the graph of $\dfrac{x^2}{a^2} + \dfrac{y^2}{b^2} = 1$ is an ellipse. Assume first that $a > b$ and let $c = \sqrt{a^2 - b^2}$. Let E be the ellipse with foci $(-c, 0)$ and $(c, 0)$ and distance sum $2a$. E consists of all points (x, y) such that

(distance from $(-c, 0)$ to (x, y)) + (distance from (x, y) to $(c, 0)$) = $2a$

(a)   Use the distance formula to show that every point on E satisfies the equation
$$\sqrt{(x + c)^2 + y^2} = 2a - \sqrt{(x - c)^2 + y^2}$$

(b)   Square both sides of the equation in part (a) and show that the resulting equation simplifies to
$$\sqrt{(x - c)^2 + y^2} = a - \frac{c}{a}x$$

(c)   Square both sides of the equation in (b) and show that the resulting equation simplifies to
$$\frac{a^2 - c^2}{a^2}x^2 + y^2 = a^2 - c^2$$

(d)   Use the fact that $c = \sqrt{a^2 - b^2}$ to show that the equation in part (c) is equivalent to
$$\frac{x^2}{a^2} + \frac{y^2}{b^2} = 1$$

(e)   Use parts (a)–(d) to verify that every point on the ellipse E is on the graph of the equation $\dfrac{x^2}{a^2} + \dfrac{y^2}{b^2} = 1$. The rest of the proof (namely, that every point on the graph is on the ellipse) is a bit tricky and is omitted here.

(f)   If $b > a$, let $c = \sqrt{b^2 - a^2}$ and show that every point on the ellipse with foci $(0, -c)$ and $(0, c)$ and distance sum $2b$ is on the graph of $\dfrac{x^2}{a^2} + \dfrac{y^2}{b^2} = 1$. (*Hint:* adapt parts (a)–(e) in the obvious fashion.)

DO IT YOURSELF!

GRAPHS OF SECOND DEGREE EQUATIONS

The following examples show how the technique of completing the square° can be used to find the graph of any second-degree equation of the form $Ax^2 + By^2 + Dx + Ey + F = 0$, where *both* A and B are nonzero. The graph is usually a conic section, but there are exceptions, as noted below.

CIRCLES

EXAMPLE To find the graph of the equation $x^2 + y^2 - 3x + 12y + 28 = 0$, we begin by rearranging the terms:

$$(x^2 - 3x) + (y^2 + 12y) = -28$$

Now we complete the square in the expression $x^2 - 3x$ by adding the square of half the coefficient of x, namely, $[\frac{1}{2}(-3)]^2 = \frac{9}{4}$, so that $(x^2 - 3x + \frac{9}{4}) = (x - \frac{3}{2})^2$. In order to leave the equation unchanged we add $\frac{9}{4}$ to *both* sides:

$$(x^2 - 3x + \tfrac{9}{4}) + (y^2 + 12y) = -28 + \tfrac{9}{4}$$
$$(x - \tfrac{3}{2})^2 + (y^2 + 12y) = -28 + \tfrac{9}{4}$$

In order to complete the square in $y^2 + 12y$ we must add $(\frac{1}{2} \cdot 12)^2 = 36$. Since we want to leave the equation unchanged, we add 36 to both sides:

$$(x - \tfrac{3}{2})^2 + (y^2 + 12y + \underline{36}) = -28 + \tfrac{9}{4} + \underline{36}$$
$$(x - \tfrac{3}{2})^2 + (y + 6)^2 = 8 + \tfrac{9}{4} = \tfrac{41}{4}$$
$$(x - \tfrac{3}{2})^2 + (y + 6)^2 = (\sqrt{\tfrac{41}{4}})^2$$

Therefore the graph is a circle with center $(\frac{3}{2}, -6)$ and radius $\sqrt{\dfrac{41}{4}} = \dfrac{\sqrt{41}}{2}$.

EXAMPLE To find the graph of $3x^2 + 3y^2 - 12x - 30y + 60 = 0$, we write the equation as

$$(3x^2 - 12x) + (3y^2 - 30y) = -60$$

Dividing both sides by 3 yields

$$(x^2 - 4x) + (y^2 - 10y) = -20$$

In order to complete the square in $x^2 - 4x$ we must add 4. In order to complete the square in $y^2 = 10y$ we must add 25:

$$(x^2 - 4x + \underline{4}) + (y^2 - 10y + \underline{25}) = -20 + \underline{4} + \underline{25}$$
$$(x - 2)^2 + (y - 5)^2 = 9 = 3^2$$

Therefore the graph is the circle with center $(2, 5)$ and radius 3.

° If you don't remember this technique, see the Algebra Review on page 560.

EXAMPLE The equation $(x - 1)^2 + (y + 4)^2 = -9$ has no solutions and *no graph*. For every real number x, the number $(x - 1)^2$ is ≥ 0. Similarly, $(y + 4)^2 \geq 0$ for *every* y. Thus $(x - 1)^2 + (y + 4)^2 \geq 0$ always, and hence $(x - 1)^2 + (y + 4)^2 = -9$ has no solutions.

EXAMPLE The only solution of $x^2 + y^2 = 0$ is $(0, 0)$, so the graph consists of the *single point* $(0, 0)$. You can consider the graph to be a "degenerate" circle with center $(0, 0)$ and radius 0.

In the preceding examples x^2 and y^2 always had the same coefficient. (That is, $A = B$ in $Ax^2 + By^2 + Dx + Ey + F = 0$.) These examples illustrate the fact that

> The graph of $Ax^2 + Ay^2 + Dx + Ey + F = 0$ *is either a circle or a single point, or there is no graph.*

ELLIPSES AND HYPERBOLAS

The next two examples are included for the sake of completeness, even though they require a bit more information about the graphs of ellipses and hyperbolas than was presented earlier in the text.

EXAMPLE To graph the equation $4x^2 + 9y^2 - 32x - 90y + 253 = 0$, we first write the equation as

$$(4x^2 - 32x) + (9y^2 - 90y) = -253$$
$$4(x^2 - 8x) + 9(y^2 - 10y) = -253$$

Now complete the square in $x^2 - 8x$ and $y^2 - 10y$:

$$4(x^2 - 8x + 16) + 9(y^2 - 10y + 25) = -253 + ? + ?$$

Be careful here: on the left side we haven't just added 16 and 25. When the left side is multiplied out we have actually added in $4 \cdot 16 = 64$ and $9 \cdot 25 = 225$. Therefore to leave the original equation unchanged, we must add these numbers on the right:

$$4(x^2 - 8x + 16) + 9(y^2 - 10y + 25) = -253 + 64 + 225$$
$$4(x - 4)^2 + 9(y - 5)^2 = 36$$
$$\frac{4(x - 4)^2}{36} + \frac{9(y - 5)^2}{36} = \frac{36}{36}$$
$$\frac{(x - 4)^2}{9} + \frac{(y - 5)^2}{4} = 1$$

The only difference between this equation and the equations of ellipses seen above is that x has been replaced by $(x - 4)$ and y by $(y - 5)$. It can be shown that the graph is actually an ellipse with center at $(4, 5)$ as shown in Figure 2-53.

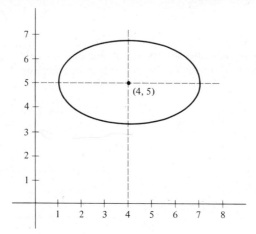

Figure 2-53

EXAMPLE The equation $2x^2 - y^2 + 4x + 6y - 11 = 0$ can be written

$$2(x^2 + 2x) - (y^2 - 6y) = 11$$

Completing the square in $x^2 - 2x$ and $y^2 - 6y$ (and adding the same amount to the right side of the equation) yields:

$$2(x^2 + 2x + \underline{1}) - (y^2 - 6y + \underline{9}) = 11 + \underline{2} + (-\underline{9})$$
$$2(x + 1)^2 - (y - 3)^2 = 4$$
$$\frac{(x + 1)^2}{2} - \frac{(y - 3)^2}{4} = 1$$

It can be shown that the graph of this equation is a hyperbola with center $(-1, 3)$, as shown in Figure 2-54.

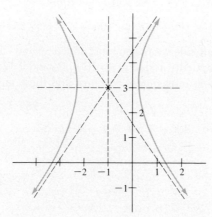

Figure 2-54

EXAMPLE Consider the equation $x^2 - y^2 = 0$. We can factor the left side and obtain $(x + y)(x - y) = 0$. Since a product is zero only when one of the factors is zero, the equation $(x + y)(x - y) = 0$ is equivalent to

$$x + y = 0 \quad \text{or} \quad x - y = 0$$

Thus the graph consists of all (x, y) that are solutions of *either* of the linear equations $x + y = 0$ or $x - y = 0$. Therefore the graph consists of two straight lines, as shown in Figure 2-55.

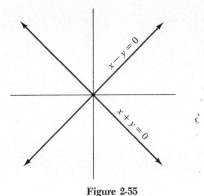

Figure 2-55

The examples above all dealt with the equations $Ax^2 + By^2 + Cxy + Dx + Ey + F = 0$ in which $C = 0$. If the coefficient of xy is nonzero, then the graph is usually a conic which is rotated from its standard position. Trigonometry is usually needed to deal effectively with rotation of axes and the graphs of such equations.

EXERCISES

A.1. Find the center and radius of each of the circles whose equations are:

(a) $x^2 + y^2 + 8x - 6y - 15 = 0$ (e) $x^2 + y^2 + 25x + 10y + 12 = 0$

(b) $15x^2 + 15y^2 = 10$ (f) $3x^2 + 3y^2 + 12 - 18y + 12 = 0$

(c) $x^2 + y^2 + 6x - 4y - 15 = 0$ (g) $2x^2 + 2y^2 + x - y - 3 = 0$

(d) $x^2 + y^2 + 10x - 75 = 0$ (h) $5x^2 + 5y^2 + 30y = 0$

B.1. (a) Find the equation, center, and radius of the circle that passes through $(1, 0)$, $(-1, 0)$, and $(0, 2)$. [*Hint:* the equation must be of the form $x^2 + y^2 + Dx + Ey + F = 0$. Since $(1, 0)$ satisfies the equation, we must have $1 + 0 + D + E \cdot 0 + F = 0$, or equivalently, $D + F = -1$. Similarly, since $(-1, 0)$ and $(0, 2)$ are also solutions, we have $-D + F = -1$ and $2E + F = -4$. Solve these three equations for D, E, and F. Then use the usual completing-the-square technique to find the center and radius.]

(b) Do the same for the circle that passes through $(0, 0)$, $(1, 0)$, and $(2, 1)$.

(c) Do the same for the circle that passes through $(0, 5)$, $(2, 5)$, and $(2, -1)$.

(d) Do the same for the circle that passes through $(-1, -2)$, $(1, -1)$, and $(0, 0)$.

B.2. Graph these equations. [*Hint:* first complete the square in x and y (if necessary), then follow the pattern illustrated in the last three examples above.]

(a) $\dfrac{(x-1)^2}{4} + \dfrac{(y-5)^2}{9} = 1$

(d) $4x^2 - y^2 + 8x - 4y - 4 = 0$

(b) $x^2 - 16y^2 = 0$

(e) $9x^2 + 4y^2 + 54x - 8y + 49 = 0$

(c) $\dfrac{(y+3)^2}{25} - \dfrac{(x+1)^2}{16} = 1$

(f) $9x^2 - 4y^2 + 54x + 8y + 45 = 0$

Functions

You have probably seen the term "function" before. Nevertheless, you may feel that you don't really understand just *what* a function is or *why* functions are important. Both the what and the why of functions are explained in detail in this chapter.

1. FUNCTIONS

There are many situations in which two numerical quantities depend on each other, or vary with each other, or determine each other. The following three examples of such situations will be used continually throughout this chapter and will be referred to by the numbers used here.

EXAMPLE 1 In a certain state, the amount of state income tax you pay depends on your income.° The income determines the tax. Here the two numerical quantities are:

(i)   the income; and
(ii) the amount of tax.

The way the income determines the tax is laid down by law.

° For simplicity we assume that this particular state bases its income tax on your entire income, regardless of number of dependents, source of income, federal taxes, or other factors.

EXAMPLE 2 Weather stations have a device which records the temperature over the 24 hours of a day in the form of a graph (Figure 3-1).

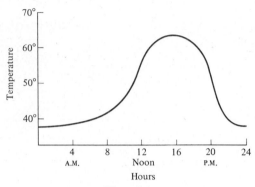

Figure 3-1

Here the two numerical quantities are

    (i)   the hours of the day, that is, the time elapsed since midnight; and
   (ii)   the temperature.

The way time and temperature are related to each other is indicated by the height of the graph over the horizontal axis. The height of the graph at 12 noon is the temperature at noon. The temperature at 6:06 P.M. (that is, 18.1 hr after midnight) is the height of the graph at 18.1, and so on.

EXAMPLE 3 You drop a rock straight down and want to know how far the rock will have fallen after 1 second, 2 seconds, 3 seconds, and so on. Here the two numerical quantities are

    (i)   the time elapsed after you drop the rock; and
   (ii)   the distance the rock has traveled.

The way in which the distance traveled depends on the time elapsed is given by a formula discovered by physicists: after t sec, the rock has traveled $16t^2$ ft.

The three preceding examples deal with things very much *different* from each other:

Income and income tax.
Times of the day and temperature.
How far a rock has traveled after so many seconds.

However, the examples also have important *common* features. The first is that

    In each example, we have two sets of numbers.

This is indicated in the following chart:

	Set 1	Set 2
Example 1	All possible incomes	All possible amounts of tax
Example 2	The various times of the day	The various possible temperatures
Example 3	The seconds elapsed after dropping the rock	The distances the rock has traveled

The second common feature of our three examples is that

> *In each example, there is a very definite way in which an element of set 1 determines an element of set 2, or, alternatively, in which an element of set 2 is associated with an element of set 1.*

For instance,

The income determines the income tax by law (Example 1).
The device measures and registers the temperature at any given time of the day (Example 2).
The formula of the physicists predicts how far the rock has traveled in any span of time (Example 3).

Let us call any such situation with these two features a **functional situation.** By the word "functional" we want to express such ideas as:

The tax is a *function* of, or *depends* on, the income (Example 1).
The location of the needle recording the temperature *varies* according to the time (Example 2).
The distance traveled is *determined* by the time elapsed (Example 3).

Instead of studying each functional situation *separately,* the mathematician focuses on the *common* features of the various functional situations. His or her goal is to design tools to handle *all* functional situations simultaneously. For this, the mathematician must strip the different functional situations of their connections with time, tax, temperature, and so on, and deal only with what is common to all functional situations.

For the mathematician,

> A **function** consists of three items:
>
> (i)   a set of numbers, called the **domain.**
> (ii) another set of numbers, called the **range.**
> (iii) a **rule** that assigns to each number in the domain one and only one *number in the range.*

The domain corresponds to Set 1 and the range to Set 2 in the three examples above.

EXAMPLE 1   The *domain* is the set of all possible incomes. The *range* is the set of all possible tax amounts. The *rule* is the law that decrees the amount of tax to be paid on each income.

EXAMPLE 2   The *domain* is the set of various times during the day (measured in hours after midnight), that is, all real numbers from 0 to 24. The *range* is the set of possible temperatures. The *rule* is the graphical temperature record, which shows the temperature at any given time during the day.

EXAMPLE 3   The *domain* consists of the possible numbers of seconds elapsed after dropping the rock, that is, the set of all nonnegative real numbers.° The *range* is the set of possible distances traveled by the rock, that is, the set of all nonnegative real numbers.° The *rule* is the physical law discovered by the physicists:

$$\text{distance} = 16t^2 \text{ ft, where } t = \text{time elapsed in seconds}$$

You should be sure you understand the meaning of the phrase *"one and only one"* in the description of the rule of a function [item (iii) in the box above]:

EXAMPLE 1   For a particular income—say, $7500—there is *exactly one* tax amount. Indeed, a tax law that said something like "the tax on $7500 is $155 or $187 or $210" would be ludicrous. For each income (number in the domain), there is *one and only one* tax amount (number in the range).

EXAMPLE 2   At a specific time during the day—say, 12:00 noon, the temperature is exactly 56°. It is impossible to have two different temperatures at the same time. In other words, for each time (number in the domain), there is *one and only one* temperature (number in the range).

Notice, however, that it is quite possible to have the *same temperature* at two *different times.* For instance, the graph on page 88 shows that the temperature is 56° at

---

° We assume here, for convenience, that the rock can drop infinitely far. In actual practice, there would be some maximum distance and hence some maximum time.

12:00 noon and at 7:30 P.M. (= 19.5 hours after midnight). In this case, two different numbers in the domain (times) are associated with the same number in the range (temperature).

Example 3 and some further aspects of this topic are discussed in Exercises B.2–B.4. For now we can summarize the discussion as follows:

> *The rule of a function assigns to each number in the domain* one and only one *number in the range. But the rule may possibly assign to two (or more)* different *numbers in the domain the* same *number in the range.*

Functions in the abstract sense just discussed (domain, range, rule) are at the core of all functional situations. Thus the *motivation* for studying functions is the fact that particular concrete functional situations (such as those in Examples 1–3) regularly occur in the real world. But the *manner* in which functions are actually studied by mathematicians is something else again.

The mathematician usually restricts his or her view to the abstract concept of a function. The typical mathematics text rarely discusses just how or why a particular function might be connected with (or arise from) some concrete functional situation. It deals almost exclusively with properties of functions that are common to all functional situations.

In this book we shall try to remind you regularly of the origin of functions in real-life situations. But you will also have to get accustomed to dealing with functions as entities in themselves, independent of possible interpretations in concrete situations.

EXAMPLE The rule of the **identity function** assigns to each real number the number itself. Thus 4 is assigned to 4, -7 to -7, $\pi + 3$ to $\pi + 3$,   and so on. The domain of the identity function is the set of all real numbers; the range is the same set.

EXAMPLE The domain of the **absolute value function** is the set of all real numbers. Its rule assigns to each real number x the number $|x|$. For instance,

$$|-6| = 6 \text{ is assigned to } -6.$$
$$|\pi - 3| = \pi - 3 \text{ is assigned to } \pi - 3.$$

Since $|x| \geq 0$ for every number x, the range is the set of all nonnegative real numbers, that is, the interval $[0, \infty)$.

EXAMPLE The function whose rule assigns to every real number the number 6 is called a **constant function.** Its domain is the set of all real numbers; its range consists of the single number 6. More generally, if c is a fixed real number, then there is a constant function whose rule assigns to each real number the number c.

EXAMPLE For each real number s let $[s]$ denote the *integer* which is closest to s on the *left* side of s on the number line. If s is itself an integer, we define $[s] = s$. Some examples are given in Figure 3-2.

$$[-4.7] = -5 \qquad [-3] = -3 \qquad [-1.5] = -2 \qquad [0] = 0 \qquad [\tfrac{5}{3}] = 1 \qquad [\pi] = 3$$

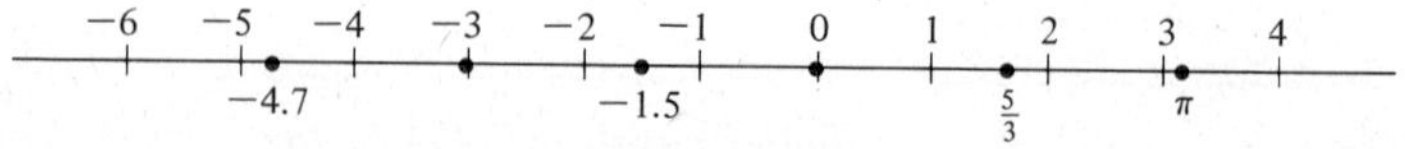

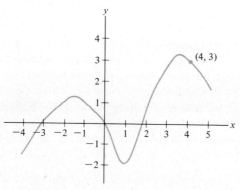

Figure 3-2

The function whose domain is the set of all real numbers and whose rule is:

assign to the real number x the integer $[x]$

is called the **greatest integer function.** The range of the greatest integer function is the set of all integers.

EXAMPLE The graph in Figure 3-3

Figure 3-3

defines a function whose domain is the interval $[-4, 5]$ and whose rule is:

Assign to a number x the number y such that (x, y) lies on the graph.

For instance, the rule assigns to 4 the number 3 since $(4, 3)$ is on the graph. Since all the second coordinates of points on the graph lie between -2 and 3.3, the range of this function is the interval $[-2, 3.3]$.

Warning Not every graph defines a function in this way. See Exercises B.6 and B.7.

Remark The definition of "function" given in this section assumes that the elements of both the domain and the range are real numbers. Consequently, such a function is sometimes called a **real-valued function of a real variable.** The vast majority of the functions you will encounter in this book and in calculus are of this type. However, the term "function" can be defined in a more general way by allowing the domain or range (or both) to consist of elements other than real numbers. Some examples of this kind of function are given in Section 9 of Chapter 6.

EXERCISES

A.1. **(a)** The notation $[r]$ used here is explained in the Example of the greatest integer function on page 92. In each of the following statements, give examples of two pairs of numbers for which the statement is true and two pairs for which it is false:

   (i) $[u + v] = [u]$;   (ii) $[r] + [s] < [r + s]$;   (iii) $[st] > [s][t]$.

   (b) Evaluate $[-\frac{4}{3}]$, $[-\frac{10}{3}]$, $[-16\frac{1}{2}]$, $[.75]$, $[6.75]$, $[-9]$, $[\frac{2}{3}]$.

B.1. Each of the following functional situations involves at least one function. Verbally describe the domain, range, and rule of each such function, providing there is sufficient information to do so. For example,

   Situation: Harry Hamburger owns a professional baseball team. He decides that the only factor to be considered in determining a player's salary for this year is his batting average last season. *Function:* Salaries are a function of batting averages. *Domain:* All possible batting averages. *Range:* All possible salaries. *Rule:* There isn't enough information given to determine the rule.

   (a) The area of a circle depends on its radius.
   (b) If you drive your car at a constant rate of 55 mph, then your distance from the starting point varies with the time elapsed.
   (c) When more is spent on advertising at Grump's Department Store, sales increase. If less is spent on advertising, sales drop.
   (d) A widget manufacturer has a contract to produce widgets at a price of $1 each. His profit per widget is determined by the cost of manufacturing the widget. (Note that if it costs more than one dollar to make a widget, he loses money; interpret such a loss as "negative profit.") Find an algebraic formula for the rule of this function.

B.2. In Example 1 of the text, suppose that the state income tax law reads as follows:

Annual Income	Amount of Tax
Less than $2000	0
$2000–$6000	2% of income over $2000
More than $6000	$80 plus 5% of income over $6000

   (a) Find the number in the range (tax amount) that is assigned to each of the following numbers in the domain (incomes):

   $500, $1509, $3754, $6783, $12,500, $55,342

   (b) Find four different numbers in the domain of this function that are associated with the same number in the range.

(c)   Explain why your answer in part (b) does *not* contradict item (iii) in the definition of a function (in the box on page 90).

(d)   Is it possible to do part (b) if all four numbers in the domain are required to be greater than 2000? Why not?

B.3.   The amount of postage required to mail a first-class letter is determined by its weight. In this situation, is weight a function of postage? or vice-versa? or both?

B.4.   Some (but not all) functions have the following property: any two different numbers in the domain are always assigned to different numbers in the range. Such functions are said to be **one-to-one** or **injective.**

(a)   Show that the function in Example 3 of the text *is* one-to-one; that is, two different numbers in the domain (times) are always assigned to two different numbers in the range (distances). (This may be intuitively clear to you, but give a mathematical argument, using the formula of the physicists.)

(b)   Give examples of two functions which are *not* one-to-one. [Hint: see Exercise B.2.(b) for one.]

(c)   Give another example of a function which *is* one-to-one.

B.5.   Could the following statement ever be the rule of a function?

Assign to a number x in the domain the number in the range whose square is x.

Why or why not? If there is a function with this rule, what is its domain and range?

B.6.   Each of the graphs in Figure 3-4 defines a function as in the Example on page 92. In each case,

(i)   State the domain and range of the function.

(ii)   By measuring as carefully as you can, state what numbers in the range the function assigns to each of the following numbers in the domain: -2, -1, 0, $\frac{1}{2}$, $\frac{3}{2}$.

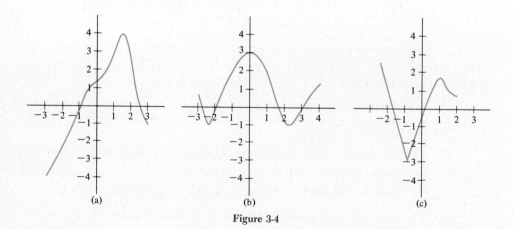

(a)   (b)   (c)

Figure 3-4

B.7.   Explain why *none* of the graphs in Figure 3-5 defines a function according to the procedure in Exercise B.6. What goes wrong?

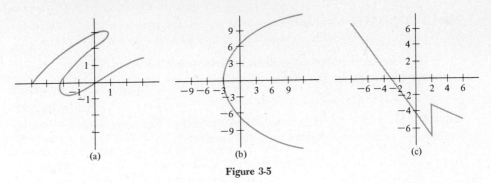

Figure 3-5

2. FUNCTIONAL NOTATION

Like most mathematical symbolism, functional notation has its origins in real-life situations. It was developed in such functional situations in order to facilitate discussion and analysis of various problems. Examples 1–3 of Section 1 provide good illustrations of the need for and development of functional notation. So we begin with them.

EXAMPLE 1—INCOME TAX   In a certain state, the income tax is determined by the following law.

If the Annual Income Is	Then the Amount of Income Tax Is
Less than $2000	0
$2000–$6000	2% of income over $2000
More than $6000	$80 plus 5% of income over $6000

Let's write i for the income (in dollars). The tax due on income i will be written $T(i)$, which is read "T of i." The T indicates "tax" and the i in parentheses indicates that the tax depends on, or varies with, the income i. In this way, the long sentence,

The income tax on an income of $5512 is $70.24,

is abbreviated as

$$T(5512) = 70.24$$

which is read "T of 5512 equals 70.24." Similarly, if your income is $12,500, then according to the table above the income tax is $405. So we write

$$T(12,500) = 405$$

which is *read* "T of 12,500 equals 405" and *means*

> The income tax on an income of $12,500 is $405.

Note 1   There is nothing that forces us to choose the letters i and T as abbreviations. We might just as well have chosen α for income and s for tax and written $s(\alpha)$:

> **Any choice of letters or symbols will do, provided we make clear what is meant by the letters or symbols chosen.**

Note 2   The symbol $T(i)$ is to be treated as *one* thing. The parentheses are *not* the parentheses of elementary algebra, as in the equation

$$3(a + b) = 3a + 3b.$$

The parentheses in $T(i)$ express the fact that the amount T of tax is forced upon us by the income i. For example,

$$T(236.14 + 8750)$$

stands for the amount of tax due on an income of $236.14 + \$8750 = \$8986.14$.

EXAMPLE 2—TEMPERATURE   Write t for the time elapsed (in hours since midnight) and tem for temperature. Then tem(t) represents the temperature at time t. The sentence,

> At 11:30 A.M. the temperature was 52°,

is now written

$$\text{tem}(11.5) = 52$$

which is read

> tem of 11.5 equals 52

Since time in this example is measured in hours after midnight, 3:15 P.M. is written as 15.25 hr after midnight. Thus the sentence,

> At 3:15 P.M. the temperature was 63°,

is the same as the sentence,

> At 15.25 hr after midnight the temperature was 63°,

which is abbreviated

$$\text{tem}(15.25) = 63.$$

EXAMPLE 3—FALLING ROCK   Let t denote the time (in seconds) after the rock is released. As we saw in Section 1, the physicists have found a formula to compute the distance (in feet) the rock has traveled after t sec, namely, $16t^2$. Let $d(t)$ denote the distance (in feet) the rock has traveled after t sec [so that $d(t) = 16t^2$]. In functional notation the sentence,

After 1 sec the rock has traveled 16 ft,

is written

$$d(1) = 16$$

Similarly, $d(4) = 256$ means that "after 4 sec the rock has traveled 256 ft." The sentence, "After 5 sec the rock has traveled 400 ft," is written $d(5) = 400$.

Note 1 The formula for computing the distance traveled $d(t)$ after t sec is

$$d(t) = 16t^2$$

Suppose we had chosen different letters, say, the Greek letter θ for time and F for distance. Then the *same* formula reads:

$$F(\theta) = 16\theta^2$$

The letters here are not important. What matters is the information they convey. Both formulas, $d(t) = 16t^2$ and $F(\theta) = 16\theta^2$, describe exactly the same function, namely, the function that assigns to any number the number obtained by squaring the original number and then multiplying this result by 16.

Note 2 Once again, the parentheses in the symbol "$d(t)$" do *not* indicate multiplication. For instance,

$$d(1 + 4) \neq d(1) + d(4)$$

For $d(1)$ is the "distance traveled after 1 sec," namely, 16 ft; $d(4)$ is the "distance traveled after 4 sec," namely, 256 ft. Thus

$$d(1) + d(4) = 16 + 256 = 272$$

But $d(1 + 4)$ is the "distance traveled after $1 + 4 = 5$ sec," namely, 400 ft. Thus

$$d(1 + 4) = 400, \quad \text{but} \quad d(1) + d(4) = 272$$

MORE FUNCTIONAL NOTATION

In Examples 1–3 above, functional notation was used to abbreviate certain sentences in the English language. We now want to adapt this convenient shorthand notation to the usual mathematical setting, where the particulars of time, temperature, distance, and so on, are eliminated. In this abstract situation we don't deal with sentences such as, "At 11:30 the temperature was 52°," but there *are* sentences which can be conveniently abbreviated by a similar sort of functional notation.

Let the letter x represent a number in the domain of a function and y a number in the range. Denote the function by the letter f. In functional notation, the sentence,

The rule of the function f assigns to the number
x in the domain the number y in the range, is written

$$f(x) = y \qquad [\text{or equivalently, } y = f(x)]$$

which is read

f of x equals y [or equivalently, y equals f of x]

EXAMPLE $f(5) = 7$ (read "f of 5 equals 7") means "the rule of the function f assigns to the number 5 in the domain the number 7 in the range."

Sometimes just the symbol $f(x)$ (read "f of x") is used alone. Consistent with the usage above,

> f(x) *denotes the number in the range which is assigned to the* *number* x *in the domain by the rule of the function* f.

EXAMPLE $f(-6)$ denotes the number in the range which is assigned to the number -6 in the domain by the rule of the function f.

The meaning of the symbols $f(x)$ and $y = f(x)$ as just explained differs slightly from the way functional notation was used in Examples 1–3 above. But the basic idea is the same:

> *Functional notation is just a convenient shorthand for phrases and* *sentences in the English language.*

Once they have made a careful definition, most mathematicians tend to become quite casual about their language. Instead of the precise sentence,

> The rule of the function f assigns to the number
> x in the domain the number y in the range,

or its symbolic abbreviation $y = f(x)$, you are quite likely to hear one or more of the following statements. Each of them is intended to mean *exactly the same thing* as the precise sentence above:

> The **value of the function** f at x is y.
> The **value** of f at x is y.
> The function f **maps** x to y.
> y is the **image** of x under (the function) f

Similarly, the number $f(x)$ in the range which is assigned to the number x in the domain by the rule of the function f is sometimes called

"the **value** (of the function) f at x," or
"the **image** of x (under the function f)."

There is nothing sacred about the letters x, y, and f. Just as in Examples 1–3 above, *we are free to choose any letters we want*, so long as we clearly explain their meanings. You can frequently reconstruct from the circumstances which letters have been used for domain, range, and rule. In particular, there are several **universally used conventions for interpreting certain algebraic expressions as functions**, as illustrated in the following examples.

EXAMPLE   An algebraic formula, such as

$$f(x) = \sqrt{x + 1}$$

is understood to denote a function f, as follows. The letter x denotes an element of the domain. The *rule* of the function f is understood to assign to the number x in the domain, the number $\sqrt{x + 1}$ in the range. For instance,

$$f(2) = \sqrt{2 + 1} = \sqrt{3} \text{ is the number assigned to } 2$$
$$f(-1) = \sqrt{-1 + 1} = 0 \text{ is the number assigned to } -1$$
$$f(\pi + 3) = \sqrt{(\pi + 3) + 1} = \sqrt{\pi + 4} \text{ is the number assigned to } \pi + 3$$

The *domain* of the function f is determined by this convention:

> *Unless specific information to the contrary is given, the domain of a function* f *is taken to be the set consisting of every real number for which the rule of* f *produces a well-defined real number.*

In this case, $\sqrt{x + 1}$ is a well-defined real number only when

$$x + 1 \geq 0 \qquad \text{or equivalently} \qquad x \geq -1$$

Consequently, the domain of this function is the set of all real numbers that are ≥ -1, that is, the interval $[-1, \infty)$. The *range* of this function is some set of real numbers. For most purposes in this course and in calculus, it is not necessary to specify the range more precisely. Consequently, we shall often omit any mention of the range. Determining the domain, however, is often essential.

EXAMPLE   The expression

$$h(u) = \frac{u^2 + 3}{u^2 - 9u + 20}$$

is understood to define a function (denoted h) whose rule is: assign to a number u in the domain the number $\dfrac{u^2 + 3}{u^2 - 9u + 20}$ in the range. For example,

$$h(-2) = \frac{(-2)^2 + 3}{(-2)^2 - 9(-2) + 20} = \frac{4 + 3}{4 + 18 + 20} = \frac{7}{42} = \frac{1}{6}$$

Since no contrary information is given, the domain consists of all real numbers for which $\dfrac{u^2 + 3}{u^2 - 9u + 20}$ is a well-defined real number. The only time when $\dfrac{u^2 + 3}{u^2 - 9u + 20}$ is *not* a real number occurs when the denominator $u^2 - 9u + 20$ is zero. Note that $u^2 - 9u + 20 = (u - 4)(u - 5)$. Clearly, this expression is 0 precisely when $u = 4$ or $u = 5$ and is nonzero for all other values of u. Therefore, the domain of the function h is the set of all real numbers, *except* 4 and 5.

EXAMPLE An expression such as

$$y = t^3 + 6t^2 - 5, \qquad t \geq 0$$

is understood to define a function whose rule is: assign to a number t in the domain the number $t^3 + 6t^2 - 5$ in the range. In this case, we are given the additional information that $t \geq 0$. So the domain of the function is taken to be the set of all real numbers t such that $t^3 + 6t^2 - 5$ is a real number *and* $t \geq 0$. Since $t^3 + 6t^2 - 5$ is a real number whenever t is, the domain consists of all nonnegative real numbers, that is, the interval $[0, \infty)$. Restrictions on the domain, such as $t \geq 0$ here, sometimes occur with functions that arise in practical problems. For instance, negative values of t might not be meaningful in a physical situation where t represents total distance traveled, even though the mathematical rule of the function makes sense for such values.

EXAMPLE The equation in the preceding example defined a function since for each number t there was one and only one number y that made the equation true. But this isn't always the case. The equation $y^2 = 4t^2$ does *not* define a function as it stands. For instance, if $t = 3$, then $4t^2 = 36$. There are two possible values of y that satisfy $y^2 = 36$, namely, $y = 6$ and $y = -6$. Thus it is not true that each number t in the domain is assigned to *one and only one* number in the range. Therefore, this equation is *not* the rule of a function. On the other hand, the statement $y^2 = 4t^2$ is true exactly when $y = \sqrt{4t^2}$ or $y = -\sqrt{4t^2}$. The equation $y = \sqrt{4t^2}$ *does* define a function whose domain is the set of all real numbers and whose rule is: assign to the number t the number $\sqrt{4t^2}$. This function assigns to the number 3 the number $\sqrt{4 \cdot 3^2} = 6$. The equation $y = -\sqrt{4t^2}$ also defines a function whose domain is all real numbers, but it is a *different* function since it has a different rule: assign to each number t the number $-\sqrt{4t^2}$. This function assigns to the number 3 the number -6.

EXAMPLE The statement

$$f(x) = \begin{cases} 2x + 3 & \text{if } x < 4 \\ x^2 - 1 & \text{if } 4 \leq x \leq 10 \end{cases}$$

defines a function f as follows. The rule of the function assigns to every number x that is *less than* 4 the number $2x + 3$. For instance,

$$f(-50) = 2(-50) + 3 = -97 \qquad \text{and} \qquad f(\tfrac{7}{4}) = 2(\tfrac{7}{4}) + 3 = \tfrac{13}{2}$$

The rule of the function f assigns to every number in the interval $[4, 10]$ the number $x^2 - 1$. For example,

$$f(\tfrac{9}{2}) = (\tfrac{9}{2})^2 - 1 = \tfrac{77}{4} \quad \text{and} \quad f(8) = 8^2 - 1 = 63$$

Since the rule of the function is defined only for numbers that are less than or equal to 10, the domain is the interval $(-\infty, 10]$.

Don't let the preceding examples mislead you into thinking that the rule of *every* function is given by an algebraic formula or by several algebraic formulas. For instance, in Example 2 above, where temperature is a function of time, the rule is given by the graph on page 88. It is very unlikely that there is *any* simple algebraic formula to describe this rule.

USING FUNCTIONAL NOTATION

Let f be the function whose rule is

$$f(x) = \sqrt{x^2 + 1}$$

The rule of f can be written in words as:

Assign to a given number the number obtained by squaring the given number, adding 1 to the result, and then taking the square root of this sum.

Keeping this fact in mind will help you to understand such symbols as:

$$f(a), \quad f(a + b), \quad f\left(\frac{1}{a}\right), \quad f(\sqrt{c}), \quad f(c^4)$$

Each of them denotes

The number obtained by applying the rule of the function f to the number in parentheses, namely: square it, add 1, then take the square root of the sum.

Thus

$$f(a) = \sqrt{a^2 + 1}$$
$$f(a + b) = \sqrt{(a + b)^2 + 1} = \sqrt{a^2 + 2ab + b^2 + 1}$$
$$f\left(\frac{1}{a}\right) = \sqrt{\left(\frac{1}{a}\right)^2 + 1} = \sqrt{\frac{1}{a^2} + 1} = \sqrt{\frac{1 + a^2}{a^2}}$$
$$f(\sqrt{c}) = \sqrt{(\sqrt{c})^2 + 1} = \sqrt{c + 1}$$
$$f(c^4) = \sqrt{(c^4)^2 + 1} = \sqrt{c^8 + 1}$$

The very same principle applies to expressions such as

$$f(x - 1), \quad f(x + h), \quad f(x^3), \quad f\left(\frac{1}{x}\right), \quad f(-x)$$

To compute these numbers, proceed as above and *apply the rule of the function to the number in parentheses.* In this case, the rule is $f(x) = \sqrt{x^2 + 1}$; that is, "square it, add 1, take the square root of the sum." Consequently,

$$f(x - 1) = \sqrt{(x - 1)^2 + 1} = \sqrt{x^2 - 2x + 1 + 1} = \sqrt{x^2 - 2x + 2}$$
$$f(x + h) = \sqrt{(x + h)^2 + 1} = \sqrt{x^2 + 2xh + h^2 + 1}$$
$$f(x^3) = \sqrt{(x^3)^2 + 1} = \sqrt{x^6 + 1}$$
$$f\left(\frac{1}{x}\right) = \sqrt{\left(\frac{1}{x}\right)^2 + 1} = \sqrt{\frac{1}{x^2} + 1} = \sqrt{\frac{1 + x^2}{x^2}}$$
$$f(-x) = \sqrt{(-x)^2 + 1} = \sqrt{x^2 + 1}$$

EXAMPLE Suppose $g(x) = \dfrac{|x + 5|}{x^3}$. Then:

$$g(t) = \frac{|t + 5|}{t^3} \qquad\qquad g(x + h) = \frac{|(x + h) + 5|}{(x + h)^3}$$

$$g(x + 4) = \frac{|(x + 4) + 5|}{(x + 4)^3} \qquad\qquad g(-x) = \frac{|-x + 5|}{(-x)^3}$$

$$g(x + \Delta x) = \frac{|(x + \Delta x) + 5|}{(x + \Delta x)^3} \qquad\qquad g\left(\frac{1}{x}\right) = \frac{|(1/x) + 5|}{(1/x)^3}$$

COMMON MISTAKES WITH FUNCTIONAL NOTATION

Consider the function f whose domain and range are the set of all real numbers and whose rule is: assign to the number x the number x^3. According to the conventions introduced earlier, all this information is contained in the notation

$$f(x) = x^3 \qquad \text{or} \qquad y = x^3$$

Note that each of the symbols

$$f(x), \qquad x^3, \qquad \text{and} \qquad y$$

represents a *number*, namely the number assigned to the number x by the rule of the function. Consequently, all of the following statements are *logically incorrect* even though their intended meaning may seem clear:

"the function $f(x)$"
"the function x^3"
"the function y"

For a *function* is *not* a number; it consists of two sets (domain and range) and a rule that assigns to each number in the domain one and only one number in the range.

Nevertheless, the use of such inaccuracies as "the function $f(x)$" or "the function x^3" or "the function $y = f(x)$" is so widespread that you may as well get accustomed to it.

So long as you know that these phrases *are* inaccurate and know the precise statements they represent, you won't have any difficulty.

Similarly, mathematicians often say things like "the function assigns $\sqrt{x + 3}$ to x," meaning that the *rule* of the function assigns $\sqrt{x + 3}$ to x. Once again, there is no problem in understanding what is intended.

Unfortunately, there are more serious mistakes in the use of functional notation. These are not just inaccuracies of language in situations where the intended meaning is clear, but out and out *falsehoods*. Usually they result from an attempt to treat functional notation as if it were ordinary algebraic notation, instead of a very specialized *shorthand language*. Here are examples of the most common errors.

Note There may be some functions for which one or more of the statements below are true. But all of these statements are *false* for most functions.

> **MISTAKES:** $f(a + b) = f(a) + f(b)$
> $f(a - b) = f(a) - f(b)$

EXAMPLE If $f(x) = x^2$, then $f(3 + 2) = (3 + 2)^2 = 25$, but $f(3) + f(2) = 3^2 + 2^2 = 9 + 4 = 13$. Hence $f(3 + 2) \neq f(3) + f(2)$. Furthermore, $f(3 - 2) = (3 - 2)^2 = 1$, but $f(3) - f(2) = 3^2 - 2^2 = 9 - 4 = 5$, Hence $f(3 - 2) \neq f(3) - f(2)$.

> **MISTAKE:** $f(ab) = f(a)f(b)$

EXAMPLE If $f(x) = x + 7$, then $f(3 \cdot 4) = (3 \cdot 4) + 7 = 12 + 7 = 19$, but $f(3)f(4) = (3 + 7)(4 + 7) = 10 \cdot 11 = 110$. Hence $f(3 \cdot 4) \neq f(3)f(4)$.

> **MISTAKES:** $f(ab) = af(b)$
> $f(ab) = bf(a)$

EXAMPLE If $f(x) = x^2 + 1$, then $f(2 \cdot 3) = (2 \cdot 3)^2 + 1 = 36 + 1 = 37$, but $2f(3) = 2(3^2 + 1) = 2(10) = 20$ and $3f(2) = 3(2^2 + 1) = 3(5) = 15$. Hence $f(2 \cdot 3) \neq 2f(3) \neq 3f(2)$.

EXERCISES

A.1. Let f be the function defined by the expression $f(x) = \sqrt{x + 3} - x + 1$. Evaluate:

(a) $f(0)$ (c) $f(\frac{5}{2})$ (e) $f(\sqrt{2})$ (g) $f(-2)$
(b) $f(1)$ (d) $f(\pi)$ (f) $f(\sqrt{2} - 1)$ (h) $f(-\frac{3}{2})$

A.2. Let h be the function defined by $h(x) = x^2 + \dfrac{1}{x} + 2$. Evaluate:

(a) $h(3)$ (c) $h(\frac{3}{2})$ (e) $h(a + k)$ (g) $h(2 - x)$

(b) $h(-4)$ (d) $h(\pi + 1)$ (f) $h(-x)$ (h) $h(x - 3)$

A.3. Let g be the function defined by $g(t) = t^2 - 1$. Evaluate:

(a) $g(3)$ (d) $g(x)$ (g) $g(-t)$

(b) $g(-2)$ (e) $g(s + 1)$ (h) $g(t + h)$

(c) $g(0)$ (f) $g(1 - r)$ (i) $g\left(\dfrac{1}{t}\right)$

A.4. Let f be the function defined by $f(x) = |2 - x| + \sqrt{x - 2} + x^2$. Evaluate:

(a) $f(2)$ (d) $f(5) + f(11)$ (g) $f(-x)$

(b) $f(11)$ (e) $f(5 - 2)$ (h) $f(5 - x)$

(c) $f(5 + 11)$ (f) $f(5) - f(2)$ (i) $f(5) - f(x)$

A.5. For each of the following functions, compute $f(2)$; $f(\frac{16}{3})$; $f(2) - f(\frac{16}{3})$; $f(r)$; $f(r) - f(x)$; and $\dfrac{f(r) - f(x)}{r - x}$ (assume $r \neq x$ and simplify your answer to the last one). For example, if $f(x) = x^2$, then

$$f(2) = 4, \quad f\left(\frac{16}{3}\right) = \frac{16^2}{3^2} = \frac{256}{9}, \quad f(2) - f\left(\frac{16}{3}\right) = 4 - \frac{256}{9} = -\frac{220}{9},$$

$$f(r) = r^2, \quad f(r) - f(x) = r^2 - x^2,$$

$$\frac{f(r) - f(x)}{r - x} = \frac{r^2 - x^2}{r - x} = \frac{(r + x)(r - x)}{r - x} = r + x \quad (r \neq x)$$

(a) $f(x) = x$ (d) $f(x) = -10x + 12$ (g) $f(x) = x - x^2$

(b) $f(x) = -10x$ (e) $f(x) = 3x + 7$ (h) $f(x) = \sqrt{x}$

(c) $f(x) = 12$ (f) $f(x) = x^3$ (i) $f(x) = \dfrac{1}{x}$

A.6. Each of the graphs in Figure 3-6 defines a function f with domain $[-3, 4]$. In each case, find $f(-3)$, $f(-\frac{3}{2})$, $f(0)$, $f(1)$, $f(\frac{5}{2})$, $f(4)$. Careful approximate answers are acceptable.

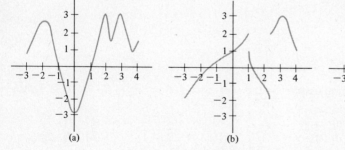

(a)   (b)   (c)

Figure 3-6

A.7. For each of the functions in Exercise A.6., find $f(-3) + f(-\frac{3}{2})$; $f(0) - f(2)$; $f(\frac{5}{2}) - f(3)$; $f(4) + 3f(-2)$.

A.8. In Washington state the sales tax $T(p)$ on an item of price p (*dollars*) equals 5% of p.

    (a) Which of the following formulas gives the correct sales tax in all cases?

        (i) $T(p) = p + 5$ (iv) $T(p) = p + (5/100)p = p + .05p$

        (ii) $T(p) = 1 + 5p$ (v) $T(p) = (5/100)p = .05p$

        (iii) $T(p) = p/20$

    (b) Find $T(3.6)$, $T(4.8)$, $T(0.6)$, and $T(0)$.

B.1. The sales tax law of state X reads as follows: "on any item up to and including 99¢, the sales tax is 5¢. On an item of \$1.00 or more, the tax is 5¢ *plus* 3% of the amount in excess of \$1.00."

    (a) Find an algebraic rule for the sales tax (in dollars) in state X on an item of price p dollars.

    (b) We want to compare the 5% Washington state sales tax (as discussed in Exercise A.8. immediately above) with the tax in state X. To keep the two taxes straight, we write $T_X(p)$, which is read "T sub X of p," for the tax in state X, and continue writing $T(p)$ for the tax in Washington. Which of the following statements are true, which false?

        (i) $T(2) > T_X(2)$ (iv) $T(2 \cdot 3) = T(2)T(3)$

        (ii) $T_X(2 + 6) = T_X(2) + T_X(6)$ (v) $T_X(1) = T(1)$

        (iii) $T_X(p) < T(p)$ for $0 \le p \le 0.99$

B.2. (a) Which of the following statements are true for *all* numbers x in the domain of the identity function [whose rule is $f(x) = x$]? If a statement is not true for some x in the domain, give an example to demonstrate this fact.

        (i) $f(x^2) = (f(x))^2$ (iii) $f(-x) = f(x)$

        (ii) $f(|x|) = |f(x)|$ (iv) $f(3x) = 3f(x)$

    (b) Do part (a) for the function $f(x) = 4x$.

    (c) Do part (a) for the absolute value function $f(x) = |x|$.

    (d) Do part (a) for the function $f(x) = x^2$.

B.3. Determine the *domain* of each of the following functions, according to the usual conventions.

    (a) $f(x) = x^2$ (g) $k(x) = |x| + \sqrt{x - 1}$

    (b) $g(x) = \dfrac{1}{x^2} + 2$ (h) $h(x) = \sqrt{(x + 1)^2}$

    (c) $h(t) = |t| - 1$ (i) $f(x) = x - [x]$

    (d) $k(u) = \sqrt{u}$ (j) $r(t) = [t] + [-t]$

    (e) $f(x) = [x]^2$ (k) $g(u) = \dfrac{|u|}{u}$

    (f) $g(t) = |t - 1|$ (l) $h(x) = \begin{cases} -x & \text{if } x < 1 \\ 4x - 5 & \text{if } x \ge 1 \end{cases}$

B.4. Give an example of two different functions f and g which have all of the following properties:
$$f(-1) = 1 = g(-1) \quad \text{and} \quad f(0) = 0 = g(0) \quad \text{and} \quad f(1) = 1 = g(1).$$

B.5. **(a)** Give an example of a function h which has the property that $h(u) = h(2u)$ for every real number u.
   (b) Give an example of a function f which has the property that $f(x) = 2f(x)$ for every real number x.

B.6. Determine the *domain* of each of the following functions, according to the usual conventions.

(a) $f(-x) = |-x|$

(b) $h(x) = \dfrac{\sqrt{x-1}}{x^2 - 1}$

(c) $g(t) = \sqrt{t^2}$

(d) $y = x^3 + 2$
(e) $g(y) = [-y]$
(f) $f(t) = \sqrt{-t}$

(g) $h(x) = \sqrt{(1-x)^3}$

(h) $y = \dfrac{1}{x} + \dfrac{1}{x-1} + \dfrac{1}{x+2}$

(i) $g(u) = \dfrac{u^2 + 1}{u^2 - u - 6}$

(j) $f(t) = \sqrt{4 - t^2}$
(k) $y = -\sqrt{9 - (x-9)^2}$

B.7. Let the symbol Δx represent a *nonzero* number. (That is, Δx represents a *single* number; it does *not* mean Δ times x.) For each of the following functions, compute $f(x + \Delta x)$. *Example:* if $f(x) = x^2 + 1$, then

$$f(x + \Delta x) = (x + \Delta x)^2 + 1$$
$$= x^2 + 2x\,\Delta x + (\Delta x)^2 + 1$$

(a) $f(x) = x$ **(d)** $h(x) = x^3$ **(g)** $f(x) = x + 5$

(b) $f(x) = x^2$ **(e)** $k(x) = \dfrac{1}{x}$ **(h)** $f(x) = 7x + 2$

(c) $g(x) = 12$ **(f)** $g(x) = -10x$ **(i)** $k(x) = x^2 + 3x - 7$

B.8. For each of the functions in Exercise B.7 compute and simplify the **difference quotient,** which is the quotient $\dfrac{f(x + \Delta x) - f(x)}{\Delta x}$. The difference quotient is used extensively in calculus. *Example:* if $f(x) = x^2 + 1$, then

$$\frac{f(x + \Delta x) - f(x)}{\Delta x} = \frac{[(x + \Delta x)^2 + 1] - (x^2 + 1)}{\Delta x}$$

$$= \frac{x^2 + 2x\,\Delta x + (\Delta x)^2 + 1 - x^2 - 1}{\Delta x} = \frac{(2x + \Delta x)\,\Delta x}{\Delta x} = 2x + \Delta x \quad (\Delta x \neq 0)$$

B.9. **(a)** Evaluate each of the difference quotients in Exercise B.8 when x is fixed and $\Delta x = 2$; $\Delta x = 1$; $\Delta x = .5$; $\Delta x = .1$;   and $\Delta x = .01$. *Example:* if $f(x) = x^2 + 1$, then

$$\frac{f(x + \Delta x) - f(x)}{\Delta x} = 2x + \Delta x \quad (\Delta x \neq 0)$$

Hence,

Δx	2	1	.5	.1	.01
$2x + \Delta x$	$2x + 2$	$2x + 1$	$2x + .5$	$2x + .1$	$2x + .01$

(b) When Δx is a nonzero number very close to 0, what is the difference quotient very close to? (In the example above, the difference quotient $2x + \Delta x$ is very close to $2x$ when Δx is very close to 0.)

C.1. As explained in Example 3 of the text, the distance traveled by a falling rock after t sec is $16t^2$. In functional notation, $d(t) = 16t^2$.
   (a) Find $d(0)$ and $d(2)$.
   (b) Find $d(1)$ and $d(3)$.
   (c) Find $d(2)$ and $d(4)$.
   (d) How far does the rock travel during the first 2 sec?
   (e) How far does the rock travel during the second and third seconds (that is, what distance is traveled from $t = 1$ to $t = 3$)?
   (f) How far does the rock travel from $t = 2$ to $t = 4$?
   (g) Use functional notation to express the distance traveled by the rock from time t_1 to time t_2.
   (h) Use functional notation to express the distance traveled by the rock during a 4-sec period beginning at time t.

C.2. This exercise deals with Example 2 in the text; use the graph shown in Figure 3-1 (p. 88).
   (a) By measuring as carefully as you can, fill the blanks in the following statements. There may be more than one correct answer in some cases.
      (i) The temperature at 8 A.M. was _____ degrees.
      (ii) The temperature at 8 P.M. was _____ degrees.
      (iii) The temperature at 8 A.M. was the same as the temperature at _____ .
      (iv) The difference between the temperatures at noon and at 6 P.M. was _____ degrees.
      (v) The temperature at _____ was greater than the temperature at 8 P.M.
      (vi) The temperature at _____ P.M. was less than the temperature at _____ A.M.
      (vii) The temperature at 10 A.M. was _____ than the temperature at 5 P.M.
      (viii) The temperature at noon was _____ than the temperature at 8 P.M. but _____ than the temperature at 4 P.M.
   (b) After filling the blanks in part (a), write each of the statements in the functional notation introduced on page 96.

DO IT YOURSELF!

BACK TO REAL LIFE

This chapter began with some real-life situations which suggested various kinds of functions and functional notation. These in turn led to the abstract concept of function in a purely mathematical setting, together with more functional notation. Now that you have some familiarity with functional notation, it's time to return to the starting point. If you want to understand mathematics fully or to apply it, you must be able to:

Recognize functional situations in real-life settings.
Determine which quantities are functions of others.
Find a mathematical description of the function involved (including, it is to be hoped, an algebraic formula for the rule of the function).

There are no hard and fast rules for doing these things. The only way to become proficient at them is to work through some examples and then do some exercises on your own.

SITUATION A 20-in. square sheet of tin is to be used to make an open-top box by cutting a small square from each corner of the sheet and bending up the sides (see Figure 3-7).

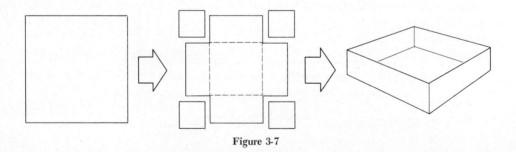

Figure 3-7

Approximately what size square should be cut out of each corner in order to obtain a box with the largest possible volume?

Analysis It is clear that the volume of the box depends on the size of the square cut from the corners. The size of the square is determined by the length of its side. Call this length x. Then the *volume is a function of the length x of the side* of the square to be cut from the corners. Since the whole sheet of tin is 20×20 in., we see that each of the dotted lines in Figure 3-8 has the length $20 - 2x$ in.

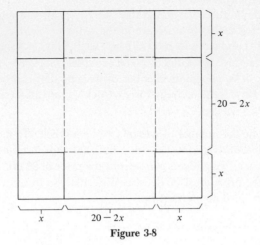

Figure 3-8

When the sides are folded up, the bottom of the box will measure $(20 - 2x)$ by $(20 - 2x)$ in. and its height will be x in. Therefore its volume will be

$$\text{height} \times \text{length} \times \text{width} = x(20 - 2x)(20 - 2x) = x(20 - 2x)^2$$

Thus the volume is a function of the length x and the *rule* of the function is:

$$V(x) = x(20 - 2x)^2$$

What is the *domain* of this function? Certainly $V(x) = x(20 - 2x)^2$ is a well-defined real number for any real number x. But in this situation the only numbers which *make sense* are those numbers x with $0 < x < 10$. For you can't cut a zero or negative amount from the corners. Nor can you cut four squares each of side length 10 in. or more from a 20×20-in. sheet (and still have something left). Consequently, we write

$$V(x) = x(20 - 2x)^2 \qquad 0 < x < 10$$

to indicate that the domain is the interval $(0, 10)$.

Here are some values of the function for different numbers:

x	1	2	3	4	5	6	9
$V(x) = x(20 - 2x)^2$	324	512	588	576	500	384	36

It appears that a box of largest possible volume will be obtained by cutting out a square of side *approximately* 3 in. This is the best we can do with purely algebraic techniques. In calculus you will learn how to find the *exact* value of x which makes the volume $V(x)$ a maximum. (It turns out to be $x = \frac{10}{3}$, which gives a box of volume $\frac{16,000}{27} = 592.59$ cu in.)

SITUATION   A manufacturer has found that she can't sell any widgets at a price of $2.00 or more. But for each 10¢ decrease in the price per widget, she can sell 1000 widgets a week (that is, she can sell 1000 at $1.90 each, 2000 at $1.80 each, and so on). It costs 50¢ to manufacture one widget. Fixed expenses (mortgage, taxes, and the like) run $2000 per week. The maximum number of widgets that can be manufactured in one week is 17,000. How is the manufacturer's weekly *profit* affected by the number of widgets she sells?

Analysis   Let x be the number of widgets sold per week. To measure the number or widgets sold *in thousands* we must divide x by 1000. For instance, 2500 is the same as $2.5 \left(= \frac{2500}{1000} \right)$ *thousands*. No widgets can be sold at a price of $2.00. For each thousand widgets to be sold, the price must be decreased by 10¢. So to sell x widgets per week the price per widget (in dollars) must be:

$$2 - \text{(number of thousands sold)}(.10) = 2 - \left(\frac{x}{1000} \right)(.10) = 2 - \frac{x}{10,000}$$

For instance, if $x = 5000$, the price is $2 - \frac{5000}{10,000} = 2 - .5 = \$1.50$. Therefore the *income* from selling x widgets is

$$\text{(price per widget)} \times \text{(number sold)} = \left(2 - \frac{x}{10,000} \right)x$$

The *cost* of manufacturing x widgets per week is

$$\text{(cost per widget)} \times \text{(number sold)} + \text{(fixed expenses)} = .50x + 2000$$

The manufacturer's profit is

$$\text{income} - \text{cost} = \left(2 - \frac{x}{10,000} \right)x - (.50x + 2000)$$

$$= 2x - \frac{x^2}{10,000} - .5x - 2000$$

$$= 1.5x - \frac{x^2}{10,000} - 2000$$

Thus *profit is a function P of the number x of widgets sold:*

$$P(x) = 1.5x - \frac{x^2}{10,000} - 2000$$

Note that $P(x)$ is a well-defined real number for any real number x. But in this situation the only numbers which make sense for x are integers from 0 to 17,000. Thus the domain of P is all integers in the interval [0, 17,000].

The preceding examples illustrate a common phenomenon:

> *A real life functional situation may lead to a function whose domain does not include all real numbers, even though the rule of the function may make sense for all real numbers.*

EXERCISES

A.1. Suppose a car travels at a constant rate of 55 mph for 2 hr, and travels at 45 mph thereafter. Show that distance traveled is a function of time and find the rule of the function.

A.2. (a) The distance between city C and city S is 2000 mi. A plane flying directly to S passes over C at noon. If the plane travels at 475 mph, express the distance of the plane from *city S* as a function of time.
    (b) Do part (a) for a plane which travels at 325 mph.

A.3. The list price of a textbook is $12. But if 10 or more copies are purchased, then the price per copy is reduced by 25¢ for every copy above 10. (That is, $11.75 per copy for 11 copies, $11.50 per copy for 12 copies, and so on.)
    (a) The price per copy is a function of the number of copies purchased. Find the rule of this function.
    (b) The total cost of a quantity purchase is (number of copies) × (price per copy). Show that the total cost is a function of the number of copies and find the rule of the function.

A.4. A potato chip factory has a daily overhead from salaries and building costs of $1800. The cost of ingredients and packaging to produce a pound of potato chips is 50¢. A pound of potato chips sells for $1.20. Show that the factory's daily profit is a function of the number of pounds of potato chips sold and find the rule of this function. (Assume that the factory sells all the potato chips it produces each day.)

B.1. (a) A rectangular region of 6000 sq ft is to be fenced in on three sides with fencing costing $3.75 per ft and on the fourth side with fencing costing $2.00 per ft. Express the cost of the fence as a function of the length x of the fourth side.
    (b) Estimate the value of x which produces the cheapest fence.
    (c) Do parts (a) and (b) assuming that the side opposite the fourth side is a river, so that no fencing is required there.

B.2. The fast food king Ray Rotgut can sell 10,000 hamburgers per day at a price of 75¢ each. Each price increase of 10¢ per hamburger results in 1000 fewer hamburgers being sold.
    (a) Express the number of hamburgers sold as a function of the price per hamburger.
    (b) Express the total income from hamburger sales as a function of the price per hamburger.
    (c) Rotgut's fixed costs (salaries, building, equipment, and so on) are $1100 per day The ingredients for one hamburger costs 40¢. Express Rotgut's daily profit as a function of the price per hamburger.
    (d) Use the function in part (c) to estimate the price Rotgut should charge per hamburger in order to maximize his daily profit.

B.3. A box with a square base measuring $t \times t$ ft is to be made of three kinds of wood. The cost of the wood for the base is 85¢ per sq ft; the wood for the sides costs 50¢ per sq ft and the wood for the top $1.15 per sq ft. The volume of the box is to be 10 cu ft.

(a) Express the total cost of the box as a function of the length t.

(b) Estimate the value of t which makes the cost as low as possible.

B.4. An open-top rectangular box is to be made from a piece of cardboard 10 in. wide and 16 in. long by cutting a square from each corner and bending up the sides. The volume of the box is a function of the length t of the side of the cut square. Find the rule of this function and *estimate* the value of t which will produce a box of maximum volume.

3. GRAPHS OF FUNCTIONS

The next step in analyzing functional situations is to draw a picture. The graph of a function is essentially just a picture of the function. We have seen such pictures before, when we considered graphs of equations in Chapter 2. The graph of an *equation*, such as $y = 7x^2$ or $y = 3x - 4$, is just the set of all points (x, y) in the coordinate plane whose coordinates, x and y, satisfy the given equation. The graph of a *function* is similar:

> The **graph of the function** f *is the graph of the equation* y = f(x).

Thus the point (x, y) is on the graph of the function f precisely when x is a number in the domain of f and y is the number $f(x)$, the value of the function at x. In other words,

> The graph of the function f *consists of all points* (x, f(x)), *where* x *is any number in the domain of* f.

The letters used are not important. For instance, if the function is denoted by g, then its graph is the set of all points $(t, g(t))$, where t is any number in the domain of the function.

Graphs allow us to apply various geometric insights to the study of function. Properties of a function, which may not be immediately apparent, often become obvious when you look at the graph. This interaction between geometry and algebra is crucial in the development of the mathematics needed to cope with numerous problems in the real world.

In theory, once you know the function f, you automatically know its graph: just plot each of the points $(x, f(x))$ for every number x in the domain. In practice, however, you can actually plot only finitely many points and usually must make an "educated guess" about the rest. For most of the functions in this course and in calculus, you should follow this general procedure:

How to Graph the Function *f*

1. *Choose a few numbers* x *in the domain of* f *and compute the corresponding numbers* f(x).
2. *Plot the points* (x, f(x)) *determined in step 1. In many cases, they will suggest a geometric pattern.*
3. *Apply any other algebraic or geometric information that may be available about the function* f *in order to determine the general shape of the entire graph.*
4. *Using the points plotted in step 2 and the information developed in step 3, sketch the graph. Unless the information from step 3 indicates otherwise, the graph will usually be an unbroken curve.*°

This general procedure, especially step 3, isn't very enlightening until you have seen it applied in examples. As you will see, it isn't necessary to be rigid about the order in which you do steps 1–3.

EXAMPLE   Let g be the **square root function,** given by $g(x) = \sqrt{x}$. We know that $\sqrt{x}$ is defined only when $x \geq 0$ and that in such cases $\sqrt{x} \geq 0$. Therefore the only points $(x, g(x)) = (x, \sqrt{x})$ on the graph are those with both coordinates ≥ 0. Thus the entire graph lines in the first quadrant. With the use of the table on page 588, or a calculator, we see that

$$g(0) = \sqrt{0} = 0, \qquad g(.25) = \sqrt{.25} = .5, \qquad g(1) = \sqrt{1} = 1,$$
$$g(2) = \sqrt{2} \approx 1.414, \qquad g(3) = \sqrt{3} \approx 1.732, \qquad g(4) = \sqrt{4} = 2,$$
$$g(6) = \sqrt{6} \approx 2.449, \qquad g(9) = \sqrt{9} = 3, \qquad g(10) = \sqrt{10} \approx 3.162$$

If we plot the corresponding points, we have Figure 3-9.

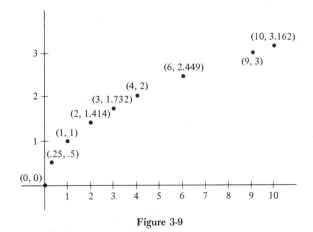

Figure 3-9

°The word "curve" here includes the possibility of straight line segments.

In this case there is some additional algebraic information:

$$\text{if } 0 \le b < c, \qquad \text{then } \sqrt{b} < \sqrt{c}$$

In graphical terms this means that;

If c lies to the *right* of b on the horizontal axis,
then the graph at c is *higher* than the graph at b

In other words, as you move to the *right*, the graph always *rises*. Consequently, the entire graph looks like the one shown in Figure 3-10 (where the arrow indicates that the graph continues upward to the right).

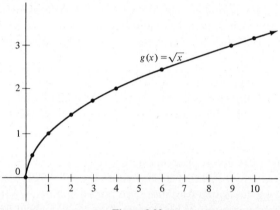

Figure 3-10

EXAMPLE The graph of $f(x) = x^2$ consists of all points (x, x^2). Since $x^2 \ge 0$ for every x, every point on the graph lies on or above the x-axis. By computing x^2 for various values of x, we are able to locate some points on the graph, as shown in Figure 3-11.

x	$f(x) = x^2$
-2.5	6.25
-2	4
-1.5	2.25
-1	1
$-.5$	.25
0	0
.5	.25
1	1
1.5	2.25
2	4
2.5	6.25

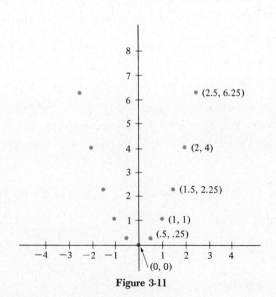

Figure 3-11

As x gets very large in absolute value, x^2 gets even larger. Consequently, the graph of $f(x) = x^2$ looks like the one shown in Figure 3-12 (where the arrows indicate that the graph continues sharply upward and outward).

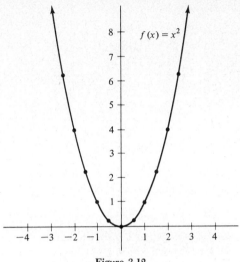

$f(x) = x^2$

Figure 3-12

A curve shaped like the graph of $f(x) = x^2$ is called a **parabola.** Its lowest point [in this case $(0, 0)$] is called the **vertex** of the parabola. Parabolas are discussed further in Section 6.

EXAMPLE   To graph the function $f(x) = \dfrac{1}{x^2 + 1}$, we first find some points on the graph by computing the value of the function at several numbers:

x	-10	-7	-3	-1	$-\frac{1}{2}$	0	$\frac{1}{3}$	1	2	3	7	10
$f(x) = \dfrac{1}{x^2 + 1}$	$\frac{1}{101}$	$\frac{1}{50}$	$\frac{1}{10}$	$\frac{1}{2}$	$\frac{4}{5}$	1	$\frac{9}{10}$	$\frac{1}{2}$	$\frac{1}{5}$	$\frac{1}{10}$	$\frac{1}{50}$	$\frac{1}{101}$

Observe that as x gets larger and larger in absolute value, $f(x) = \dfrac{1}{x^2 + 1}$ is a positive number that gets closer and closer to zero. In geometric terms this means that as you move to the far left or far right ($|x|$ large), the graph lies above the x-axis [$f(x)$ positive] but gets closer and closer to the x-axis. Furthermore, $x^2 + 1 \geq 1$ for every number x, so that $f(x) = \dfrac{1}{x^2 + 1} \leq 1$. Geometrically, this means that the graph always lies below the horizontal line $y = 1$. Using these facts and the points obtained from the table above, we see that the entire graph looks like the one shown in Figure 3-13.

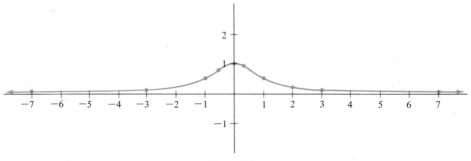

Figure 3-13

EXAMPLE No such elaborate analysis is needed to graph **linear functions** such as

$$f(x) = 3x - 4, \qquad g(t) = 7t + 2, \qquad h(x) = -6$$

For instance, the graph of $f(x) = 3x - 4$ is, by definition, the graph of the equation $y = f(x)$, that is, $y = 3x - 4$. From Section 4 of Chapter 2 we know that the graph of any linear equation, such as $y = 3x - 4$, $y = 7t + 2$, and so on, is just a straight line. Hence we need only plot two points in each case to determine the entire graph, as shown in Figure 3-14.

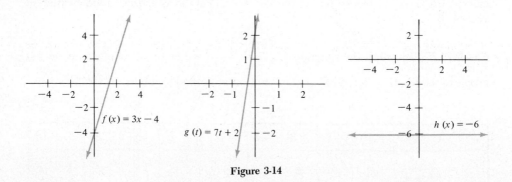

Figure 3-14

EXAMPLE Let f be the absolute value function, given by $f(x) = |x|$. If we write down the definition of $|x|$, we see that the rule of this function is in two parts:

$$f(x) = |x| = \begin{cases} x & \text{if } x \geq 0 \\ -x & \text{if } x < 0 \end{cases}$$

Thus we first consider those points on the graph with $x \geq 0$. If $x \geq 0$, then $|x| = x$ and the equation $y = f(x)$ becomes $y = |x| = x$. But we know from Section 4 of Chapter 2 that the graph of $y = x$ (with $x \geq 0$) is just half of the straight line $y = x$, as shown in Figure 3-15.

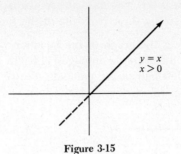

Figure 3-15

Similarly, when $x < 0$, then $|x| = -x$, and the equation $y = f(x)$ becomes $y = |x| = -x$. Once again we know that the graph of $y = -x$ (with $x < 0$) is just half of the straight line $y = -x$, as shown in Figure 3-16.

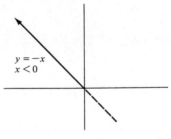

Figure 3-16

Combining this information, we see that the entire graph of $f(x) = |x|$ looks like the one in Figure 3-17.

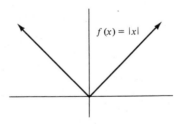

Figure 3-17

EXAMPLE Just plotting points may not be enough to find the graph of the function h given by

$$h(t) = \begin{cases} t^2 & \text{if } t < 0 \\ t & \text{if } 0 \leq t < 5 \\ -2t + 11 & \text{if } t \geq 5 \end{cases}$$

If you simply compute $h(t)$ for some selected values of t, plot the corresponding points, and join them by a single curve, you'll probably get the graph wrong. To obtain the correct graph in cases such as this, you need a careful analysis of the separate parts of the graph, corresponding to the three-part rule of the function.

$t < 0$ The graph of h is the graph of the equation $y = h(t)$. When $t < 0$, this equation is $y = t^2$. From the example on pages 114–115 we know that the graph of $y = t^2$ with $t < 0$ is the left half of the parabola $y = t^2$, as plotted in Figure 3-18.

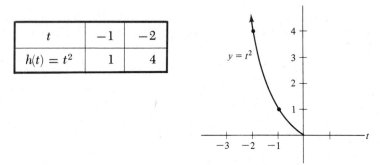

t	-1	-2
$h(t) = t^2$	1	4

Figure 3-18

$0 \le t < 5$ For these values of t the rule of the function is $h(t) = t$. So the graph of $y = h(t)$ is just the graph of $y = t$. Since the entire graph of $y = t$ is a straight line, the graph of the function h over the interval $[0, 5)$ is just a segment of that line, as plotted in Figure 3-19.

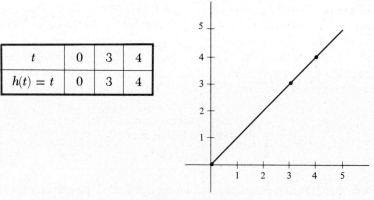

t	0	3	4
$h(t) = t$	0	3	4

Figure 3-19

$t \ge 5$ Over the interval $[5, \infty)$, the graph of h is the graph of $y = h(t) = -2t + 11$, namely, a half line, as plotted in Figure 3-20.

t	5	6	7
$h(t) = -2t + 11$	1	-1	-3

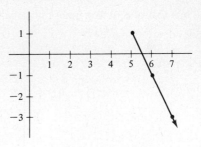

Figure 3-20

Combining the above information yields the entire graph of the function h, as shown in Figure 3-21.

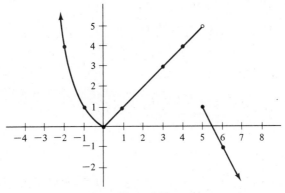

Figure 3-21

The open circle indicates that the point $(5, 5)$ is *not* on the graph of h (it is on the graph of $y = t$, of course). The closed circle indicates that the point $(5, 1)$ *is* on the graph of h. Note that this graph, unlike the others discussed so far, is *not* an *unbroken* curve.

EXAMPLE Whenever you want to graph a function whose rule involves the greatest integer function, such as $g(x) = x - [x]$, the best procedure is to consider what the function and graph look like between each two consecutive integers. For instance,

If	Then $[x] =$	So That $x - [x] =$
$-2 \leq x < -1$	-2	$x - (-2) = x + 2$
$-1 \leq x < 0$	-1	$x - (-1) = x + 1$
$0 \leq x < 1$	0	x
$1 \leq x < 2$	1	$x - 1$
$2 \leq x < 3$	2	$x - 2$

Thus the rule of the function g really consists of many parts:

$$g(x) = \begin{cases} \vdots \\ x + 2 & \text{if } -2 \leq x < -1 \\ x + 1 & \text{if } -1 \leq x < 0 \\ x & \text{if } 0 \leq x < 1 \\ x - 1 & \text{if } 1 \leq x < 2 \\ x - 2 & \text{if } 2 \leq x < 3 \\ \vdots \end{cases}$$

Its graph can be found by considering the function over each of the intervals . . . , $[-2, -1), [-1, 0), [0, 1), [1, 2), \ldots$, and so on. Over each such interval the graph is a segment of a straight line. For instance, when $1 \leq x < 2$, we have the graph shown in Figure 3-22.

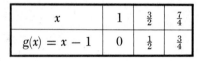

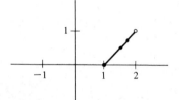

Figure 3-22

The open circle indicates that the point $(2, 1)$ is not on the graph.

A similar analysis shows that for $-2 \leq x < 3$, the graph of g looks like the one in Figure 3-23.

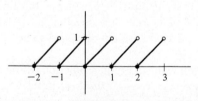

Figure 3-23

You should convince yourself that this same pattern continues both to the left and the right.

A WARNING AND A PREVIEW

The examples above are a bit misleading. It is not *always* possible to find the graph simply by plotting a few points and making a simple algebraic analysis of the function. Even when plotting points *does* suggest a general pattern, you may not be able to answer questions such as these:

(i) When the graph rises from one point to another, which way does it bend?

(ii) Does the graph wiggle between two points?

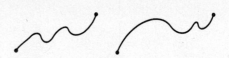

(iii) When the graph appears to change from rising to falling between points P and Q, exactly where does it change?

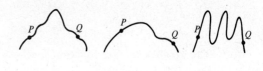

(iv) Is the graph a continuous, unbroken curve between two points? Or are there gaps, jumps, holes, or isolated points?

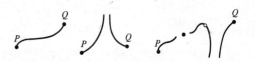

The last of these four questions is especially important. In many of the examples above, it was possible only to plot a finite number of points. Unless there was specific information to the contrary, we *assumed* that the graph between two such points was a continuous, unbroken curve. Although this assumption may seem *reasonable*, we can't be absolutely *sure* it is true without some proof. The techniques needed to provide this proof, as well as to answer all of the questions above, are developed in calculus. Until then, you should continue to make such reasonable assumptions and apply the general graphing procedure outlined above as carefully as possible.

EXERCISES

A.1 Sketch the graph of the function d in Example 3 (page 96).

A.2. Sketch the graph of $g(x) = x^3$.

A.3. Sketch the graph of $f(x) = -x^3 + 1$.

A.4. Sketch the graph of the cube root function h, given by $h(x) = \sqrt[3]{x}$. (*Hint:* the table of cube roots on p. 588 may be helpful.)

A.5. Graph each of the following functions.

(a) $f(x) = \begin{cases} x^2 & \text{if} \quad x \geq -1 \\ 2x + 3 & \text{if} \quad x < -1 \end{cases}$

(b) $g(x) = \begin{cases} |x| & \text{if} \quad x < 1 \\ -3x + 4 & \text{if} \quad x \geq 1 \end{cases}$

(c) $h(x) = \begin{cases} x - 2 & \text{if} \quad x < 4 \\ x & \text{if} \quad x \geq 4 \end{cases}$

(d) $f(t) = \begin{cases} t & \text{if} \quad t < -3 \\ t^2 & \text{if} \quad -3 \leq t \leq 3 \\ 2t & \text{if} \quad t > 3 \end{cases}$

(e) $k(u) = \begin{cases} -2u - 2 & \text{if} \quad u < -3 \\ u - [u] & \text{if} \quad -3 \leq u \leq 1 \\ 2u^2 & \text{if} \quad u > 1 \end{cases}$

(f) $f(x) = \begin{cases} x^2 & \text{if} \quad x < -2 \\ x & \text{if} \quad -2 \leq x < 4 \\ \sqrt{x} & \text{if} \quad x \geq 4 \end{cases}$

B.1. The greatest integer function f, whose rule is $f(x) = [x]$, is called a **step function**. To see why, carefully sketch the graph of f. (*Hint:* what does the graph look like on the interval $[0, 1)$? on $[1, 2)$? on $[-1, 0)$? on $[-2, -1)$? and so forth.)

B.2. (a) Graph $f(x) = -[x]$ for $-3 \leq x \leq 3$.
(The hint for Exercise B.1 also applies here.)
(b) Graph $g(x) = [-x]$ for $-3 \leq x \leq 3$. (*Note:* this *not* the same as $-[x]$.)
(c) Graph $h(x) = [x] + [-x]$ for $-3 \leq x \leq 3$.

B.3. The amount of postage required for each ounce of first-class mail, or fraction thereof, is k cents. (At this writing $k = 15$.) The *number* of k-cent stamps required to mail a first-class letter is a function of the weight of the letter (in ounces). Call this function the postage stamp function.
(a) Describe the rule of the postage stamp function algebraically.
(b) Graph the postage stamp function.
(c) Graph the function f whose rule is $f(x) = p(x) - [x]$, where p is the postage stamp function.

B.4. Sketch the graph of $f(x) = 1/x$. (*Hints:* the function is not defined when $x = 0$; what does the graph look like *near* $x = 0$? What does the graph look like when $|x|$ is very large?)

B.5. (a) Sketch the graph of the function given by $y = x + |x|$.
(b) Sketch the graph of the function given by $y = x|x|$.

B.6. Sketch the graph of $f(x) = x^3 - x$. [*Hints:* determine whether $f(x)$ is always positive or always negative in each of the intervals $(-\infty, -1), (-1, 0), (0, 1), (1, \infty)$. What happens at $x = -1, x = 0, x = 1$?]

B.7. Sketch the graphs of the following functions. (*Hint:* think circular.)
(a) $f(x) = \sqrt{1 - x^2}$
(b) $g(t) = -\sqrt{16 - t^2}$
(c) $f(x) = \sqrt{4 - (x - 1)^2} + 2$
(d) $g(t) = \sqrt{16 - (t + 3)^2} - 5$

B.8. In each part, graph the two given functions (one of them was graphed in the text). What is the geometric relationship between the two graphs?

(a) $f(x) = |x|$ and $h(x) = |x| + 1$ (e) $g(x) = \sqrt{x}$ and $k(x) = -\sqrt{x}$

(b) $f(x) = |x|$ and $k(x) = |x| - 4$ (f) $f(x) = |x|$ and $h(x) = |x - 5|$

(c) $g(x) = \sqrt{x}$ and $r(x) = \sqrt{x} - 1$ (g) $f(x) = |x|$ and $v(x) = |x + 5|$

(d) $g(x) = \sqrt{x}$ and $s(x) = \sqrt{x} + 2$

B.9. When you have done Exercises A.2 and A.4, answer this question: how is the graph of $g(x) = x^3$ related geometrically to the graph of $h(x) = \sqrt[3]{x}$?

B.10. Let f be a *function*. Suppose x and r are two *different* numbers in the domain of f. Then the points $(r, f(r))$ and $(x, f(x))$ are on the graph of f.

(a) Why is it *impossible* for $(r, f(r))$ and $(x, f(x))$ to be on the same *vertical* straight line?

(b) Draw three different graphs which are *not* the graphs of functions.

B.11. Graph each of these functions as best you can (some of them may be harder than the examples and other exercises).

(a) $f(x) = 2x^2 - 3x + 2$ (d) $f(u) = \dfrac{1}{u^3 + 1}$

(b) $g(x) = -3x^2 + 4x - 7$ (e) $k(x) = x^3 - 3x + 1$

(c) $h(t) = \dfrac{t}{1 - t}$ (f) $h(x) = \dfrac{x}{x^2 + 1}$

4. MORE GRAPH READING

In order to understand the interaction of algebra and geometry in the study of functions, or to apply this knowledge to practical problems, you must be able to "translate" statements from one to another of three different "languages":

The English language.
Formula language (algebraic and functional notation).
Graphical language (graphs).

The emphasis in this section is on translating graphical information into equivalent statements in English or functional notation; in short, graph reading. We begin with a familiar example from Sections 1 and 2.

EXAMPLE 2—TEMPERATURE The recording device at the weather bureau in city F directly produces the graph of the function which relates the time of day (measured in hours after midnight) to the temperature, as shown below (Figure 3-24). Using functional notation we write tem(t) for the temperature at t hours after midnight. The graph consists of all points with coordinates $(t, \text{tem}(t))$. For instance, since the point (4, 39) is on the graph,° we know that the temperature at 4:00 A.M. was 39°. To determine the time period during which the temperature was below 50°, we must *first* find every point $(t, \text{tem}(t))$ on the graph whose second coordinate, $\text{tem}(t)$ is < 50, that is, every point which lies *below* the horizontal line through 50, as shown in Figure 3-24.

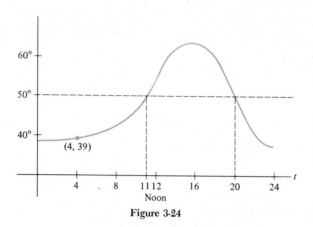

Figure 3-24

Having found all such points, we now must determine their first coordinates, for these first coordinates are precisely the times when the temperature is less than 50°. A careful examination of the graph above suggests that $\text{tem}(t) < 50$ whenever

$$0 \le t \le 11 \qquad \text{or} \qquad 20 \le t \le 24$$

In other words, the temperature was below 50° from midnight to 11:00 A.M. and again from 8:00 P.M. $(t = 20)$ to midnight.

EXAMPLE 2 (PART 2)   A slightly more complicated problem is to determine the time period *before* 4 P.M. during which the temperature was *at least* 60°. To do so, we translate these requirements into functional notation and graphical terms. Remember that 4:00 P.M. is 16 hours after midnight:

Statement	Functional Notation	Graph
The time is before 4:00 P.M.	$t < 16$	$(t, \text{tem}(t))$ lies to the left of the vertical line through $t = 16$
The temperature is at least 60°	$\text{tem}(t) \ge 60$	$(t, \text{tem}(t))$ lies on or above the horizontal line through 60

° Here and below our results are only as accurate as our measuring ability. But the basic idea should be clear.

Using this information and examining the graph in Figure 3-25

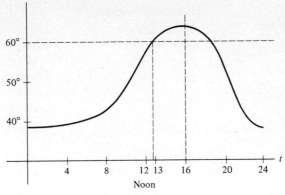

Figure 3-25

shows that the points $(t, \text{tem}(t))$ with $t < 16$ and $\text{tem}(t) \geq 60$ are those with first coordinates between 12.8 and 16. Thus the temperature was at least 60° from 12:48 P.M. $(t = 12.8)$ to 4:00 P.M. $(t = 16)$.

EXAMPLE 2 (PART 3) Suppose the temperature graph for a second city (city S) is superimposed on the temperature graph given above for city F, as shown in Figure 3-26.

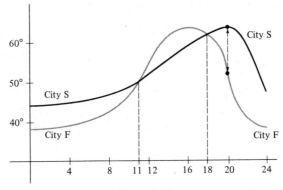

Figure 3-26

We continue to denote the temperature function for our original city (city F) by $\text{tem}(t)$. We shall denote the temperature function for the second city (city S) by $\text{tem}_S(t)$. By examining the graphs of $\text{tem}(t)$ and $\text{tem}_S(t)$ carefully, we can answer such questions as:

(i)   During what time period was it warmer in city F than in city S?
(ii) Was there any time during the day when it was at least 10 degrees warmer in city S than in city F?

Once again it is a matter of three-way translation from English language to functional notation to graphical terms:

Statement	Functional Notation	Graph
City F is warmer than city S at time t	$\text{tem}(t) > \text{tem}_S(t)$	The point $(t, \text{tem}(t))$ lies directly above $(t, \text{tem}_S(t))$
City S is at least 10 degrees warmer than city F at time t	$\text{tem}_S(t) \geq \text{tem}(t) + 10$	The point $(t, \text{tem}_S(t)$ is at least 10 units above the point $(t, \text{tem}(t))$

Examination and careful measurement on the graph on page 125 shows that

(i) $\text{tem}(t) > \text{tem}_S(t)$ for all t in the interval $(11, 18)$.
(ii) $\text{tem}_S(t) \geq \text{tem}(t) + 10$ for many values of t, including $t = 20$.

In other words,

(i)   It was warmer in city F than in city S between 11:00 A.M. and 6:00 P.M. $(t = 18)$.
(ii) It was at least 10 degrees warmer in city S at 8:00 P.M. $(t = 20)$ and at other times.

Of course, in the usual mathematical setting for graph reading, there is no reference to times, temperatures, and so on. You are simply required to use the graphs of functions to determine information about the functions and vice-versa.

EXAMPLE   Figure 3-27 shows the graphs of functions g and h over the interval $[-5, 7]$.

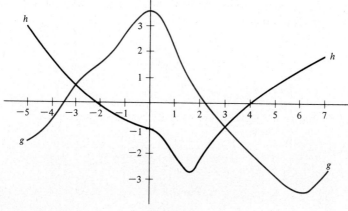

Figure 3-27

We can use these graphs to solve such problems as:

(i)   Find all numbers x in the interval $[-3, 3]$ such that $g(x) = 2$.
(ii) Find the largest interval over which the graph of g is rising, the graph of h is falling,° and $h(x) \geq g(x)$ for every number x in the interval.

° Here and below "rising" and "falling" refer to movement from left to right.

In problem (i), for instance, the numbers x with x in $[-3, 3]$ and $g(x) = 2$ are the first coordinates of all points $(x, g(x))$ on the graph of g which lie on or between the vertical lines through -3 and 3, and on the horizontal line through 2, as shown in Figure 3-28.

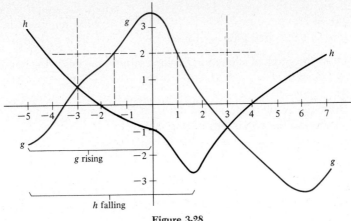

Figure 3-28

Thus the answer is $x = -1.5$ and $x = 1$.

To solve problem (ii), observe first that the graph of h is falling over the interval $[-5, 1.5]$. Over *that* interval, the graph of g is rising only when $-5 \le x \le 0$. Thus the only interval over which the graph of g is rising *and* the graph of h is falling is $[-5, 0]$. Clearly, $h(x) \ge g(x)$ exactly when the point $(x, h(x))$ lies on or above the point $(x, g(x))$. The only time this occurs over the interval $[-5, 0]$ is when $-5 \le x \le -3$, as shown above. Therefore the interval asked for in (ii) is the interval $[-5, -3]$.

EXERCISES

B.1.   Let g be the function with the graph shown in Figure 3-29.

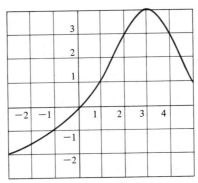

Figure 3-29

(a) If $t = 1.5$, then $g(2t) = ?$

(b) If $t = 1.5$, then $2g(t) = ?$

(c) If $y = 2$, then $g(y + 1.5) = ?$

(d) If $y = 2$, then $g(y) + g(1.5) = ?$

(e) If $y = 2$, then $g(y) + 1.5 = ?$

(f) $g(0) = ?$

(g) If $v = 1.5$, then $g(3v - 1.5) = ?$

(h) If $s = 2$, then $g(-s) = ?$

(i) For what values of z is $g(z) = 1$?

(j) For what values of z is $g(z) = -1$?

(k) What is the largest interval over which the graph is rising?

(l) At what number t in the interval $[-1, 2]$ is $g(t)$ largest?

B.2. (a) Draw the graph of a function f, which satisfies the following four conditions:
(i) domain $f = [-2, 4]$; (ii) range $f = [-5, 6]$; (iii) $f(-1) = f(2)$;
(iv) $f(\tfrac{1}{2}) = 0$.

(b) Draw the graph of a function different from the one in part (a), which also satisfies all the conditions of part (a).

B.3. Figure 3-30 shows the entire graph of a function f.

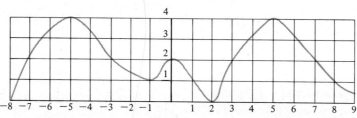

Figure 3-30

(a) What is the domain of f?

(b) What is the range of f?

(c) Find all numbers x such that $f(x) = 2$.

(d) Find all numbers x such that $f(x) > 2$.

(e) Find at least three numbers x such that $f(x) = f(-x)$.

(f) Find a number x such that $f(x + 1) = 0$.

(g) Find two numbers x such that $f(x - 2) = 4$.

(h) Find a number x such that $f(x + 1) = f(x - 2)$.

(i) Find a number x such that $f(x) + 1 = f(x - 4)$.

B.4. Figure 3-31 shows the entire graphs of functions f and g.

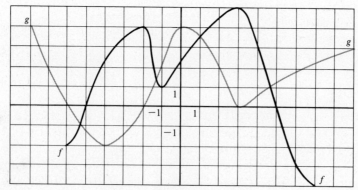

Figure 3-31

(a) What is the domain of f? the domain of g?
(b) What is the range of f? the range of g?
(c) Find all numbers x in the interval $[-3, 1]$ such that $f(x) = 2$.
(d) Find all numbers x in the interval $[-3, 3]$ such that $g(x) \geq 2$.
(e) Find the number x for which $f(x) - g(x)$ is largest.
(f) For how many values of x is it true that $f(x) = g(x)$?
(g) Find all intervals over which both functions are defined, the graph of f is falling *and* the graph of g is rising (from left to right).
(h) Find all intervals over which the graph of $g(x)$ is falling *and* $0 \leq f(x) \leq 2$.

B.5. The owners of the Rieben & Tabares Deluxe Widget Works have determined that both their weekly manufacturing expenses and their weekly sales income are functions of the number of widgets manufactured each week. Figure 3-32 shows the graphs of these two functions.

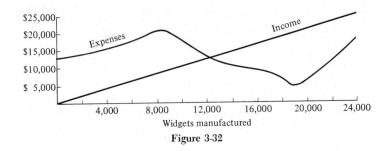

Figure 3-32

(a) Use careful measurement on the graph and the fact that profit = income − expenses to determine the weekly profit if 5000 widgets are manufactured.
(b) Do the same if 10,000, 14,000, 18,000, or 22,000 widgets are manufactured.
(c) What is the smallest number of widgets that can be manufactured each week without losing money?
(d) What is the largest number of widgets that can be manufactured without losing money?
(e) The owners build a new lounge and swimming pool for their employees. This raises their expenses by approximately $5000 per week. Draw the graph of the new "expense function."
(f) Due to competitive pressure, widget prices cannot be increased and the income function remains the same. Answer parts (a)–(d) with the new expense function in place of the old.

B.6. A weather bureau device records the graph of the temperature as a function of time as shown in Figure 3-33.

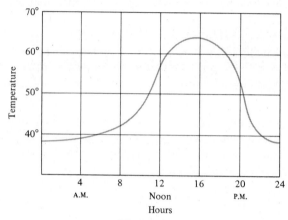

Figure 3-33

We write tem(t) for the temperature at t hours after midnight.

(a) Find tem(10). Find tem(3 + 12).

(b) Is tem(6) bigger than, equal to, or less than tem(18)?

(c) At which time is the temperature 50°?

(d) Find a 4-hr period for which tem(h) > 40 for all h in this 4-hr period.

(e) Find the difference in temperature at 10 and 16 hours. Identify this difference in the graph; that is, express difference in terms of points and their location on the graph.

(f) Is it true that tem(6) = tem(8)? Explain in terms of the graph.

(g) Is it true that tem(6 · 2) = 6 · tem(2)?

(h) The temperature graph above was recorded in city F. City B is 500 miles to the south of city F, and its temperature is 7° higher all day long. Sketch into the above drawing the temperature for city B during the same day.

(i) Find an hour h at which the temperature in city B is the same as tem(12) in city F.

5. THE SHAPE OF A GRAPH

We now consider various *geometric* properties a graph might have. Each of them is easily understood simply by looking at the *shape* of the graph. But a careful analysis is needed to understand just what these geometric properties mean in terms of the *algebraic* behavior of the function.

The ability to translate graphical properties into algebraic terms, and vice-versa, can be very helpful in graphing. For instance, by using algebra we may be able to tell *in advance* that the graph of a given function will have certain geometric features. This will not only make it easier to graph the function but will often result in a much more accurate graph as well.

SYMMETRY WITH RESPECT TO THE *y*-AXIS

Geometric Description   Figure 3-34 shows some graphs which are **symmetric with respect to the *y*-axis.**

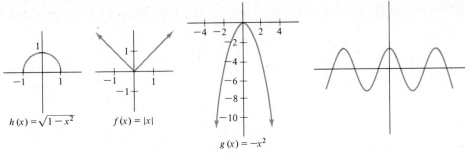

$$h(x) = \sqrt{1 - x^2} \qquad f(x) = |x| \qquad\qquad g(x) = -x^2$$

Figure 3-34

In each case the part of the graph on the right side of the *y*-axis is the *mirror image* of the part of the graph on the left side of the *y*-axis (with the *y*-axis being the mirror). In more technical terms, the left half of the graph can be obtained by *reflecting* the right half in the *y* axis. This means that if you fold the plane along the *y*-axis, so that the right-hand half of the plane goes on top of the left-hand half, then the right-hand half of the graph will *fit exactly* on the left-hand half. A point on the right half of the graph will be folded onto its mirror image point on the left-hand half of the graph. For example, in each of the graphs shown in Figure 3-35, all of which are symmetric with respect to the *y*-axis, *P* is the mirror image of *Q*:

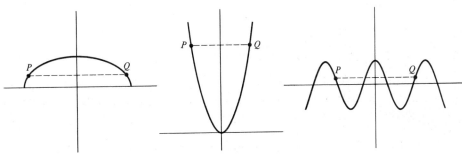

Figure 3-35

Observe that in each case:

(i)   The straight line from P to Q is perpendicular to the y-axis (and hence parallel to the x-axis).

(ii) The distance along this line from P to the y-axis is the same as the distance from Q to the y-axis.

Similar statements apply to any pair of mirror-image points.

Algebraic Description We need only translate the geometric statements (i) and (ii) above into algebraic terms in order to obtain an algebraic description of two mirror-image points. Figure 3-36 is a typical example of mirror-image points on the graph of a function f which is symmetric with respect to the y-axis. The point Q has coordinates $(c, f(c))$.

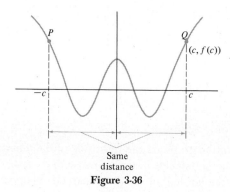

Figure 3-36

Statement (ii) says P and Q are the same distance from the y-axis. Therefore the first coordinate of P must be $-c$.

Since P lies on the graph of the function f, its second coordinate must be $f(-c)$. But statement (i) says that P and Q lie on the same horizontal line and hence must have the *same* second coordinate, as shown in Figure 3-37.

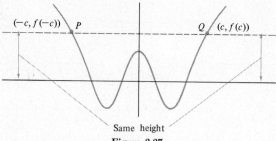

Figure 3-37

But to say the second coordinates are the same means that $f(-c) = f(c)$. A similar argument works for any pair of mirror-image points on the graph of f. Consequently,

> The graph of a function f is **symmetric with respect to the y-axis** provided
>
> $$f(-x) = f(x)$$
>
> for every number x in the domain of f. Such a function is said to be an **even function.**

When the rule of the function f is given by an algebraic formula involving x, then the condition that $f(-x) = f(x)$ means that *the formula remains the same when x is replaced by* $-x$.

EXAMPLE   The function $f(x) = \dfrac{1}{x^2 + 1}$ is an even function, since for every real number x

$$f(-x) = \frac{1}{(-x)^2 + 1} = \frac{1}{x^2 + 1} = f(x)$$

This explains algebraically why the graph of f (which is sketched on p. 116) is symmetric with respect to the y-axis.

In order to find the *graph of an even function f,* you need only consider points $(x, f(x))$ with $x \geq 0$, for once you have this part of the graph, you can immediately obtain the other half by using symmetry with respect to the y-axis.

PERIODIC FUNCTIONS

Geometric Description   A function whose graph repeats the same pattern at regular intervals is said to be **periodic.** If the shortest length in which the graph completes a pattern is c units, then the function is said to have **period** c. For instance, see Figure 3-38.

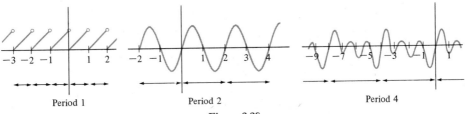

Period 1   Period 2   Period 4

Figure 3-38

The critical fact about the graph of a periodic function of period c is that you need only determine the graph over a single interval of length c. For instance, you can consider only points with $0 \leq x \leq c$. The rest of the graph will just be a repetition of this piece.

Algebraic Description   Suppose $(b, f(b))$ is a point on the graph of a periodic function f with period c. We can think of the basic graph pattern as starting at $(b, f(b))$ and extending to the right for c units, as shown in Figure 3-39.

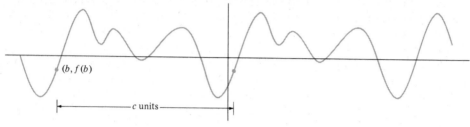

Figure 3-39

The point at which the pattern begins to repeat is c units to the *right* of $(b, f(b))$ and therefore has first coordinate $b + c$. Thus the point is $(b + c, f(b + c))$. But since the pattern begins at $(b, f(b))$ and starts to repeat at $(b + c, f(b + c))$, these two points must have the same second coordinate, as shown in Figure 3-40.

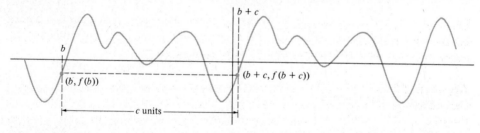

Figure 3-40

In other words, $f(b) = f(b + c)$. This analysis applies to *any* point $(x, f(x))$ on the graph. The second coordinate of $(x, f(x))$ is the same as the second coordinate of the point c units to the right $(x + c, f(x + c))$. Consequently,

---

A function f *is **periodic with period** c provided c is the smallest positive number such that*

$$f(x) = f(x + c)$$

for every number x in the domain of f.

The most important examples of periodic functions are the trigonometric functions to be discussed in Chapter 6. The function $g(x) = x - [x]$, whose graph is on page 120, is periodic.

INCREASING FUNCTIONS

Geometric Description   A function is said to be **increasing on an interval** if the graph of the function always *rises* as you move from left to right in the interval. For example, each of the functions shown in Figure 3-41 is increasing on the stated intervals.

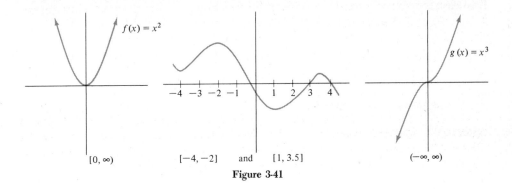

$[0, \infty)$ $[-4, -2]$ and $[1, 3.5]$ $(-\infty, \infty)$

Figure 3-41

Obviously, if a function is increasing on an interval I, it is also increasing on any interval contained in I. A function, such as the function g in Figure 3-41, which is increasing on *every* interval in its domain is called an **increasing function.**

Algebraic Description   Suppose the function f is increasing on an interval and that c and d are numbers in the interval with $c < d$. Since $c < d$, the point $(d, f(d))$ lies to the *right* of the point $(c, f(c))$ on the graph. Since f is increasing, its graph is rising from left to right. Therefore the right-hand point $(d, f(d))$ must be *higher* than the left-hand point $(c, f(c))$, as shown in Figure 3-42.

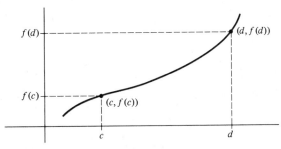

Figure 3-42

Consequently, the second coordinate of $(d, f(d))$ must be *larger* than the second coordinate of $(c, f(c))$; that is, $f(c) < f(d)$. Thus

---

> *A function* f *is **increasing on an interval** provided that for any numbers* c *and* d *in the interval,*
>
> $$\text{whenever } c < d, \qquad \text{then } f(c) < f(d)$$

EXAMPLE We know from the graph of the function $f(x) = x^2$ that f is increasing on $[0, \infty)$. Since $f(c) = c^2$ and $f(d) = d^2$, the algebraic meaning of this fact is:

$$\text{whenever } 0 \leq c < d, \qquad \text{then } c^2 < d^2$$

DECREASING FUNCTIONS

Geometric Description A function is said to be **decreasing on an interval** if the graph of the function always *falls* as you move from left to right in the interval. A function which is decreasing on *every* interval in its domain is called a **decreasing function.**

Algebraic Description Suppose the function f is decreasing on an interval and that c and d are numbers in the interval with $c < d$. Then the point $(d, f(d))$ lies to the *right* of the point $(c, f(c))$ on the graph. Since f is decreasing, its graph is *falling* from left to right. Therefore, the right-hand point $(d, f(d))$ must be *lower* than the left-hand point $(c, f(c))$, as shown in Figure 3-43.

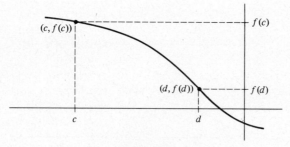

Figure 3-43

Consequently, $f(c) > f(d)$ and we have:

> *The function* f *is **decreasing on an interval** provided that for any numbers* c *and* d *in the interval,*
>
> *whenever* c < d, *then* f(c) > f(d)

SUMMARY

The following chart summarizes the preceding discussion; it also mentions "odd functions," which are explained in the DO IT YOURSELF! segment at the end of this section.

Geometric Description	Algebraic Description	Name of Property
Graph of f is *symmetric with respect to y-axis*	$f(-x) = f(x)$ for all x in domain of f	f is *even*
Graph of f repeats its pattern every c units	$f(x) = f(x + c)$ for all x in domain of f	f is *periodic with period c*
Graph of f *rises* from left to right over the interval I	Whenever $b < c$, then $f(b) < f(c)$ for all b, c in I	f is *increasing on I*
Graph of f *falls* from left to right over the interval I	Whenever $b < c$, then $f(b) > f(c)$ for all b, c in I	f is *decreasing on I*
Graph of f is *symmetric with respect to the origin*	$f(-x) = -f(x)$ for all x in the domain of f	f is *odd*

EXERCISES

A.1. Without graphing, decide which of the following functions have graphs which are symmetric to the y-axis (that is, which functions are *even*).

(a) $f(x) = |x| + 3$

(b) $g(x) = |x - 1|$

(c) $y = x^2 - |x|$

(d) $k(t) = t^4 - 6t^2 + 5$

(e) $f(u) = (u + 2)^2 + u^4 + 2$

(f) $g(x) = x^3(3x + x^5) + 6x^2 + 3$

(g) $h(x) = x^2(x^3 - 3x^5) + 5$

(h) $y = \sqrt{t^2 - 5}$

A.2. Each of the graphs in Figure 3-44 is the graph of a function. For each of these graphs answer the following questions:

(i) Is the function even?   (iii) Is the function increasing on $[1,5]$?

(ii) Is the function periodic?   (iv) Is the function decreasing on $[-3,2]$?

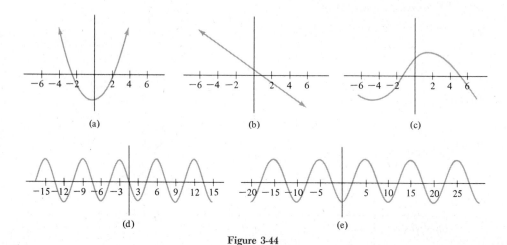

Figure 3-44

A.3. For each of the functions whose graphs appear in Figure 3-45, state the intervals on which the function is increasing and the intervals on which it is decreasing.

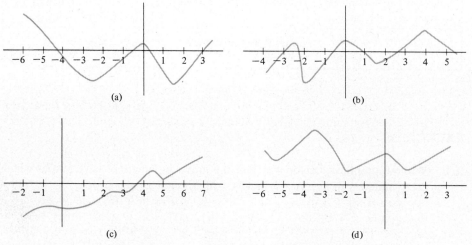

Figure 3-45

B.1. Sketch the graph of a function f which satisfies these five conditions:

    (i) $f(-1) = 2$ (iii) $f(x)$ starts decreasing when $x = 1$

    (ii) $f(x) \geq 2$ when x is in   (iv) $f(3) = 3 = f(0)$

       the interval $(-1, \tfrac{1}{2})$ (v) $f(x)$ starts increasing when $x = 5$

Note: the function whose graph you sketch need not be given by an algebraic formula.

B.2. Examine Figure 3-46.

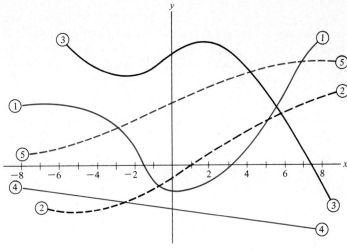

Figure 3-46

(a) For *each* of the following *five* statements, find one (or more) functions (among the five whose graphs appear in Figure 3-46) for which the statement is true.

    (i) $f(2) < f(1)$

    (ii) $f(x)$ is negative and increasing from $x = 1$ to $x = 2$

    (iii) $f(0) < 0$ but $f(2) > 0$

    (iv) $f(x)$ is negative and decreasing from $x = -4$ to $x = -3$

    (v) $f(x) > f(-x)$ for some number x with $|x| \leq 5$

(b) For each of the following six statements, find a pair of functions (or several pairs) from among the five whose graphs appear in Figure 3-46 for which the statement is true.

    (i) $f(1) - g(1) > 0$ (iv) $\dfrac{f(x)}{g(x)} < 1$ for some $x < 0$

    (ii) $f(x) < g(3)$ for $0 \leq x \leq 2$ (v) $f(x) = g(x)$ for some x with $|x| \geq 2$

    (iii) $f(x)g(x) < 0$ for $x > 4$ (vi) $f(x) \leq g(-x)$ for some x with $|x| < 4$

B.3. (a) Draw some coordinate axes and plot the points $(0, 1)$, $(1, -3)$, $(-5, 2)$, $(-3, -5)$, $(2, 3)$, and $(4, 1)$.

(b)   Suppose the points in part (a) lie on the graph of an *even* function f. Plot the points $(0, f(0))$, $(-1, f(-1))$, $(5, f(5))$, $(3, f(3))$, $(-2, f(-2))$,   and $(-4, f(-4))$.

B.4.   Draw the graph of an *even* function which includes the points $(0, -3)$, $(-3, 0)$, $(2, 0)$, $(1, -4)$, $(2.5, -1)$, $(-4, 3)$ and $(-5, 3)$. (*Note:* there are many possible correct answers here.)

B.5.   Suppose f is a periodic function with period c and that for some number x, $f(x) = k$.
   (a)   Find $f(x + c)$, $f(x + 2c)$, $f(x + 3c)$, $f(x + 4c)$, $f(x + 5c)$. (*Hint:* how far apart are $x + c$ and $x + 2c$? How far apart are $x + 2c$ and $x + 3c$?)
   (b)   Find $f(x - c)$, $f(x - 2c)$, $f(x - 3c)$, $f(x - 4c)$, $f(x - 5c)$. (*Hint:* how far apart are $x - c$ and x? How far apart are $x - 2c$ and $x - c$?)
   (c)   Find $f(x + nc)$ for any integer n. [See part (a) when $n = 1, 2, 3, 4,$ or 5 and part (b) when $n = -1, -2, -3, -4,$ or -5.]

B.6.   Figure 3-47 shows the graphs of some periodic functions. What is the period of each?

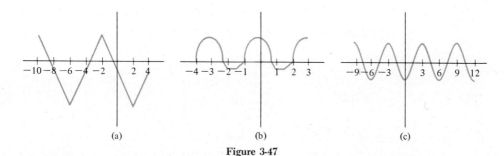

(a)   (b)   (c)

Figure 3-47

B.7.   Draw the graph of an even function which is periodic of period 3. (There are many possible correct answers.)

B.8.   Use algebra to show that each of the functions below is increasing on the interval $(0, 10]$. Some of them may also be increasing on other intervals. You may assume the usual facts about inequalities, including the following:

$$\text{if } 0 \le c < d, \quad \text{then } c^2 < d^2;$$
$$\text{if } c < d, \quad \text{then } c^3 < d^3;$$
$$\text{if } 0 \le c < d, \quad \text{then } \sqrt{c} < \sqrt{d}.$$

(a) $f(x) = x^2 + 3$ **(e)** $h(x) = \sqrt{x + 3}$
(b) $g(x) = x^3 - 10{,}000$ **(f)** $g(x) = x^3 + x^2$
(c) $h(t) = t^2 + t + 5$ **(g)** $k(x) = \sqrt{x^3 + 2x - 1}$
(d) $k(u) = u^3 + u^2 + 1$ **(h)** $f(x) = -\dfrac{1}{x}$

B.9. Suppose an equation in x and y has the following property: the equation remains the same when y is replaced by $-y$. For example, each of these equations has this property:

$$x = y^2, \quad |x| + |y| = 1, \quad x^2 + y^2 - 2x + 1 = 0, \quad 16x^2 + 9y^2 + 16x = 20.$$

    (a) If the point (c, d) is on the graph of the equation, what other point is *necessarily* on the graph as well?

    (b) How does the part of the graph of the equation *above* the x-axis look in comparison with the part of the graph *below* the x-axis? (*Hint:* see the analogous, but different, discussion on pp. 130–133.)

    (c) In view of your answer to part (b), what would be a good name for the property defined above?

    (d) Is the graph of such an equation *ever* the same as a graph of some function?

C.1. (a) Suppose x is a real number and $[x] = k$. Show that $[x + 1] = k + 1$. (Note that k and $k + 1$ are consecutive integers.)

    (b) Let $g(x) = x - [x]$. Use part (a) to give an algebraic proof that $g(x) = g(x + 1)$ for every number x. Hence g is periodic of period (at most) 1.

    (c) Use an algebraic argument to show that the function $f(x) = [x] + [-x]$ is periodic of period 1.

DO IT YOURSELF!

SYMMETRY WITH RESPECT TO THE ORIGIN

Geometric Description Figure 3-48 shows some graphs which are **symmetric with respect to the origin** O:

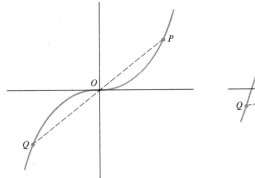
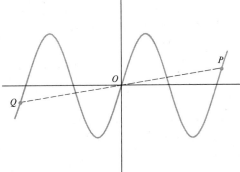

Figure 3-48

In each case, the following condition holds:

> Whenever P is on the graph and $P \neq O$, then the straight line through P and O intersects the graph at a third point Q such that the distance from P to O is the same as the distance from Q to O.

Points P and Q related in this fashion are said to be symmetric with respect to the origin.

Algebraic Description   Suppose points P and Q (neither of which is the origin) are on the graph of a function f and are symmetric with respect to the origin. P has coordinates $(c, f(c))$ for some nonzero number c, as shown in Figure 3-49. Consequently, the straight line L through $P = (c, f(c))$ and $O = (0, 0)$ has slope $\dfrac{f(c) - 0}{c - 0} = \dfrac{f(c)}{c}$ and equation $y = \dfrac{f(c)}{c} x$.

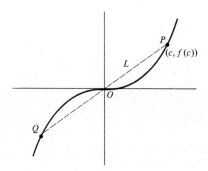

Figure 3-49

Observe that the point M with coordinates $(-c, -f(c))$ also lies on the line L since it satisfies the equation of L:

$$-f(c) = \frac{f(c)}{c} \cdot (-c)$$

Furthermore,

distance from P to $O = \sqrt{(c - 0)^2 + (f(c) - 0)^2} = \sqrt{c^2 + f(c)^2}$,

distance from M to $O = \sqrt{(-c - 0)^2 + (-f(c) - 0)^2} = \sqrt{c^2 + f(c)^2}$

But by symmetry, Q is the only point on L whose distance from O is the same as the distance from P to O. Therefore M must be the point Q; that is, Q has coordinates $(-c, -f(c))$. But a point on the graph of f with the first coordinate $-c$ has second coordinate $f(-c)$. Thus

$$f(-c) = -f(c)$$

Since this analysis applies to any point $(x, f(x))$ and its symmetric point on the graph, we conclude:

> The graph of the function f is **symmetric with respect to the origin** provided
>
> $$f(-x) = -f(x)$$
>
> for every number x in the domain of f. Such a function is said to be an **odd function.**

EXAMPLE   The function $f(x) = x^3$ is an odd function since

$$f(-x) = (-x)^3 = -x^3 = -f(x)$$

This is an algebraic verification that the graph of f (p. 135) is indeed symmetric with respect to the origin.

EXERCISES

A.1.   Which of the graphs in Figure 3-50 are symmetric with respect to the origin?

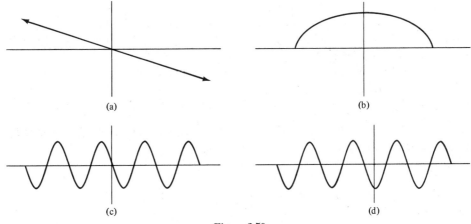

(a)   (b)

(c)   (d)

Figure 3-50

A.2.   Without graphing, decide which of the following functions have graphs which are symmetric with respect to the origin (that is, which functions are *odd*).

(a) $f(x) = 4x$ (d) $k(t) = -5t$ (g) $h(x) = \dfrac{1}{x}$

(b) $g(x) = 4x^3 - 3x$ (e) $y = \sqrt{5 - x^2}$ (h) $y = x(x^4 - x^2) + 4$

(c) $h(u) = |3u|$ (f) $g(t) = -t^3 + 1$

A.3.   Determine which of the functions, whose graphs appear in Figure 3-51, are even, which are odd, and which are neither.

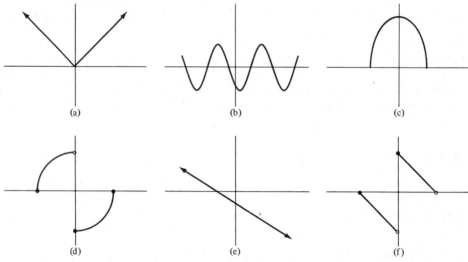

Figure 3-51

B.1. **(a)** Plot the points $(0, 0)$; $(2, 3)$; $(3, 4)$; $(5, 0)$; $(7, -3)$; $(-1, -1)$; $(-4, -1)$; $(-6, 1)$.

   (b) Suppose the points in part (a) lie on the graph of an *odd* function f. Plot the points $(-2, f(-2))$; $(-3, f(-3))$; $(-5, f(-5))$; $(-7, f(-7))$; $(1, f(1))$; $(4, f(4))$; $(6, f(6))$.

   (c) Draw the graph of an odd function f which includes all of the points plotted in parts (a) and (b).°

B.2.   Draw the graph of an odd function which includes the points $(-3, 5)$, $(-1, 1)$, $(2, -6)$, $(4, -9)$, and $(5, -5)$.°

B.3.   Draw the graph of an odd function which is periodic of period 3.°

6. NEW GRAPHS FROM OLD

In this section we consider the question:

> If the rule of a function is changed algebraically, so
> as to produce a new function, how is the graph of the
> new function related to the graph of the original function?

In other words, what do *algebraic* manipulations on "formulas" mean in *geometric* terms? If we know the graph of a particular function f, then the answer to these

° There are many possible correct answers.

questions will provide techniques for easily graphing many of the functions whose rules are obtained from the rule of f by various algebraic manipulations.

NEW FUNCTIONS FROM OLD

Here are some common ways in which new functions are constructed from old ones, via algebraic manipulation.

Adding or Subtracting a Constant   If f is a function and c is a positive real number, then we can form two new functions g and h whose rules are:

$$g(x) = f(x) + c, \qquad h(x) = f(x) - c$$

EXAMPLE   If $f(x) = x^2$ and $c = 3$, we have the functions

$$g(x) = f(x) + 3 = x^2 + 3 \qquad \text{and} \qquad h(x) = f(x) - 3 = x^2 - 3$$

EXAMPLE   The function $g(x) = x^3 + x^2 + 2$ may be thought of as $f(x) + 2$, where $f(x) = x^3 + x^2$.

Multiplying by a Constant   If f is a function and c is a real number, then a new function g can be formed by the rule

$$g(x) = cf(x)$$

EXAMPLE   If $f(x) = x^2$ and $c = 16$, then we can form the function g whose rule is: $g(x) = 16f(x) = 16x^2$.

EXAMPLE   The function $g(x) = -4x^2 + 8x + 4$ may be thought of as $g(x) = (-4)f(x)$, where $f(x) = x^2 - 2x - 1$.

Change of Variable   If f is a function and c is a real number, then we can construct two new functions g and h, whose rules are:

$$g(x) = f(x + c) \qquad h(x) = f(x - c)$$

EXAMPLE   If $f(x) = x^2$ and $c = 3$, then

$$g(x) = f(x + 3) = (x + 3)^2 = x^2 + 6x + 9$$

and

$$h(x) = f(x - 3) = (x - 3)^2 = x^2 - 6x + 9$$

NEW GRAPHS FROM OLD

If you begin with a function f, the methods outlined above provide numerous ways of constructing new functions from f. As we shall now see, the graph of each of these new functions can easily be obtained from the graph of f, by performing some simple geometric procedure.

Adding or Subtracting a Constant Suppose f is a function, c is a positive real number, and g is the function given by $g(x) = f(x) + c$. Then f and g have the same domain. For every number x in this domain:

$$(x, f(x)) \text{ is on the graph of } f,$$
$$(x, g(x)) = (x, f(x) + c) \text{ is on the graph of } g$$

But the point $(x, f(x) + c)$ lies c units directly *above* the point $(x, f(x))$, as shown in Figure 3-52.

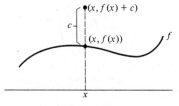

Figure 3-52

Thus for every number x the graph of $g(x) = f(x) + c$ lies exactly c units above the graph of f, as shown in the following box. In other words,

---

If c $>$ 0, *then the graph of* g(x) = f(x) + c *is the graph of* f *shifted upward* c *units.*

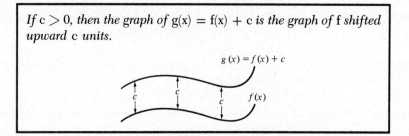

EXAMPLE If $f(x) = x^2$, then the graph of the function $g(x) = x^2 + 2$ [that is, $g(x) = f(x) + 2$] is just the graph of f shifted 2 units upward, as shown in Figure 3-53.

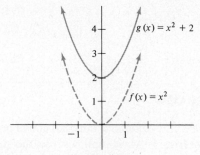

Figure 3-53

If $c > 0$ and h is the function given by $h(x) = f(x) - c$, then for each x in the domain of f and h,

$(x, f(x))$ is on the graph of f,
$(x, h(x)) = (x, f(x) - c)$ is on the graph of h.

But the point $(x, f(x) - c)$ lies c units directly *below* the point $(x, f(x))$. So,

> *If* c > 0, *then the graph of* h(x) $=$ f(x) $-$ c *is the graph of* f *shifted downward* c *units.*

EXAMPLE The graph of $h(x) = x^2 - 3$ is just the graph of $f(x) = x^2$ shifted 3 units downward, as shown in Figure 3-54.

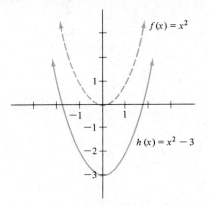

Figure 3-54

EXAMPLE The graph of $f(x) = \sqrt{x}$ was determined on page 114. We can now use this graph to find the graphs of the functions

$$g(x) = \sqrt{x} + 5, \qquad h(x) = \sqrt{x} + 1, \qquad k(x) = \sqrt{x} - 3, \qquad r(x) = \sqrt{x} - \tfrac{11}{2}$$

without plotting any points, as shown in Figure 3-55.

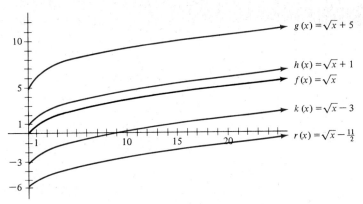

Figure 3-55

Multiplying by a Positive Constant This situation is most easily understood by considering a specific example. Suppose $f(x) = x^2 - 3$. Then we know what the graph of f looks like (see $h(x)$, p. 147). Suppose $c = 2$, so that the function $g(x) = cf(x)$ is just the function $g(x) = 2f(x) = 2(x^2 - 3) = 2x^2 - 6$. If $x = 3$, then

$$(3, f(3)) = (3, 6) \text{ is on the graph of } f,$$
$$(3, g(3)) = (3, 12) \text{ is on the graph of } g$$

Observe that $(3, 12)$ lies *directly above* $(3, 6)$ and is *twice as far from the x-axis;* as shown in Figure 3-56.

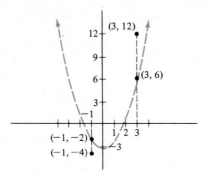

Figure 3-56

Similarly, if $x = -1$, then

$$(-1, f(-1)) = (-1, -2) \text{ is on the graph of } f,$$
$$(-1, g(-1)) = (-1, -4) \text{ is on the graph of } g$$

The point $(-1, -4)$ lies *directly below* $(-1, -2)$ and is *twice as far from the x-axis,* as shown in Figure 3-56. A similar argument for any number x shows that the point $(x, g(x)) = (x, 2f(x))$ on the graph of g lies

> directly above or below,
> on the same side of the x-axis,
> *twice as far* from the x-axis as

the point $(x, f(x))$ on the graph of f. Consequently, the graph of g looks like the one in Figure 3-57.

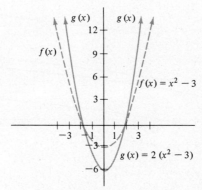

Figure 3-57

Thus the graph of g is the graph of f *stretched vertically away from the x-axis* by a factor of 2.

A similar analysis enables us to obtain the graph of $h(x) = \frac{1}{4}f(x) = \frac{1}{4}(x^2 - 3)$ from the graph of $f(x) = x^2 - 3$. But now the factor is $c = \frac{1}{4}$ instead of $c = 2$. If we plot a few points on both graphs, as shown in Figure 3-58, we can see what happens.

x	$f(x) = x^2 - 3$	$h(x) = \frac{1}{4}(x^2 - 3)$
-3	6	$\frac{1}{4}(6) = \frac{3}{2}$
-1	-2	$\frac{1}{4}(-2) = -\frac{1}{2}$
0	-3	$\frac{1}{4}(-3) = -\frac{3}{4}$
2	1	$\frac{1}{4}(1) = \frac{1}{4}$
3	6	$\frac{1}{4}(6) = \frac{3}{2}$

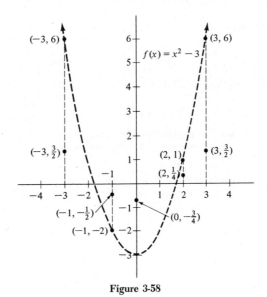

Figure 3-58

In each case, we see that the point $(x, h(x)) = (x, \frac{1}{4}f(x))$ on the graph of h lies

> directly above or below,
> on the same side of the x-axis,
> *one-fourth as far from the x-axis as*

the point $(x, f(x))$ on the graph of f. Thus the graph of h is the graph of f *shrunk vertically toward the x-axis* by a factor of $\frac{1}{4}$; as shown in Figure 3-59.

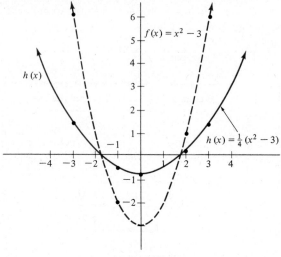

Figure 3-59

Analogous arguments work for any function f and positive real number c:

> If $c > 1$, *then the graph of* $g(x) = cf(x)$ *is the graph of* f *stretched vertically away from the x-axis by a factor of* c.
>
> If $0 < c < 1$, *then the graph of* $h(x) = cf(x)$ *is the graph of* f *shrunk vertically toward the x-axis by a factor of* c.

Multiplying by −1   Suppose f is a function and g is the function given by $g(x) = -f(x)$. If $(x, f(x))$ is on the graph of f, then the point $(x, g(x)) = (x, -f(x))$ is on the graph of g, as shown in Figure 3-60.

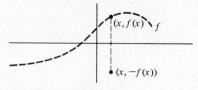

Figure 3-60

Observe that $(x, f(x))$ and $(x, -f(x))$ lie on the same vertical line, on *opposite sides* of the x-axis, at the *same distance* from the x-axis. Thus if we fold the plane along the x-axis, the point $(x, f(x))$ will land on the point $(x, -f(x))$. Consequently,

The graph of g(x) = −f(x) *is the graph of* f *reflected in the x-axis, as shown here.*

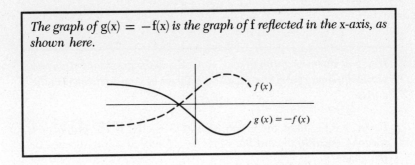

EXAMPLE   The graph of $g(x) = -(x^2 - 3)$ is obtained from the graph of $f(x) = x^2 - 3$ by reflecting the graph of f in the x-axis (as if the axis were a mirror) (see Figure 3-61).

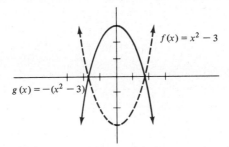

Figure 3-61

Multiplying by a Negative Constant   Suppose f is a function, c is a *positive* real number, and g is the function given by $g(x) = -cf(x)$. Then the graph of g can be obtained in two steps from the graph of f. First, the graph of $cf(x)$ is just the graph of f stretched or shrunk by a factor of c (depending on whether $c > 1$ or $c < 1$). Second, the graph of $-cf(x)$ is just the graph of $cf(x)$ reflected in the x-axis. Figure 3-62 shows the process in pictures (with $c > 1$).

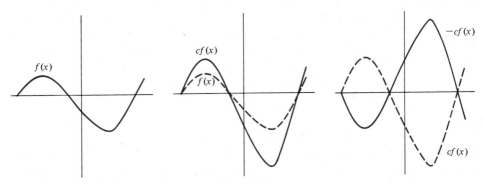

Figure 3-62

In summary,

> If $c > 1$, then the graph of $g(x) = -cf(x)$ is the graph of f stretched vertically away from the x-axis by a factor of c and reflected in the x-axis.
>
> If $0 < c < 1$, then the graph of $h(x) = -cf(x)$ is the graph of f shrunk vertically toward the x-axis by a factor of c and reflected in the x-axis.

EXAMPLE In order to graph $h(x) = -\frac{1}{2}(x^2 - 3)$, we begin with the known graph of $f(x) = x^2 - 3$ and then proceed to the graph of $\frac{1}{2}f(x) = \frac{1}{2}(x^2 - 3)$. Finally, reflecting the latter graph in the x-axis yields the graph of h, as shown in Figure 3-63.

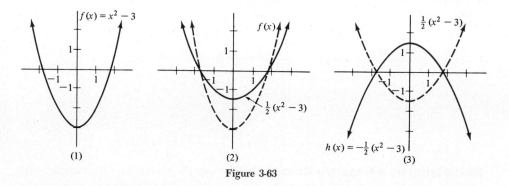

Figure 3-63

Change of Variable Suppose f is a function, c is a positive real number, and $g(x) = f(x + c)$. For each number x in the domain of g, $x + c$ is in the domain of f and

$$(x + c, f(x + c)) \text{ is on the graph of } f,$$
$$(x, g(x)) = (x, f(x + c)) \text{ is on the graph of } g$$

Note that $(x, g(x)) = (x, f(x + c))$ lies on the same horizontal line as the point $(x + c, f(x + c))$, exactly c units to the *left*, as shown in Figure 3-64.

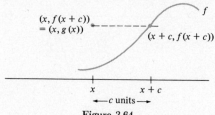

Figure 3-64

Thus for each point on the graph of f, there is a point c units to the *left* which is on the graph of g. Consequently,

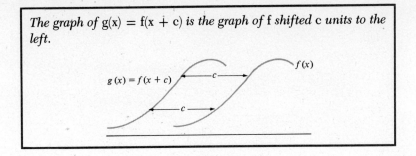

The graph of g(x) = f(x + c) *is the graph of* f *shifted* c *units to the* *left.*

EXAMPLE The function $g(x) = (x + 4)^2$ is just $f(x + 4)$, where $f(x) = x^2$. Therefore the graph of g is just the well-known graph of $f(x) = x^2$ shifted 4 units to the left, as shown in Figure 3-65.

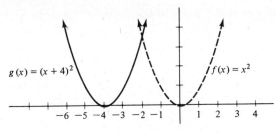

Figure 3-65

A similar analysis can be used to obtain the graph of $h(x) = f(x - c)$ from the graph of f when $c > 0$. In this case, for each x in the domain of h,

$$(x - c, f(x - c)) \text{ is on the graph of } f,$$
$$(x, h(x)) = (x, f(x - c)) \text{ is on the graph of } h$$

The point $(x, h(x)) = (x, f(x - c))$ lies on the same horizontal line as $(x - c, f(x - c))$, exactly c units to the *right*, as shown in Figure 3-66.

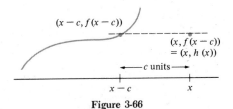

Figure 3-66

Consequently, for every point on the graph of f, there is a point c units to the *right* which is on the graph of h and

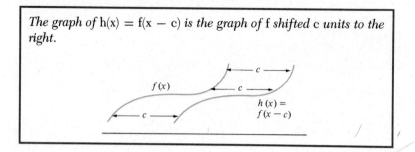

The graph of $h(x) = f(x - c)$ *is the graph of* f *shifted* c *units to the right.*

EXAMPLE   The graph of the function $f(x) = x^3 - 3x$ is shown by the dotted line in Figure 3-67. The graph of the function h whose rule is

$$h(x) = f(x - 2) = (x - 2)^3 - 3(x - 2)$$
$$= x^3 - 6x^2 + 9x - 2$$

is just the graph of f shifted 2 units to the right, as shown in Figure 3-67.

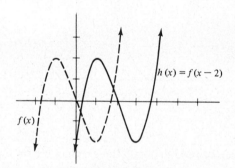

Figure 3-67

The techniques developed above can be used to find the graph of *any* quadratic function, such as

$$g(x) = 3x^2 - 2x + 1, \qquad h(x) = -7x^2 + x, \qquad k(x) = -4x^2 + 7x - 5$$

For details, see the DO IT YOURSELF! segment at the end of this section.

EXERCISES

A.1.   Use the graphs of $g(x) = \sqrt{x}$ and $f(x) = |x|$ (given in Section 3) to graph:
    (a) $h(x) = |x| - 5$ (d) $s(x) = |x - 5|$ (g) $k(x) = \sqrt{x} + 3$
    (b) $k(x) = 3|x|$ (e) $t(x) = |x + 5|$ (h) $r(x) = 5\sqrt{x}$
    (c) $r(u) = \frac{1}{2}|u|$ (f) $h(u) = -4|u|$ (i) $s(x) = -\sqrt{x}$

A.2.   Plot the graphs of these four functions on the same set of axes:
(a) $y = x^2$ (c) $g(x) = 4(x + 3)^2$
(b) $f(x) = (x + 3)^2$ (d) $h(x) = 4(x + 3)^2 - 2$

A.3.   Use the graph of $g(x) = x^3$ on page 201 to sketch the graphs of:
(a) $f(x) = \frac{1}{3}x^3$ (d) $f(x) = -\frac{1}{3}x^3$
(b) $h(x) = (x - 2)^3$ (e) $h(x) = -\frac{1}{3}(x + 5)^3$
(c) $k(x) = 5(x - 2)^3$ (f) $y = -\frac{1}{3}(x + 5)^3 - 3$

A.4.   Use the graph of $f(x) = \dfrac{1}{x^2 + 1}$ on page 116 to graph:

(a) $g(x) = \dfrac{1}{x^2 + 1} + 5$ (b) $h(x) = \dfrac{10}{x^2 + 1}$ (c) $\dfrac{10}{(x - 4)^2 + 1}$

A.5.   Graph $f(x) = -g(x)$, where $g(x) = x - [x]$ (see p. 120).

A.6.   Use the graph of the function f in Figure 3-68 to graph each of these functions:

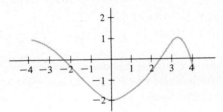

Figure 3-68

(a) $g(x) = 2f(x)$ (c) $k(x) = \frac{1}{2}f(x)$
(b) $h(x) = 5f(x)$ (d) $r(x) = \frac{1}{3}f(x)$

B.1.   Use the graphs of $g(x) = \sqrt{x}$ and $f(x) = |x|$ (given in Section 3) to graph:
(a) $h(x) = 3|x| + 5$ (e) $h(x) = -2\sqrt{x} + 3$

(b) $k(x) = -2|x| - 2$ (f) $r(x) = \dfrac{\sqrt{x}}{2} - 2$

(c) $r(x) = \frac{1}{3}|x| + \frac{5}{3}$ (g) $k(u) = -\sqrt{u + 1} + 2$
(d) $h(u) = 2|u - 3|$ (h) $h(x) = 4\sqrt{x - 2} + \frac{3}{2}$

B.2.   Figure 3-69 is the graph of a function f (whose rule is not given by an algebraic formula). Use this graph to sketch the graphs of the functions listed below.

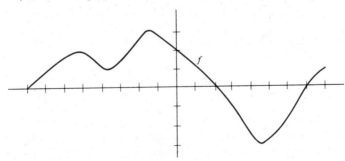

Figure 3-69

(a) $g(x) = f(x) + 3$

(b) $h(x) = f(x) - 1$

(c) $k(x) = f(x) + \frac{7}{4}$

(d) $r(x) = f(x) + \sqrt{2}$

(e) $g(x) = 5f(x)$

(f) $h(x) = \frac{1}{2}f(x)$

(g) $k(t) = -f(t)$

(h) $g(u) = -4f(u)$

(i) $h(z) = -\frac{1}{3}f(z)$

(j) $g(x) = f(x - 2)$

(k) $h(x) = f(x + 3)$

(l) $k(x) = f(x - \frac{5}{2})$

B.3. Let f be the function whose graph is given in Exercise B.2. Sketch the graphs of

(a) $g(x) = f(x - 1) + 2$

(b) $h(x) = f(x + 3) - 5$

(c) $g(x) = 3f(x + 1) - 4$

(d) $h(x) = \frac{1}{2}f(x - 2) + 3$

B.4. Let h be the function whose graph is given on page 126. Graph

(a) $f(x) = 3h(x)$

(b) $g(x) = 3h(x) - 2$

(c) $k(x) = 3h(x + 1) + 2$

(d) $f(x) = -2h(x)$

(e) $g(x) = -2h(x - 2)$

(f) $k(x) = -2h(x - 2) - 2$

C.1. If f is a function and g is the function given by $g(x) = |f(x)|$, then by the definition of absolute value,

$$g(x) = \begin{cases} f(x) & \text{if} \quad f(x) \geq 0 \\ -f(x) & \text{if} \quad f(x) < 0 \end{cases}$$

Therefore the graph of g will be the *same* as the graph of f for all numbers x for which $f(x) \geq 0$. For numbers x with $f(x) < 0$, the graph of g will be the *reflection* of the graph of f in the x-axis. Figure 3-70 is the graph of a function f.

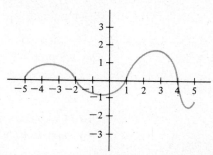

Figure 3-70

Use it to sketch the graphs of the following functions for $-5 \leq x \leq 5$:

(a) $h(x) = -f(x)$

(b) $k(x) = |f(x)|$

(c) $J(x) = |f(x)| + 2$

(d) $R(x) = |f(x - 3)|$

C.2. Graph these functions. (The introduction to Exercise C.1 may be helpful.)

(a) $f(x) = |3x|$

(b) $g(x) = |x + 5|$

(c) $h(x) = |3x - 6|$

(d) $f(x) = |x^2 - 3|$

(e) $g(x) = |x^2 + 1| - 5$

(f) $h(x) = |-(x^2 + 1)|$

DO IT YOURSELF!

QUADRATIC FUNCTIONS

A function f whose rule is of the form $f(x) = ax^2 + bx + c$, where a, b, c are real numbers with $a \neq 0$, is called a **quadratic function.** For example,

$$f(x) = 3x^2 - 2x + 6 \qquad \text{(here } a = 3, \, b = -2, \, c = 6\text{)}$$
$$h(x) = -2x^2 + x \qquad \text{(here } a = -2, \, b = 1, \, c = 0\text{)}$$

The examples below show how the technique of completing the square° can be used to write the rule of any quadratic function f in the form

$$f(x) = d(x + k)^2 + s$$

for some real numbers d, k. and s.

EXAMPLE We want to find numbers d, k, and s such that $g(x) = 3x^2 + 30x + 77$ can be written as $g(x) = d(x + k)^2 + s$. First we write

$$g(x) = 3x^2 + 30x + 77 = 3(x^2 + 10x) + 77$$

The next step is to complete the square in $x^2 + 10x$ by *adding* $(\frac{1}{2} \cdot 10)^2 = 5^2 = 25$, so that $x^2 + 10x + 25 = (x + 5)^2$. But we don't want the rule of the function to change, so we must also *subtract* 25:

$$g(x) = 3(x^2 + 10x) + 77 = 3(x^2 + 10x + 25 - 25) + 77$$

Using the distributive law on the quantity in parentheses yields

$$g(x) = 3(x^2 + 10x + 25) - 3 \cdot 25 + 77,$$
$$g(x) = 3(x + 5)^2 - 75 + 77 = 3(x + 5)^2 + 2$$

Thus $g(x) = d(x + k)^2 + s$ with $d = 3$, $k = 5$, and $s = 2$.

EXAMPLE If $f(x) = -4x^2 + 12x - 8$, then $f(x) = -4(x^2 - 3x) - 8$. In order to complete the square in $x^2 - 3x$, we must *add* $(-\frac{3}{2})^2 = \frac{9}{4}$. In order to leave the rule of f unchanged we must also *subtract* $\frac{9}{4}$:

$$f(x) = -4(x^2 - 3x) - 8 = -4\left(x^2 - 3x + \frac{9}{4} - \frac{9}{4}\right) - 8,$$

$$f(x) = -4\left(x^2 - 3x + \frac{9}{4}\right) - 4\left(-\frac{9}{4}\right) - 8,$$

$$f(x) = -4\left(x - \frac{3}{2}\right)^2 + 9 - 8 = -4\left(x - \frac{3}{2}\right)^2 + 1$$

Thus $f(x) = d(x + k)^2 + s$ with $d = -4$, $k = -\frac{3}{2}$, and $s = 1$.

° This technique is discussed in the Algebra Review on page 560.

GRAPHS OF QUADRATIC FUNCTIONS

By using the techniques discussed earlier in this section, the graph of *any* quadratic function can be easily obtained from the graph of $f(x) = x^2$, as illustrated in the following examples.

EXAMPLE To graph $g(x) = 3x^2 + 30x + 77$, we complete the square, as shown in the example above, and write $g(x) = 3(x + 5)^2 + 2$ (see Figure 3-71).

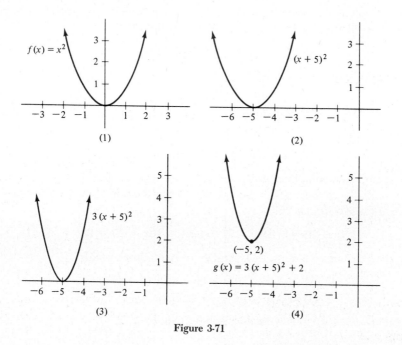

Figure 3-71

Observe that the graph of g is a **parabola** which opens upward. Its **vertex** (lowest point) is $(-5, 2)$. Notice how the coordinates of the vertex are related to the algebraic rule of the function:

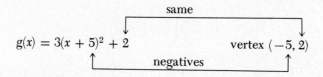

EXAMPLE To graph $f(x) = -4x^2 + 12x - 8$, we complete the square, as shown in the example above, and write $f(x) = -4(x - \frac{3}{2})^2 + 1$. The graph is obtained in Figure 3-72.

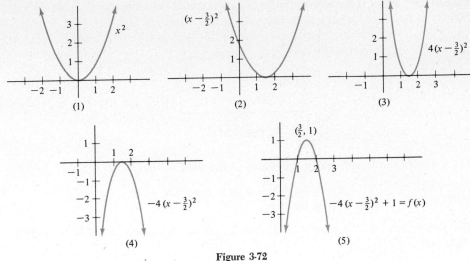

Figure 3-72

The graph of f is a parabola which opens downward. Its vertex (highest point) is $(\frac{3}{2}, 1)$. The coordinates of the vertex are closely related to the graph of the function:

$$f(x) = -4(x - \tfrac{3}{2})^2 + 1 \qquad \text{vertex } (\tfrac{3}{2}, 1)$$

with "same" connecting the $\frac{3}{2}$ to the vertex and "negatives" connecting below.

The techniques illustrated above can be used to obtain the graph of *any* quadratic function $f(x) = ax^2 + bx + c$ (with $a \neq 0$) from the graph of $y = x^2$. The preceding examples illustrate these facts:

> The graph of the quadratic function $f(x) = ax^2 + bx + c$ *is a parabola; it opens upward if* $a > 0$ *and downward if* $a < 0$. *If the rule of* f *is rewritten in the form* $f(x) = d(x + k)^2 + s$, *then the vertex of the parabola is the point* $(-k, s)$.

APPLICATIONS

If the graph of a quadratic function is a *downward*-opening parabola with vertex $(r, f(r))$, then the number $f(r)$ is the **maximum value** of the function f. In other words, $f(x) \leq f(r)$ for every number x. Similarly, if the graph of f opens *upward* and has vertex $(r, f(r))$, then $f(r)$ is the **minimum value** of the function f. In other words, $f(x) \geq f(r)$ for every number x.

EXAMPLE What is the *area* of the largest rectangular field that can be enclosed with 3000 ft of fence and what are its *dimensions* (see Figure 3-73)?

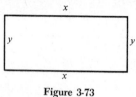

Figure 3-73

Let x denote the length and y the width of the field so that:

$$\text{perimeter} = x + y + x + y = 2x + 2y$$
$$\text{area} = xy$$

But the perimeter is precisely 3000 (the length of the fence), so that $2x + 2y = 3000$ and

$$2y = 3000 - 2x, \quad \text{and hence} \quad y = 1500 - x$$

Consequently, the area is

$$A = xy = x(1500 - x) = 1500x - x^2$$

The largest possible area is just the maximum value of the quadratic function $A(x) = 1500x - x^2 = -x^2 + 1500x$. The graph of $A(x)$ is a parabola which opens downward (why?). Its vertex can be found by completing the square:

$$A(x) = -x^2 + 1500x = -1(x^2 - 1500x) = -1(x^2 - 1500x + (750)^2 - (750)^2),$$
$$A(x) = -(x^2 - 1500x + (750)^2) + (750)^2 = -(x - 750)^2 + (750)^2$$

Therefore the vertex is $(750, 750^2)$, as explained in the last box above. This means that the maximum area is $750^2 = 562{,}000$ sq ft and it occurs when the length $x = 750$. Since the width $y = 1500 - x = 1500 - 750 = 750$, the enclosed area is a square measuring 750 by 750 ft.

EXAMPLE Suppose c and d are real numbers whose difference is 5. What is the smallest possible value for cd and in this case, what are c and d? Since $c - d = 5$, we have $c = 5 + d$. We want the product

$$cd = (5 + d)d = 5d + d^2$$

to be a minimum. Since the graph of $f(d) = d^2 + 5d$ is a parabola which opens upward, the minimum value of $f(d)$ occurs at the vertex of the parabola. To find the vertex we complete the square:

$$f(d) = d^2 + 5d = d^2 + 5d + (\tfrac{5}{2})^2 - (\tfrac{5}{2})^2 = (d^2 + 5d + (\tfrac{5}{2})^2) - (\tfrac{5}{2})^2,$$
$$f(d) = (d + \tfrac{5}{2})^2 - \tfrac{25}{4}$$

Therefore the vertex is $(-\tfrac{5}{2}, -\tfrac{25}{4})$. It occurs when $d = -\tfrac{5}{2}$ and $c = 5 + d = 5 - \tfrac{5}{2} = \tfrac{5}{2}$. The smallest value for $cd = f(d)$ is $-\tfrac{25}{4} = (-\tfrac{5}{2})(\tfrac{5}{2})$.

EXERCISES

A.1. *Without graphing,* determine the vertex of each of the following parabolas and state whether it opens upward or downward.

(a) $f(x) = 3(x - 5)^2 + 2$

(b) $g(x) = -6(x - 2)^2 - 5$

(c) $y = -(x - 1)^2 + 2$

(d) $h(x) = -x^2 + 1$

(e) $y = -\frac{3}{2}(x + \frac{3}{2})^2 + \frac{3}{2}$

(f) $v = 656(t - 590)^2 + 7284$

A.2. Do the same as in Exercise A.1 for these parabolas.

(a) $f(x) = -3x^2 + 4x + 5$

(b) $g(x) = 2x^2 - x - 1$

(c) $y = -x^2 + x$

(d) $h(t) = \frac{1}{2}t^2 - \frac{3}{2}t + \frac{5}{2}$

(e) $y = t^2 + t + 1$

(f) $g(x) = x^2 - 9x$

(g) $h(t) = -2t^2 + 2t - 1$

(h) $y = 3x^2 + x - 4$

A.3. In the example which begins on page 110, $P(x)$ is the weekly profit obtained from selling x widgets. $P(x)$ is given by $P(x) = 1.5x - \dfrac{x^2}{10,000} - 2000$. Find the number x which makes $P(x)$ as large as possible.

B.1. Graph each of these quadratic functions.

(a) $p(x) = x^2 - 4x - 1$

(b) $q(x) = x^2 + 8x + 6$

(c) $y = x^2 - 10x + 20$

(d) $h(x) = x^2 + 3x + 1$

(e) $f(x) = 2x^2 - 4x + 1$

(f) $r(x) = -3x^2 + 9x - 5$

B.2. What must the number b be in order that the vertex of the parabola $y = x^2 + bx + c$ lie on the y-axis?

B.3. For what value of c does the vertex of the parabola $y = x^2 + 8x + c$ lie on the x-axis?

C.1. A ball is thrown upward from the top of a tower. After t sec, its height h above the ground is given by the formula $h = -16t^2 + 80t + 96$. *When* does the ball reach its maximum height and *how high* is it at that time? [*Hint:* consider h as a function of t. To say that a point (a, b) is on the graph of h means that the ball is b ft high at time $t = a$. Thus the maximum height and the time at which it occurs can be determined by finding the highest point on the graph.]

C.2. A rocket is fired upward from ground level. At t sec after blast-off its height h is $-16t^2 + 1600t$ ft. When does it attain its maximum height and what is it? (See the hint for C.1.)

C.3. A projectile is fired at an angle of $45°$ upward. Exactly t sec after firing, its vertical height above the ground is $500t - 16t^2$. What is the greatest height the projectile reaches and at what time does this occur? (See the hint for C.1.)

C.4. What is the minimum product of two numbers whose difference is 4? What are the numbers?

C.5. The sum of the height h and base b of a triangle is 30. What height and base will produce a triangle of maximum area?

C.6. A field bounded on one side by a river is to be fenced on three sides so as to form a rectangular enclosure. If the total amount of fence to be used is 200 ft, what dimensions will yield an enclosure of the largest possible area?

C.7. A rectangular box (with top) has a square base. The sum of the lengths of its 12 edges is 8 ft. What dimensions should the box have in order that its surface area be as large as possible?

7. OPERATIONS ON FUNCTIONS

In this section we explore more ways of creating new functions from given ones. In each case we begin with *two* functions and use them to create a new function. Unlike Section 6, the stress here is primarily on the algebraic aspect of the subject.

COMPOSITION OF FUNCTIONS

Once again we begin in the "real world" with an example of a functional situation (which is numbered 4 to avoid confusion with Examples 1–3 in earlier sections).

EXAMPLE 4—AIR CONDITIONERS The power company has to estimate the additional load on the system when people start turning on their air conditioners. Such an estimate rests on two basic facts:

Fact 1: As the temperature rises, more air conditioners are turned on.

Fact 2: Weather bureau records of past years provide a reasonably good prediction of what the temperature will be at a specific time on a given day.

To predict the additional load due to air conditioners on a particular day—say, July 24—the power company reasons as follows. On the one hand, the weather bureau predicts the expected temperature T on July 24 at all times of the day. Write $T(h)$ for the expected temperature at h hr after midnight. In other words, the weather bureau furnishes a function $T = T(h)$.

On the other hand, a telephone survey gives an indication of how many people turn on air conditioners when the temperature is 75°, 76°, 77°, and so on. Write $A(T)$ for the number of air conditioners in operation when the temperature is reading $T°$. In other words, the company which conducts the telephone survey provides a function $A(T)$.

In order to find out how many air conditioners are likely to be in operation at 4:00 P.M. (= 16 hr after midnight), the power company

first finds $T(16)$, the temperature at $h = 16$ hr after midnight and, second, finds the value of the function $A(T)$ for the temperature $T = T(16)$.

In short, at 4:00 P.M. the company expects

$$A(T(16))$$

(which is read "A of T of 16") air conditioners to be in operation.

What works for 4:00 P.M. works for any time h:

At h hr, the temperature is expected to be $T(h)$ degrees.

At a temperature of $T(h)$ degrees, there will be $A(T(h))$ air conditioners in operation.

From the two functions $T(h)$ and $A(T)$ we obtain in two steps a new function, namely, the function which assigns to h the number $A(T(h))$. This new function is called the **composite** of the functions $T(h)$ and $A(T)$ because $T(h)$ and $A(T)$ are put together, or "composed," to form a new function. In the language of functions,

> The number of air conditioners operating at h hours is given by the composition $A(T(h))$ of the functions $T(h) =$ temperature at h hours and $A(T) =$ number of air conditioners in operation when the temperature is $T°$

In Example 4 the functions $T(h)$ and $A(T)$ were composed to yield the new function $A(T(h))$. Similarly, given any two functions we can construct a new composite function as follows:

Let f(x) *and* g(t) *be functions. The* **composite function** *of* f *and* g *is the function which assigns to the number* x, *the number* g(f(x)). *The composite function of* f *and* g *is denoted* g ∘ f.

The symbol $g \circ f$ is read "f followed by g" (note the order carefully). Thus the rule of the composite function may be written

$$(g \circ f)(x) = g(f(x))$$

that is, $(g \circ f)(x)$ and $g(f(x))$ represent the *same* number, the value of the composite function at x.

The domain of the composite function $g \circ f$ is determined by the usual conventions. Since $g(f(x))$ only makes sense when the number x is in the domain of f *and* $f(x)$ is in the domain of g, we have:

The domain of the composite function g ∘ f *is the set of all real numbers* x *such that* x *is in the domain of* f *and* f(x) *is in the domain of* g.

EXAMPLE Suppose $f(x) = 4x^2 + 1$ and $g(t) = \dfrac{1}{t + 2}$. Then $(g \circ f)(2)$ is the number $g(f(2))$. Since $f(2) = 4 \cdot 2^2 + 1 = 17$ and $g(17) = \dfrac{1}{17 + 2} = \dfrac{1}{19}$, we have:

$$(g \circ f)(2) = g(f(2)) = g(17) = \frac{1}{19}$$

Similarly, we can compute $(g \circ f)(x)$ for any number x by evaluating $g(f(x))$. This means that whenever t appears in the formula for $g(t)$, we must replace it by $f(x) = 4x^2 + 1$:

$$(g \circ f)(x) = g(f(x)) = \frac{1}{f(x) + 2} = \frac{1}{(4x^2 + 1) + 2} = \frac{1}{4x^2 + 3}$$

For every real number x, both $f(x) = 4x^2 + 1$ and $g(f(x)) = \dfrac{1}{4x^2 + 3}$ are well defined real numbers. Consequently, the domain of $g \circ f$ is the set of all real numbers.

EXAMPLE If $f(x) = x - 5$ and $g(t) = 3t + \sqrt{t}$, then

$$(g \circ f)(x) = g(f(x)) = 3f(x) + \sqrt{f(x)} = 3(x - 5) + \sqrt{x - 5} = 3x - 15 + \sqrt{x - 5}$$

Thus $(g \circ f)(7) = 3 \cdot 7 - 15 + \sqrt{7 - 5} = 6 + \sqrt{2}$ and $(g \circ f)(9) = 3 \cdot 9 - 15 + \sqrt{9 - 5} = 12 + \sqrt{4} = 14$. Observe that $f(x)$ is defined for every real number x. Since $\sqrt{t}$ is a real number only when $t \geq 0$, the domain of g is the interval $[0, \infty)$. Consequently, $g(f(x))$ will be defined only when $f(x) \geq 0$, that is, when $x - 5 \geq 0$, or equivalently, $x \geq 5$. Therefore the domain of $g \circ f$ is the interval $[5, \infty)$.

In the preceding examples we began with functions f and g and constructed the composite function $g \circ f$. Sometimes it's necessary to reverse this process and write a *given* function as the composite of two others.

EXAMPLE If $h(x) = \sqrt{3x^2 + 1}$, then h may be considered as the composite $g \circ f$, where $f(x) = 3x^2 + 1$ and $g(u) = \sqrt{u}$:

$$(g \circ f)(x) = g(f(x)) = g(3x^2 + 1) = \sqrt{3x^2 + 1} = h(x)$$

EXAMPLE If $k(x) = (x^2 - 2x + \sqrt{x})^3$, then k is $g \circ f$, where $f(x) = x^2 - 2x + \sqrt{x}$ and $g(t) = t^3$:

$$(g \circ f)(x) = g(f(x)) = g(x^2 - 2x + \sqrt{x}) = (x^2 - 2x + \sqrt{x})^3 = k(x)$$

As you may have noticed, there are two possible ways to form a composite function. If f and g are functions, we can consider either

$$(g \circ f)(x) = g(f(x)), \quad \text{the composite of } f \text{ and } g$$
$$(f \circ g)(x) = f(g(x)), \quad \text{the composite of } g \text{ and } f$$

The *order is important,* as we shall now see:

> g ∘ f *and* f ∘ g *usually are not the same function.*

EXAMPLE If $f(x) = x^2$ and $g(x) = x + 3$,° then

$$(g \circ f)(x) = g(f(x)) = g(x^2) = x^2 + 3$$

° Up to now we have used different letters in the rules of two functions being composed. Now that you have the idea, we shall use the same letter for the variable in both functions.

but

$$(f \circ g)(x) = f(g(x)) = f(x + 3) = (x + 3)^2 = x^2 + 6x + 9$$

Obviously, $g \circ f$ and $f \circ g$ are different functions (for example, they have different values at $x = 0$). In other words, $g \circ f \neq f \circ g$.

Although $g \circ f$ and $f \circ g$ are usually different functions, there are some instances where they turn out to be the same function; see Exercise A.6.

ARITHMETIC OPERATIONS ON FUNCTIONS

Although we know how to add, subtract, multiply, and divide *numbers,* we have not yet given a meaning to addition, subtraction, multiplication, and division of *functions.* We do so now, and thus produce several arithmetic ways of creating a new function from two given functions.

Suppose f and g are functions. With one exception, noted below, all of the functions to be constructed will have the same *domain:*

The set of all real numbers x that are in
both the domain of f and the domain of g

The first new function to be constructed is the sum function. The **sum** of f and g is the function h defined by the rule:

$$h(x) = f(x) + g(x)$$

EXAMPLE If $f(x) = 3x^2 + 2$ and $g(x) = x^2 + \dfrac{1}{\sqrt{x}} - 5$, then their sum is the function

given by

$$h(x) = f(x) + g(x) = (3x^2 + 2) + \left(x^2 + \frac{1}{\sqrt{x}} - 5\right) = 4x^2 + \frac{1}{\sqrt{x}} - 3$$

Instead of using a different letter h for the sum function of f and g, we shall usually denote the sum function by $f + g$. Thus the sum $f + g$ is defined by the rule:

$$(f + g)(x) = f(x) + g(x)$$

It is important to realize that this rule is *not* just a formal manipulation of symbols. If x is a number, then so are $f(x)$ and $g(x)$. The plus sign in $f(x) + g(x)$ is addition of *numbers.* (The result is a number.) But the plus sign in $f + g$ is addition of *functions:* the result is the function which assigns to the number x the number $f(x) + g(x)$.

The **difference** of two functions f and g is defined similarly. It is the function denoted $f - g$ whose rule is: assign to the number x the number $f(x) - g(x)$. In symbols,

$$(f - g)(x) = f(x) - g(x)$$

The minus sign on the right side indicates subtraction of *numbers.* The minus sign on the left denotes subtraction of *functions.*

EXAMPLE   If $f(x) = 3x^2$ and $g(x) = \sqrt{x - 1}$, then $f - g$ is the function given by:

$$(f - g)(x) = f(x) - g(x) = 3x^2 - \sqrt{x - 1}$$

The **product** of two functions f and g is the function denoted fg, whose rule is: assign to the number x the number $f(x)g(x)$. In symbols,

$$(fg)(x) = f(x)g(x)$$

Once again, $f(x)g(x)$ is a product of numbers, while fg is a product of functions.

Warning   Do not confuse the *product* fg of f and g with the *composite* function $f \circ g$ (g followed by f). They are *different*. For example, if $f(x) = 2x^2$ and $g(x) = x - 3$, then the product function fg is given by

$$(fg)(x) = f(x)g(x) = 2x^2(x - 3) = 2x^3 - 6x^2$$

But the composite function $f \circ g$ is given by

$$(f \circ g)(x) = f(g(x)) = f(x - 3) = 2(x - 3)^2 = 2(x^2 - 6x + 9) = 2x^2 - 12x + 18$$

The functions fg and $f \circ g$ are clearly different since they take different values at $x = 0$.

EXAMPLE   A special case of product functions occurs when one of the functions is a constant function. For instance, suppose $f(x) = x^3 + 2x^2 - 1$ and $g(x) = 6$. In such a situation, we let $6f$ denote the product function gf whose rule is:

$$(6f)(x) = g(x)f(x) = 6f(x) = 6(x^3 + 2x^2 - 1) = 6x^3 + 12x^2 - 6$$

Similarly, for any constant c, the function cf is given by

$$(cf)(x) = cf(x) = c(x^3 + 2x^2 - 1) = cx^3 + 2cx^2 - c$$

Note   If f is any function, then the product ff of f with itself is denoted f^2. The product $f(f^2)$ is denoted f^3, and so on.

The **quotient** of two functions f and g is the function denoted f/g. whose rule is

$$\left(\frac{f}{g}\right)(x) = \frac{f(x)}{g(x)}$$

Since $f(x)/g(x)$ is defined only when $g(x) \neq 0$, the domain of f/g is

> The set of all numbers x in both the domain of f and the domain of g with $g(x) \neq 0$

EXAMPLE   If $f(x) = \sqrt{x}$ and $g(x) = x^2 - 1$, then f/g is given by the rule

$$\left(\frac{f}{g}\right)(x) = \frac{f(x)}{g(x)} = \frac{\sqrt{x}}{x^2 - 1}$$

The domain of g is the set of all real numbers, and $g(x) = 0$ only when $x = 1$ or $x = -1$. The domain of f is the interval $[0, \infty)$. Consequently, the domain of f/g is the set of all numbers in $[0, \infty)$ *except* $x = 1$.

A COMPREHENSIVE EXAMPLE

The most common way that the operations introduced above are used in calculus is to consider a fairly complicated function as being built up from simple parts. For example, the function $f(x) = \sqrt{\dfrac{3x^2 - 4x + 5}{x^3 + 1}}$ may be considered as the composite $f = g \circ h$, where

$$h(x) = \frac{3x^2 - 4x + 5}{x^3 + 1} \qquad \text{and} \qquad g(x) = \sqrt{x}$$

since

$$(g \circ h)(x) = g(h(x)) = g\left(\frac{3x^2 - 4x + 5}{x^3 + 1}\right) = \sqrt{\frac{3x^2 - 4x + 5}{x^3 + 1}} = f(x)$$

The function $h(x) = \dfrac{3x^2 - 4x + 5}{x^3 + 1}$ is the quotient $\dfrac{p}{q}$, where

$$p(x) = 3x^2 - 4x + 5 \qquad \text{and} \qquad q(x) = x^3 + 1$$

The function $p(x) = 3x^2 - 4x + 5$ may be written $p = k - s + r$, where

$$k(x) = 3x^2, \qquad s(x) = 4x, \qquad r(x) = 5$$

The function k, in turn, can be considered as the product $3I^2$, where I is the identity function [whose rule is $I(x) = x$]:

$$(3I^2)(x) = 3(I^2(x)) = 3(I(x)I(x)) = 3 \cdot x \cdot x = 3x^2 = k(x)$$

Similarly, $s(x) = (4I)(x) = 4I(x) = 4x$. The function $q(x) = x^3 + 1$ may be "decomposed" in the same way.

Thus the complicated function f is just the result of performing suitable operations on the identity function I and various constant functions.

EXERCISES

A.1. Given the functions $g(t) = t^2 - t$ and $f(x) = 1 + |x|$, evaluate the following:
(a) $g(f(0))$ (c) $g(f(2) + 3)$ (e) $(g \circ f)(1 + 2 + 3)$
(b) $(f \circ g)(3)$ (d) $f(2g(1))$ (f) $(f \circ g)(0)$

A.2. In each part, find $(g \circ f)(3)$, $(f \circ g)(1)$, and $(f \circ f)(0)$.
(a) $f(x) = 3x - 2$, $g(x) = x^2$ (d) $f(x) = x$, $g(x) = -3$
(b) $f(x) = -x + 7$, $g(x) = 7x - 1$ (e) $f(x) = [x]$, $g(x) = 2x - 1$
(c) $f(x) = |x + 2|$, $g(x) = -x^2$ (f) $f(x) = x^2 - 1$, $g(x) = \sqrt{x}$

A.3. Let $f(x) = x + 3$ and $g(u) = u^2 - 1$. Find:
(a) $(g \circ f)(x)$ (b) $(f \circ g)(u)$ (c) $(f \circ f)(x)$ (d) $(g \circ g)(u)$

A.4. In each part, find formulas for $(g \circ f)(x)$ and $(f \circ g)(x)$.

(a) $f(x) = 2x^2 + 2x - 1$ (c) $f(x) = -3x + 2$ (e) $f(x) = \sqrt[3]{x}$
    $g(x) = |x - 1| + 2$ $g(x) = x^3$ $g(x) = x^2 - 1$

(b) $f(x) = 4x^2 + x^4$ (d) $f(x) = x^2 + 1$ (f) $f(x) = \dfrac{1}{x}$

    $g(x) = \sqrt{x^2 + 1}$ $g(x) = -5x + \pi$ $g(x) = \sqrt{x}$

A.5. If f is any function and I is the identity function, what are $f \circ I$ and $I \circ f$?

A.6. Verify that in each part below, $f \circ g = I$ and $g \circ f = I$, where I is the identity function. Functions f and g with this property are said to be **inverses** of each other.

(a) $f(x) = 9x + 2$ (d) $f(x) = \sqrt[3]{\dfrac{7 - x}{3}}$

    $g(x) = \dfrac{x - 2}{9}$ $g(x) = 7 - 3x^3$

(b) $f(x) = \sqrt[3]{x - 1}$ (e) $f(x) = \sqrt[3]{x} + 2$

    $g(x) = x^3 + 1$ $g(x) = (x - 2)^3$

(c) $f(x) = 6x + 2$ (f) $f(x) = 2x^3 - 5$

    $g(x) = \dfrac{x}{6} - \dfrac{1}{3}$ $g(x) = \sqrt[3]{\dfrac{x + 5}{2}}$

A.7. In each part of Exercises A.4, find $(f + g)(x)$, $(f - g)(x)$ and $(g - f)(x)$.

A.8. In each part of Exercise A.4, find $(f + g)(x + h)$ and $(fg)(x + h)$ (where h is a fixed constant).

A.9. In each part of Exercise A.4, find $(fg)(x)$, $\left(\dfrac{f}{g}\right)(x)$ and $\left(\dfrac{g}{f}\right)(x)$.

B.1. Figure 3-74 is the graph of a function f.

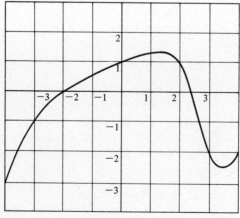

Figure 3-74

Let g be the composite function $f \circ f$ [that is, $g(x) = (f \circ f)(x) = f(f(x))$].

(a) Use the graph of f to fill in the following table (approximate where necessary).

x	$f(x)$	$g(x) = f(f(x))$
-4		
-3		
-2	0	1
-1		
0		
1		
2		
3		
4		

(b) Use the information obtained in part (a) to sketch the graph of the function g.

B.2. Here are tables which show the values of functions f and g at certain numbers.

x	$f(x)$
1	3
2	5
3	1
4	2
5	3

t	$g(t)$
1	5
2	4
3	4
4	3
5	2

Fill in each of the following tables.

(a)

x	$(g \circ f)(x)$
1	4
2	
3	5
4	
5	

(b)

t	$(f \circ g)(t)$
1	
2	2
3	
4	
5	

(c)

x	$(f \circ f)(x)$
1	
2	
3	3
4	
5	

(d)

t	$(g \circ g)(t)$
1	
2	
3	
4	4
5	

B.3. Write each of the following functions as the composite of two functions, neither of which is the identity function. (There may be more than one way to do this.)
Example: $f(x) = \sqrt{x^2 + 1}$ can be written $f = g \circ h$ with $g(x) = \sqrt{x}$ and $h(x) = x^2 + 1$.

(a) $f(x) = \sqrt[3]{x^2 + 2}$

(b) $g(x) = \sqrt{x + 3} - \sqrt[3]{x + 3}$

(c) $h(x) = (7x^3 - 10x + 17)^7$

(d) $k(x) = \sqrt[3]{(7x - 3)^2}$

(e) $f(x) = |x^2 - \sqrt{x} + 2|$

(f) $r(t) = (16t^2)^3$

(g) $h(t) = (t + 2)\sqrt{(t + 2)^2 - 5}$

(h) $k(u) = u^2 + 4u + 4$

(i) $f(x) = \dfrac{1}{3x^2 + 5x - 7}$

(j) $g(t) = \dfrac{3}{\sqrt{t - 3}} + 7$

B.4. If $f(x) = x + 1$ and $g(t) = t^2$, then
$$(g \circ f)(x) = g(f(x)) = g(x + 1) = (x + 1)^2 = x^2 + 2x + 1$$
Find two other functions $h(x)$ and $k(t)$ such that $(k \circ h)(x) = x^2 + 2x + 1$.

B.5. Determine whether the functions $f \circ g$ *and* $g \circ f$ are defined. If a composite function *is* defined, find its domain.

(a) $f(x) = x^3$
   $g(x) = \sqrt{x}$

(b) $f(x) = x^2 + 1$
   $g(x) = \sqrt{x}$

(c) $f(x) = \sqrt{x + 10}$
   $g(x) = 5x$

(d) $f(x) = -x^2$
   $g(x) = \sqrt{x}$

(e) $f(x) = \dfrac{1}{x}$
   $g(x) = x^2 + 1$

(f) $f(x) = x^2 + x + 1$
   $g(x) = x^3 - x + 2$

B.6. Use the graphs of f and g in Figure 3-75 to sketch the graph of the functions $f + g$ and $f - g$.

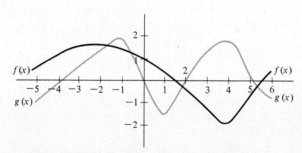

Figure 3-75

B.7. (a) If $f(x) = 2x^3 + 5x - 1$, find $f(x^2)$.

(b) If $f(x) = 2x^3 + 5x - 1$, find $(f(x))^2$.

(c) Are the answers in parts (a) and (b) the same? What can you conclude about $f(x^2)$ and $(f(x))^2$?

B.8. Give an example of a function f such that $f\left(\dfrac{1}{x}\right) \neq \dfrac{1}{f(x)}$.

C.1. This problem deals with the situation described in Example 4 of the text. The telephone survey yields the graph (shown in Figure 3-76), of the number A of air conditioners in operation when the outside temperature is T degrees.

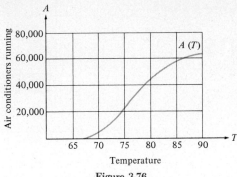

Figure 3-76

On the other hand, on a typical July 24, the graph of the temperature T at h hr after midnight looks like the one in Figure 3-77.

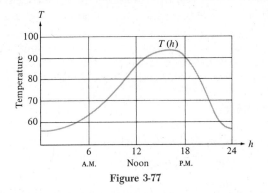

Figure 3-77

(a) Use these graphs to estimate the following quantities: $A(81)$; the temperature at 11:30 A.M.; the number of air conditioners in operation at 4:00 P.M.; $T(8.5)$; $A(T(12))$.

(b) At which time of the day are approximately 21,000 air conditioners in operation? At which time is the number of air conditioners in operation at a maximum?

(c) Find a 2-hr period for which $A(T(h))$ is between 20,000 and 60,000 for all h (in the 2-hr period).

(d) The power company announces a very steep price increase. Sketch into Figure 3-76 what you think the graph of $A(T)$ will look like *after* the rate increase. Sketch the graphs of $A(T(h))$ based on the old and the new power rates.

Polynomial and Rational Functions

The most frequently seen functions in many parts of mathematics are the polynomial functions. Since their values can be computed using only the simple arithmetic operations of addition, subtraction, and multiplication, they are ideally suited for high-speed computers. This fact is crucial, since many complicated functions which arise in applied mathematics can be approximated by polynomial functions or rational functions (which are just quotients of polynomials).

1. POLYNOMIALS

A **polynomial** is an algebraic expression such as

$$x^3 - 6x^2 + 2x - 7, \qquad 11, \qquad 37x^{15} + 12x^9 - 6x^8 + x^7 - 4, \qquad y^2 + 2, \qquad 4 + x$$

Although any letter may be used when writing polynomials, we shall usually stick with x in this discussion. A more formal definition of a polynomial is that a polynomial is an algebraic expression that can be written in the form

$$a_n x^n + a_{n-1}x^{n-1} + a_{n-2}x^{n-2} + \cdots + a_3 x^3 + a_2 x^2 + a_1 x + a_0$$

where $a_0, a_1, a_2, \ldots, a_{n-1}, a_n$ are fixed real numbers (constants), and n is a nonnegative integer.

EXAMPLE In the polynomial

$$7x^4 - 6x^3 + 17x^2 + 5$$

we have $n = 4$, $n - 1 = 3$, and so on, and

$$a_4 = 7, \qquad a_3 = -6, \qquad a_2 = 17, \qquad a_1 = 0, \qquad a_0 = 5$$

The numbers a_0, a_1, ..., a_n in the polynomial $a_n x^n + a_{n-1}x^{n-1} + \cdots + a_2 x^2 + a_1 x + a_0$ are called **coefficients.** We denote the coefficients of a typical polynomial in this manner rather than a, b, c, d, and so on, because it is then easier to keep track of which coefficient goes with which power of x.

It is important to be able to distinguish polynomials from algebraic expressions which are not polynomials.

EXAMPLE $x^3 + 2^x + 3$ is *not* a polynomial. All the exponents in a polynomial must be nonnegative integers. But $2^x + x^3 - 3x - 2^x$ *is* a polynomial since it can be written as $x^3 - 3x$.

EXAMPLE $\sqrt{x^3 + 2x - 1}$ and $x^7 + 2x^3 - \sqrt{x} + 2$ are *not* polynomials since they include square roots involving x. On the other hand, $\sqrt{2}x^3 - 6x^2 + \sqrt{3}x - \sqrt{7}$ *is* a polynomial since no square roots of x are involved.

EXAMPLE $1/x^2$ and $3x^4 + 2x^3 - 4/x^2 + 7/x - 6$ are *not* polynomials. The coefficients of a polynomial are *multiplied* by powers of x, not *divided* by powers of x.

When a polynomial is written so that the powers of x appear in *descending* order, as in $6x^7 + 4x^3 + 5x^2 - 7x + 10$ or $x^4 + 2x^3 + 3x^2 + 4x + 5$, the *nonzero coefficient* of the highest power of x is called the **leading coefficient,** and the last term is the **constant term.** For example,

Polynomial	Leading Coefficient	Constant Term
$6x^7 + 4x^3 + 5x^2 - 7x + 10$	6	10
$-x^4 + 2x^3 + 3x^2 + 4x + 5$	-1	5
$0x^4 + 5x^3 - 6x^2 + 2x - 1$	5 (note well)	-1
$7x^5 - 3x^3 + x^2 + 4x$	7	0 (note well)
$2x^6 + 3x^7 + x^8 - 2x - 4 + 4x^2$	1 (be careful)	-4 (ditto)
$a_n x^n + a_{n-1}x^{n-1} + \cdots + a_2 x^2 + a_1 x + a_0$ $(a_n \neq 0)$	a_n	a_0

A polynomial which consists only of a constant term, such as 11 or -7 or 0, is called a **constant polynomial.**

The **degree** of a polynomial is the highest power to which x is raised, provided this power of x has a nonzero coefficient.

EXAMPLE $6x^3 + 2x^2 + 1$ has degree 3. $7x^3 - 6x^4 + x^8 - 14x + 2$ has degree 8. $0x^5 + 5x^4 - 6x + 2$ has degree 4.

EXAMPLE A *nonzero* constant polynomial, such as 11 or -7, has degree 0. One way to think of this is to consider the constant term of a polynomial as the coefficient of x^0 and follow the usual custom that $x^0 = 1$.

EXAMPLE The **zero polynomial** is the constant polynomial 0. The degree of the zero polynomial is *not defined*. This is done for technical convenience in stating certain theorems and properties. It need not worry us here.

POLYNOMIAL FUNCTIONS

A **polynomial function** is a function whose rule is given by a polynomial. For example,

$$f(x) = x^2, \qquad g(y) = -y^3 + 1, \qquad h(t) = 3t^4 - 6t^3 + 14t^2 - 6t + 7$$

Many of the functions examined in Chapter 3 were in fact polynomial functions.

When dealing with a polynomial function, such as $f(x) = 7x^4 + 4x^2 - 6x + 5$, most mathematicians are pretty casual about their language. They may refer to "the polynomial $f(x)$" or "the function $7x^4 + 4x^2 - 6x + 5$."

POLYNOMIAL ARITHMETIC

We shall assume that you know how to *add, subtract,* and *multiply* polynomials. If you think you need some practice, do Exercises A.9–A.15 and B.1–B.4 in the Algebra Review at the end of the book. You should also know how to *divide* one polynomial by another. Before giving examples of division of polynomials, it will be helpful to recall how long division is carried out with numbers.

EXAMPLE   Divide 4509 by 31:

$$
\begin{array}{r}
145 \\
31\overline{)4509} \\
\underline{31} \quad \leftarrow 1 \cdot 31 \\
140 \quad \leftarrow \text{subtract} \\
\underline{124} \quad \leftarrow 4 \cdot 31 \\
169 \quad \leftarrow \text{subtract} \\
\underline{155} \quad \leftarrow 5 \cdot 31 \\
14 \quad \leftarrow \text{subtract}
\end{array}
$$

check:

$$
\begin{array}{r}
145 \text{ quotient} \\
\underline{\times 31} \text{ divisor} \\
145 \\
435 \\
\underline{} \\
4495 \\
\underline{+14} \text{ remainder} \\
4509 \text{ dividend}
\end{array}
$$

The number 4509 is the **dividend,** 31 is the **divisor,** 145 is the **quotient,** and 14 is the **remainder.** The division process stops when we reach a remainder (namely, 14) which is *less than* the divisor 31. The procedure shown above for "checking" the answer may be summarized in one line by

$$(31) \cdot (145) + 14 = 4509$$

or in words,

$$(\text{divisor})(\text{quotient}) + (\text{remainder}) = \text{dividend}$$

Long division of polynomials is similar in many ways to long division of numbers, as shown in the following examples.

EXAMPLE   To divide $x^3 + 2x^2 - 4x + 1$ by $x - 2$ we first write

$$x - 2\overline{)x^3 + 2x^2 - 4x + 1}$$

$x - 2$ is called the **divisor,** $x^3 + 2x^2 - 4x + 1$ the **dividend.** The **quotient** of the division is written above the horizontal line, as shown below. To refresh your memory, each step in the process is explained in the right-hand column. Note the similarity with the numerical example above.

$$
\begin{array}{r}
x^2 + 4x\ \ + 4 \qquad\qquad \textit{quotient} \\
x - 2\,\overline{)x^3 + 2x^2 - 4x + 1} \\
\underline{x^3 - 2x^2} \qquad\qquad \leftarrow x^2 \cdot (x - 2) \\
4x^2 - 4x + 1 \ \leftarrow \text{subtract}° \\
\underline{4x^2 - 8x} \qquad\ \ \leftarrow 4x(x - 2) \\
4x + 1 \ \leftarrow \text{subtract} \\
\underline{4x - 8} \ \leftarrow 4 \cdot (x - 2) \\
9 \ \leftarrow \text{subtract}
\end{array}
$$

The number 9 is called the **remainder.** The division process always stops when the remainder is zero or has *smaller degree* than the divisor (here, the divisor $x - 2$ has degree 1 and the remainder is the constant polynomial 9 which has degree 0). Observe that the product of the divisor and quotient is

$$(x - 2)(x^2 + 4x + 4) = x^3 + 2x^2 - 4x - 8 \qquad \text{(verify!)}$$

Adding the remainder 9 to this result yields the original dividend:

$$(x^3 + 2x^2 - 4x - 8) + 9 = x^3 + 2x^2 - 4x + 1$$

In other words,

$$(\text{divisor})(\text{quotient}) + (\text{remainder}) = \text{dividend}$$

just as in the case of division of numbers.

EXAMPLE   To divide $2x^5 + 5x^4 - 4x^3 + 8x^2 + 1$ by $2x^2 - x + 1$, we follow the same procedure:

$$
\begin{array}{r}
x^3 + 3x^2 -\ \ x\ \ + 2 \qquad\qquad\qquad \textit{quotient} \\
2x^2 - x + 1\,\overline{)2x^5 + 5x^4 - 4x^3 + 8x^2 \qquad\quad + 1} \\
\underline{2x^5 -\ \ x^4 +\ \ x^3} \qquad\qquad\qquad \leftarrow x^3 \cdot (2x^2 - x + 1) \\
6x^4 - 5x^3 + 8x^2 \qquad\quad + 1 \ \leftarrow \text{subtract} \\
\underline{6x^4 - 3x^3 + 3x^2} \qquad\qquad\qquad \leftarrow 3x^2 \cdot (2x^2 - x + 1) \\
- 2x^3 + 5x^2 \qquad\quad + 1 \ \leftarrow \text{subtract} \\
\underline{- 2x^3 +\ \ x^2 -\ \ x} \qquad\quad \leftarrow (-x) \cdot (2x^2 - x + 1) \\
4x^2 +\ \ x + 1 \ \leftarrow \text{subtract} \\
\underline{4x^2 - 2x + 2} \ \leftarrow 2 \cdot (2x^2 - x + 1) \\
3x - 1 \ \leftarrow \text{subtract}
\end{array}
$$

° If this subtraction is confusing, write it out "horizontally" and watch the signs carefully: $(x^3 + 2x^2 - 4x + 1) - (x^3 - 2x^2) = x^3 + 2x^2 - 4x + 1 - x^3 + 2x^2 = 4x^2 - 4x + 1$, as shown above in "vertical" fashion.

The process now ends since the remainder $3x - 1$ (degree 1) has smaller degree than the divisor $2x^2 - x + 1$ (degree 2). You should verify that once again:

$$\text{(divisor)(quotient)} \qquad + \text{(remainder)} = \qquad \text{dividend}$$
$$(2x^2 - x + 1)(x^3 + 3x^2 - x + 2) + \ (3x - 1) \ = 2x^5 + 5x^4 - 4x^3 + 8x^2 + 1$$

A convenient shorthand method of performing certain polynomial division problems (called synthetic division) is explained in the DO IT YOURSELF! segment at the end of this section.

PROPERTIES OF POLYNOMIAL DIVISION

The facts illustrated in the preceding examples, namely, that

$$\text{(divisor)(quotient)} + \text{(remainder)} = \text{dividend}$$

and that the remainder is zero or has smaller degree than the divisor, are important enough to be given a special name and a formal statement.

THE DIVISION ALGORITHM

If a polynomial f(x) *is divided by a nonzero polynomial* h(x), *then there is a quotient polynomial* q(x) *and a remainder polynomial* r(x) *such that*

$$h(x)q(x) + r(x) = f(x)$$

where either r(x) $= 0$ *or* r(x) *has degree less than the degree of the divisor* h(x).

Look what happens to the Division Algorithm when the divisor $h(x)$ is a *first*-degree polynomial of the form $x - c$, such as $x - 2$ or $x + 3 = x - (-3)$ or $x = x - 0$:

$$(*) \qquad\qquad f(x) = h(x)q(x) + r(x) = (x - c)q(x) + r(x)$$

where either $r(x) = 0$ or $r(x)$ has smaller degree than $h(x) = x - c$. But the only polynomials with smaller degree than $x - c$ are the nonzero constants (that is, the polynomials of degree 0). Therefore the remainder $r(x)$ is just a real number d (possibly 0), and statement (*) above becomes

$$f(x) = (x - c)q(x) + d$$

If we evaluate the polynomial function $f(x)$ at the number $x = c$, we obtain

$$f(c) = (c - c)q(c) + d = 0 \cdot q(c) + d = d$$

Thus the remainder d is just the number $f(c)$, the value of $f(x)$ at $x = c$. We have proved

> ## THE REMAINDER THEOREM
>
> *If a polynomial* f(x) *is divided by* x − c, *then the remainder is precisely the number* f(c).

EXAMPLE What is the remainder when $f(x) = x^3 + 2x^2 - 4x + 1$ is divided by $x - 2$? Now $x - 2$ is just $x - c$ with $c = 2$, so the Remainder Theorem states that the remainder is

$$f(c) = f(2) = 2^3 + 2 \cdot 2^2 - 4 \cdot 2 + 1 = 8 + 8 - 8 + 1 = 9$$

If you have any doubts, refer to page 175, where the entire division of $f(x)$ by $x - 2$ is written out: the remainder is indeed 9.

EXAMPLE If you divide $f(x) = x^3 + 4x^2 + 2x - 3$ by $x + 3$, the remainder is the number 0. You can obtain the same conclusion by writing $x + 3 = x - (-3)$ and using the Remainder Theorem with $c = -3$: the remainder is

$$f(c) = f(-3) = (-3)^3 + 4(-3)^2 + 2(-3) - 3 = -27 + 36 - 6 - 3 = 0$$

FACTORS

Recall that $h(x)$ is said to be a *factor* of the polynomial $f(x)$ provided that $f(x) = h(x)q(x)$ for some polynomial $q(x)$. For instance, $2x^2 + 1$ is a factor of $6x^3 - 4x^2 + 3x - 2$ since $6x^3 - 4x^2 + 3x - 2 = (2x^2 + 1)(3x - 2)$.

If $h(x)$ is a factor of $f(x)$, then

$$f(x) = h(x)q(x) = h(x)q(x) + 0$$

The Division Algorithm tells us that

> *The remainder is zero when* f(x) *is divided by one of its factors*

Thus to test whether or not a given polynomial *is* a factor of $f(x)$, we just divide $f(x)$ by the polynomial and see if the remainder is 0.

Testing for factors by long division can be quite tedious. But in one case there is an easier way. If $f(x)$ is a polynomial and c is any real number, we can determine whether or not $x - c$ is a factor of $f(x)$ by using the following theorem.

> ## THE FACTOR THEOREM
>
> x − c *is a factor of the polynomial* f(x) *exactly when* f(c) = 0.

To prove this statement, just divide $f(x)$ by $x - c$. We know from the Division Algorithm that

$$f(x) = (x - c)q(x) + r(x)$$

where $r(x)$ is the remainder. But the Remainder Theorem says that in this case the remainder $r(x)$ is just the number $f(c)$, so that

(*) $$f(x) = (x - c)q(x) + f(c)$$

If $f(c) = 0$, then statement (*) becomes $f(x) = (x - c)q(x)$, so that $x - c$ is a factor of $f(x)$. Conversely, if $x - c$ is a factor of $f(x)$, we must have remainder 0 [that is, $f(c) = 0$] in order for statement (*) to be true. This proves the Factor Theorem.

EXAMPLE In order to check whether $x - 2$ is a factor of $f(x) = x^3 - 4x^2 + 5x - 2$, we need only compute the number $f(2) = 2^3 - 4 \cdot 2^2 + 5 \cdot 2 - 2 = 8 - 16 + 10 - 2 = 0$. Since $f(2) = 0$, the Factor Theorem tells us that $x - 2$ *is* a factor of $x^3 - 4x^2 + 5x - 2$.

EXAMPLE Is $x + 1$ a factor of the polynomial $f(x) = x^4 + x^3 + 5x^2 + x - 6$? We have $x + 1 = x - (-1)$, so we apply the Factor Theorem with $c = -1$:

$$f(c) = f(-1) = (-1)^4 + (-1)^3 + 5(-1)^2 + (-1) - 6$$
$$= 1 - 1 + 5 - 1 - 6 = -2$$

Since $f(-1) = -2 \neq 0$, $x + 1$ is *not* a factor of $f(x)$.

EXERCISES

A.1. Which of the following are polynomial functions? For those that are, list their leading coefficient, constant term, and degree. For those that aren't, tell why not.
   (a) $h(x) = 1 + x^3$
   (b) $g(t) = (t - 1)(t^2 + 1)$
   (c) $p(x) = -7$
   (d) $f(u) = \dfrac{u^2 + 1}{u}$
   (e) $g(t) = 10^t$
   (f) $r(x) = (x - 1)^k$ (k a fixed positive integer)
   (g) $f(x) = \sqrt{x - 7}$
   (h) $f(x) = (x + \sqrt{3})(x - \sqrt{3})$

A.2. Perform the indicated division and determine the quotient and remainder [*Note*: $f(x) \div h(x)$ means "divide $f(x)$ by $h(x)$"; that is, $h(x)$ is the divisor.] Check your division by calculating (divisor)(quotient) + (remainder).
   (a) $(3x^4 + 2x^2 - 6x + 1) \div (x + 1)$
   (b) $(x^5 - x^3 + x - 5) \div (x - 2)$
   (c) $(x^5 + 2x^4 - 6x^3 + x^2 - 5x + 1) \div (x^3 + 1)$
   (d) $(3x^4 - 2x^3 - 11x^2 + 6x - 1) \div (x^3 + x^2 - 2)$
   (e) $(5x^4 + 5x^2 + 5) \div (x^2 - x + 1)$
   (f) $(x^3 + 2x^2 - 6x + 3) \div (x - 2)$

A.3. Use long division to determine whether the first polynomial is a factor of the second.
(a) $x^2 + 3x - 1$ and $x^3 + 2x^2 - 5x - 6$
(b) $x^2 + 9$ and $x^5 + x^4 - 81x - 81$
(c) $x^2 + 3x - 1$ and $x^4 + 3x^3 - 2x^2 - 3x + 1$
(d) $x^2 - 5x + 7$ and $x^3 - 3x^2 - 3x + 9$

A.4. Find the remainder when $f(x)$ is divided by $g(x)$ *without* using long division.
(a) $f(x) = x^{10} + x^8; g(x) = x - 1$
(b) $f(x) = 3x^4 - 6x^3 + 2x - 1; g(x) = x + 1$
(c) $f(x) = x^5 - 3x^2 + 2x - 1; g(x) = x - 2$
(d) $f(x) = x^3 - 2x^2 + 5x - 4; g(x) = x + 2$
(e) $f(x) = x^6 - 10; g(x) = x - 2$
(f) $f(x) = 10x^{75} - 8x^{65} + 6x^{45} + 4x^{32} - 2x^{15} + 5; g(x) = x - 1$

A.5. Use the Factor Theorem to determine whether or not $h(x)$ is a factor of $f(x)$.
(a) $h(x) = x - 1; f(x) = x^5 + 1$
(b) $h(x) = x - \frac{1}{2}; f(x) = 2x^4 + x^3 + x - \frac{3}{4}$
(c) $h(x) = x + 2; f(x) = x^3 - 3x^2 - 4x - 12$
(d) $h(x) = x + 1; f(x) = x^3 - 4x^2 + 3x + 8$
(e) $h(x) = x - 1; f(x) = 14x^{99} - 65x^{56} + 51$
(f) $h(x) = x - 2; f(x) = x^3 + x^2 - 4x + 4$

B.1. (a) Compute these polynomial products:
(i) $(x + 2)(x - 1)$ (iii) $x(x^2 + 2x + 1)$.
(ii) $(3x^2 - 2)(2x^3 - 3x^2 + x - 1)$
(b) If $f(x)$ is a polynomial of degree p and $g(x)$ is a polynomial of degree q, what is the degree of the product polynomial $f(x)g(x)$? [*Hint:* part (a) may provide a clue.]
(c) Prove that your answer in part (b) is correct. (You will need to use the fact that the product of two *nonzero* real numbers is necessarily nonzero.)

B.2. Use the Factor Theorem to show that for every real number c, $x - c$ is *not* a factor of $x^4 + x^2 + 1$.

B.3. Let c be a real number and n a positive integer.
(a) Show that $x - c$ is a factor of $x^n - c^n$.
(b) If n is even, show that $x + c$ is a factor of $x^n - c^n$. [Remember $x + c = x - (-c)$.]
(c) If n is odd, give an example to show that $x + c$ may not be a factor of $x^n - c^n$.
(d) If n is odd, show that $x + c$ is a factor of $x^n + c^n$.

B.4. Find a number k such that
(a) $x + 2$ is a factor of $x^3 + 3x^2 + kx - 2$
(b) $x - 3$ is a factor of $x^4 - 5x^3 + kx^2 + 18k + 18$
(c) $x - 1$ is a factor of $k^2x^4 - 2kx^2 + 1$
(d) $x + 2$ is a factor of $x^3 - kx^2 + 3x + 7k$

DO IT YOURSELF!

SYNTHETIC DIVISION

Synthetic division is a fast method of doing polynomial division when the divisor is a first-degree polynomial of the form $x - c$ for some real number c. We begin with an example of ordinary long division. We shall see that the calculations can be written in a brief shorthand notation. This shorthand version will then lead to the method of synthetic division.

The usual long division process for dividing $3x^4 - 8x^2 - 11x + 1$ by $x - 2$ goes like this:

$$
\begin{array}{r}
3x^3 + 6x^2 + 4x - 3 \qquad \leftarrow \text{quotient} \\
\text{divisor} \rightarrow x - 2 \overline{)\, 3x^4 \qquad - 8x^2 - 11x + 1} \quad \leftarrow \text{dividend} \\
\underline{3x^4 - 6x^3} \\
6x^3 - 8x^2 \\
\underline{6x^3 - 12x^2} \\
4x^2 - 11x \\
\underline{4x^2 - 8x} \\
-3x + 1 \\
\underline{-3x + 6} \\
-5 \quad \leftarrow \text{remainder}
\end{array}
$$

This calculation obviously involves a lot of repetitions. If we keep the various powers of x aligned properly, we can eliminate many of these repetitions without causing any confusion:

$$
\begin{array}{r}
3x^3 + 6x^2 + 4x - 3 \qquad \leftarrow \text{quotient} \\
\text{divisor} \rightarrow x - 2 \overline{)\, 3x^4 \qquad - 8x^2 - 11x + 1} \quad \leftarrow \text{dividend} \\
\underline{-6x^3} \\
6x^3 \\
\underline{-12x^2} \\
4x^2 \\
\underline{-8x} \\
-3x \\
\underline{+6} \\
-5 \quad \leftarrow \text{remainder}
\end{array}
$$

As long as we keep the coefficients in the proper columns, there is no need to keep writing out all the powers of x. But to avoid confusion here we must insert 0 coefficients for those terms that don't appear above, such as the x^3 term in the dividend:

$$
\begin{array}{r}
\;3\quad 6\quad 4\;\; -3 \quad\leftarrow \text{quotient}\\
\text{divisor}\rightarrow 1-2\overline{)3\quad 0\;\; -8\;\; -11\quad\; 1}\;\leftarrow\text{dividend}\\
\underline{-6}\\
6\\
\underline{-12}\\
4\\
\underline{-8}\\
-3\\
\underline{+6}\\
-5\;\leftarrow\text{remainder}
\end{array}
$$

Since it's easy to remember that the coefficient of x in the divisor $x-2$ is the number 1, there's really no need to write the 1 in this last calculation. Furthermore, we can save some space by moving the various numbers upward, thus obtaining this shorthand version of the calculation:

$$
\begin{array}{r}
3\quad 6\quad 4\;\; -3 \quad\leftarrow \text{quotient}\\
\text{divisor}\rightarrow -2\overline{)3\quad 0\;\; -8\;\; -11\quad 1}\;\leftarrow\text{dividend}\\
\underline{-6\;\;-12\;\;-8\quad 6}\\
6\quad 4\;\; -3\;\lfloor -5\;\leftarrow\text{remainder}
\end{array}
$$

But there are still some repetitions: the last three entries in the quotient row are the same as the first three entries in the last row. By inserting a 3 at the beginning of the last row, we can omit the top row and still preserve all the essential information:

$$
\begin{array}{r}
\text{divisor}\rightarrow -2\overline{)3\quad 0\;\; -8\;\; -11\quad 1}\;\leftarrow\text{dividend}\\
\underline{-6\;\;-12\;\;-8\quad 6}\\
3\quad 6\quad 4\;\; -3\;\lfloor -5\;\leftarrow\text{remainder}\\
\underbrace{}_{\text{quotient}}
\end{array}
$$

For reasons that will soon be apparent, we shall now rewrite this shorthand version in a slightly different manner. We replace -2 by 2 and change all the signs in the *second* row (and leave everything else alone):

$$
\begin{array}{r}
\text{divisor}\rightarrow 2\overline{)3\quad 0\;\; -8\;\; -11\quad 1}\;\leftarrow\text{dividend}\\
\underline{6\quad 12\quad 8\;\; -6}\\
3\quad 6\quad 4\;\; -3\;\lfloor -5\;\leftarrow\text{remainder}\\
\underbrace{}_{\text{quotient}}
\end{array}
$$

Providing that we remember the sign change on the 2, this three-line array of numbers still preserves all the essential information (divisor, dividend, quotient, and remainder). Note that the three rows are related to each other as follows:

(i) The second row can be obtained by multiplying each entry in the last row (except the remainder -5) by 2. For instance, $2 \cdot 3 = 6$, $2 \cdot 6 = 12$, and so on, as shown by the arrows below:

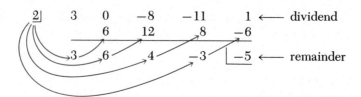

(ii) The last row can be obtained by *adding* the corresponding entries in the first and second rows. For instance, the 4 in the last row is the sum of the two numbers directly above it, $-8 + 12$.

The three-line array of numbers just obtained is a shorthand summary of the division of $3x^4 - 8x^2 - 11x + 1$ by $x - 2$. **Synthetic division** is just a method of producing this summary directly, without having to go through the entire long division process first. It is based on properties (i) and (ii) above.

Here is a step-by-step explanation of the division of $3x^4 - 8x^2 - 11x + 1$ by $x - 2$ by means of synthetic division:

Step 1. In the first row list the constant term of the divisor (namely, 2) and the coefficients of the dividend in order of decreasing powers of x (insert 0 coefficients for missing powers of x).

$$\underline{2|} \quad 3 \quad 0 \quad -8 \quad -11 \quad 1$$

Step 2. Bring down the first dividend coefficient (namely, 3) to the third row.

$$\underline{2|} \quad 3 \quad 0 \quad -8 \quad -11 \quad 1$$
$$3$$

Step 3. Multiply $2 \cdot 3$ and insert the answer 6 in the second row, in the position shown here.

$$\underline{2|} \quad 3 \quad 0 \quad -8 \quad -11 \quad 1$$
$$6$$
$$3$$

Step 4. Add $0 + 6$ and write the answer 6 in the third row.

$$\underline{2|} \quad 0 \quad -8 \quad -11 \quad 1$$
$$6$$
$$3 \quad 6$$

Step 5. Multipy $2 \cdot 6$ and insert the answer 12 in the second row.

$$\underline{2|} \quad 3 \quad 0 \quad -8 \quad -11 \quad 1$$
$$6 \quad 12$$
$$3 \quad 6$$

Step 6. Add $-8 + 12$ and write the answer 4 in the third row.

$$\begin{array}{r|rrrrr} 2 & 3 & 0 & -8 & -11 & 1 \\ & & 6 & 12 & & \\ \hline & 3 & 6 & 4 & & \end{array}$$

Step 7. Multiply $2 \cdot 4$ and insert the answer 8 in the second row.

$$\begin{array}{r|rrrrr} 2 & 3 & 0 & -8 & -11 & 1 \\ & & 6 & 12 & 8 & \\ \hline & 3 & 6 & 4 & & \end{array}$$

Step 8. Add $-11 + 8$ and write the answer -3 in the third row.

$$\begin{array}{r|rrrrr} 2 & 3 & 0 & -8 & -11 & 1 \\ & & 6 & 12 & 8 & \\ \hline & 3 & 6 & 4 & -3 & \end{array}$$

Step 9. Multiply $2 \cdot (-3)$ and insert the answer -6 in the second row.

$$\begin{array}{r|rrrrr} 2 & 3 & 0 & -8 & -11 & 1 \\ & & 6 & 12 & 8 & -6 \\ \hline & 3 & 6 & 4 & -3 & \end{array}$$

Step 10. Add $1 + (-6)$ and write the answer -5 in the third row.

$$\begin{array}{r|rrrrr} 2 & 3 & 0 & -8 & -11 & 1 \\ & & 6 & 12 & 8 & -6 \\ \hline & 3 & 6 & 4 & -3 & -5 \end{array}$$

We have arrived at the same three-line summary that we obtained earlier via long division. The divisor and dividend are listed in the first row. The quotient and remainder can be read directly from the third row:

The last number in the third row is the remainder.
The other numbers in the third row are coefficients of the quotient (arranged in order of decreasing powers of x).

Since we are dividing the *fourth-degree* polynomial $3x^4 - 8x^2 - 11x + 1$ by the *first*-degree polynomial $x - 2$, the quotient must be the polynomial of degree *three* with coefficients 3, 6, 4, -3, namely, $3x^3 + 6x^2 + 4x - 3$. The remainder is -5.

Warning Synthetic division can be used *only* when the divisor is of the form $x - c$. The constant c can be either positive or negative. In the example above, the divisor was $x - 2$, which is of the form $x - c$ with $c = 2$. On the other hand, $x + 5$ can be written as $x - (-5)$, which is of the form $x - c$ with $c = -5$. So synthetic division can be used with the divisor $x + 5$. But synthetic division *cannot* be used with divisors such as x^2 or $x^3 + 5$ or $6x - 7$ since these cannot be put in the form $x - c$.

EXAMPLE The first step in dividing $x^5 + 5x^4 + 6x^3 - x^2 + 4x + 29$ by $x + 3$ is to write the divisor in the form $x - c$: $x + 3 = x - (-3)$. The first row of the synthetic division consists of -3 and the coefficients of the dividend:

$$\begin{array}{r|rrrrrr} -3 & 1 & 5 & 6 & -1 & 4 & 29 \end{array}$$

We now proceed as in the preceding example and obtain:

$$\begin{array}{r|rrrrrr} -3 & 1 & 5 & 6 & -1 & 4 & 29 \\ & & -3 & -6 & 0 & 3 & -21 \\ \hline & 1 & 2 & 0 & -1 & 7 & \underline{8} \end{array}$$

Since we are dividing a fifth-degree polynomial by a first-degree polynomial, the quotient is a fourth-degree polynomial. The last row above shows that the quotient is $1x^4 + 2x^3 + 0x^2 - 1x + 7$, that is, $x^4 + 2x^3 - x + 7$ and that the remainder is 8.

EXAMPLE Show that $x - 7$ is a factor of $f(x) = 8x^5 - 52x^4 + 2x^3 - 198x^2 - 86x + 14$ and find the other factor. We know that $x - 7$ will be a factor exactly when division by $x - 7$ leaves remainder 0. The other factor will be the quotient of this division. Using synthetic division, we have:

$$\begin{array}{r|rrrrrr} 7 & 8 & -52 & 2 & -198 & -86 & 14 \\ & & 56 & 28 & 210 & 84 & -14 \\ \hline & 8 & 4 & 30 & 12 & -2 & \underline{0} \end{array}$$

The last row shows that the remainder is 0 and the quotient is $8x^4 + 4x^3 + 30x^2 + 12x - 2$. Therefore $f(x)$ factors as:

$$8x^5 - 52x^4 + 2x^3 - 198x^2 - 86x + 14 = (x - 7)(8x^4 + 4x^3 + 30x^2 + 12x - 2)$$

EXAMPLE Let $f(x)$ be as in the preceding example and find $f(6)$. The Remainder Theorem states that the number $f(6)$ is just the remainder after division by $x - 6$. It can be found quickly by synthetic division:

$$\begin{array}{r|rrrrrr} 6 & 8 & -52 & 2 & -198 & -86 & 14 \\ & & 48 & -24 & -132 & -1980 & -12{,}396 \\ \hline & 8 & -4 & -22 & -330 & -2066 & \underline{-12{,}382} \end{array}$$

Therefore $f(6) = \text{remainder} = -12{,}382$. The arithmetic involved in the synthetic division here is much easier than the arithmetic needed to find $f(6)$ by substituting $x = 6$ in $f(x) = 8x^5 - 52x^4 + 2x^3 - 198x^2 - 86x + 14$.

EXERCISES

A.1. Use synthetic division to find the quotient and remainder.
(a) $(3x^4 - 8x^3 + 9x + 5) \div (x - 2)$
(b) $(4x^3 - 3x^2 + x + 7) \div (x - 2)$
(c) $(2x^4 + 5x^3 - 2x - 8) \div (x + 3)$
(d) $(3x^3 - 2x^2 - 8) \div (x + 5)$
(e) $(5x^4 - 3x^2 - 4x + 6) \div (x - 7)$
(f) $(3x^4 - 2x^3 + 7x - 4) \div (x - 3)$
(g) $(x^4 - 6x^3 + 4x^2 + 2x - 7) \div (x - 2)$
(h) $(x^6 - x^5 + x^4 - x^3 + x^2 - x + 1) \div (x + 3)$

A.2. Use synthetic division and the Remainder Theorem to find $f(c)$ when:

(a) $f(x) = 2x^5 - 3x^4 + x^3 - 2x^2 + x - 8$ and $c = 10$

(b) $f(x) = x^3 + 8x^2 - 29x + 44$ and $c = -11$

(c) $f(x) = 2x^5 - 3x^4 + 2x^3 - 8x - 8$ and $c = 20$

(d) $f(x) = x^5 - 10x^4 + 20x^3 - 5x - 95$ and $c = -10$

(e) $f(x) = 2x^5 + x^3 - 3x^2 + 4$ and $c = \frac{1}{2}$

B.1. Use synthetic division to find the quotient and remainder. In each divisor $x - c$, the number c is not an integer, but the same technique will work.

(a) $(3x^4 - 2x^2 + 2) \div (x - \frac{1}{4})$

(b) $(2x^4 - 3x^2 + 1) \div (x - \frac{1}{2})$

(c) $(2x^4 - 5x^3 - x^2 + 3x + 2) \div (x + \frac{1}{2})$

(d) $(10x^5 - 3x^4 + 14x^3 + 13x^2 - \frac{4}{3}x + \frac{7}{3}) \div (x + \frac{1}{5})$

(e) $(x^4 + (5 - \sqrt{3})x^3 - (6 + 5\sqrt{3})x^2 + (1 + 6\sqrt{3})x + (1 - \sqrt{3})) \div (x - \sqrt{3})$

B.2. Use synthetic division to show that $f(x)$ is a factor of $g(x)$ and find the other factor.

(a) $f(x) = x + 4$ and $g(x) = 3x^3 + 9x^2 - 11x + 4$

(b) $f(x) = x - 5$ and $g(x) = x^5 - 8x^4 + 17x^2 + 293x - 15$

(c) $f(x) = x - \frac{1}{2}$ and $g(x) = 2x^5 - 7x^4 + 15x^3 - 6x^2 - 10x + 5$

(d) $f(x) = x + \frac{1}{3}$ and $g(x) = 3x^6 + x^5 - 6x^4 + 7x^3 + 3x^2 - 15x - 5$

B.3. Use a calculator and synthetic division to find the quotient and remainder.

(a) $(x^3 - 5.27x^2 + 10.708x - 10.23) \div (x - 3.12)$

(b) $(2.79x^4 + 4.8325x^3 - 6.73865x^2 + .9255x - 8.125) \div (x - 1.35)$

2. ROOTS OF POLYNOMIALS

A great deal of time is spent in high school algebra solving polynomial equations, such as

$$x^2 - 5x + 6 = 0, \qquad x^2 + 7x - 2 = 0, \qquad x^4 + 2x^2 + 1 = 0$$

In this section we shall review some of this high school material and discuss the solutions of higher degree polynomial equations.

If $f(x)$ is a polynomial, then a solution of the equation $f(x) = 0$ is called a **root** (or **zero**) of the polynomial $f(x)$. Thus a number c is a root of $f(x)$ provided that $f(c) = 0$.

EXAMPLE Determine whether or not 2 and 3 are roots of the polynomial $f(x) = x^3 + x^2 - 4x - 4$. The roots of $f(x)$ are the solutions of the equation $f(x) = 0$, that is, solutions of

$$x^3 + x^2 - 4x - 4 = 0$$

In order to test whether or not 2 and 3 are roots we need only substitute each of them in this equation to see if they are solutions:

$$f(2) = 2^3 + 2^2 - 4 \cdot 2 - 4 = 0 \qquad \text{and} \qquad f(3) = 3^3 + 3^2 - 4 \cdot 3 - 4 = 20$$

Since $f(2) = 0$, 2 is a solution of $f(x) = 0$, whence 2 *is* a root. Since $f(3) = 20 \neq 0$, 3 is *not* a root.

FIRST-DEGREE POLYNOMIALS

A polynomial function of degree one has the form $f(x) = ax + b$ for some constants a and b, with $a \neq 0$. *Every first-degree polynomial has exactly one root*, and it is easily found.

EXAMPLE To find the roots of the polynomial $-4x + 1$, we need only solve the equation $-4x + 1 = 0$:

$$-4x + 1 = 0$$
$$-4x = -1$$
$$x = \tfrac{1}{4}$$

Therefore $\tfrac{1}{4}$ is the only root of the polynomial $-4x + 1$.

SECOND-DEGREE POLYNOMIALS

A second-degree (or **quadratic**) polynomial function has the form $f(x) = ax^2 + bx + c$ for some constants a, b, c with $a \neq 0$. Its real roots° (if any) are the solutions of the quadratic equation $ax^2 + bx + c = 0$. Depending on just what the numbers a, b, and c are, there may be several ways to solve this equation. The most common method of solving quadratic equations is by factoring.

EXAMPLE In order to solve the equation $x^2 - 5x + 6 = 0$, we first factor the left side:

$$x^2 - 5x + 6 = 0$$
$$(x - 3)(x - 2) = 0$$

Remember that the only time a product of real numbers is zero occurs when one of the factors is zero. Consequently, the last equation above is equivalent to:

$$x - 3 = 0 \quad \text{or} \quad x - 2 = 0$$

But $x - 3 = 0$ means $x = 3$ and $x - 2 = 0$ means $x = 2$. Therefore the solutions of $x^2 - 5x + 6 = 0$ are the numbers 2 and 3, since either of these numbers satisfies the equation.

EXAMPLE Solve the equation $3x^2 - x - 10 = 0$. Once again, the left side can be factored

$$3x^2 - x - 10 = 0$$
$$(3x + 5)(x - 2) = 0$$

° Here and below, the term "real root" means "a root which is a real number." A polynomial may have no real roots, but still have roots which are *complex* numbers. Until we reach Chapter 8 we shall consider only the real number system and real roots.

This last equation is equivalent to

$$3x + 5 = 0 \qquad \text{or} \qquad x - 2 = 0$$
$$3x = -5 \qquad\qquad\qquad x = 2$$
$$x = \tfrac{-5}{3}$$

Therefore the solutions are $\tfrac{-5}{3}$ and 2.

Another method of solving quadratic equations uses the technique of **completing the square.**°

EXAMPLE   In order to solve $x^2 + 6x + 1 = 0$, we first rewrite it as

$$x^2 + 6x = -1$$

The idea now is to add a suitable number to both sides of the equation, so that the left side will factor as a perfect square $(x + ?)^2$. But we know how to find the number needed to complete the square in the expression $x^2 + 6x$: *take half the coefficient of* x—namely, $\tfrac{6}{2} = 3$—*and square it*, thus obtaining $3^2 = 9$. Adding 9 to both sides of the equation above and factoring the left side yields:

$$x^2 + 6x + 9 = -1 + 9$$
$$(x + 3)^2 = 8$$

Thus $x + 3$ must be a number whose square is 8. The only two numbers whose square is 8 are $\sqrt{8}$ and $-\sqrt{8}$. Therefore

$$x + 3 = \sqrt{8} \qquad \text{or} \qquad x + 3 = -\sqrt{8}$$
$$x = \sqrt{8} - 3 \qquad\qquad\qquad x = -\sqrt{8} - 3$$

Consequently, the solutions are $\sqrt{8} - 3$ and $-\sqrt{8} - 3$, or in more compact notation $\pm\sqrt{8} - 3$.

EXAMPLE   In order to solve $3x^2 + 11x + 5 = 0$ by completing the square, we must have 1 as the coefficient of x^2. So we divide both sides of the equation by 3, and *then* proceed as above:

$$x^2 + \tfrac{11}{3}x + \tfrac{5}{3} = 0$$

$$x^2 + \tfrac{11}{3}x = \tfrac{-5}{3} \qquad \text{(subtract } \tfrac{5}{3} \text{ from both sides)}$$

$$x^2 + \tfrac{11}{3}x + (\tfrac{11}{6})^2 = (\tfrac{11}{6})^2 - \tfrac{5}{3} \qquad \text{(add the square of half the coefficient of } x, \text{ namely, } (\tfrac{1}{2} \cdot \tfrac{11}{3})^2 = (\tfrac{11}{6})^2, \text{ to } both \text{ sides)}$$

$$(x + \tfrac{11}{6})^2 = (\tfrac{11}{6})^2 - \tfrac{5}{3} \qquad \text{(factor left side as a perfect square)}$$

$$(x + \tfrac{11}{6})^2 = \tfrac{121}{36} - \tfrac{5}{3} = \tfrac{61}{36} \qquad \text{(put right side over a common denominator)}$$

Therefore we must have

$$x + \tfrac{11}{6} = \sqrt{\tfrac{61}{36}} \qquad \text{or} \qquad x + \tfrac{11}{6} = -\sqrt{\tfrac{61}{36}}$$

° This technique is explained in the Algebra Review on page 560.

or in more compact notation

$$x + \frac{11}{6} = \pm\sqrt{\frac{61}{36}} = \pm\frac{\sqrt{61}}{6}$$

Subtracting $\frac{11}{6}$ from both sides yields

$$x = \frac{-11}{6} \pm \frac{\sqrt{61}}{6} = \frac{-11 \pm \sqrt{61}}{6}$$

Therefore the solutions are $\dfrac{-11 + \sqrt{61}}{6}$ and $\dfrac{-11 - \sqrt{61}}{6}$.

The technique just described can be used to find a formula for the solution of *any* quadratic equation. Here is a step-by-step solution of the general equation $ax^2 + bx + c = 0$. The reason for each step is given at the right. Although it may look complicated, it is *exactly* what was done in the preceding example, where we had $a = 3, b = 11,$ and $c = 5$. If you have trouble following this argument, just compare it step for step with the preceding example.

$$ax^2 + bx + c = 0$$

$$x^2 + \frac{b}{a}x + \frac{c}{a} = 0 \qquad \text{(divide both sides by } a\text{)}$$

$$x^2 + \frac{b}{a}x \qquad = -\frac{c}{a} \qquad \text{(subtract } \frac{c}{a} \text{ from both sides)}$$

Now we want to complete the square in the expression $x^2 + \dfrac{b}{a}x$. To do this we *take one-half the coefficient of x*—namely $\dfrac{1}{2} \cdot \dfrac{b}{a}$, *and square it:* $\left(\dfrac{1}{2} \cdot \dfrac{b}{a}\right)^2 = \left(\dfrac{b}{2a}\right)^2$. Adding this quantity to both sides of the last equation above yields:

$$x^2 + \frac{b}{a}x + \left(\frac{b}{2a}\right)^2 = \left(\frac{b}{2a}\right)^2 - \frac{c}{a}$$

$$\left(x + \frac{b}{2a}\right)^2 = \left(\frac{b}{2a}\right)^2 - \frac{c}{a} \qquad \text{(factor left side as a perfect square)}$$

$$\left(x + \frac{b}{2a}\right)^2 = \frac{b^2}{4a^2} - \frac{c}{a} = \frac{b^2 - 4ac}{4a^2} \qquad \begin{array}{l}\text{(put right side over}\\\text{common denominator)}\end{array}$$

Thus the *square* of $x + \dfrac{b}{2a}$ is $\dfrac{b^2 - 4ac}{4a^2}$, so that

$$x + \frac{b}{2a} = \sqrt{\frac{b^2 - 4ac}{4a^2}} \qquad \text{or} \qquad x + \frac{b}{2a} = -\sqrt{\frac{b^2 - 4ac}{4a^2}}$$

or in more compact notation

$$x + \frac{b}{2a} = \pm\sqrt{\frac{b^2 - 4ac}{4a^2}} = \pm\frac{\sqrt{b^2 - 4ac}}{2a}$$

Subtracting $\dfrac{b}{2a}$ from both sides yields:

$$x = -\frac{b}{2a} \pm \frac{\sqrt{b^2 - 4ac}}{2a} = \frac{-b \pm \sqrt{b^2 - 4ac}}{2a}$$

We have proved

THE QUADRATIC FORMULA

The solutions of the quadratic equation $ax^2 + bx + c = 0$ *are the numbers*

$$x = \frac{-b \pm \sqrt{b^2 - 4ac}}{2a}$$

You should memorize the quadratic formula and be able to use it easily.

EXAMPLE In order to solve $x^2 + 8x + 3 = 0$, apply the quadratic formula with $a = 1$, $b = 8$, and $c = 3$:

$$x = \frac{-b \pm \sqrt{b^2 - 4ac}}{2a} = \frac{-8 \pm \sqrt{8^2 - 4 \cdot 1 \cdot 3}}{2 \cdot 1} = \frac{-8 \pm \sqrt{64 - 12}}{2} =$$

$$\frac{-8 \pm \sqrt{52}}{2} = \frac{-8 \pm \sqrt{4 \cdot 13}}{2} = \frac{-8 \pm \sqrt{4}\sqrt{13}}{2} = \frac{-8 \pm 2\sqrt{13}}{2} = -4 \pm \sqrt{13}$$

The solutions are $-4 + \sqrt{13}$ and $-4 - \sqrt{13}$. In other words, the polynomial $x^2 + 8x + 3$ has *two distinct real roots*, $-4 + \sqrt{13}$ and $-4 - \sqrt{13}$.

EXAMPLE In order to solve $\frac{1}{3}x^2 - 2x + 3 = 0$, apply the quadratic formula with $a = \frac{1}{3}$, $b = -2$, and $c = 3$:

$$x = \frac{-b \pm \sqrt{b^2 - 4ac}}{2a} = \frac{-(-2) \pm \sqrt{(-2)^2 - 4(\frac{1}{3})3}}{2(\frac{1}{3})}$$

$$= \frac{2 \pm \sqrt{4 - 4}}{\frac{2}{3}} = \tfrac{3}{2}(2 \pm \sqrt{0}) = 3$$

The only solution is 3. So the polynomial $\frac{1}{3}x^2 - 2x + 3$ has just *one real root*.

EXAMPLE To solve $2x^2 + x + 3 = 0$, apply the quadratic formula with $a = 2$, $b = 1$, and $c = 3$:

$$x = \frac{-b \pm \sqrt{b^2 - 4ac}}{2a} = \frac{-1 \pm \sqrt{1 - 4 \cdot 2 \cdot 3}}{2 \cdot 2} = \frac{-1 \pm \sqrt{1 - 24}}{4}$$

But $\sqrt{b^2 - 4ac} = \sqrt{1 - 24} = \sqrt{-23}$ is *not* a real number. There are *no solutions* in the real number system. In other words, the polynomial $2x^2 + x + 3$ has *no real roots*.

The term $b^2 - 4ac$ in the quadratic formula is called the **discriminant**. As the three preceding examples illustrate, the discriminant determines the number of real roots of the polynomial $ax^2 + bx + c$:

Discriminant $b^2 - 4ac$	Number of Real Roots of $ax^2 + bx + c$	Example
$b^2 - 4ac > 0$	Two distinct real roots	$x^2 + 8x + 3$
$b^2 - 4ac = 0$	One real root	$\frac{1}{3}x^2 - 2x + 3$
$b^2 - 4ac < 0$	No real roots	$2x^2 + x + 3$

POLYNOMIALS OF HIGHER DEGREE

It is often difficult to solve polynomial equations of degree 3 or more. But there are some cases where this can be done. For instance, the rational number roots of a polynomial with integer coefficients can always be found. A method for doing so is explained in the DO IT YOURSELF! segment at the end of this section. Here are some other situations in which roots of polynomials can be found.

EXAMPLE The roots of the polynomial $f(x) = 3(x - 2)^3(x + 7)(x - \frac{5}{2})^2(x + \sqrt{6})^3$ are the solutions of the equation

$$3(x - 2)^3(x + 7)(x - \tfrac{5}{2})^2(x + \sqrt{6})^3 = 0$$
$$3(x - 2)(x - 2)(x - 2)(x + 7)(x - \tfrac{5}{2})(x - \tfrac{5}{2})(x + \sqrt{6})(x + \sqrt{6})(x + \sqrt{6}) = 0$$

Since a product is zero only when at least one of the factors is zero, this equation is equivalent to:

$$x - 2 = 0 \quad \text{or} \quad x + 7 = 0 \quad \text{or} \quad x - \tfrac{5}{2} = 0 \quad \text{or} \quad x + \sqrt{6} = 0$$

which is the same as

$$x = 2 \quad \text{or} \quad x = -7 \quad \text{or} \quad x = \tfrac{5}{2} \quad \text{or} \quad x = -\sqrt{6}$$

Therefore the roots of $f(x)$ are 2, -7, $\frac{5}{2}$, and $-\sqrt{6}$.

EXAMPLE The argument used in the preceding example also applies to the polynomial $5(x + 1)^{10}(x - 1)^{17}(x + \frac{3}{2})^8(x - \frac{17}{3})^4$. Its roots are -1, 1, $\frac{-3}{2}$, and $\frac{17}{3}$.

Occasionally, a polynomial which is not presented in factored form can be factored and its roots determined.

EXAMPLE To find the roots of the polynomial $x^7 - 7x^5 - 18x^3$ we factor the left side of the equation

$$x^7 - 7x^5 - 18x^3 = 0$$
$$x^3(x^4 - 7x^2 - 18) = 0$$
$$x^3(x^2 - 9)(x^2 + 2) = 0$$
$$x^3(x + 3)(x - 3)(x^2 + 2) = 0$$

The last equation is equivalent to

$x^3 = 0$	or	$x + 3 = 0$	or	$x - 3 = 0$	or	$x^2 + 2 = 0$
$x = 0$	or	$x = -3$	or	$x = 3$	or	$x^2 = -2$

Since $x^2 = -2$ has no solution in the real number system, the only real roots of $x^7 - 7x^5 - 18x^3$ are 0, 3, and -3.

In the last section we proved the Factor Theorem, which can be restated in terms of roots as follows:

FACTOR THEOREM

The number c is a root of the polynomial f(x) *exactly when* x − c *is a factor of* f(x).

If you know at least one root of a polynomial, the Factor Theorem can sometimes be used to obtain a factorization of the polynomial and hence other roots.

EXAMPLE Suppose we know that one of the roots of the polynomial $x^3 - 4x^2 + 2x + 4$ is the number 2 (verify that this is indeed the case). Then $x - 2$ must be a factor of $x^3 - 4x^2 + 2x + 4$. We can use synthetic or long division to determine the other factor. We find that $x^3 - 4x^2 + 2x + 4 = (x - 2)(x^2 - 2x - 2)$. We use this factorization to solve the equation:

$$x^3 - 4x^2 + 2x + 4 = 0$$
$$(x - 2)(x^2 - 2x - 2) = 0$$

Since a product is zero only when at least one of its factors is zero, this last equation is equivalent to

$$x - 2 = 0 \quad \text{or} \quad x^2 - 2x - 2 = 0$$

We can use the quadratic formula on $x^2 - 2x - 2 = 0$ and conclude that

$$x = 2 \quad \text{or} \quad x = \frac{-(-2) \pm \sqrt{(-2)^2 - 4 \cdot 1(-2)}}{2}$$

$$= \frac{2 \pm \sqrt{12}}{2} = \frac{2 \pm \sqrt{4 \cdot 3}}{2} = \frac{2 \pm 2\sqrt{3}}{2} = 1 \pm \sqrt{3}$$

Therefore the roots of $x^3 - 4x^2 + 2x + 4$ are 2, $1 + \sqrt{3}$, and $1 - \sqrt{3}$.

MULTIPLICITY OF ROOTS

If $f(x)$ is a polynomial and the number c is a root of $f(x)$, then as we have just seen, $x - c$ is a factor of $f(x)$. Now it may happen that $(x - c)^2$ or $(x - c)^3$ or some higher power of $x - c$ is also a factor of $f(x)$. Suppose the highest power of $x - c$ which is a factor of $f(x)$ is k; that is,

$$(x - c)^k \text{ is a factor of } f(x), \text{ but } (x - c)^{k+1} \text{ is not a factor}$$

Then we say that c is a **root of multiplicity** k.

EXAMPLE The roots of the polynomial $f(x) = (x - 3)^5(x + \frac{1}{2})^2(x - \frac{7}{3})^4(x + 9)$ are 3, $-\frac{1}{2}$, and $\frac{7}{3}$, and -9. Since $(x - 3)^5$ is a factor of f but $(x - 3)^6$ is not, 3 is a root of multiplicity 5. Since $(x + \frac{1}{2})^2 = (x - (-\frac{1}{2}))^2$ is a factor of f, but $(x - (-\frac{1}{2}))^3$ is not, $-\frac{1}{2}$ is a root of multiplicity 2. Similarly, $\frac{7}{3}$ is a root of multiplicity 4 and -9 is a root of multiplicity 1.

THE NUMBER OF ROOTS

The discussion of the discriminant on page 190 shows that a polynomial of degree 2 may have two, one, or no real roots. Similarly, in the preceding examples, we saw that the polynomial $x^7 - 7x^5 - 18x^3$ of degree 7 has exactly three real roots and the polynomial $x^3 - 4x^2 + 2x + 4$ of degree 3 has exactly three real roots. These examples are illustrations of the following useful fact:

> *A polynomial of degree* n *has at most* n *distinct real roots.*

Actually, a bit more is true, as we shall see in Chapter 8. If you count *all* the roots of a polynomial, including the complex ones, and if you count each root the same number of times as its multiplicity, then the total number is precisely the degree of the polynomial.

EXERCISES

A.1. (a) Which of the numbers 7, 3, -7, -3, 2 are roots of $f(x) = x^2 - 4x - 21$?
(b) Which of 2, -2, -1, 0 are roots of $g(x) = x^4 - 16$?
(c) Which of 2, 3, 0, -1 are roots of $h(x) = x^4 + 6x^3 - x^2 - 30x$?
(d) Which of -1, 1, 2, 3 are roots of $k(x) = 2x^3 - 3x^2 + 2x - 1$?

A.2. Find the roots of these first-degree polynomials:
(a) $f(x) = 10x - 3$ (c) $g(t) = 4t + 7$ (e) $f(x) = 3(x + 1) - 5$
(b) $h(x) = 9x - \frac{2}{3}$ (d) $h(x) = x$ (f) $g(x) = 11x + 2$

A.3. List the roots of these factored polynomials and state the multiplicity of each root.
    (a) $k(x) = x^{127}(x - 4)^{15}$
    (b) $h(t) = (4t - 1)^3(t + 7)^2$
    (c) $f(x) = (x + 100)^7(x + 7)^{100}(x - 2)^7(x - 7)^2$
    (d) $g(t) = (t + 5)(t - 5)(t - 5)^2(t + 10)^5$

A.4. Find the roots of these polynomials by *factoring*.
    (a) $f(x) = x^2 - x - 12$ **(e)** $f(x) = 2x^2 + 5x - 3$
    (b) $g(x) = 15x^2 + 7x - 2$ **(f)** $g(u) = 3u^2 + u - 4$
    (c) $h(x) = x^2 + 9x + 14$ **(g)** $h(x) = 5x^2 + 26x + 5$
    (d) $k(t) = 4t^2 + 9t + 2$ **(h)** $k(x) = 3x^2 - 10x - 8$

A.5. Find the roots of these polynomials by *completing the square*.
    (a) $f(x) = x^2 - 2x - 15$ **(c)** $h(x) = x^2 - 4x - 32$
    (b) $g(x) = x^2 - x - 1$ **(d)** $k(x) = 2x^2 + x - 15$

A.6. Use the *quadratic formula* to find the *real* roots of these polynomials.
    (a) $f(x) = x^2 + 8x + 15$ **(e)** $f(u) = 5u^2 + 8u + 2$
    (b) $g(x) = 4x^2 - 8x + 1$ **(f)** $g(u) = \frac{2}{3}u^2 - u + \frac{2}{3}$
    (c) $h(t) = 2t^2 + 4t + 1$ **(g)** $h(x) = 4x^2 - 3x - 5$
    (d) $k(t) = 3t^2 + 4t + 2$ **(h)** $k(x) = \frac{3}{4}x^2 + \frac{1}{4}x + \frac{1}{2}$

A.7. Compute the discriminant of each of these polynomials and state the *number* of real roots of the polynomial.
    (a) $f(x) = x^2 + 4x + 1$ **(d)** $k(t) = 9t^2 - 30t + 15$
    (b) $g(x) = 4x^2 - 4x - 3$ **(e)** $f(t) = 25t^2 - 70t + 49$
    (c) $h(x) = 9x^2 - 12x - 1$ **(f)** $g(t) = 49t^2 - 42t + 5$

B.1. Find all *real* roots of these polynomials. (*Note:* the methods of factoring cubics and higher degree polynomials presented in the Algebra Review at the end of the book may be helpful here.)
    (a) $f(x) = x^3 - 1$ **(e)** $f(u) = u^4 - 2u^2 - 3$
    (b) $h(x) = x^3 + 8$ **(f)** $g(u) = 24u^4 - 14u^2 - 5$
    (c) $g(x) = x^4 - 125x$ **(g)** $k(x) = 12x^5 + 2x^4 - 2x^3$
    (d) $k(x) = x^3 + 64$ **(h)** $f(t) = t^8 + t^4 - 2$

B.2. One root of each of these polynomials is given. Find all its real roots.
    (a) $f(x) = x^3 - 5x^2 + 7x - 2$; root 2
    (b) $g(x) = 3x^3 - 10x^2 - 23x - 10$; root -1
    (c) $f(x) = 8x^3 + 22x^2 - 7x - 3$; root -3
    (d) $h(x) = x^3 - 7x^2 + 9x + 5$; root 5
    (e) $f(x) = x^4 + 6x^3 - 7x^2$; root 0
    (f) $h(x) = 2x^3 - x^2 - 3x - 1$; root $-\frac{1}{2}$

B.3. Find a number k such that the given polynomial has exactly *one* real root.
    (a) $f(x) = x^2 + kx + 25$ **(d)** $f(x) = kx^2 + 24x + 16$
    (b) $g(x) = x^2 - kx + 49$ **(e)** $f(x) = 9x^2 - 30x + k$
    (c) $h(x) = kx^2 + 8x + 1$ **(f)** $g(u) = 4u^2 + ku + 9$

B.4. Find a number k such that:
  (a) 4 and 1 are the roots of $f(x) = x^2 - 5x + k$
  (b) 3 is a root of $f(x) = x^3 + x^2 + kx + 1$

B.5. Find a polynomial with the given degree n and the given roots and no other roots.
  (a) $n = 3$; roots 1, 7, -4 (c) $n = 6$; roots 1, 2, 3
  (b) $n = 3$; roots 1, -1 (d) $n = 5$; root 2

B.6. Find a polynomial $f(x)$ of degree 3 such that $f(10) = 17$ and the roots of $f(x)$ are 0, 5, and 8.

B.7. Given the equation $3x^2 + xy - 4y^2 = 9$, use the quadratic formula to
  (a) solve for x in terms of y (b) solve for y in terms of x

B.8. Find two consecutive integers whose product is 272. (*Hint:* if x is an integer, then $x + 1$ is the next consecutive integer.)

B.9. Find two consecutive integers the sum of whose squares is 313.

B.10. The diameter of a circle is 16 in. By what amount must the radius be decreased in order to decrease the area of the circle by 48π sq in.?

B.11. Let $ax^2 + bx + c$ be a polynomial with $a \neq 0$ and $b^2 - 4ac > 0$.
  (a) Use the quadratic formula to show that $ax^2 + bx + c = a(x - r)(x - s)$, where r and s are the roots of $ax^2 + bx + c$.
  (b) If $b^2 - 4ac = 0$, show that $ax^2 + bx + c = a(x - r)^2$ where r is the one real root of $ax^2 + bx + c$.

C.1. A favorite topic of science fiction writers is the possibility of animals either shrinking to an incredibly small size or growing to an incredibly large size. One reason these stories are scientifically unsound involves a certain polynomial function dealing with heat.

  Any living thing generates heat as a result of its metabolism. In addition, any animal that is warmer than its surroundings (as is usually the case with mammals) loses heat through its body's surface. In the long run, of course, these two processes must balance out.

  Let x be a number measuring the "general size" of an animal in inches. For a small shrew x might be .1, for a cat x might be 20, for a human 70, for a hippopotamus 150. The amount of heat generated by an animal is proportional to the mass of the animal, which is proportional to its volume, which is proportional to x^3. Consequently, the amount of heat generated is given by ax^3, where a is a positive number adjustable within certain bounds by the animal in question. The amount of heat lost is proportional to the amount of body surface, and is hence given by bx^2, where b is a positive number which is *fixed* for each animal. Thus the "heat balance formula" for a mammal of size x is given by $h(x) = ax^3 - bx^2$, where a and b must be chosen for each specific mammal. If the animal is to survive, x must be a root of $h(x)$, or at least close to a root (so that heat generated, ax^3, will be approximately equal to heat lost, bx^2). Otherwise the animal will bake itself or freeze to death.

(a) Which type of mammal has the hardest time staying warm in cold weather, very large ones or very small ones? Which type of mammal has the hardest time keeping cool in hot weather?

(b) Do you think that fur should cause more or less heat to be lost through the body surface? Do smaller or larger animals need fur the most? Does this fit the facts of which animals are the furriest? (*Note:* humans are quite large, as mammals go.)

(c) Suppose that $b = \frac{1}{2}$ for furry mammals and $b = 1$ for mammals with less hair. Suppose that a is given by the fraction of its own weight that an animal eats in a 1-day period. For each case given below, find the function $h(x)$ by inserting the proper values for a and b, and find the root of $h(x)$ to determine what size (in inches) this animal should be.

  (i) A furry creature that eats its own weight in food four times a day.

  (ii) A furry creature that eats its own weight every 5 days.

  (iii) A nonfurry creature that eats its own weight every 60 days.

(d) If a furry mammal were of size $x = .001$ in., how often would it have to eat its own weight to stay alive? (*Hint:* substitute the given values for b and x into the "heat balance formula," set it equal to 0, and solve for a. Then interpret your answer for a). Is it likely that such a creature could eat this much?

(e) If a hairless mammal were of size $x = 600$ in., how often would it be allowed to eat its own weight in order to maintain heat balance? Do you think such a creature could survive on this diet? If it could, would it be likely to undertake any energetic activities?

(f) As is always the case when mathematics is applied to a practical situation, many simplifying assumptions have been made in this problem. What are some factors that have been ignored in this problem? Can you think of problems other than diet that a 50-ft mammal might have? Highly readable discussions of these topics can be found in volume 2 of *The World of Mathematics* (ed. James R. Newman) in the article "On Being the Right Size" by J.B.S. Haldane and in *The Solar System and Back* by Isaac Asimov in the article "Just Right."

DO IT YOURSELF!

RATIONAL ROOTS

The real roots of a polynomial are either rational numbers or irrational numbers. A root which is a rational number is called a **rational root.** Here is a result that can be used to find the rational roots of certain polynomials.

RATIONAL ROOT TEST

Let $f(x) = a_n x^n + a_{n-1} x^{n-1} + \cdots + a_1 x + a_0$ *be a polynomial with* integer *coefficients* $a_n, a_{n-1}, \ldots, a_0$. *If a rational number* r/s *(expressed in lowest terms) is a root of* $f(x)$, *then* r *is a factor of the constant term* a_0 *and* s *is a factor of the leading coefficient* a_n.

The proof of the Rational Root Test depends on some elementary number theory. Since the proof sheds little light on how the result is actually *used,* we shall omit it.

EXAMPLE The polynomial $f(x) = 2x^4 + x^3 - 21x^2 - 14x + 12$ may or may not have any roots that are rational numbers. But *if* some rational number r/s is a root of $f(x)$, then the number s must be a factor of the leading coefficient 2. Since the only integer factors of 2 are $1, -1, 2,$ and $-2, s$ must be one of these numbers. Similarly, the number r must be a factor of the constant term 12. Therefore r *must* be one of $1, -1, 2, -2, 3, -3, 4, -4, 6, -6, 12,$ or -12. Consequently, the only possibilities for the quotient r/s (in lowest terms) are:

$$1, \quad -1, \quad 2, \quad -2, \quad 3, \quad -3, \quad 4, \quad -4, \quad 6, \quad -6, \quad 12, \quad -12, \quad \tfrac{1}{2}, \quad -\tfrac{1}{2}, \quad \tfrac{3}{2}, \quad -\tfrac{3}{2}$$

Obviously all 16 of these numbers can't be roots. But *if* there is any rational root, it *must* be one of these 16 numbers. By evaluating $f(x)$ at each of these numbers we can find out which, if any, actually *are* roots. Although this is a tedious procedure, it can be done. If you do it, you will find that

$$f(-3) = 2(-3)^4 + (-3)^3 - 21(-3)^2 - 14(-3) + 12 = 0$$
$$f(\tfrac{1}{2}) = 2(\tfrac{1}{2})^4 + (\tfrac{1}{2})^3 - 21(\tfrac{1}{2})^2 - 14(\tfrac{1}{2}) + 12 = 0$$

and that none of the other numbers on the list are roots. Therefore -3 and $\tfrac{1}{2}$ are the only *rational* roots of $f(x)$.

What about other roots of $f(x)$? By the Factor Theorem we know that $x - (-3) = x + 3$ and $x - \tfrac{1}{2}$ are factors of $f(x)$. Therefore $f(x) = q(x)(x + 3)(x - \tfrac{1}{2})$ for some polynomial $q(x)$. We can determine $q(x)$ by dividing $f(x) = 2x^4 + x^3 - 21x^2 - 14x + 12$ by $(x + 3)(x - \tfrac{1}{2}) = x^2 + \tfrac{5}{2}x - \tfrac{3}{2}$:

$$
\begin{array}{r}
2x^2 - 4x - 8 \\
x^2 + \tfrac{5}{2}x - \tfrac{3}{2} \overline{\smash{\big)}\, 2x^4 + x^3 - 21x^2 - 14x + 12} \\
\underline{2x^4 + 5x^3 - 3x^2 } \\
- 4x^3 - 18x^2 - 14x + 12 \\
\underline{- 4x^3 - 10x^2 + 6x } \\
- 8x^2 - 20x + 12 \\
\underline{- 8x^2 - 20x + 12} \\
0
\end{array}
$$

Thus

$$f(x) = 2x^4 + x^3 - 21x^2 - 14x + 12 = (x^2 + \tfrac{5}{2}x - \tfrac{3}{2})(2x^2 - 4x - 8)$$
$$= (x + 3)(x - \tfrac{1}{2})\,2(x^2 - 2x - 4)$$

The roots of $f(x)$ are the solutions of $2(x + 3)(x - \tfrac{1}{2})(x^2 - 2x - 4) = 0$, which is equivalent to

$$x + 3 = 0 \quad \text{or} \quad x - \tfrac{1}{2} = 0 \quad \text{or} \quad x^2 - 2x - 4 = 0$$

Using the quadratic formula on the last equation, we see that

$$x = -3 \quad \text{or} \quad x = \tfrac{1}{2} \quad \text{or} \quad x = \frac{2 \pm \sqrt{4 - 4 \cdot 1(-4)}}{2}$$

$$= \frac{2 \pm \sqrt{20}}{2} = \frac{2 \pm 2\sqrt{5}}{2} = 1 \pm \sqrt{5}$$

Therefore all the roots of $f(x)$ are -3, $\tfrac{1}{2}$, $1 + \sqrt{5}$, and $1 - \sqrt{5}$.

EXAMPLE To find the rational roots of $f(x) = x^5 + 4x^4 + x^3 - x^2$ (or any polynomial with constant term 0), begin by factoring out the highest possible power of x from *all* terms:

$$f(x) = x^5 + 4x^4 + x^3 - x^2 = x^2(x^3 + 4x^2 + x - 1)$$

Obviously, $x = 0$ is a rational root. The other roots of $f(x)$ are the roots of the factor $h(x) = x^3 + 4x^2 + x - 1$. Now if $h(x)$ has a rational root r/s, then

r is a factor of the constant term -1
s is a factor of the leading coefficient 1

Therefore the only possibilities for both r and s are 1 and -1. Hence the only possibilities for r/s are 1 and -1 as well. But

$$h(1) = 1^3 + 4 \cdot 1^2 + 1 - 1 = 5 \quad \text{and} \quad h(-1) = (-1)^3 + 4(-1)^2 + (-1) - 1 = 1$$

So neither of the possible rational roots is actually a root. If $h(x)$ has any real roots, they must be *irrational* numbers. Therefore the original polynomial $f(x)$ has $x = 0$ as a rational root and no other rational roots.

APPROXIMATION OF ROOTS

The roots of polynomials of high degree can be quite difficult to find, even *with* the help of a computer. Consequently, it is often necessary to *approximate* the roots of a given polynomial to some stated degree of accuracy. The following theorem provides one method of doing this.

LOCATION THEOREM

If f(x) *is a polynomial function and* c *and* d *are real numbers such that* f(c) *and* f(d) *have opposite signs, then* f(x) *has at least one root between* c *and* d.

A rigorous proof of this fact is beyond the scope of this book, but the intuitive idea underlying the proof is quite clear. The fact that $f(c)$ and $f(d)$ have opposite signs (one positive, the other negative) means that on the graph of $f(x)$, the points $(c, f(c))$ and $(d, f(d))$ lie on *opposite sides* of the x-axis, as shown in Figure 4-1.

$$f(c) < 0 \text{ and } f(d) > 0 \qquad\qquad f(c) > 0 \text{ and } f(d) < 0$$

Figure 4-1

The graph of $f(x)$ is known to be a smooth unbroken curve, passing through $(c, f(c))$ and $(d, f(d))$. There is no way that an unbroken curve (which is the graph of a function) can travel from $(c, f(c))$ to $(d, f(d))$ *without* crossing the x-axis at least once between $x = c$ and $x = d$. (If you doubt it, try to draw one.) Therefore the graph of $f(x)$ crosses the x-axis for at least one point $(b, f(b))$, with b between c and d. Since the second coordinate of any point on the x-axis is 0, we must have $f(b) = 0$; that is, b is a root of $f(x)$.

EXAMPLE   If we apply the Rational Root Test to the polynomial $f(x) = x^3 + 2x^2 - 7x + 1$, we find that the only possibilities (± 1) are *not* roots. On the other hand, we have

x	-4	-3	-2	-1	0	1	2
$f(x) = x^3 + 2x^2 - 7x + 1$	-3	13	15	9	1	-3	3

Since $f(-4) = -3$ and $f(-3) = 13$ have opposite signs, the Location Theorem guarantees that $f(x)$ has a root betweeen -4 and -3. Similarly, $f(0) = 1$ and $f(1) = -3$ have opposite signs and hence there is a root between 0 and 1. Finally, since $f(1) = -3$ and $f(2) = 3$ have opposite signs, there is a root between 1 and 2. Since $f(x)$ has at most three roots, there is exactly one root between -4 and -3, one between 0 and 1, and one between 1 and 2.

If you want a better approximation of one of these roots —say the one between 0 and 1—you need only compute $f(x)$ for several values of x between 0 and 1. Using a calculator, we evaluate f at $x = .1$, $x = .2$, $x = .3$, and so on. We immediately find that $f(.1) = .321$ and $f(.2) = -.312$. Since $f(.1)$ and $f(.2)$ have opposite signs, the root actually lies between .1 and .2. For a better approximation, we compute values of $f(x)$ in increments of .01:

x	.11	.12	.13	.14	.15
$f(x) = x^3 + 2x^2 - 7x + 1$	.255531	.190528	.125997	.061944	$-.001625$

Since $f(.14)$ and $f(.15)$ have opposite signs, the root lies between .14 and .15. Further calculation shows that

$$f(.1497) \approx .00027 \quad \text{and} \quad f(.1498) \approx -.00036$$

Thus the root lies between .1497 and .1498. This means that .1497 is within .0001 unit of the root. If necessary, even greater accuracy could be obtained by further calculation.

EXERCISES

A.1. Find all rational roots of the given polynomial.
  (a) $x^3 + 3x^2 - x - 3$
  (b) $x^3 - x^2 - 3x + 3$
  (c) $x^3 + 5x^2 - x - 5$
  (d) $3x^3 + 8x^2 - x - 20$
  (e) $2x^3 + 5x^2 - 11x + 4$
  (f) $2x^3 - 3x^2 - 7x - 6$

A.2. Find all real roots of these polynomials. (*Hint:* first find all the rational roots.)
  (a) $2x^3 - 5x^2 + x + 2$
  (b) $t^4 - t^3 + 2t^2 - 4t - 8$
  (c) $6x^3 - 11x^2 + 6x - 1$
  (d) $z^3 + z^2 + 2z + 2$
  (e) $x^4 - x^3 - x^2 - x - 2$
  (f) $3x^5 + 2x^4 - 7x^3 + 2x^2$
  (g) $3y^3 - 2y^2 + 3y - 2$
  (h) $x^5 - x^3 + x$

A.3. Find all rational roots of the given polynomial.
  (a) $f(x) = \frac{1}{12}x^3 - \frac{1}{12}x^2 - \frac{2}{3}x + 1$ [*Hint:* the equation $f(x) = 0$ has the same solutions as $(12)f(x) = 0$ (Why?). But $(12)f(x) = x^3 - x^2 - 8x + 12$ has integer coefficients; use the Rational Root Test.]
  (b) $\frac{2}{3}x^4 + \frac{1}{2}x^3 - \frac{5}{4}x^2 - x - \frac{1}{6}$
  (c) $\frac{1}{3}x^4 - x^3 - x^2 + \frac{13}{3}x - 2$
  (d) $\frac{1}{3}x^3 - \frac{1}{2}x^2 - \frac{1}{6}x + \frac{1}{6}$
  (e) $\frac{2}{3}x^3 - \frac{1}{2}x^2 + \frac{2}{3}x - \frac{1}{2}$

A.4. Locate *one* (irrational) root of each of the following polynomials between two consecutive integers.
  (a) $g(x) = 12x^3 - 28x^2 - 7x - 10$
  (b) $f(x) = x^5 + x^2 - 7$
  (c) $h(x) = x^3 + 4x^2 + 10x + 15$
  (d) $k(x) = x^3 - 3x^2 - 6x + 9$
  (e) $f(t) = t^4 + 2t^3 - 10t^2 - 6t + 1$
  (f) $g(u) = u^4 + 8u^3 + 17u^2 + 4u - 2$

B.1. Use a calculator to approximate at least one irrational root of each of the polynomials in Exercise A.4 to within .01 unit.

C.1. Prove that $\sqrt{2}$ is not a rational number. (*Hint:* use the Rational Root Test on $x^2 - 2$.)

3. GRAPHS OF POLYNOMIAL FUNCTIONS

As we saw in Section 4 of Chapter 2, the graph of a first-degree polynomial function, such as $f(x) = 3x - 7$ or $g(x) = \frac{7}{3}x + 5$, is just a straight line. The graph of every second degree polynomial is a parabola. For example, see Figure 4-2.

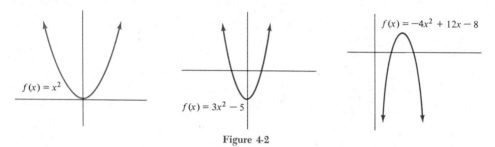

Figure 4-2

A complete discussion of such graphs was given in Section 6 of Chapter 3. If you have not already done so, you may wish to read that section now. However, it is not necessary to read that section in order to understand the material below.

THE FUNCTION $f(x) = x^n$

The simplest kind of higher degree polynomial function is one whose rule is given by a power of x, such as x^3, x^4, x^5, x^6, and so on. The graphs of such functions are easily determined.

EXAMPLE In order to obtain the graph of $g(x) = x^4$, we plot some points and find that the graph looks like the one in Figure 4-3.

x	$g(x) = x^4$
-3	81
-2	16
-1	1
$-\frac{1}{2}$	$\frac{1}{16}$
0	0
1	1
2	16
3	81

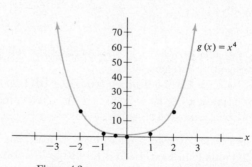

Figure 4-3

This graph is *not* a parabola, but it does have an upward-opening cup shape which is similar to the graph of the parabola $f(x) = x^2$ shown in Figure 4-2. The graphs of $h(x) = x^6$ and other *even* powers of x all have this same general shape. See Exercise A.1 for more details.

EXAMPLE   The graph of $g(x) = x^3$ may be obtained, as usual, by plotting some points. (See Figure 4-4.)

x	$g(x) = x^3$
-4	-64
-3	-27
-2	-8
-1	-1
$-\frac{1}{2}$	$-\frac{1}{8}$
0	0
$\frac{1}{2}$	$\frac{1}{8}$
1	1
2	8
3	27
4	64

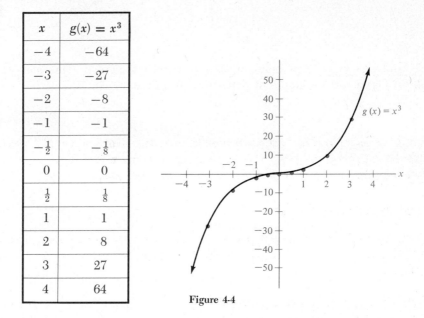

Figure 4-4

The graphs of $h(x) = x^5$ and other *odd* powers of x have the same general shape as the graph of $g(x) = x^3$. See Exercise A.2 for more details.

BASIC PROPERTIES OF POLYNOMIAL GRAPHS

For more complicated polynomial functions, the best we can do without calculus is to present some general principles and suggest a basic procedure to follow. We begin with a discussion of certain common properties shared by *all* graphs of polynomial functions.

Extent   Since the domain of a polynomial function is always the set of all real numbers, the graph extends on forever both to the left and to the right.

Continuity The graph of a polynomial function is a smooth, unbroken, continuous curve, such as the ones shown in Figure 4-5.

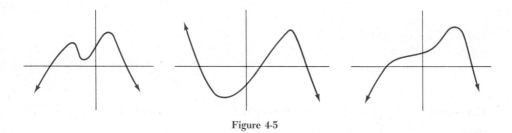

Figure 4-5

There can be no jumps, gaps, holes, or sharp corners on the graph. Thus *none* of the graphs in Figure 4-6 are graphs of polynomial functions.

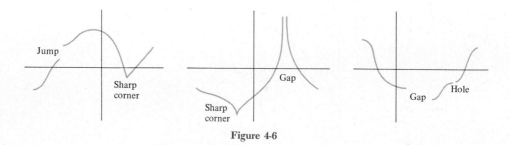

Figure 4-6

We have tacitly assumed these facts in the previous examples where we plotted only a few points and then connected them by a curve.

Behavior When |x| Is Large When you move very far to the right or very far to the left along the x-axis, the graph of a polynomial function begins to move farther and farther *away* from the x-axis. For example, see Figure 4-7.

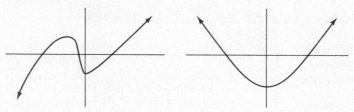

Figure 4-7

Graphs such as the ones in Figure 4-8 *cannot* be the graphs of polynomial functions.

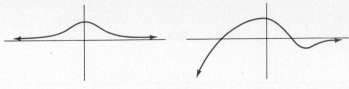

Figure 4-8

The reason why polynomial graphs behave this way is discussed in the DO IT YOUR-SELF! segment at the end of this section.

Maxima and Minima   The graph of a typical polynomial function *may* have several "peaks" and "valleys" as shown in Figure 4-9.

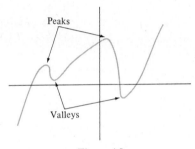

Figure 4-9

The technical term for a peak is a **relative maximum** (plural, maxima). The technical term for a valley is a **relative minimum** (plural, minima). The word "relative" is often omitted. It refers to the fact that a peak or relative maximum is not necessarily the highest point on the entire graph, but only the highest of all nearby points. Similarly, a valley or relative minimum is the lowest point of all the points near it. In calculus it is proved that

> *The total number of relative maxima and minima on the graph of a polynomial function of degree* n *is at most* n − 1.

We have already seen (in Chapter 3) one common example of this fact. The graph of every polynomial of degree 2 is a parabola. As Figure 4-2 shows, every parabola has exactly *one* maximum or exactly *one* minimum. More generally, the graph of the fifth-degree polynomial function $f(x) = x^5 - 6x^4 + 2x^2 + 1$ has a total of at most *four* relative maxima and minima.

Meeting the x-axis The points on the x-axis are precisely the points with second coordinate zero. If a point $(c, f(c))$ on the graph of a function $f(x)$ is also on the x-axis, then the second coordinate of $(c, f(c))$ must be zero; that is, $f(c) = 0$. Thus the graph of a polynomial function meets the x-axis at $x = c$ exactly when c is a *root* of the polynomial. As we saw in Section 2, a polynomial of degree n has at most n real roots. Therefore

> *The graph of a polynomial function of degree* n *meets the x-axis at most* n *times.*

To say that the graph of a function *meets* the x-axis does not necessarily mean that it *crosses* the axis. For instance, the graph of $f(x) = x^2$ shown above meets the x-axis at $x = 0$ but does not cross the axis. On the other hand, the graph of $g(x) = x^3$ given above does cross the x-axis at $x = 0$. One difference between these two functions is that $x = 0$ is a root of $f(x) = x^2$ of multiplicity 2 (an even number), while $x = 0$ is a root of $g(x) = x^3$ of multiplicity 3 (an odd number). More generally, suppose the number c is a root of a polynomial function:

> *If the multiplicity of the root* c *is an* odd *number, then the graph of the function crosses the* x-axis *at* x = c.
>
> *If the multiplicity of the root* c *is an* even *number, then the graph of the function meets, but does not cross, the* x-axis *at* x = c.

The reason why this fact is true will become clear in the examples below. For now we just illustrate some of the possibilities in Figure 4-10.

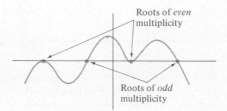

Figure 4-10

Suppose c and d are both roots of the polynomial $f(x)$ and that there are *no* roots *between* c and d. Then the graph of $f(x)$ meets the x-axis at $x = c$ and $x = d$, but *not* at any point in between. Since the graph of $f(x)$ between $x = c$ and $x = d$ is a continuous, unbroken curve that does *not* meet the *x*-axis, *it must lie entirely on one side of the x-axis,* as illustrated in Figure 4-11.

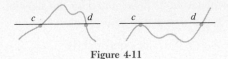

Figure 4-11

This fact is so useful that it's worth repeating:

> *If* c *and* d *are roots of the polynomial function* f, *and* f *has no roots between* x = c *and* x = d, *then the graph of* f *between* x = c *and* x = d *lies either completely above the* x-*axis or completely below the* x-*axis.*

GRAPHING A POLYNOMIAL FUNCTION

Until the techniques of calculus are available, we suggest that you use the four-step procedure outlined in the following examples for graphing polynomial functions. It won't work well every time, but will usually be satisfactory whenever the polynomial can be completely factored.

EXAMPLE   The *first step in graphing* $f(x) = 2x^3 - x^2 - 15x$ is to *find all its real roots* and *write* $f(x)$ *in factored form:*

$$f(x) = 2x^3 - x^2 - 15x = x(2x^2 - x - 15),$$
$$f(x) = x(x - 3)(2x + 5) = x(x - 3)2(x + \tfrac{5}{2}),$$
$$f(x) = 2x(x - 3)(x + \tfrac{5}{2})$$

The roots are obviously $x = 0$, $x = 3$, and $x = -\tfrac{5}{2}$. They divide the *x*-axis into four intervals, as shown in Figure 4-12.

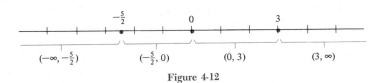

Figure 4-12

We know that in the interval between any two adjacent roots the graph either lies entirely above the *x*-axis or entirely below it. Likewise, the graph lies entirely on one side of the *x*-axis over the intervals $(-\infty, -\tfrac{5}{2})$ and $(3, \infty)$, since if it crossed over, there would be another root.

   The second step is to determine on which side of the x-axis the graph lies for each of these intervals. To do this, we shall use the fact that the point $(x, f(x))$ lies *above* the

x-axis exactly when its second coordinate is *positive;* that is, $f(x) > 0$. Similarly, $(x, f(x))$ lies *below* the x-axis when $f(x) < 0$.

Observe that for each x, $f(x) = 2x(x - 3)(x + \frac{5}{2})$ is a product of three factors. The sign of $f(x)$ (positive or negative) is determined by the signs of the three factors. For example, the factor $x + \frac{5}{2}$ is positive exactly when

$$x + \tfrac{5}{2} > 0 \qquad \text{or equivalently} \qquad x > -\tfrac{5}{2}$$

Thus $(x + \frac{5}{2})$ is positive when x is in any one of the intervals $(-\frac{5}{2}, 0)$, $(0, 3)$ or $(3, \infty)$ and $(x + \frac{5}{2})$ is negative when x is in the interval $(-\infty, -\frac{5}{2})$, as shown in Figure 4-13.

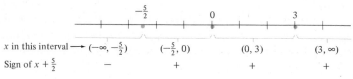

Figure 4-13

Similarly, $2x$ is positive whenever $x \geq 0$, that is, when x is in $(0, 3)$ or $(3, \infty)$. An analogous argument works for the factor $x - 3$, as summarized in the chart shown in Figure 4-14.

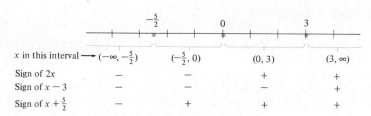

Figure 4-14

This chart shows, for example, that in the interval $(-\frac{5}{2}, 0)$, the function $f(x) = 2x(x - 3)(x + \frac{5}{2})$ is a product of one positive and two negative factors, so that $f(x)$ is positive. Similar arguments work in the other intervals, as summarized in the chart shown in Figure 4-15.

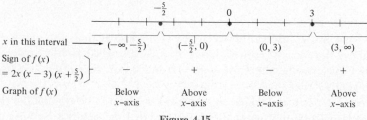

Figure 4-15

The third step is to use the information above to make a rough sketch of the graph. Since $f(x)$ changes sign at each of its roots, the graph must *cross* the x-axis at each root. (Note that the multiplicity of each root is the odd number 1). So we obtain the rough sketch shown in Figure 4-16.

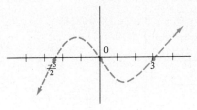

Figure 4-16

We claim that this sketch has the correct basic shape, except for the exact location and height of the maxima and minima (peaks and valleys). For we noted above that the graph of $f(x)$ is an *unbroken* curve passing through $(-\frac{5}{2}, 0)$ and $(0, 0)$ and lying *above* the x-axis between these points. Thus it must rise and then fall at least once between these points as shown in the rough sketch above. Similarly, since the graph passes through $(0, 0)$ and $(3, 0)$ and lies *below* the x-axis, it must fall, then rise at least once between these points, as shown in the rough sketch. This accounts for at least one relative maximum (peak) and one relative minimum (valley). But $f(x) = 2x^3 - x^2 - 15x$ has degree 3 and therefore has *at most* a total of two relative maxima and minima. Hence there can be *no more* maxima or minima.

Finally, if $x > 3$ our rough sketch indicates that the graph continues to rise forever. This must be the case, for if the graph fell even briefly, we would have a situation such as the one in Figure 4-17.

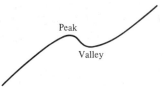

Peak

Valley

Figure 4-17

In this case there would be additional maxima and minima. Similarly, the graph must always fall as x moves left from $-\frac{5}{2}$. Likewise, there can be no other "wiggles" in the graph since any wiggle, such as $\sim$, would result in additional maxima or minima.

The fourth step is to sketch the graph accurately. This can often be done by plotting only 6 to 8 additional points, once you know the correct basic shape of the graph (see Figure 4-18).

x	$f(x) = 2x^3 - x^2 - 15x$
-3	-18
$-\frac{5}{2}$	0
-2	10
-1	12
0	0
1	-14
2	-18
3	0
4	52

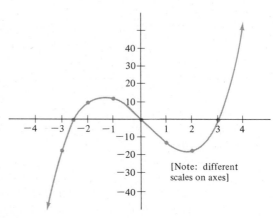

[Note: different scales on axes]

Figure 4-18

One problem with this graph is the location of the maxima and minima. As sketched above, they are very near $(-1, 12)$ and $(2, -18)$, respectively. As you will see in calculus, the actual relative maximum occurs when $x = (1 - \sqrt{91})/6$ (roughly, the point on the graph is $(-1.42, 13.56)$); the actual relative minimum occurs when $x = (1 + \sqrt{91})/6$ (roughly, the point on the graph is $(1.76, -18.59)$).

EXAMPLE If the function $g(x) = (x + 3)^2(x^2 - 2)(x - 2)$ were multiplied out, it would be a fifth-degree polynomial. However, it is convenient to leave it as is. *First*, we factor $g(x)$ further:

$$g(x) = (x + 3)^2(x^2 - 2)(x - 2) = (x + 3)^2(x + \sqrt{2})(x - \sqrt{2})(x - 2)$$

The roots of $g(x)$ are $-3, -\sqrt{2}, \sqrt{2}$, and 2. They divide the x-axis into five intervals, as shown in Figure 4-19.

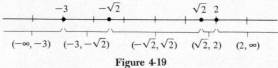

Figure 4-19

Second, we determine whether the graph of g lies above or below the x-axis over each of these intervals by finding the sign of $g(x)$. As above, this is done by examining the sign of each of the factors. Note that for *any* number x, the factor $(x + 3)^2$ is always positive. The signs of the other factors are easily determined, as indicated in Figure 4-20.

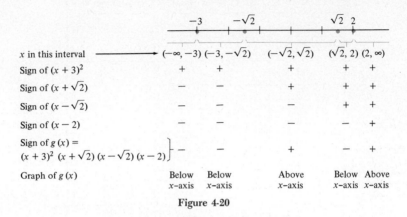

	$(-\infty, -3)$	$(-3, -\sqrt{2})$	$(-\sqrt{2}, \sqrt{2})$	$(\sqrt{2}, 2)$	$(2, \infty)$
Sign of $(x + 3)^2$	+	+	+	+	+
Sign of $(x + \sqrt{2})$	−	−	+	+	+
Sign of $(x - \sqrt{2})$	−	−	−	+	+
Sign of $(x - 2)$	−	−	−	−	+
Sign of $g(x) =$ $(x + 3)^2 (x + \sqrt{2}) (x - \sqrt{2}) (x - 2)$	−	−	+	−	+
Graph of $g(x)$	Below x–axis	Below x–axis	Above x–axis	Below x–axis	Above x–axis

x in this interval ⟶

Figure 4-20

Third, we use this information to make a rough sketch of the graph, as shown in Figure 4-21.

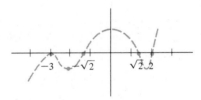

Figure 4-21

This sketch *does* show the correct basic shape since $g(x)$ is a fifth-degree polynomial and thus has at most four relative maxima or minima. Therefore there cannot be additional peaks, valleys, or wiggles in the graph.

The graph does *not* cross the x-axis at $x = -3$ since the analysis above shows that $g(x)$ is negative on *both* sides of $x = -3$. Note that $x = -3$ is a root of *even multiplicity* (2) whereas the roots at which the graph crosses the x-axis (namely, $x = -\sqrt{2}$, $x = \sqrt{2}$, and $x = 2$) are roots of *odd multiplicity* (1).

The *final step* is to plot some additional points to obtain a more accurate sketch of the graph, as shown in Figure 4-22.

x	$g(x) = (x + 3)^2(x^2 - 2)(x - 2)$
-4	-84
-3	0
-2	-8
$-\sqrt{2}$	0
-1	12
0	36
1	16
$\sqrt{2}$	0
1.5	-2.53125
1.8	-5.71392
2	0
3	252

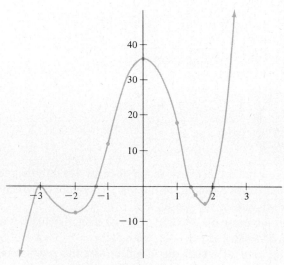

Figure 4-22

If you can't factor a given polynomial completely (that is, you don't know all its roots), then the procedure outlined in the preceding examples may not be of much help. In such cases, all that you can do is to plot a reasonable number of points and use these, together with the basic properties of polynomial functions discussed above, to make an educated guess about the shape of the graph.

EXERCISES

A.1. (a) Sketch careful graphs of $f(x) = x^2$, $g(x) = x^4$, $h(x) = x^6$, for $-3 \leq x \leq 3$ on the same set of coordinate axes.

(b) Use part (a) to estimate what the graph of $k(x) = x^8$ will look like for $-3 \leq x \leq 3$. Sketch the graph of $k(x)$ without plotting any points first.

A.2. (a) Sketch careful graphs of $f(x) = x^3$, $g(x) = x^5$, $h(x) = x^7$, for $-3 \leq x \leq 3$ on the same set of coordinate axes.

(b) Use part (a) to estimate what the graph of $k(x) = x^9$ will look like for $-3 \leq x \leq 3$. Sketch the graph of $k(x)$ without plotting any points first.

A.3. Sketch the graph of these functions. (*Hint:* the quickest way is to use the techniques of Section 6 of Chapter 3 and the graphs on pages 200–201.)

(a) $f(x) = x^3 + 5$ (c) $h(x) = -x^3 - 3$ (e) $k(x) = (x - 2)^4$

(b) $g(x) = -2x^3$ (d) $k(x) = \dfrac{x^3}{2} + 1$ (f) $f(x) = (x + 5)^3$

A.4. Figure 4-23 shows the graphs of several functions. Which functions are definitely not polynomial functions? Which functions could *possibly* be polynomial functions?

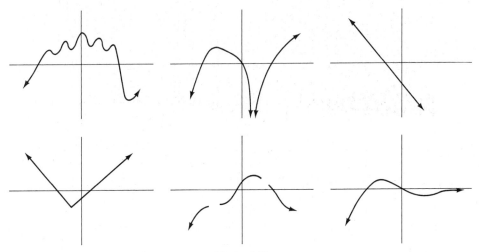

Figure 4-23

A.5.   Which of the graphs in Figure 4-24 could possibly be the graph of a polynomial function of degree 3? of degree 4? of degree 5?

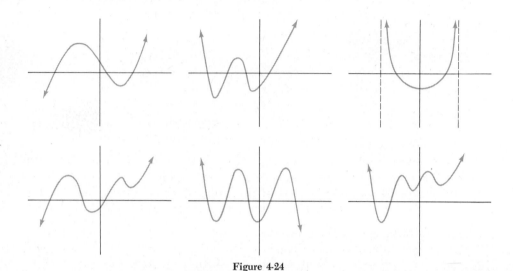

Figure 4-24

B.1.   Do a rough sketch of the graph of each of these functions. (Check signs, but don't plot any points except the roots. Then see Exercise B.2.)
(a) $f(x) = (x - 1)(x + 5)(x + 2)$
(b) $g(x) = (x - 2)^2(x - 4)$
(c) $h(x) = -x(x - 3)^2(2x + 3)$
(d) $k(x) = x^2(x + 1)(3x - 7)$
(e) $f(x) = \frac{1}{6}(x - 1)(x - 2)(x - 3)(x - 4)$
(f) $g(x) = -(x - 1)(x + 2)^2x$

B.2.   Plot 6–8 points on the graph of each of the functions in Exercise B.1 to see how well the rough sketch you did there corresponds to the actual graph.

B.3.   Is the given function increasing (= graph rising from left to right) or decreasing (= graph falling from left to right) on the given interval?
(a) $f(x) = (x - 1)^2(x + 2)(x - 3)$ on $(3, \infty)$
(b) $f(x) = (x - 1)^2(x + 2)(x - 3)$ on $(-\infty, -2)$
(c) $g(x) = (x - 6)(x + 5)^2(x - 3)x$ on $(-\infty, -5)$
(d) $g(x) = (x - 6)(x + 5)^2(x - 3)x$ on $(6, \infty)$

B.4.   (a)   Show that $x^2 - 6x + 7$ is a factor of $x^4 - 6x^3 + 6x^2 + 6x - 7$.
(b)   Graph the function $f(x) = x^4 - 6x^3 + 6x^2 + 6x - 7$.

B.5. Use the four-step method presented above to sketch a reasonably accurate graph of each of these functions.
  (a) $f(x) = -x^3 - x^2 + 2x$
  (b) $g(x) = x^2 - 6x + 5$
  (c) $h(x) = x^3 - 4x$
  (d) $k(x) = x^4 - 5x^2 + 5$
  (e) $f(x) = x^4 - 6x^2 + 8$
  (f) $g(x) = x^4 - x^2$

B.6. Graph these functions.
  (a) $f(x) = x^3 - 3x^2 - 4x$
  (b) $g(x) = x^3 - 3x^2 - 4x + 2$ (*Hint:* see part (a) and p. 146.)
  (c) $h(x) = x^3 - 3x^2 - 4x - 6$.

B.7. Graph these functions. [*Hint:* in each case, first use the Rational Root Test (p. 196) to find all rational roots; then use the Factor Theorem (p. 191) to factor and find the rest of the roots.]
  (a) $f(x) = x^3 - 3x^2 - x + 3$
  (b) $g(x) = x^3 - 3x^2 + x + 1$
  (c) $f(x) = x^3 - 4x^2 + 2x + 3$
  (d) $h(x) = x^4 - 4x^3 + 4x - 1$
  (e) $k(x) = x^4 - 3x^3 - 6x^2 + 14x + 12$
  (f) $g(x) = 4x^4 + 12x^3 - x^2 - 3x$

DO IT YOURSELF!

WHAT HAPPENS FOR LARGE x?

It is important to realize just how *explosive* the growth of the function $f(x) = x^n$ is when $x > 1$. Figure 4-25 shows a combined graph of

$$p(x) = x, \qquad q(x) = x^2, \qquad r(x) = x^3, \qquad s(x) = x^4$$

for positive values of x.

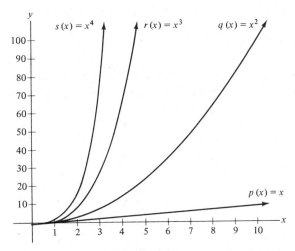

Figure 4-25

The situation only gets worse when x gets still larger. For instance, Figure 4-26 is a combined graph of these same functions from 0 to 100, where the scale has been adjusted to accommodate x^4.

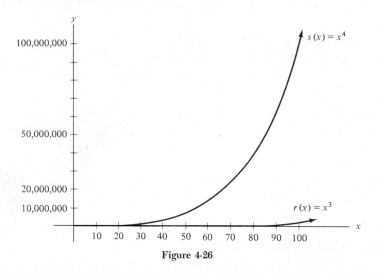

Figure 4-26

Here the values of x^3, x^2 and x are so insignificant in comparison to x^4 that their graphs hardly even show up! We can barely see the graph of x^3 beginning to creep above the x-axis. It would take a microscope to distinguish the graph of x^2 from the x-axis, and an electron microscope to find the graph of $p(x) = x$ on this scale.

The moral here is that each power of x makes lower powers look sickly and is humbled in turn by the next higher power. Thus for large values of x, the highest power of x in a polynomial is extremely important.

Analogous remarks hold when x is a negative number which is large in absolute value. For instance, when $|x|$ is large, then $|x^4|$ is very much larger than $|x^3|$, which is in turn very much larger than $|x^2|$.

EXAMPLE When x is large in absolute value, the graph of $f(x) = x^3 - 3x^2 + 5x + 4$ will be almost the same as that of x^3, turning sharply upwards to the right, sharply downwards to the left. The reason is that when x is large in absolute value, the terms $-3x^2$, $5x$, and 4 will be insignificant *compared* with x^3, as shown in the table.

x	100	500	1,000	$-1{,}500$
$5x$	500	2,500	5,000	$-7{,}500$
$-3x^2$	$-30{,}000$	$-750{,}000$	$-3{,}000{,}000$	$-6{,}750{,}000$
x^3	1,000,000	125,000,000	1,000,000,000	$-3{,}375{,}000{,}000$
$f(x) = x^3 - 3x^2 + 5x + 4$	970,504	124,252,504	997,005,004	$-3{,}381{,}757{,}496$

EXERCISES

A.1. **(a)** Make a table of values for the functions $f(x) = x^3$ and $g(x) = 10x^2$ to show that x^3 dominates $10x^2$ for large values of x.

(b) Repeat part (a) for $f(x)$ and $L(x) = 200x^2$. How large (approximately) does x have to go before x^3 exceeds $200x^2$?

(c) Make a table of values for $k(x) = 50x^3$ and $t(x) = \frac{1}{100}x^4$. Which function dominates the other for large values of x? (Make sure you take x large enough to see what's going on.)

A.2. Calculate a table of values for the following polynomials for *large* values of x. Your table should demonstrate that for large values of x, the value of a polynomial of degree n is approximately equal to the value of x^n.

(a) $x^3 - 7x^2 + 10x - 1$ **(b)** $x^4 + 10x^3 + x^2 - x + 5$ **(c)** $x^5 - x^4 + 1$

B.1. Do the graphs of $f(x) = 10x^2$ and $g(x) = \frac{1}{10}x^3$ cross at any point on the right side of the y-axis? Explain.

4. RATIONAL FUNCTIONS

A **rational function** is a function whose rule is a quotient of two polynomials, such as

$$f(x) = \frac{1}{x}, \qquad g(x) = \frac{4x - 3}{2x + 1}, \qquad h(x) = \frac{5}{x - 3},$$

$$k(x) = \frac{2x^2 + 5x + 2}{2x + 7}, \qquad t(x) = \frac{x^3 - 2}{x - 1}$$

In particular, every polynomial function is also a rational function. For example, $f(x) = x^2 + 3$ can be written as $f(x) = (x^2 + 3)/1$.

Let $f(x) = g(x)/h(x)$ be a rational function [with $g(x)$ and $h(x)$ polynomials]. If c is a real number, then $f(c)$ is a well-defined real number whenever

$$f(c) = \frac{g(c)}{h(c)} \text{ is a real number}$$

The only time $g(c)/h(c)$ is *not* a real number occurs when the denominator $h(c)$ is 0. In other words, $f(c) = g(c)/h(c)$ is not defined when c is a root of the polynomial $h(x)$. Therefore

> The domain of a rational function $f(x) = g(x)/h(x)$ *is the set of all real numbers that are* not *roots of the denominator* $h(x)$.

EXAMPLE The domain of the function $k(x) = \dfrac{x - 2}{x^2(x - 3)(x + 1)}$ is the set of all real numbers *except* 0, 3, and -1, the roots of the denominator.

EXAMPLE You are probably accustomed to canceling factors in algebraic expressions, such as

$$\frac{x^2 - 1}{x - 1} = \frac{(x + 1)(x - 1)}{x - 1} = x + 1$$

Nevertheless, the *function* with rule $p(x) = x + 1$ and the *function* with rule $q(x) = \frac{x^2 - 1}{x - 1}$ are *not* the same function. The difference occurs when $x = 1$. Then

$$p(1) = 1 + 1 = 2, \quad \text{but} \quad q(1) = \frac{1^2 - 1}{1 - 1} = \frac{0}{0}$$

Thus $q(1)$ is not defined so that the number 1 is *not* in the domain of the function $q(x)$. But 1 *is* in the domain of the function $p(x) = x + 1$. It is true that for any number c *except* 1, the two functions have the same value; for in this case $c - 1 \neq 0$, and the factor $c - 1$ *can* be canceled:

$$q(c) = \frac{c^2 - 1}{c - 1} = \frac{(c + 1)(c - 1)}{c - 1} = c + 1 \quad \text{and} \quad p(c) = c + 1$$

But the difference for $x = 1$ makes $p(x)$ and $q(x)$ *different functions.*

GRAPHS OF LINEAR RATIONAL FUNCTIONS

The simplest rational functions to graph are functions such as

$$f(x) = \frac{5}{x - 3}, \quad g(x) = \frac{4x - 3}{2x + 1}, \quad h(x) = \frac{7}{5}, \quad k(x) = \frac{6x - 9}{3} = 2x - 3$$

in which both numerator and denominator are either first-degree polynomials or constants. Such functions are called **linear rational functions.** We already know that the graphs of linear rational functions such as $h(x)$ and $k(x)$ above are straight lines. The key to graphing other linear rational functions, such as $f(x)$ and $g(x)$ above, is this simple fact from arithmetic:

THE BIG-LITTLE PRINCIPLE

The farther the number c is from 0, the closer the number 1/c is to 0. Conversely, the closer the number c is to 0, the farther the number 1/c is from 0. In less precise but more suggestive terms:

$$\frac{1}{big} = little \quad and \quad \frac{1}{little} = big$$

For example, each of the numbers 100, 750, 1500, and 5000 is bigger (farther from 0) than the preceding one, but each of the numbers

$$\frac{1}{100}, \quad \frac{1}{750}, \quad \frac{1}{1500}, \quad \frac{1}{5000}$$

is smaller (closer to 0) than the preceding one. Similarly, each of the numbers

$$\frac{-2}{10}, \quad \frac{-4}{100}, \quad \frac{-5}{1000}, \quad \frac{-2}{10{,}000}$$

is closer to 0 (smaller in absolute value) than the preceding one, but each of the numbers

$$\frac{1}{\frac{-2}{10}} = -5, \qquad \frac{1}{\frac{-4}{100}} = -25, \qquad \frac{1}{\frac{-5}{1000}} = -200, \qquad \frac{1}{\frac{-2}{10{,}000}} = -5000$$

is farther from 0 (bigger in absolute value) than the preceding one.

EXAMPLE In order to graph $f(x) = 5/(x - 3)$, we first observe that there is one *bad point* at which the function is not defined, namely, $x = 3$. It is important to determine what the graph looks like *near* this bad point. When x is a number bigger than 3, but very close to 3, then $x - 3$ is a positive number very close to 0. By the Big-Little Principle $1/(x - 3)$ must be a very large positive number. Thus $f(x) = 5\left(\dfrac{1}{x-3}\right)$ is also a large positive number. In fact, the closer x is to 3, the smaller $x - 3$ is, so that

The closer x is to 3, the *larger* $f(x) = \dfrac{5}{x - 3}$ is

In graphical terms this means that the closer x is to 3 on the right side of 3, the farther the point $(x, f(x))$ is above the x-axis. So the graph rises sharply on the right side of $x = 3$, as shown in Figure 4-27.

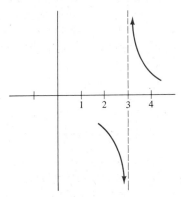

Figure 4-27

Similarly, if x is a bit smaller than 3, but very close to 3, then $x - 3$ is a negative number very close to 0 and $f(x) = 5/(x - 3)$ is a negative number that is very far from 0 (large in absolute value). So the graph of f falls very sharply on the left side of $x = 3$, as shown in Figure 4-27. The graph gets closer and closer to the vertical line $x = 3$, but never touches it. We say that the line $x = 3$ is a **vertical asymptote**° of the graph. Informally, we sometimes say that the graph *"blows up"* at $x = 3$.

Since $f(x) = 5/(x - 3)$ is positive when $x > 3$ and negative when $x < 3$, the graph of f lies above the x-axis to the right of $x = 3$ and below the x-axis to the left of $x = 3$. In order to determine what the graph looks like at the far right and the far left, we must examine $f(x)$ when x is a number very large in absolute value (that is, far from 0). As x gets bigger and bigger in absolute value, so does $x - 3$, and the Big-Little Principle tells us that $1/(x - 3)$ gets closer and closer to 0. Consequently, $f(x) = \dfrac{5}{x - 3} = 5 \cdot \dfrac{1}{x - 3}$ gets closer and closer to 0 as well. Thus the corresponding points $(x, f(x))$ on the graph get closer and closer to the x-axis without touching it, as shown in Figure 4-28.

Figure 4-28

We say that the x-axis is a **horizontal asymptote** of the graph.°

By plotting a few points and using the information developed above, we see that the entire graph looks like the one shown in Figure 4-29.

x	$f(x) = \dfrac{5}{x - 3}$
-12	$-\frac{1}{3}$
-2	-1
0	$-\frac{5}{3}$
1	$-\frac{5}{2}$
2	-5
2.5	-10
3.5	10
4	5
6	$\frac{5}{3}$
8	1
13	$\frac{1}{2}$

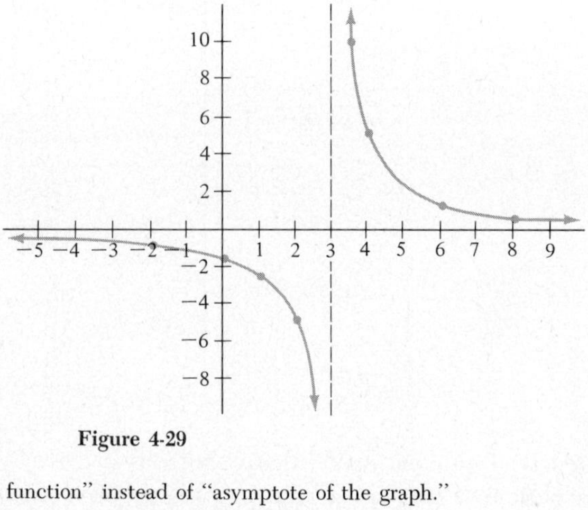

Figure 4-29

° Sometimes we say "asymptote of the function" instead of "asymptote of the graph."

The kind of analysis used in the preceding example works for any rational function of the form $f(x) = b/(cx + d)$ (where b, c, d are fixed real numbers and $c \neq 0$). The graph has a vertical asymptote at the number which is the root of the denominator. The graph lies above the x-axis on one side of the vertical asymptote and below the x-axis on the other side. The x-axis is a horizontal asymptote of the graph. These facts are sufficient to determine the general shape of the graph. A reasonably accurate sketch of the graph can then be obtained by plotting a few points. Some examples are shown in Figure 4-30.

$$f(x) = \frac{1}{x} \quad \text{vertical asymptote } x = 0$$

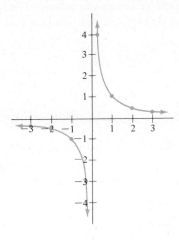

x	$f(x)$
-3	$-\frac{1}{3}$
-2	$-\frac{1}{2}$
-1	-1
$-\frac{1}{3}$	-3
$\frac{1}{4}$	4
1	1
2	$\frac{1}{2}$
3	$\frac{1}{3}$

$$k(x) = \frac{-3}{2x + 5} \quad \text{vertical asymptote } x = -\frac{5}{2}$$

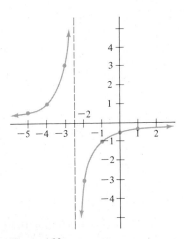

x	$k(x)$
-5	$\frac{3}{5}$
-4	1
-3	3
-2	-3
-1	-1
0	$-\frac{3}{5}$
1	$-\frac{3}{7}$

Figure 4-30

EXAMPLE The graph of the function $f(x) = \dfrac{4x - 3}{2x + 1}$ can be obtained by similar techniques. Since $f(x) = \dfrac{4x - 3}{2x + 1} = \dfrac{4x - 3}{2(x + \frac{1}{2})}$, the function is not defined at $x = -\frac{1}{2}$. To determine the shape of the graph near this bad point, note that when x is a number very close to $-\frac{1}{2}$, then the denominator of $f(x) = \dfrac{4x - 3}{2(x + \frac{1}{2})}$ is very close to 0, and the numerator $4x - 3$ is very close to $4(-\frac{1}{2}) - 3 = -5$. By the Big-Little Principle $f(x)$ must be a number that is very large in absolute value (since $\dfrac{-5}{\text{little}}$ must be far from 0). In graphical terms this means that when x is close to $-\frac{1}{2}$, then $f(x)$ is far from the x-axis. As in the previous examples, we must have a vertical asymptote at $x = -\frac{1}{2}$. In order to determine the shape of the graph when $|x|$ is large, we note that for any *nonzero* number x, we can divide both the numerator and denominator of $f(x)$ by x:

$$f(x) = \frac{4x - 3}{2x + 1} = \frac{\dfrac{4x - 3}{x}}{\dfrac{2x + 1}{x}} = \frac{4 - \dfrac{3}{x}}{2 + \dfrac{1}{x}}$$

Now when $|x|$ is large we know by the Big-Little Principle that both $-3/x$ and $1/x$ are very close to 0. Therefore when $|x|$ is large, $f(x) = \dfrac{4 - (3/x)}{2 + (1/x)}$ is very close to $\dfrac{4 - 0}{2 + 0} = 2$. We can see what this means graphically by plotting some points, as shown in Figure 4-31.

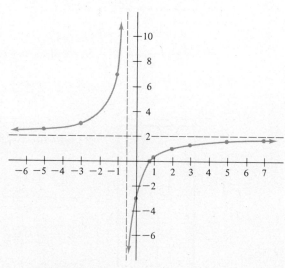

x	$f(x) = \dfrac{4x - 3}{2x + 1}$
-5	$\frac{23}{9}$
-3	3
-1	7
0	-3
$\frac{3}{4}$	0
1	$\frac{1}{3}$
2	1
3	$\frac{9}{7}$
5	$\frac{17}{11}$
7	$\frac{5}{3}$

Figure 4-31

As x gets larger and larger in absolute value, the graph of $f(x)$ gets closer and closer to the horizontal line $y = 2$. This line is called a **horizontal asymptote** of the graph.

The kind of analysis used in the preceding example works for any linear rational function of the form

$$f(x) = \frac{ax + b}{cx + d} \quad \text{(where } a, b, c, d \text{ are fixed real numbers and } a \neq 0, c \neq 0\text{)}$$

The graph has a vertical asymptote at the number which is a root of the denominator $cx + d$. The horizontal line $y = a/c$ is a horizontal asymptote of the graph since

$$f(x) = \frac{ax + b}{cx + d} = \frac{\dfrac{ax + b}{x}}{\dfrac{cx + d}{x}} = \frac{a + \dfrac{b}{x}}{c + \dfrac{d}{x}} \quad (x \neq 0)$$

so that $f(x)$ is very close to a/c when $|x|$ is large. Some examples are shown in Figure 4-32.

$$f(x) = \frac{-5x + 12}{2x - 4}$$

vertical asymptote $x = 2$
horizontal asymptote $y = -\frac{5}{2}$

x	$f(x)$
-1	$-\frac{17}{6}$
0	-3
1	$-\frac{7}{2}$
$\frac{12}{5}$	0
3	$-\frac{3}{2}$
4	-2
6	$-2\frac{1}{4}$

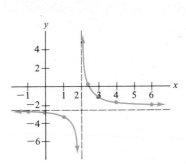

$$k(x) = \frac{3x + 6}{x}$$

vertical asymptote $x = 0$
horizontal asymptote $y = \frac{3}{1} = 3$

x	$k(x)$
-3	1
-2	0
-1	-3
1	9
2	6
3	5
4	$4\frac{1}{2}$

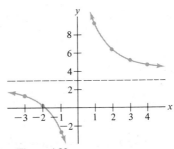

Figure 4-32

BASIC PROPERTIES OF RATIONAL GRAPHS

The graphs of more complicated rational functions share a number of common properties. We summarize them here so that you'll have some idea what to expect when graphing such functions.

Maxima and Minima   The graph of a rational function may have some relative maxima and minima (peaks and valleys). The total *number* of relative maxima and minima is much less important for graphing rational functions than it was for graphing polynomial functions.

Roots   If the graph of the rational function $f(x) = g(x)/h(x)$ touches the x-axis at $x = c$, then $(c, 0)$ is on the graph. Since the only point on the graph of f with first coordinate c is $(c, f(c))$, we must have $f(c) = 0$. In order for the number $f(c) = g(c)/h(c)$ to be 0, we must have:

$$h(c) \neq 0 \quad \text{(so that the fraction } \frac{g(c)}{h(c)} \text{ is a well-defined real number)}$$

$$g(c) = 0 \quad \text{(since a fraction is 0 only when its numerator is 0)}$$

In other words,

> *The points at which the graph of the rational function* f(x) = g(x)/h(x) *touches the* x-axis *correspond to the numbers that are roots of the numerator* g(x), *but* not *roots of the denominator* h(x).

A number that is a root of the numerator $g(x)$, but not of the denominator $h(x)$, is called a **root of the rational function** $f(x) = g(x)/h(x)$. For example, the numerator of the function $f(x) = \dfrac{x^2 - 1}{x - 1}$ has 1 and -1 as roots. But 1 is also a root of the denominator, so that -1 is the only root of the rational function $f(x)$.

Bad Points   The rational function $f(x) = g(x)/h(x)$ is not defined at those numbers that are *roots of the denominator* $h(x)$. If $x = d$ is a root of $h(x)$ that is also a root of $g(x)$, then the graph of $f(x)$ will have either a *hole* or a *vertical asymptote* at $x = d$, as shown in Figure 4-33.

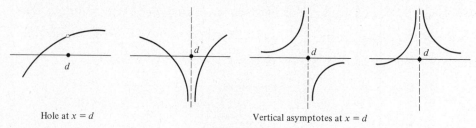

Hole at $x = d$ Vertical asymptotes at $x = d$

Figure 4-33

As we saw above, the function $f(x) = 5/(x - 3)$ has a vertical asymptote at $x = 3$. Observe that $x = 3$ is a root of the denominator $x - 3$ which is *not* a root of the numerator 5. An analogous argument works for any rational function and shows that

> Whenever x = d *is a root of the denominator* h(x), *but not a root of the numerator* g(x), *then the rational function* f(x) = g(x)/h(x) *has a vertical asymptote at* x = d.

Extent and Continuity As just noted, the graph of a rational function may have several breaks in it (holes or places where there is a vertical asymptote). These occur *only* where the function is not defined. Wherever the function *is* defined, however, its graph is always a continuous, smooth, unbroken curve.

Behavior When $|x|$ Is Large Consider the function $f(x) = \dfrac{3x^4 - 2x}{2x^4 + 5}$ in which the numerator and denominator polynomials have the *same* degree. When $x \neq 0$, we can divide both the numerator and denominator by the highest power of x that occurs, namely, x^4, without affecting the value of $f(x)$:

$$f(x) = \frac{3x^4 - 2x}{2x^4 + 5} = \frac{\dfrac{3x^4 - 2x}{x^4}}{\dfrac{2x^4 + 5}{x^4}} = \frac{3 - \dfrac{2}{x^3}}{2 + \dfrac{5}{x^4}}.$$

As $|x|$ gets larger and larger, the terms $2/x^3$ and $5/x^4$ get closer and closer to 0 by the Big-Little Principle. So $f(x)$ gets closer and closer to $\dfrac{3 - 0}{2 + 0} = \dfrac{3}{2}$. Thus as $|x|$ gets larger, the graph of f gets closer and closer to the horizontal line $y = \frac{3}{2}$. Hence this line is a horizontal asymptote of the graph. The same argument works in the general case:

> If f(x) = $\dfrac{\text{ax}^n + \cdots}{\text{cx}^n + \cdots}$ *is a rational function whose numerator and denominator have the same degree* n, *then the line* y = a/c *is a horizontal asymptote of the graph.*

Now suppose f is a rational function in which the denominator has *larger* degree than the numerator, such as $f(x) = \dfrac{x^2 + 7}{x^3 - x^2}$. Dividing both numerator and denominator by the highest power of x that occurs, namely, x^3, shows that for $x \neq 0$

$$f(x) = \frac{x^2 + 7}{x^3 - x^2} = \frac{\dfrac{x^2 + 7}{x^3}}{\dfrac{x^3 - x^2}{x^3}} = \frac{\dfrac{1}{x} + \dfrac{7}{x^3}}{1 - \dfrac{1}{x}}$$

As $|x|$ gets larger and larger, the Big-Little Principle shows that $f(x)$ gets closer and closer to $(0 + 0)/(1 - 0) = 0/1 = 0$. In geometric terms, the graph of f gets closer and closer to the x-axis. So the x-axis is a horizontal asymptote of the graph. This argument also carries over to the general case:

> If f(x) = g(x)/h(x) *is a rational function in which the denominator* h(x) *has larger degree than the numerator* g(x), *then the* x-axis *is a horizontal asymptote of the graph.*

The asymptotes of rational functions in which the degree of the denominator is *less* than the degree of the numerator are somewhat more complicated. Some examples of such functions are discussed at the end of this section.

Change of Sign   If the graph of a function f crosses the x-axis at the root $x = c$, then $f(x)$ is a positive number when x is on one side of c and $f(x)$ is a negative number when x is on the other side of c, as shown in Figure 4-34.

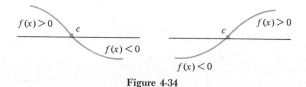

Figure 4-34

We say that f **changes sign** at the root $x = c$. There are two other situations in which a rational function f may possibly change sign. If the graph has a vertical asymptote or a hole on the x-axis at $x = d$, then it *may* happen that $f(x)$ is positive when x is on one side of d and negative when x is on the other side of d, as shown in the examples in Figure 4-35.

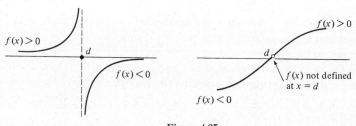

Figure 4-35

Since the graph of a rational function is a continuous, unbroken curve wherever the function *is* defined, it can be proved that

> A *rational function can change sign only at a root, or a vertical*
> *asymptote, or a hole on the x-axis.*

This statement does not mean that a rational function *must* change sign at *every* root, vertical asymptote, or hole on the x-axis. But *if* it *does* change sign, such a change can occur only at these places.

GRAPHING A RATIONAL FUNCTION

The basic procedure is to examine the behavior of the function near its bad points and roots to obtain an idea of the general shape of the graph. Then plot a few points and obtain a reasonably accurate graph.

EXAMPLE   The first step in graphing $f(x) = \dfrac{x-1}{x^2-x-6}$, is to write $f(x)$ in factored form in order to *find all the real roots and bad points:*

$$f(x) = \frac{x-1}{x^2-x-6} = \frac{x-1}{(x+2)(x-3)}$$

Clearly, $x = 1$ is a root of $f(x)$, and $f(x)$ is not defined when $x = -2$ or $x = 3$. Since neither $x = -2$ nor $x = 3$ is a root of the numerator $x - 1$, the graph must have vertical asymptotes at $x = -2$ and $x = 3$. These asymptotes and the root divide the x-axis into four intervals, as shown in Figure 4-36.

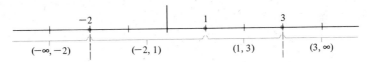

Figure 4-36

Since this rational function can *change* sign only at a root or a vertical asymptote, the sign of $f(x)$ remains the same throughout any one of these intervals. This means that the graph of $f(x)$ must lie *entirely above* the x-axis ($f(x) > 0$) or *entirely below* the x-axis ($f(x) < 0$) on each of these intervals.

In order to *determine the sign of $f(x)$ on each of the intervals* above, we need only examine the signs of each of the factors $x - 1$, $x + 2$, and $x - 3$. For example, $x - 1$ is positive exactly when

$$x - 1 > 0 \qquad \text{or equivalently} \qquad x > 1$$

Thus $x - 1$ is positive when x is in the intervals $(1, 3)$ and $(3, \infty)$, and negative when x is in the intervals $(-\infty, -2)$ and $(-2, 1)$. Similar arguments work for the other factors, as summarized in the chart shown in Figure 4-37.

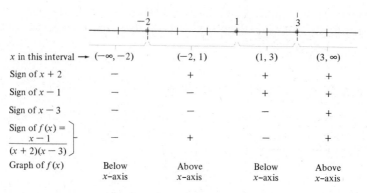

x in this interval $\rightarrow$	$(-\infty, -2)$	$(-2, 1)$	$(1, 3)$	$(3, \infty)$
Sign of $x + 2$	−	+	+	+
Sign of $x - 1$	−	−	+	+
Sign of $x - 3$	−	−	−	+
Sign of $f(x) = \dfrac{x - 1}{(x + 2)(x - 3)}$	−	+	−	+
Graph of $f(x)$	Below x-axis	Above x-axis	Below x-axis	Above x-axis

Figure 4-37

Since $f(x)$ changes sign at the root $x = 1$, the graph must cross the x-axis there. Since the numerator of $k(x) = \dfrac{x - 1}{(x + 2)(x - 3)}$ has degree 1 and the denominator $(x + 2)(x - 3) = x^2 - x - 6$ has strictly larger degree 2, the x-axis must be a horizontal asymptote of the graph. Using all the information we now have about roots, asymptotes, and signs, we can make a *rough sketch* of the general shape of most of the graph, as shown in Figure 4-38.

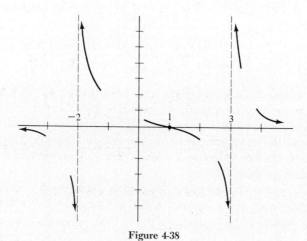

Figure 4-38

It is now easy to plot a few points and obtain a reasonably accurate sketch of the graph, as shown in Figure 4-39.

x	$f(x) = \dfrac{x-1}{x^2-x-6}$
-4	$-\frac{5}{14}$
-3	$-\frac{2}{3}$
-2.2	$-\frac{80}{26} \approx -3.08$
-1.9	$\frac{290}{49} \approx 5.92$
-1	$\frac{1}{2}$
0	$\frac{1}{6}$
1	0
2	$-\frac{1}{4}$
2.8	$-\frac{15}{8} = -1.875$
3.1	$\frac{210}{51} \approx 4.12$
4	$\frac{1}{2}$

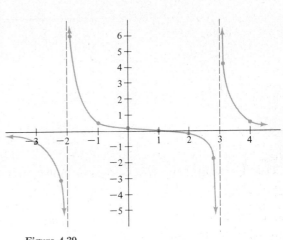

Figure 4-39

EXAMPLE In order to graph $k(x) = \dfrac{x^2-3x+2}{(x^2-4)(x-3)}$ we first factor:

$$k(x) = \frac{x^2-3x+2}{(x^2-4)(x-3)} = \frac{(x-2)(x-1)}{(x-2)(x+2)(x-3)}$$

Since $x = 2$ is a root of *both* numerator and denominator, $k(2) = \frac{0}{0}$ is *not* defined. However, for any number x *except* 2, we have $x - 2 \neq 0$ so that

$$k(x) = \frac{(x-2)(x-1)}{(x-2)(x+2)(x-3)} = \frac{x-1}{(x+2)(x-3)} \qquad \text{when } x \neq 2$$

Therefore the graph of $k(x)$ looks exactly like the graph of $f(x) = \dfrac{x-1}{(x+2)(x-3)}$ with *one point deleted,* namely, the point where $x = 2$. The graph of $f(x) = \dfrac{x-1}{(x+2)(x-3)}$

was obtained in the preceding example. All we have to do is to delete the point where $x = 2$ in order to get the graph of $k(x)$, as shown in Figure 4-40.

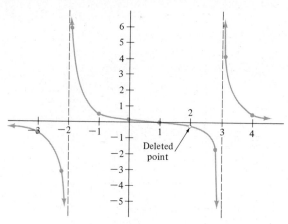

Figure 4-40

Thus the graph of $k(x)$ has a vertical asymptote at $x = -2$ and $x = 3$ and a *hole* at $x = 2$.

EXAMPLE In order to graph $f(x) = \dfrac{x - 4}{(2x^2 - x - 6)(x - 2)}$ we factor:

$$f(x) = \frac{x - 4}{(2x^2 - x - 6)(x - 2)} = \frac{x - 4}{(2x + 3)(x - 2)(x - 2)} = \frac{x - 4}{2(x + \frac{3}{2})(x - 2)^2}$$

Clearly, $f(x)$ has vertical asymptotes at $x = -\frac{3}{2}$ and $x = 2$ and a root at $x = 4$. The root and asymptotes divide the x-axis into four intervals. We can determine the sign of $f(x)$ when x is in each of these intervals by examining the signs of the factors $2(x + \frac{3}{2})$, $(x - 2)^2$, and $x - 4$, as summarized in the chart shown in Figure 4-41.

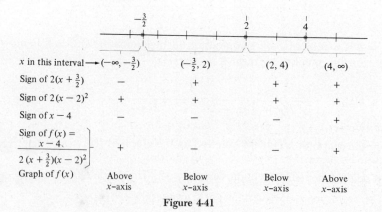

	$(-\infty, -\frac{3}{2})$	$(-\frac{3}{2}, 2)$	$(2, 4)$	$(4, \infty)$
x in this interval →				
Sign of $2(x + \frac{3}{2})$	$-$	$+$	$+$	$+$
Sign of $2(x - 2)^2$	$+$	$+$	$+$	$+$
Sign of $x - 4$	$-$	$-$	$-$	$+$
Sign of $f(x) = \dfrac{x - 4}{2(x + \frac{3}{2})(x - 2)^2}$	$+$	$-$	$-$	$+$
Graph of $f(x)$	Above x-axis	Below x-axis	Below x-axis	Above x-axis

Figure 4-41

Since f changes sign at the root $x = 4$, the graph crosses the x-axis there. Note that f does *not* change sign at the asymptote $x = 2$. Since the numerator of $f(x) = \dfrac{x-4}{(2x^2 - x - 6)(x - 2)}$ has smaller degree than the denominator, the x-axis is a horizontal asymptote of the graph. It is now easy to plot some points and obtain a reasonably accurate sketch of the graph, as shown in Figure 4-42.

x	$f(x) = \dfrac{x-4}{(2x^2 - x - 6)(x - 2)}$
-3	$\frac{7}{75}$
-2	$\frac{3}{8}$
-1	$-\frac{5}{9}$
0	$-\frac{1}{3}$
1	$-\frac{3}{5}$
1.5	$-\frac{5}{3}$
3	$-\frac{1}{9}$
4	0
5	$\frac{1}{117}$

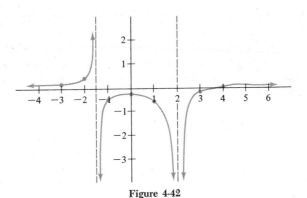

Figure 4-42

Without calculus we can't precisely locate the top of the peak between $x = -\frac{3}{2}$ and $x = 2$ or the tiny peak (molehill?) to the right of $x = 4$. Nor can we guarantee that there aren't some further bends or wiggles in the graph.

In all the examples up to now, the degree of the denominator has always been greater than or equal to the degree of the numerator. We have seen that the graphs of such rational functions have only vertical or horizontal straight lines as asymptotes. But when $f(x) = g(x)/h(x)$ and the degree of the denominator $h(x)$ is strictly less than the degree of $g(x)$, the situation is different. The graph may have oblique asymptotes (that is, straight lines that are neither vertical nor horizontal). Or it may have curves other than straight lines as asymptotes.

EXAMPLE   In order to graph $f(x) = \dfrac{2x^2 + 5x + 2}{2x + 7}$, we first factor:

$$f(x) = \frac{2x^2 + 5x + 2}{2x + 7} = \frac{(2x + 1)(x + 2)}{2x + 7} = \frac{2(x + \frac{1}{2})(x + 2)}{2(x + \frac{7}{2})}$$

We see that $x = -\frac{1}{2}$ and $x = -2$ are roots of f and that there is a vertical asymptote at $x = -\frac{7}{2}$. The usual analysis of signs produces the chart in Figure 4-43.

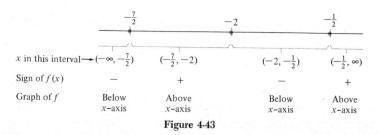

	$\left(-\infty, -\frac{7}{2}\right)$	$\left(-\frac{7}{2}, -2\right)$	$\left(-2, -\frac{1}{2}\right)$	$\left(-\frac{1}{2}, \infty\right)$
x in this interval →				
Sign of $f(x)$	$-$	$+$	$-$	$+$
Graph of f	Below x-axis	Above x-axis	Below x-axis	Above x-axis

Figure 4-43

By plotting some points we see that near the vertical asymptote and roots the graph looks like the one shown in Figure 4-44.

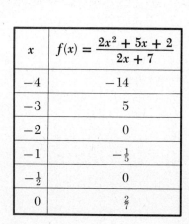

x	$f(x) = \dfrac{2x^2 + 5x + 2}{2x + 7}$
-4	-14
-3	5
-2	0
-1	$-\frac{1}{5}$
$-\frac{1}{2}$	0
0	$\frac{2}{7}$

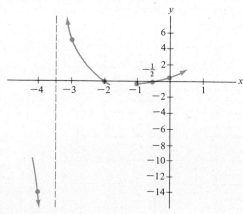

Figure 4-44

In order to determine the rest of the graph, we divide the numerator $2x^2 + 5x + 2$ of $f(x)$ by the denominator $2x + 7$ and find that the quotient is $x - 1$ and the remainder is 9 (verify this!). According to the division algorithm,

$$\text{dividend} = (\text{quotient})(\text{divisor}) + (\text{remainder})$$
$$2x^2 + 5x + 2 = (x - 1)(2x + 7) \quad + 9$$

Dividing both sides of this last equation by $2x + 7$, we obtain

$$\underbrace{\frac{2x^2 + 5x + 2}{2x + 7}}_{f(x)} = \frac{(x - 1)(2x + 7)}{2x + 7} + \frac{9}{2x + 7}$$

$$f(x) = (x - 1) + \frac{9}{2x + 7}$$

Now when $|x|$ is large, the Big-Little Principle tells us that $9/(2x + 7)$ is very close to 0. Therefore when x is large, $f(x) = (x - 1) + \dfrac{9}{2x + 7}$ is very close to $(x - 1) +$ $0 = x - 1$. Thus for $|x|$ large the graph of f is very close to the graph of $y = x - 1$, as shown in Figure 4-45.

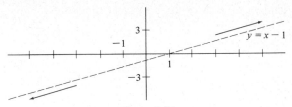

Figure 4-45

The straight line $y = x - 1$ is an **asymptote** of the graph. Note that the graph of f lies above the asymptote $y = x - 1$ when x is a large positive number. The reason is that when x is large positive, $9/(2x + 7)$ is a small positive number, so that $f(x) = (x - 1) +$ $\dfrac{9}{2x + 7}$ is slightly larger than $x - 1$. Similarly, when x is a negative number far to the left of 0, $f(x)$ is slightly smaller than $x - 1$.

We now have enough information to plot a reasonably accurate graph, as shown in Figure 4-46.

x	$f(x) = \dfrac{2x^2 + 5x + 2}{2x + 7}$
-6	$-\frac{44}{5}$
-5	-9
-4	-14
-3	5
-2	0
-1	$-\frac{1}{5}$
$-\frac{1}{2}$	0
0	$\frac{2}{7}$
1	1
2	$\frac{20}{11}$

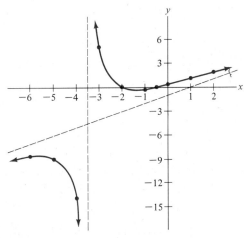

Figure 4-46

Once again, we cannot determine the exact location of the relative maxima and minima (peaks and valleys) without calculus.

EXAMPLE The graph of $f(x) = \dfrac{x^3 - 2}{x - 1}$ is discussed in Exercise B.6. It has the parabola $y = x^2 + x + 1$ as an asymptote.

EXERCISES

A.1. State the domain of each of these rational functions.

(a) $f(x) = \dfrac{1}{x + 2}$

(d) $q(x) = \dfrac{x^2 + x + 1}{x^2 - 3}$

(b) $g(x) = \dfrac{x - 3}{x^2 - 4}$

(e) $r(x) = \dfrac{2x^3 - 3x^2 + 1}{x^3 - x}$

(c) $h(x) = \dfrac{x^2 + 5}{x^2 + 2x + 1}$

(f) $f(x) = \dfrac{x^2 - 9}{2x^3 - 5x^2 - 3x}$

A.2. Give examples of three different rational functions, each of which has as its domain the set of all real numbers except $-\tfrac{1}{2}$, 1, and 2.

A.3. Graph these linear rational functions.

(a) $f(x) = \dfrac{1}{x + 5}$

(e) $k(x) = \dfrac{-3}{2x + 5}$

(i) $f(x) = \dfrac{2 - x}{x - 3}$

(b) $q(x) = \dfrac{-7}{x - 6}$

(f) $g(x) = \dfrac{-4}{2 - x}$

(j) $g(x) = \dfrac{3x - 2}{x + 3}$

(c) $g(x) = \dfrac{10}{x + 1}$

(g) $f(x) = \dfrac{3x}{x - 1}$

(k) $q(x) = \dfrac{7x - 3}{-4x + 12}$

(d) $h(x) = \dfrac{2}{3x - 7}$

(h) $p(x) = \dfrac{x - 2}{x}$

(l) $f(x) = \dfrac{-5x + 1}{-2x - 1}$

A.4. Find the roots of these rational functions.

(a) $f(x) = \dfrac{x^2 - 5x + 6}{x^2 + 1}$

(d) $q(x) = \dfrac{2x^3 + 3x^2 - x}{x^2 + 5x - 3}$

(b) $g(x) = \dfrac{x^2 + 2x - 3}{x^2 - 1}$

(e) $k(x) = \dfrac{(x^2 - 9)(x^2 + 6x + 9)}{x^2 + x - 6}$

(c) $k(x) = \dfrac{x^2 - 7x + 10}{x^2 + 2x}$

(f) $f(x) = \dfrac{x^3 - 4x^2 + 2x}{x^4 - 8x}$

B.1. Find at least one linear rational function whose graph passes through the points $(0, 1)$, $(1, 0)$, and $(2, -2)$.

B.2. Find a linear rational function whose asymptotes are the lines $x = -1$ and $y = 2$ and whose graph contains the point $(1, 3)$.

B.3. Graph these rational functions, in which the degree of the denominator is always greater than the degree of the numerator.

(a) $f(x) = \dfrac{1}{x(x+1)^2}$

(e) $p(x) = \dfrac{x^2 - 9}{(x+3)(x-5)(x+4)}$

(b) $g(x) = \dfrac{x}{2x^2 - 5x - 3}$

(f) $q(x) = \dfrac{x^3 + 3x^2}{x^4 - 4x^2}$

(c) $f(x) = \dfrac{x-3}{x^2 + x - 2}$

(g) $h(x) = \dfrac{(x^2 + 6x + 5)(x+5)}{(x+5)^3(x-1)}$

(d) $g(x) = \dfrac{x+2}{x^2 - 1}$

(h) $f(x) = \dfrac{x^2 - 1}{x^3 - 2x^2 + x}$

B.4. Graph these functions, using the same general procedure as in Exercise B.3. The only difference here is that both numerator and denominator have the same degree. So instead of the x-axis, some other horizontal straight line will be a horizontal asymptote (see the last box on p. 223).

(a) $f(x) = \dfrac{-4x^2 + 1}{x^2}$

(c) $q(x) = \dfrac{x^2 + 2x}{x^2 - 4x - 5}$

(b) $k(x) = \dfrac{x^2 + 1}{x^2 - 1}$

(d) $F(x) = \dfrac{x^2 + x}{x^2 - 2x + 4}$

B.5. Graph these functions. (The next to last example in the text may be helpful as a guide.)

(a) $f(x) = \dfrac{x^2 - x - 6}{x - 2}$

(d) $P(x) = \dfrac{x^2 + 1}{x}$

(b) $k(x) = \dfrac{x^2 + x - 2}{x}$

(e) $Q(x) = \dfrac{4x^2 + 4x - 3}{2x - 5}$

(c) $p(x) = \dfrac{x^2 + 1}{x - 1}$

(f) $K(x) = \dfrac{3x^2 - 12x + 15}{3x + 6}$

B.6. Graph the function $f(x) = \dfrac{x^3 - 2}{x - 1}$ as follows:

(a) Determine the roots, vertical asymptotes, signs, and sign changes as usual.

(b) Divide $x^3 - 2$ by $x - 1$ and use the division algorithm (as in the next to last example in the text) to show that $f(x) = (x^2 + x + 1) + (-1)/(x - 1)$. When $|x|$ is large, $-1/(x - 1)$ is very close to 0 (why?), so that $f(x)$ is very close to $x^2 + x + 1$. Thus the curve $y = x^2 + x + 1$ is an asymptote of the graph of f. (*Note:* $y = x^2 + x + 1$ is a parabola—see p. 159)

(c) Plot some points and use parts (a) and (b) to sketch the graph of f.

B.7. Graph these functions. (*Hint:* all have parabolic asymptotes; see Exercise B.6.)

(a) $p(x) = \dfrac{x^3 + 8}{x + 1}$

(c) $f(x) = \dfrac{x^4 - 1}{x^2}$

(b) $q(x) = \dfrac{x^3 - 1}{x - 2}$

(d) $k(x) = \dfrac{(x+1)(x-1)(x+3)}{x + 2}$

B.8.   Find the rule of a rational function f which has these properties:
   (i)   the curve $y = x^3 - 8$ is an asymptote of the graph of f.
   (ii) $f(2) = 1$.
   (iii) $x = 1$ is a vertical asymptote of the graph.

C.1.   The formula for the gravitational acceleration of an object (relative to the earth) is $g(r) = (4 \cdot 10^{14})/r^2$, where r is the distance of the object from the center of the earth (in meters).
   (a)   What is the gravitational acceleration at the earth's surface? [The radius of the earth is approximately $(6.4)(10^6)$ m.]
   (b)   Graph the function $g(r)$.
   (c)   Does the function $g(r)$ have any roots? Does the gravitational acceleration ever vanish for any value of r? Can you ever "escape the pull of gravity"?

C.2.   Graph these functions. [Their graphs are somewhat trickier than those discussed in the text. Some involve peaks and valleys whose existence may not be evident unless you plot many points (or use calculus). Furthermore, you may not be able to find all of the roots of these functions exactly.]
   (a) $k(x) = \dfrac{2x^3 + 1}{x^2 - 1}$ **(c)** $f(x) = \dfrac{3x^3 - 11x - 1}{x^2 - 4}$
   (b) $p(x) = \dfrac{(x - 2)(x^2 + 1)}{x^2 - 1}$

5. POLYNOMIAL AND RATIONAL INEQUALITIES

In this section we discuss the solutions of inequalities involving polynomial and rational expressions, such as

$$2x^3 - 15x < x^2, \qquad \frac{x + 3}{x - 1} \geq -2, \qquad \left| \frac{3x + 2}{x - 2} \right| \leq 2, \qquad |2x^2 + 3x - 1| \geq 3$$

There are many instances in mathematics and applied problems where it is necessary to solve these kinds of inequalities. The strategy for solving inequalities that do not involve absolute values is exactly the same as that used to solve linear inequalities such as $7 - 2x \leq 3$. So you should begin by reviewing the basic principles for solving inequalities listed on page 21.

   Two inequalities that have the same solutions are said to be **equivalent.** The fundamental idea is to use the basic principles to transform a given inequality into an equivalent one, whose solutions we know how to find. As we shall now see, the algebraic procedures needed to find such solutions are some of the same techniques used to graph polynomial and rational functions.

EXAMPLE   The first step in solving the inequality $2x^3 - 15x < x^2$ is to collect all the terms on the same side of the inequality sign and to simplify and factor the resulting expressions:

$$2x^3 - 15x < x^2$$

$2x^3 - x^2 - 15x < 0$ (subtract x^2 from both sides)

$(2x^2 - x - 15)x < 0$ (factor x out of left side)

$(2x + 5)(x - 3)x < 0$ (factor left side further)

$2(x + \frac{5}{2})(x - 3)x < 0$ (factor 2 out of $2x + 5$)

$(x + \frac{5}{2})(x - 3)x < 0$ (divide both sides by 2)

Because of the steps used to obtain this last inequality from the original one, we know that it is equivalent to the original one. To solve this last inequality, consider the polynomial function f whose rule is given by the left side of the last inequality: $f(x) = (x + \frac{5}{2})(x - 3)x$. A solution of the last inequality is any number x such that $f(x) < 0$. For such a number x, the point $(x, f(x))$ on the graph of f has a *negative* second coordinate and therefore lies *below* the x-axis, as shown in Figure 4-47.

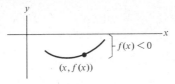

Figure 4-47

In other words,

> The solutions of the inequality are just the numbers x for
> which the graph of the function f lies below the x-axis

But we have already seen in Section 3 how to determine the places where the graph of a polynomial function lies below, on, or above the x-axis. The roots of the polynomial $f(x) = (x + \frac{5}{2})(x - 3)x$ are $x = -\frac{5}{2}$, $x = 0$, and $x = 3$. These numbers divide the x-axis into four intervals, as shown in Figure 4-48. On any one of these intervals, the sign of $f(x)$ is always the same (it can only change at a root). The sign of $f(x)$ on each of the intervals can be determined by inspecting the signs of the factors (see Figure 4-48).

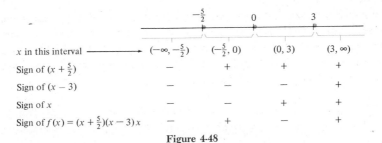

	$(-\infty, -\frac{5}{2})$	$(-\frac{5}{2}, 0)$	$(0, 3)$	$(3, \infty)$
x in this interval				
Sign of $(x + \frac{5}{2})$	$-$	$+$	$+$	$+$
Sign of $(x - 3)$	$-$	$-$	$-$	$+$
Sign of x	$-$	$-$	$+$	$+$
Sign of $f(x) = (x + \frac{5}{2})(x - 3)x$	$-$	$+$	$-$	$+$

Figure 4-48

The last line of the chart in Figure 4-48 gives us all the information we need:

$$f(x) < 0 \quad \text{for every number } x \text{ in the intervals } (-\infty, -\tfrac{5}{2}) \text{ and } (0, 3)$$

In other words, the solutions of the inequality $f(x) < 0$, and hence of the original inequality $2x^3 - 15x < x^2$, are all numbers x such that

$$x < -\tfrac{5}{2} \quad \text{or} \quad 0 < x < 3$$

EXAMPLE In order to solve $\dfrac{x+3}{x-1} \geq -2$, we begin by collecting all terms on one side, writing this side as a rational expression, factoring, and simplifying:

$$\frac{x+3}{x-1} \geq -2$$

$$\frac{x+3}{x-1} + 2 \geq 0 \qquad \text{(add 2 to both sides)}$$

$$\frac{x+3}{x-1} + \frac{2(x-1)}{x-1} \geq 0 \qquad \text{(put left side over common denominator)}$$

$$\frac{x+3+2x-2}{x-1} \geq 0 \qquad \text{(multiply out left side)}$$

$$\frac{3x+1}{x-1} \geq 0 \qquad \text{(simplify)}$$

$$\frac{3(x+\tfrac{1}{3})}{x-1} \geq 0 \qquad \text{(factor 3 out of } 3x+1)$$

$$\frac{x+\tfrac{1}{3}}{x-1} \geq 0 \qquad \text{(divide both sides by 3)}$$

The original inequality is equivalent to the last inequality above. In order to solve it, let h be the rational function given by $h(x) = \dfrac{x + \tfrac{1}{3}}{x - 1}$. We want to find the numbers x for which $h(x) \geq 0$. The function h has a root at $x = -\tfrac{1}{3}$ and a vertical asymptote at $x = 1$. It can change sign only at these two numbers. These numbers divide the x-axis into three intervals, as we see in Figure 4-49.

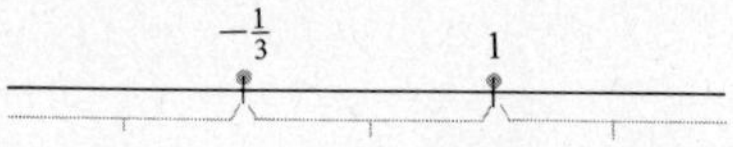

x in this interval	$(-\infty, -\tfrac{1}{3})$	$(-\tfrac{1}{3}, 1)$	$(1, \infty)$
Sign of $x + \tfrac{1}{3}$	−	+	+
Sign of $x - 1$	−	−	+
Sign of $h(x) = \dfrac{x+\tfrac{1}{3}}{x-1}$	+	−	+

Figure 4-49

From the last line of the chart in Figure 4-49, together with the root $x = -\frac{1}{3}$ of h, we see that the solutions of $h(x) \geq 0$, and hence of the original inequality $\dfrac{x + 3}{x - 1} \geq -2$, are all numbers x such that

$$x \leq -\tfrac{1}{3} \quad \text{or} \quad x > 1$$

that is, all numbers in the intervals $(-\infty, -\frac{1}{3}]$ and $(1, \infty)$.

QUADRATIC INEQUALITIES

Inequalities involving second-degree (quadratic) polynomials appear frequently enough that they deserve separate mention. In formal terms, a **quadratic inequality** is an inequality that is equivalent to an inequality of one of these four forms:

$$ax^2 + bx + c > 0 \qquad ax^2 + bx + c \geq 0$$
$$ax^2 + bx + c < 0 \qquad ax^2 + bx + c \leq 0$$

where a, b, c are fixed real numbers. For example, each of the following is a quadratic inequality:

$$2x^2 + 5x + 7 > 0 \qquad \text{(here, } a = 2, b = 5, c = 7\text{)}$$
$$3x^2 \geq -11x + 4 \qquad \text{(equivalent to } 3x^2 + 11x - 4 \geq 0\text{)}$$

A slight variation on the method presented above can be used to solve *any* quadratic inequality quickly, without factoring. This quick method depends on these two facts:

Fact 1.   *The graph of the function* $f(x) = ax^2 + bx + c$ *is a parabola that opens upward if* $a > 0$ *and downward if* $a < 0$. *(See pp. 158–159 for details.)*

Fact 2.   *All real roots of the polynomial* $ax^2 + bx + c$ *can always be found by using the quadratic formula (given in Section 2 of this chapter).*

EXAMPLE   In order to solve $2x^2 + 3x - 4 < 0$, we let f be the function whose rule is $f(x) = 2x^2 + 3x - 4$. The solutions of the inequality are just those numbers x for which $f(x) < 0$, that is, for which the graph of f lies *below* the x-axis. Since the coefficient of x^2 is positive, the graph of $f(x) = 2x^2 + 3x - 4$ is a parabola that opens upward (Fact 1). If the graph touches or crosses the x-axis, it must do so at the numbers that are the real roots of $f(x) = 2x^2 + 3x - 4$. Using the quadratic formula, we see that f has two real roots:

$$x = \frac{-3 \pm \sqrt{3^2 - 4 \cdot 2(-4)}}{4} = \frac{-3 \pm \sqrt{41}}{4}$$

Consequently, even without plotting any points, we know that the graph of $f(x) = 2x^2 + 3x - 4$ must have the form shown in Figure 4-50.

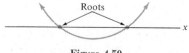

Figure 4-50

Clearly, the numbers x such that $f(x) < 0$ are precisely the numbers *between* the roots of f, that is, between $\dfrac{-3 - \sqrt{41}}{4}$ and $\dfrac{-3 + \sqrt{41}}{4}$. Therefore, the solutions of the inequality are all real numbers in the interval $\left(\dfrac{-3 - \sqrt{41}}{4}, \dfrac{-3 + \sqrt{41}}{4} \right)$.

With the information developed above, it is now easy to solve the inequality

$$2x^2 + 3x - 4 \geq 0$$

as well. The solutions are the numbers x for which $f(x) \geq 0$, that is, the numbers for which the graph of f lies on or *above* the x-axis. As Figure 4-50 shows, these are the numbers that lie to the left of the left-hand root and to the right of the right-hand root (including these roots). Therefore the solutions of $2x^2 + 3x - 4 \geq 0$ are all numbers in the intervals $\left(-\infty, \dfrac{-3 - \sqrt{41}}{4} \right]$ and $\left[\dfrac{-3 + \sqrt{41}}{4}, \infty \right)$.

EXAMPLE Solve each of these inequalities:

$$-4x^2 + 4x - 1 < 0, \qquad -4x^2 + 4x - 1 > 0$$

Using the quadratic formula, see that the polynomial function $f(x) = -4x^2 + 4x - 1$ has exactly one real root:

$$x = \frac{-4 \pm \sqrt{4^2 - 4(-4)(-1)}}{-8} = \frac{-4}{-8} = \frac{1}{2}$$

Therefore the graph of f touches the x-axis only once at the point where $x = \frac{1}{2}$. Since the coefficient of x^2 in $f(x)$ is negative, the graph of f is a downward-opening parabola (Fact 1). Consequently, the graph of f must look like the one in Figure 4-51.

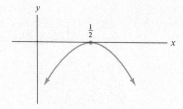

Figure 4-51

The solutions of the first inequality above are the numbers x for which $f(x) < 0$, that is, for which the graph of f lies *below* the x-axis. This is the case for every number *except* $x = \frac{1}{2}$. So the solutions of $-4x^2 + 4x - 1 < 0$ are all numbers in the intervals $(-\infty, \frac{1}{2})$ and $(\frac{1}{2}, \infty)$.

The solutions of the second inequality are all numbers x for which the graph lies *above* the x-axis. Since the graph is *never* above the x-axis, the inequality $-4x^2 + 4x - 1 > 0$ has *no solutions*.

ABSOLUTE VALUE INEQUALITIES

A great many inequalities involving absolute values, such as

$$\left|\frac{x + 3}{x - 1}\right| < 2, \qquad \left|\frac{3x + 2}{x - 2}\right| \leq 2, \qquad |2x^2 + 3x - 1| > 3, \qquad |x^2 - 5| < 4$$

can be solved by using the methods presented earlier in this section, together with certain fundamental facts about absolute value. These are the same facts that were used in Section 4 of Chapter 1 to solve linear inequalities, such as $|2x - 3| < 7$. For convenience, we repeat those facts here (they are also true with $<$ and $>$ in place of $\leq$ and $\geq$):

Let k *be a fixed positive number. Then for any real number* r:

$$|r| \leq k \textit{ exactly when } -k \leq r \leq k$$
$$|r| \geq k \textit{ exactly when } r \geq k \textit{ or } r \leq -k$$

EXAMPLE In order to solve $\left|\dfrac{x + 3}{x - 1}\right| < 2$, we observe that for any real number x (except $x = 1$), $(x + 3)/(x - 1)$ is also a real number. Using $(x + 3)/(x - 1)$ in place of r and 2 in place of k in the box above, we see that

$$\left|\frac{x + 3}{x - 1}\right| < 2 \qquad \text{exactly when} \qquad -2 < \frac{x + 3}{x - 1} < 2$$

As usual, $-2 < \dfrac{x + 3}{x - 1} < 2$ is shorthand for the statement

$$\frac{x + 3}{x - 1} > -2 \qquad and \qquad \frac{x + 3}{x - 1} < 2$$

Consequently, we need only find all the numbers that are solutions of *both* of these inequalities. The first of these inequalities was solved on page 236, where we found that its solutions are all numbers in the intervals $(-\infty, -\frac{1}{3})$ and $(1, \infty)$. The second inequality can also be solved by reducing it to an equivalent one and checking signs:

$$\frac{x + 3}{x - 1} < 2$$

$$\frac{x + 3}{x - 1} - 2 < 0 \qquad \text{(subtract 2 from both sides)}$$

$$\frac{x + 3 - 2(x - 1)}{x - 1} < 0 \qquad \text{(put left side over common denominator)}$$

$$\frac{-x + 5}{x - 1} < 0 \qquad \text{(simplify)}$$

$$\frac{(-1)(x - 5)}{x - 1} < 0 \qquad \text{(factor } -1 \text{ out of } -x + 5\text{)}$$

$$\frac{x - 5}{x - 1} > 0 \qquad \begin{array}{l}\text{(divide both sides by } -1 \text{ and } \textit{reverse} \\ \textit{direction} \text{ of inequality)}\end{array}$$

Since $(x - 5)/(x - 1)$ can only change sign at the root $x = 5$ or at the vertical asymptote $x = 1$, we have the results shown in Figure 4-52.

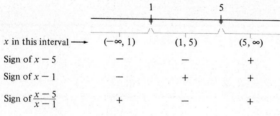

Figure 4-52

Therefore the solutions of $\dfrac{x - 5}{x - 1} > 0$, and hence of $\dfrac{x + 3}{x - 1} < 2$, are all numbers in the intervals $(-\infty, 1)$ and $(5, \infty)$.

It will be helpful to sketch a graphical picture of the solutions of these two inequalities in Figure 4-53.

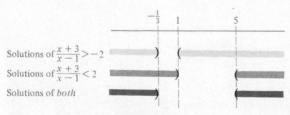

Figure 4-53

The solutions of the original inequality are the numbers that are solutions of both the inequalities above. As the picture shows, x is a solution of *both* inequalities precisely when

$$x < -\tfrac{1}{3} \quad \text{or} \quad x > 5$$

Therefore the solutions of the original inequality $\left|\dfrac{x+3}{x-1}\right| < 2$ are all numbers in the intervals $(-\infty, -\tfrac{1}{3})$ and $(5, \infty)$.

EXAMPLE To solve $|2x^2 + 3x - 2| \geq 2$, we put $2x^2 + 3x - 2$ in place of r and 2 in place of k in the box on page 239 and see that

$$|2x^2 + 3x - 2| \geq 2 \quad \text{exactly when} \quad 2x^2 + 3x - 2 \leq -2 \text{ or } 2x^2 + 3x - 2 \geq 2$$

Thus the solutions of $|2x^2 + 3x - 2| \geq 2$ are all numbers that are solutions of *either one* of the two inequalities on the right above. These two inequalities are equivalent to:

$$2x^2 + 3x \leq 0 \quad \text{or} \quad 2x^2 + 3x - 4 \geq 0$$

To solve the first of these quadratic inequalities we note that the roots of $f(x) = 2x^2 + 3x = x(2x + 3)$ are $x = 0$ and $x = -\tfrac{3}{2}$. Since the graph of f is an upward-opening parabola, the solutions of $f(x) \leq 0$ (that is, $2x^2 + 3x \leq 0$) are all numbers between these two roots. Thus the solutions are all numbers in the interval $[-\tfrac{3}{2}, 0]$. The second of the two inequalities above was solved on page 238; its solutions are all numbers in the intervals $\left(-\infty, \dfrac{-3 - \sqrt{41}}{4}\right]$ and $\left[\dfrac{-3 + \sqrt{41}}{4}, \infty\right)$. Sketching all of these solutions on the same axis, we have Figure 4-54.

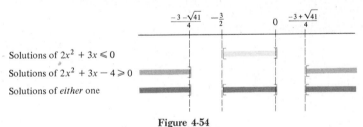

$$\begin{array}{cccc} \dfrac{-3-\sqrt{41}}{4} & -\tfrac{3}{2} & 0 & \dfrac{-3+\sqrt{41}}{4} \end{array}$$

Solutions of $2x^2 + 3x \leq 0$

Solutions of $2x^2 + 3x - 4 \geq 0$

Solutions of *either* one

Figure 4-54

Therefore the solutions of $|2x^2 + 3x - 2| \geq 2$ are all numbers in the intervals

$$\left(-\infty, \dfrac{-3 - \sqrt{41}}{4}\right], \left[-\dfrac{3}{2}, 0\right] \text{ and } \left[\dfrac{-3 + \sqrt{41}}{4}, \infty\right).$$

EXERCISES

A.1. Solve these polynomial inequalities.

(a) $x^2 + x - 2 > 0$

(b) $2x^2 - 5x - 3 < 0$

(c) $x^3 - x \geq 0$

(d) $x^2 \leq 9$

(e) $x^2 > 7$

(f) $x^3 + 2x^2 > -x$

(g) $x^2(x - 4) \leq 5x$

(h) $4x^2 + 10x > -4x^2 - 10x + 12$

B.1. Solve these inequalities.

(a) $\dfrac{x-2}{x-1} < 1$

(d) $\dfrac{-x+5}{2x+3} \geq 2$

(b) $\dfrac{1}{x} < 3$

(e) $\dfrac{x-3}{x+3} \leq 5$

(c) $\dfrac{-1}{x+1} > 2$

(f) $\dfrac{2x+1}{x-4} > 3$

B.2. Solve these inequalities.

(a) $(x-1)(x-2)(x+3) \geq 0$

(d) $\dfrac{x^2-9}{x^2+3x-10} > 0$

(b) $(x^2-1)(x^2-4) < 0$

(e) $\dfrac{x^2+6x+9}{(x-1)(x-2)(x+5)} \leq 0$

(c) $\dfrac{(x+3)(x+2)}{x(x-1)} \geq 0$

(f) $(x-2)^2(x^2-1)(x-3) > 0$

B.3. Solve these inequalities. (*Hint:* begin by collecting all nonzero terms on one side; find a common denominator, and express this side as a single fraction. Then proceed as in Exercise B.2.)

(a) $\dfrac{2}{x+3} \geq \dfrac{1}{x-1}$

(c) $\dfrac{1}{x-1} < \dfrac{1}{x}$

(b) $\dfrac{1}{x+1} > \dfrac{2}{x}$

(d) $\dfrac{1}{x-1} < \dfrac{-1}{x+2}$

B.4. Solve these inequalities. (See the hint for Exercise B.3.)

(a) $\dfrac{9}{x+3} > \dfrac{4}{x+2} + 1$

(c) $\dfrac{1}{2x-2} + \dfrac{9}{2x-6} \geq \dfrac{4}{x-2}$

(b) $1 + \dfrac{1}{x-2} \geq \dfrac{1}{x+2}$

B.5. Solve these quadratic inequalities without factoring.

(a) $x^2 + 7x + 3 > 0$

(d) $-x^2 + 7x + 2 < 0$

(b) $x^2 - 3x + 4 \geq 0$

(e) $-2x^2 + x + 4 \geq 0$

(c) $9x^2 + 30x + 25 \leq 0$

(f) $x^2 - 1 > -3x$

B.6. Solve these inequalities.

(a) $\left| \dfrac{x-1}{x+2} \right| \leq 3$

(d) $\left| \dfrac{x+1}{x+2} \right| \geq 2$

(b) $\left| \dfrac{x+1}{3x+5} \right| < 2$

(e) $\left| \dfrac{1-4x}{2+3x} \right| < 1$

(c) $\left| \dfrac{2x-1}{x+5} \right| > 1$

(f) $\left| \dfrac{3x+1}{1-2x} \right| \geq 2$

B.7. Solve these quadratic inequalities.

(a) $|x^2 - 2| < 1$ (d) $|x^2 - 5| > 4$

(b) $|x^2 - 4| \leq 3$ (e) $|x^2 - 2| > 4$

(c) $|x^2 - 4| < 5$ (f) $\left| \dfrac{1}{x^2 - 1} \right| \leq 2$

B.8. Solve these inequalities.

(a) $|x^2 + x - 2| \leq 4$ (c) $|x^2 - x - 1| \leq 1$

(b) $|x^2 + x - 1| \geq 1$ (d) $|x^2 + 3x - 2| > 3$

C.1. Suppose $ax^2 + bx + c$ is a quadratic polynomial such that $b^2 - 4ac < 0$. Show that:

(a) if $a > 0$, then for any real number x, $|ax^2 + bx + c| = ax^2 + bx + c$;

(b) if $a < 0$, then for any real number x, $|ax^2 + bx + c| = -(ax^2 + bx + c)$.

(*Hint:* consult the facts in the box on p. 237, the information about the discriminant in the box on p. 190, and the definition of absolute value on p. 14.)

C.2. Suppose $ax^2 + bx + c$ is a quadratic polynomial such that $b^2 - 4ac < 0$. Show that:

(a) if $a > 0$, then the inequality $ax^2 + bx + c < 0$ has no solutions;

(b) if $a > 0$, then every real number is a solution of $ax^2 + bx + c > 0$;

(c) if $a < 0$, then $ax^2 + bx + c > 0$ has no solutions:

(d) if $a < 0$, then every real real number is a solution of
$$ax^2 + bx + c < 0.$$

(*Hint:* consult the facts in the box on p. 237 and the facts about the discriminant in the box on p. 190.)

C.3. Use Exercise C.2 to solve each of these inequalities quickly.

(a) $x^2 + x + 1 > 0$ (c) $-x^2 + 3x - 5 \leq 0$

(b) $2x^2 - 4x + 5 < 0$ (d) $-x^2 + 5x - 7 \geq 0$

Exponential and Logarithmic Functions

Exponential functions are necessary for the mathematical description of certain physical situations. The exponential function $t(x) = e^x$, which is introduced here, is one of the most important functions in calculus. The technical problem of actually defining exponents is dealt with first. Then exponential functions and their characteristic pattern of growth and decay are discussed at length.

Logarithms, which were originally invented as a computational tool, lead to logarithmic functions. These functions are closely related to exponential functions and play an equally important role in calculus. In a certain technical sense, explained below, exponential and logarithmic functions are inverses of one another.

Note on Calculators Up to now there have not been very many opportunities for *efficient* use of calculators. Now, however, we shall deal with functions whose rules are not given by simple algebraic formulas. Evaluating these functions by hand is very difficult, except for a few numbers. It will often be necessary to use either tables or calculators. *All the needed tables are provided in the text.* If you do not now have one, it is *not* necessary to buy a calculator.

If you *do* have a calculator with appropriate capabilities, you are encouraged to use it. But remember three things: (i) A calculator is not a substitute for learning the computational material in this chapter. In those cases when computations can be done by hand, you will be expected to do them without a calculator. (ii) In order to use a calculator with maximum efficiency and to interpret certain answers obtained from one, you must have a reasonably complete knowledge of the "theoretical" properties of the functions involved. (iii) Even with a knowledgeable operator, a calculator often provides rational approximations rather than exactly correct answers. For most practical purposes this restriction causes no difficulty, but you should be aware of it.

1. RADICALS AND EXPONENTS

We already know the meaning of square root ($\sqrt{c}$), cube root ($\sqrt[3]{c}$), and exponent notation (c^n, with n a positive integer). In this section we shall extend these ideas. We shall define nth roots, for any positive integer n. These will then be used to provide a meaning for c^r, where r is a real number.

Although there are a number of technicalities involved in the presentation of this material, the vast majority of it consists simply of *new notation for old ideas*.

ROOTS AND RADICALS

You will recall that the square root (or 2nd root) of a nonnegative real number c is the unique nonnegative number $\sqrt{c}$ such that $(\sqrt{c})^2 = c$. Similarly, the cube root (or 3rd root) of *any* real number is the unique number $\sqrt[3]{c}$ such that $(\sqrt[3]{c})^3 = c$. We shall assume the following fact without proof:

> If n *is a positive integer and* c *is a nonnegative real number, then there is a unique nonnegative real number* d *such that* $d^n = c$. *This number* d *is called the* **nth root** *of* c *and is denoted* $\sqrt[n]{c}$.

In other words, $\sqrt[n]{c}$ is the unique *nonnegative* number such that $(\sqrt[n]{c})^n = c$.

EXAMPLE If $n = 2$, then the 2nd root of a nonnegative number c is just the square root of c, since $(\sqrt{c})^2 = c$. We shall continue to write $\sqrt{c}$ rather than $\sqrt[2]{c}$.

EXAMPLE Suppose $n = 4$ and $c = 16$. Then $\sqrt[4]{16} = 2$ since $2^4 = 2 \cdot 2 \cdot 2 \cdot 2 = 16$. It is also true that $(-2)^4 = 16$. But $\sqrt[4]{c}$ is defined to be *nonnegative*. Hence $-2 \neq \sqrt[4]{16}$. Just as with square roots, however, we can say $-2 = -\sqrt[4]{16}$.

Despite the examples above, the nth root of a rational number need *not* be rational. For instance, it can be shown that $\sqrt{2}, \sqrt[3]{2}, \sqrt[4]{2}, \sqrt[5]{2}, \sqrt[6]{2}$, and so on, are all irrational. Of course, if you use a calculator to find one of these numbers—say, $\sqrt[3]{2}$—it will register something like $\sqrt[3]{2} = 1.26$ or $\sqrt[3]{2} = 1.259921$. The numbers 1.26 and 1.259921 are not actually $\sqrt[3]{2}$, but just rational numbers that are good approximations of $\sqrt[3]{2}$. If you must carry a number such as $\sqrt[3]{2}$ through a series of calculations by hand, it is almost always better to write it as $\sqrt[3]{2}$ rather than 1.26.

It is also possible to define the nth root $\sqrt[n]{c}$ of a *negative* number c, provided n is *odd*. For instance, we write:

$$\sqrt[3]{-125} = -5 \quad \text{since} \quad (-5)^3 = -125$$
$$\sqrt[7]{-128} = -2 \quad \text{since} \quad (-2)^7 = -128$$

On the other hand, the square of *every* real number is nonnegative. Similarly, if n is *even*, then $d^n \geq 0$ for every number d. For instance,

$$(-2)^6 = 64, \qquad (-3)^4 = 81, \qquad 5^4 = 625, \qquad (-10)^8 = 100,000,000$$

Consequently, when n is even and $c < 0$, there is *no* real number d such that $d^n = c$; that is, *no negative number has an nth root when* n *is even.* Thus $\sqrt{-3}$, $\sqrt[4]{-17}$, $\sqrt[20]{-673}$, and so on, are *meaningless* expressions in the real number system.

Because the nth root $\sqrt[n]{c}$ is not always defined when c is negative, *we shall deal only with nth roots of nonnegative numbers hereafter,* unless stated otherwise. It is easy to verify that nth roots obey the same kind of laws as do square roots and cube roots:

> *For every real number* $c \geq 0$,
>
> $$(\sqrt[n]{c})^n = c \qquad\qquad \sqrt[n]{c^n} = c$$
>
> $$\sqrt[n]{cd} = \sqrt[n]{c}\,\sqrt[n]{d} \qquad\qquad \sqrt[n]{\frac{c}{d}} = \frac{\sqrt[n]{c}}{\sqrt[n]{d}} \qquad (d \neq 0)$$
>
> $$\sqrt[m]{\sqrt[n]{c}} = \sqrt[mn]{c}$$

EXAMPLES $\sqrt[3]{\sqrt[2]{64}} = \sqrt[3]{8} = 2 = \sqrt[6]{64}$ and $\sqrt{\sqrt[4]{81}} = \sqrt{9} = 3 = \sqrt[4]{81}$.

Warning   It is *not* true that $\sqrt[n]{c + d} = \sqrt[n]{c} + \sqrt[n]{d}$. For instance,

$$\sqrt[4]{16 + 81} = \sqrt[4]{97}, \qquad \text{but} \qquad \sqrt[4]{16} + \sqrt[4]{81} = 2 + 3 = 5 \qquad \text{and} \qquad 5^4 = 625 \neq 97$$

INTEGER EXPONENTS

For some time we have used the exponent notation c^n, where n is a nonnegative integer: $c^1 = c$, $c^2 = c \cdot c$, $c^3 = c \cdot c \cdot c$, and so on. For $c \neq 0$, c^0 was defined to be the number 1. 0^0 is *not* defined. You should recall three basic laws about such exponents.[*]

> $$c^m c^n = c^{m+n} \qquad \frac{c^m}{c^n} = c^{m-n} \qquad (c^m)^n = c^{mn}$$
>
> *for all nonnegative integers* m,n.

We shall now extend this exponent notation, so that c^n will have a meaning when n is a negative integer, such as $-1, -2, -3$, and so on. We would like to do this in such

---

[*] For more details, see the Algebra Review.

a way that the exponent laws in the box above remain true, even for negative exponents. In particular, when $c \neq 0$ we want to have:

$$c^{-1}c^1 = c^{-1+1} = c^0 = 1 \qquad c^{-2}c^2 = c^{-2+2} = c^0 = 1$$
$$c^{-3}c^3 = c^{-3+3} = c^0 = 1, \text{ and so on}$$

But for $c \neq 0$, we *already* have:

$$\frac{1}{c} \cdot c = 1, \qquad \frac{1}{c^2} \cdot c^2 = 1, \qquad \frac{1}{c^3} \cdot c^3 = 1, \qquad \text{and so on}$$

These facts suggest the following definition of negative exponents. For any nonzero number c.

$$c^{-1} \text{ denotes the number } \frac{1}{c}$$

$$c^{-2} \text{ denotes the number } \frac{1}{c^2}$$

$$c^{-3} \text{ denotes the number } \frac{1}{c^3}$$

And in general:

> For any nonzero real number c and positive integer k,
>
> c^{-k} *denotes the number* $\dfrac{1}{c^k}$

Note that negative powers of 0 are not defined since 0^{-1} would be $1/0$, which is *not* a real number.

EXAMPLES $6^{-3} = 1/6^3 = 1/216;$ $(-2)^{-5} = 1/(-2)^5 = 1/{-32};$

$$(1/3)^{-4} = \frac{1}{(1/3)^4} = \frac{1}{(1/3^4)} = \frac{1}{(1/81)} = 81.$$

RATIONAL EXPONENTS

The symbol c^n now has a meaning whenever n is an *integer*. The next step is to extend the exponent notation, so that such expressions as $c^{1/2}$, $c^{3/4}$, $c^{9/4}$, $c^{17/3}$, and so on, will have meaning. Our guide for defining c^r, with c nonnegative and r rational, is the basic law of integer exponents: $c^m c^n = c^{m+n}$. We would like this to remain true even when m and n are rational numbers. Thus we should define $c^{1/2}$ in such a way that

$$c^{1/2}c^{1/2} = c^{(1/2)+(1/2)} = c^1 = c$$

and $c^{1/3}$ so that

$$c^{1/3}c^{1/3}c^{1/3} = c^{(1/3)+(1/3)+(1/3)} = c^1 = c$$

and so on. In other words, we want

$$(c^{1/2})^2 = c, \qquad (c^{1/3})^3 = c, \qquad \text{and so on}$$

But for $c \geq 0$ we already know that

$$(\sqrt{c})^2 = c, \qquad (\sqrt[3]{c})^3 = c, \qquad \text{and so on}$$

Therefore it is reasonable to make the following definition. For any positive number c,

$$c^{1/2} \text{ denotes the number } \sqrt{c}$$
$$c^{1/3} \text{ denotes the number } \sqrt[3]{c}$$

and in general:

> *For any nonnegative real number* c *and positive integer* k,
>
> $c^{1/k}$ *denotes the number* $\sqrt[k]{c}$

EXAMPLES $4^{1/2} = \sqrt{4} = 2; \qquad 27^{1/3} = \sqrt[3]{27} = 3;$
$(32/7)^{1/5} = \sqrt[5]{32/7} = \sqrt[5]{32}/\sqrt[5]{7} = 2/\sqrt[5]{7}.$

We also want the second law of integer exponents, $(c^m)^n = c^{mn}$, to hold when m and n are *rational*. In particular, since $t/k = (1/k)t$ and $t/k = t(1/k)$, we want to define $c^{t/k}$ in such a way that

$$c^{t/k} = (c^{1/k})^t = (\sqrt[k]{c})^t \qquad \text{and} \qquad c^{t/k} = (c^t)^{1/k} = \sqrt[k]{c^t}$$

But when c, k, and t are positive, we have

$$(\sqrt[k]{c})^t = \sqrt[k]{c} \cdot \sqrt[k]{c} \cdots \sqrt[k]{c} = \sqrt[k]{c \cdot c \cdots c} = \sqrt[k]{c^t}$$

A similar argument works when t is negative. Therefore for any integers t, k, with $k > 0$ and any positive real number c,

$$(\sqrt[k]{c})^t = \sqrt[k]{c^t}$$

It is now reasonable to make this definition:

> *For every positive real number* c *and integers* t *and* k *with* k > 0,
>
> $c^{t/k}$ *denotes the number* $(\sqrt[k]{c})^t = \sqrt[k]{c^t}$

EXAMPLES $8^{2/3} = \sqrt[3]{8^2} = \sqrt[3]{64} = 4$ and $(\sqrt{2})^{4/5} = \sqrt[5]{(\sqrt{2})^4} = \sqrt[5]{(\sqrt{2})^2(\sqrt{2})^2} = \sqrt[5]{4}.$

EXAMPLE $5^{-2/3} = \sqrt[3]{5^{-2}} = \sqrt[3]{1/5^2} = \sqrt[3]{1}/\sqrt[3]{5^2} = 1/\sqrt[3]{25}.$

EXAMPLE Since every finite decimal is a rational number, expressions such as

$$10^{1.41}, \quad 5^{.3}, \quad (\tfrac{2}{3})^{7.591}$$

now have a meaning. For instance, since $1.41 = 1\tfrac{41}{100} = \tfrac{141}{100}$, we have $10^{1.41} = 10^{141/100} = \sqrt[100]{10^{141}}$. Writing rational exponents in decimal rather than fraction form is especially convenient when using a calculator.

There is a subtle point in the definition of $c^{t/k}$: the same rational number can be expressed many ways. For example,

$$\tfrac{1}{2} = \tfrac{2}{4} = \tfrac{3}{6} = \tfrac{4}{8} \quad \text{and} \quad \tfrac{2}{3} = \tfrac{6}{9} = \tfrac{8}{12} = \tfrac{10}{15}$$

Consequently, we should verify that if $t/k = a/b$, then $c^{t/k} = c^{a/b}$. Rather than present a detailed proof, we shall simply illustrate this fact with an example.

EXAMPLE $5^{1/2} = \sqrt{5}$ and $5^{3/6} = \sqrt[6]{5^3} = \sqrt[6]{125} = \sqrt{\sqrt[3]{125}} = \sqrt{5}$.

LAWS OF EXPONENTS

Our definition $c^{t/k} = \sqrt[k]{c^t}$ was motivated by a desire to have the same laws that hold for integer exponents also hold for rational exponents. In making this definition we guaranteed that some of these exponent laws would hold (such as $c^{-1}c^1 = 1$ and $(c^{1/k})^k = c$). But we have not yet verified that *all* the exponent laws are valid in *all* possible cases. Since the proofs of these various facts are rather tedious and unenlightening, we shall omit them and simply illustrate each one with some examples.

If c *and* d *are positive real numbers and* r, s *are rational numbers, then*

(i) $\quad c^r c^s = c^{r+s}$ $\qquad\qquad$ (iv) $\quad (cd)^r = c^r d^r$

(ii) $\quad \dfrac{c^r}{c^s} = c^{r-s}$ $\qquad\qquad$ (v) $\quad \left(\dfrac{c}{d}\right)^r = \dfrac{c^r}{d^r}$

(iii) $\quad (c^r)^s = c^{rs}$ $\qquad\qquad$ (vi) $\quad c^{-r} = \dfrac{1}{c^r}$

EXAMPLE (i) $(8^{2/3})(8^{1/3}) = 8^{(2/3)+(1/3)} = 8^1$ since $(8^{2/3})(8^{1/3}) = (\sqrt[3]{8^2})(\sqrt[3]{8}) = (\sqrt[3]{64})(2) = 4 \cdot 2 = 8$. Similarly, $(4^{3/2})(4^{-1/2}) = 4^{(3/2)+(-1/2)} = 4^1$ since $(4^{3/2})(4^{-1/2}) = \sqrt{4^3}\sqrt{4^{-1}} = \sqrt{64}\sqrt{\tfrac{1}{4}} = 8 \cdot \tfrac{1}{2} = 4$.

EXAMPLE (ii) $\dfrac{3^{-2}}{3^{3/2}} = 3^{-2-(3/2)} = 3^{-7/2}$ since $\dfrac{3^{-2}}{3^{3/2}} = \dfrac{1/3^2}{\sqrt{3^3}} = \dfrac{1}{3^2} \cdot \dfrac{1}{\sqrt{27}} = \dfrac{1}{9} \cdot \dfrac{1}{3\sqrt{3}} = \dfrac{1}{27\sqrt{3}}$ and $3^{-7/2} = \sqrt{3^{-7}} = \sqrt{\dfrac{1}{3^7}} = \dfrac{1}{\sqrt{3^6}\sqrt{3}} = \dfrac{1}{3^3\sqrt{3}} = \dfrac{1}{27\sqrt{3}}$.

EXAMPLE   (iii) $(4^{3/2})^2 = (\sqrt{4^3})^2 = (\sqrt{64})^2 = 64$ and $4^{(3/2)2} = 4^3 = 64$,   so $(4^{3/2})^2 = 4^{(3/2)2}$.

EXAMPLE (iv) $(8 \cdot 27)^{2/3} = \sqrt[3]{(8 \cdot 27)^2} = \sqrt[3]{8^2 \cdot 27^2} = \sqrt[3]{8^2} \sqrt[3]{27^2} = (8^{2/3})(27^{2/3})$.

EXAMPLE (v) $\left(\dfrac{7}{3}\right)^{1/5} = \sqrt[5]{\dfrac{7}{3}} = \dfrac{\sqrt[5]{7}}{\sqrt[5]{3}} = \dfrac{7^{1/5}}{3^{1/5}}$.

EXAMPLE (vi) $3^{-2/5} = \sqrt[5]{3^{-2}} = \sqrt[5]{\dfrac{1}{3^2}} = \dfrac{\sqrt[5]{1}}{\sqrt[5]{3^2}} = \dfrac{1}{3^{2/5}}$.

The exponent laws listed above can often be used to simplify complicated expressions, as illustrated in the following examples.

EXAMPLE   Express $\dfrac{3^{-2}x^{-3}}{3^{-3}y^3}$ without negative exponents.

$$\frac{3^{-2}x^{-3}}{3^{-3}y^3} = \frac{\dfrac{1}{3^2} \cdot \dfrac{1}{x^3}}{\dfrac{1}{3^3} \cdot y^3} = \frac{1}{3^2 x^3} \cdot \frac{3^3}{y^3} = \frac{3}{x^3 y^3}$$

EXAMPLE   Let k be a positive rational number and express $\sqrt[10]{c^{5k}} \sqrt{(c^{-k})^{1/2}}$ without radicals, using only positive exponents:

$$\sqrt[10]{c^{5k}} \sqrt{(c^{-k})^{1/2}} = (c^{5k})^{1/10}[(c^{-k})^{1/2}]^{1/2} = c^{k/2}c^{-k/4} = c^{2k/4}c^{-k/4} = c^{k/4}$$

EXAMPLE   Express $\sqrt[5]{\dfrac{\sqrt[4]{7^3 \cdot x^5}}{\sqrt[3]{7^2 y^4}}}$ without radicals.

$$\sqrt[5]{\frac{\sqrt[4]{7^3 \cdot x^5}}{\sqrt[3]{7^2 \cdot y^4}}} = \left(\frac{(7^3 \cdot x^5)^{1/4}}{(7^2 \cdot y^4)^{1/3}}\right)^{1/5} = \frac{[(7^3 \cdot x^5)^{1/4}]^{1/5}}{[(7^2 \cdot y^4)^{1/3}]^{1/5}} = \frac{(7^3 \cdot x^5)^{1/20}}{(7^2 \cdot y^4)^{1/15}} = \frac{7^{3/20}x^{5/20}}{7^{2/15}y^{4/15}}$$

$$= \frac{7^{9/60}x^{1/4}}{7^{8/60}y^{4/15}} = \frac{7^{1/60}x^{1/4}}{y^{4/15}}$$

IRRATIONAL EXPONENTS

The expression c^r now has a meaning whenever r is a rational number and $c > 0$. Defining c^t when t is an *irrational* number, however, is not so simple. Without resorting to limits and other topics covered in calculus it is impossible to give a precise definition of c^t. The best we can do at this stage is to make it *plausible* that:

> For each positive real number c and each irrational number t, there is a well-defined real number, denoted c^t

The general idea can best be understood in a specific example. Suppose $c = 10$ and we wish to define $10^{\sqrt{2}}$. The key fact is that *every irrational number*, such as $\sqrt{2}$, *has an infinite decimal expansion*.° For instance,

$$\sqrt{2} = 1.4142135623 \cdots$$

Consequently, every irrational number may be approximated to any desired degree of accuracy by a suitable finite decimal (that is, a rational number). For instance, there is an infinite list of decimal approximations of $\sqrt{2}$, each more accurate than the preceding:

$$1.4, \quad 1.41, \quad 1.414, \quad 1.4142, \quad 1.41421, \quad 1.414213, \quad \cdots$$

Each of these rational numbers is slightly larger than the preceding one, and *all* of them lie between 1.4 and 1.5.

We know how to raise 10 to each of the rational powers: $10^{1.14} = 10^{14/10}$, $10^{1.41} = 10^{141/100}$, and so on. It seems *reasonable*, therefore, that $10^{\sqrt{2}}$ be defined to be the real number for which the numbers

$$10^{1.4}, \quad 10^{1.41}, \quad 10^{1.414}, \quad 10^{1.4142}, \quad 10^{1.41421}, \quad 10^{1.414213}, \quad \cdots$$

are better and better approximations. Using a calculator, we find that°°

$$10^{1.4} \approx 25.1189$$
$$10^{1.41} \approx 25.7040$$
$$10^{1.414} \approx 25.9418$$
$$10^{1.4142} \approx 25.9537$$
$$10^{1.41421} \approx 25.9543$$
$$10^{1.414213} \approx 25.9545$$
$$\vdots$$

It appears that $10^{1.4}$, $10^{1.41}$, and so on, are better and better approximations of a specific real number, whose infinite decimal expansion begins $25.954 \cdots$. This number is defined to be $10^{\sqrt{2}}$.

A similar procedure can be used to define c^t for any positive real number c and any irrational number t. Although a proof is beyond the scope of this book, we shall assume this and use the fact that the exponent laws are valid for *all* real exponents, rational or irrational.

EXERCISES

A.1. Compute (without a calculator):

(a) $(.001)^{-5/3}$

(b) $(\frac{4}{49})^{-3/2}$

(c) $16^{-5/4}$

(d) $(1000)^{5/3}$

(e) $(1,000,000)^{5/6}$

(f) $(\frac{1}{5})^{-2} \cdot 5^{-3}$

(g) $\dfrac{3^0 - 0^3}{3 - 0}$

(h) $2^{-2} + 2^{-3} + 3^{-2}$

(i) $(-2)^2 + 2^{-2}$

° See pages 9–10 for more details.
°° $\approx$ means "approximately equal."

A.2. Compute carefully:

(a) $2^4 - 2^7 = ?$ (c) $(2^{1/2} + 2)^2 = ?$ (e) $\dfrac{1}{2^3} + \dfrac{1}{2^4} = ?$

(b) $3^3 - 3^{-7} = ?$ (d) $2^2 \cdot 3^3 - 3^2 \cdot 2^3 = ?$ (f) $3^2\left(\dfrac{1}{3} + \dfrac{1}{3^{-4}}\right) = ?$

A.3. Express each of the given numbers as a power of 2. (*Example:* $\sqrt{8} = \sqrt{2^3} = 2^{3/2}$.)

(a) 64   (d) $\sqrt[4]{8}$ (g) $\dfrac{\sqrt{8}}{\sqrt[3]{4}}$

(b) $(2^4 \cdot 16^{-2})^3$ (e) $(\tfrac{1}{2})^{-8}(\tfrac{1}{4})^4(\tfrac{1}{8})^{-3}$ (h) $\dfrac{1024}{\sqrt[4]{64}}$

(c) $\sqrt[3]{16}$ (f) $\sqrt[3]{256}$ (i) $\dfrac{\sqrt{128}}{2048}$

A.4. Find an equivalent expression, involving at most one radical sign. Assume all letters represent positive numbers.

(a) $\sqrt{c\sqrt{c^{10}}}$ (c) $\sqrt[5]{2}\sqrt[5]{ab}\sqrt[5]{7cd}$ (e) $\sqrt{2}\sqrt[3]{2}$

(b) $\dfrac{\sqrt{xy}}{\sqrt[4]{x^2y}}$ (d) $\left(\sqrt[3]{\dfrac{16rs}{t}}\right)(\sqrt[3]{4s^2t^7})$ (f) $\sqrt[3]{40} + 2\sqrt[3]{135} - 4\sqrt[3]{320}$

A.5. Simplify and express the following without negative exponents. (Assume r, s, t are positive integers and a, b, c are positive real numbers.)

(a) $\dfrac{3^{-r}}{3^{-s}}$ (d) $\dfrac{4^{-(t+1)}}{9r^{-2}}$ (g) $(4c^{3/2})(2c^{-1/2})$

(b) $a^{-2}b^3$ (e) $\dfrac{r^{-t}}{(6s)^{-s}}$ (h) $a^2(a^{-1} + a^{-3})$

(c) $(a^2a^{-5}a^4)^{-3}$ (f) $\dfrac{a^{-2}(b^2c^3)^{-3}}{(a^{-5}b^{-4})^2c^{-7}}$ (i) $\dfrac{a^{-2}}{b^{-2}} + \dfrac{b^2}{a^2}$

A.6. Express without radicals:

(a) $\sqrt[3]{8a^3b^3}$ (d) $\sqrt[4]{81x^8y^8}$ (g) $\sqrt[4]{\sqrt[4]{a^3}}$

(b) $\sqrt[3]{27a^6b^3}$ (e) $\sqrt[3]{a^2 + b^2}$ (h) $\sqrt{\sqrt[3]{a^3b^8}}$

(c) $\sqrt[5]{32c^{10}d^{15}}$ (f) $\sqrt[4]{a^3 - b^3}$ (i) $\sqrt{x}\sqrt[3]{x^2}\sqrt[4]{x^3}$

B.1. Compute and simplify your answer as much as possible. (Assume x and y are positive real numbers).

(a) $x^{1/2}(x^{2/3} - x^{4/3})$ (d) $(x^{1/3} + y^{1/2})(2x^{1/3} - y^{3/2})$

(b) $x^{1/2}(3x^{3/2} + 2x^{-1/2})$ (e) $(x + y)^{1/2}[(x + y)^{1/2} - (x + y)]$

(c) $(x^{1/2} + y^{1/2})(x^{1/2} - y^{1/2})$ (f) $(x^{1/3} + y^{1/3})(x^{2/3} - x^{1/3}y^{1/3} + y^{2/3})$

B.2. Factor these expressions. [For example, $x - x^{1/2} - 2 = (x^{1/2} - 2)(x^{1/2} + 1)$.]

(a) $x^{2/3} + x^{1/3} - 6$ (c) $x + 4x^{1/2} + 3$ (e) $x^{4/5} - 81$

(b) $x^{2/5} + 11x^{1/5} + 30$ (d) $x^{1/3} + 7x^{1/6} + 10$ (f) $x^{2/3} - 6x^{1/3} + 9$

B.3. *Errors to avoid.* In each part, give an example to show that the given statement may be *false* for some numbers.

    (a) $a^r + b^r = (a + b)^r$ (c) $a^r b^s = (ab)^{r+s}$ (e) $c^{-r} = -c^r$

    (b) $a^r a^s = a^{rs}$ (d) $\dfrac{c^r}{c^s} = c^{r/s}$ (f) $a^r + b^s = (a + b)^{r+s}$

B.4. Simplify the following expressions. (Assume a, b, n, t, x, y are positive rational numbers.)

    (a) $\dfrac{2^{11} \cdot 2^{-7} \cdot 2^{-5}}{2^3 \cdot 2^{-3}}$ (d) $(x^{1/2}y^3)(x^0 y^7)^{-2}$ (g) $(a^{x^2})^{1/x}$

    (b) $\sqrt{x^7} \cdot x^{5/2} \cdot x^{-3/2}$ (e) $\dfrac{(7a)^2(5b)^3}{(5a)^3(7b)^4}$ (h) $\dfrac{(x^2 y^7)^{-2}}{x^{-3}y^4}$

    (c) $\dfrac{(3^2)^{-1/2}(9^4)^{-1}}{27^{-3}}$ (f) $\sqrt[5]{t\,16t^4}$ (i) $\sqrt[6]{b^{3x}}\sqrt{(b^{-x})^{1/2}}$

B.5. Express without radicals, using only positive exponents:

    (a) $(\sqrt[3]{xy^2})^{-3/5}$ (c) $\dfrac{c}{(c^{5/6})^{42}(c^{51})^{-2/3}}$ (e) $(c^{5/6} - c^{-5/6})^2$

    (b) $(\sqrt[4]{r^{14}s^{-21/5}})^{-3/7}$ (d) $\sqrt[5]{\dfrac{(ab^2)^{-10/3}}{(a^2 b)^{-15/7}}}$ (f) $(\sqrt{a} + b^{-1/3})^{-2}$

B.6. Since $(\sqrt{2} + \sqrt{3})^2 = 2 + 2\sqrt{2} \cdot \sqrt{3} + 3 = 5 + 2\sqrt{6}$ and $\sqrt{2} + \sqrt{3} \geq 0$, we conclude $\sqrt{5 + 2\sqrt{6}} = \sqrt{2} + \sqrt{3}$. Use similar methods to verify each of the following statements.

    (a) $\sqrt{9 - 2\sqrt{14}} = \sqrt{7} - \sqrt{2}$ (c) $\sqrt[3]{-17\sqrt{2} + 11\sqrt{5}} = \sqrt{5} - \sqrt{2}$

    (b) $\sqrt{8 + 2\sqrt{15}} = \sqrt{3} + \sqrt{5}$ (d) $\sqrt[4]{49 + 20\sqrt{6}} = \sqrt{2} + \sqrt{3}$

B.7. Use a calculator to find a two-place decimal approximation of each of these numbers. (Also see Exercise B.8.)

    (a) $\sqrt[5]{176}$ (c) $(\tfrac{2}{3})^{.65}$ (e) $(\sqrt[3]{17})^{5/7}$

    (b) $(\tfrac{3}{2})^{7.56}$ (d) $(\sqrt{6})^{4/5}$ (f) $2^{936/125}$

B.8. Use a calculator to find a *six*-place decimal approximation of $(311)^{-4.2}$. Explain why your answer cannot possibly *be* the number $(311)^{-4.2}$. What goes wrong?

B.9. Express each of these numbers as a product of an integer power of 10 and a number between 1 and 10. For example, $793.1 = (7.931)10^2$ and $.0076 = (7.6)10^{-3}$. A number written in this way is said to be written in **scientific notation.** [*Note:* some calculators are equipped with scientific notation. See the *Calculator Note* on page 285.]

    (a) 79,327   (e) 5,963,000,000,000

    (b) 5,200,000   (f) .0000000035

    (c) .002   (g) $\dfrac{.0032}{160,000,000}$

    (d) .00000079   (h) $\dfrac{(1,000,000)^2\,\sqrt{.00000004}}{(8,000,000,000)^{2/3}}$

DO IT YOURSELF!

RADICAL EQUATIONS

If two numbers are equal—say, $a = b$—then it is certainly true that $a^2 = b^2$ and $a^3 = b^3$. In fact,

$$\text{if } a = b, \qquad \text{then } a^r = b^r \text{ for every number } r$$

This statement is also valid for algebraic expressions that represent real numbers. For example,

$$\text{if } x - 2 = 3, \qquad \text{then } (x - 2)^2 = 3^2$$

In particular, every solution of the equation $x - 2 = 3$ is also a solution of $(x - 2)^2 = 9$. But *be careful*. This usually works only in *one* direction. In this case, it is *not* true that every solution of $(x - 2)^2 = 9$ is a solution of $x - 2 = 3$. You can easily verify that $x = -1$ is a solution of $(x - 2)^2 = 9$, but $x = -1$ is *not* a solution of $x - 2 = 3$. This example is a good illustration of the following principle:

POWER PRINCIPLE

Every solution of the equation A = B *is also a solution of the equation* $A^r = B^r$, *but not necessarily vice-versa.*

Here are some examples that show how the Power Principle can be used to solve various equations involving radicals and fractional exponents.

EXAMPLE In order to solve the equation $\sqrt[3]{2x^2 + 7x - 6} = 3^{2/3}$, we cube both sides and obtain

$$(\sqrt[3]{2x^2 + 7x - 6})^3 = (3^{2/3})^3$$
$$2x^2 + 7x - 6 = 9$$

Any solution of the original equation must also be a solution of this one by the Power Principle. This last equation can easily be solved.

$$2x^2 + 7x - 6 = 9$$
$$2x^2 + 7x - 15 = 0$$
$$(2x - 3)(x + 5) = 0$$

$$2x - 3 = 0 \qquad \text{or} \qquad x + 5 = 0$$
$$2x = 3 \qquad \text{or} \qquad x = -5$$
$$x = \tfrac{3}{2} \qquad \text{or} \qquad x = -5$$

Therefore the only *possible* solutions of the original equation are $x = \tfrac{3}{2}$ and $x = -5$. By substituting these numbers in the original equation, we can determine its solutions.

Since

$$\sqrt[3]{2(-5)^2 + 7(-5) - 6} = \sqrt[3]{50 - 35 - 6} = \sqrt[3]{9} = \sqrt[3]{3^2} = 3^{2/3}$$
$$\sqrt[3]{2(\tfrac{3}{2})^2 + 7(\tfrac{3}{2}) - 6} = \sqrt[3]{2(\tfrac{9}{4}) + \tfrac{21}{2} - 6} = \sqrt[3]{\tfrac{9}{2} + \tfrac{21}{2} - \tfrac{12}{2}} = \sqrt[3]{\tfrac{18}{2}} = \sqrt[3]{9} = 3^{2/3}$$

both $x = \tfrac{3}{2}$ and $x = -5$ are solutions of the equation $\sqrt[3]{2x^2 + 7x - 6} = 3^{2/3}$.

EXAMPLE In order to solve $3 + \sqrt{3 - x} - x = 0$ we rearrange terms and square:

$$\sqrt{3 - x} = x - 3$$
$$(\sqrt{3 - x})^2 = (x - 3)^2$$
$$3 - x = x^2 - 6x + 9$$

According to the Power Principle, the only *possible* solutions of the original equation are the solutions of this last equation. It can be solved as follows:

$$3 - x = x^2 - 6x + 9$$
$$0 = x^2 - 5x + 6$$
$$0 = (x - 3)(x - 2)$$
$$x = 3 \quad \text{or} \quad x = 2$$

Substituting these numbers in the left side of the original equation $3 + \sqrt{3 - x} - x = 0$ yields:

$$3 + \sqrt{3 - 3} - 3 = 3 + \sqrt{0} - 3 = 0$$
$$3 + \sqrt{3 - 2} - 2 = 3 + \sqrt{1} - 2 = 4 - 2 = 2$$

Therefore, $x = 2$ is *not* a solution of the original equation; its only solution is $x = 3$.

EXAMPLE In order to solve the equation $\sqrt{2x - 3} - \sqrt{x + 7} = 2$, the first step is to rearrange terms so that one side of the equation contains only a *single* radical term:

$$\sqrt{2x - 3} = \sqrt{x + 7} + 2$$

Squaring both sides and simplifying yields

$$(\sqrt{2x - 3})^2 = (\sqrt{x + 7} + 2)^2$$
$$2x - 3 = (\sqrt{x + 7})^2 + 2 \cdot 2\sqrt{x + 7} + 2^2$$
$$2x - 3 = x + 7 + 4\sqrt{x + 7} + 4$$
$$x - 14 = 4\sqrt{x + 7}$$

In order to eliminate the radical, we square both sides again and simplify:

$$(x - 14)^2 = (4\sqrt{x + 7})^2$$
$$x^2 - 28x + 196 = 4^2 \cdot (\sqrt{x + 7})^2$$
$$x^2 - 28x + 196 = 16(x + 7)$$
$$x^2 - 28x + 196 = 16x + 112$$
$$x^2 - 44x + 84 = 0$$

According to the Power Principle, applied *twice*, the solutions of the original equation must also be solutions of this last equation. Its solutions are found as follows:

$$x^2 - 44x + 84 = 0$$
$$(x - 2)(x - 42) = 0$$

$$x - 2 = 0 \quad \text{or} \quad x - 42 = 0$$
$$x = 2 \quad \text{or} \quad x = 42$$

Substituting $x = 2$ and $x = 42$ in the left side of the original equation, $\sqrt{2x - 3} - \sqrt{x + 7} = 2$, we find that $x = 42$ *is* a solution, but $x = 2$ is *not*. (Verify this!)

EXAMPLE   A different technique is needed to solve $x^{2/3} - 2x^{1/3} - 15 = 0$. The fact that $x^{2/3} = (x^{1/3})^2$ suggests that we let $u = x^{1/3}$ so that the equation becomes:

$$x^{2/3} - 2x^{1/3} - 15 = 0$$
$$(x^{1/3})^2 - 2x^{1/3} - 15 = 0$$
$$u^2 - 2u - 15 = 0$$
$$(u - 5)(u + 3) = 0$$

$$u - 5 = 0 \quad \text{or} \quad u + 3 = 0$$
$$u = 5 \quad \text{or} \quad u = -3$$
$$x^{1/3} = 5 \quad \text{or} \quad x^{1/3} = -3$$

Since $x = (x^{1/3})^3$, the only possible solutions are $5^3 = 125$ and $(-3)^3 = -27$. Substituting $x = 125$ and $x = -27$ in the left side of the original equation, $x^{2/3} - 2x^{1/3} - 15 = 0$, shows that *both* numbers are solutions. (Verify this!)

EXERCISES

A.1.   Use the Power Principle to solve these equations. Be sure to *check your answers* in the original equation.

(a) $\sqrt[3]{x^2 - 1} = 2$
(b) $\sqrt{4x + 9} = 5$
(c) $\sqrt[3]{4 - 11x} = 3$
(d) $\sqrt[3]{4x^2 + 1} = 5$
(e) $\sqrt{x + 7} = x - 5$
(f) $\sqrt{x + 5} = x - 1$

B.1.   Solve these equations. [*Hint:* first express fractional exponents in terms of radicals.]

(a) $(x - 3)^{2/3} = 2$
(b) $(x^2 + 2x + 1)^{3/2} = 8$
(c) $(x^2 - x + 123)^{1/5} = 5^{3/5}$
(d) $(x^2 - 4x + 3)^{1/4} = 2^{3/4}$
(e) $(x + 5)^{1/3} = (x + 3)^{2/3}$
(f) $(2x^2 + 17x + 17)^{1/7} = (2x + 4)^{2/7}$

B.2.   Solve these equations.

(a) $\sqrt{5x + 6} = 3 + \sqrt{x + 3}$
(b) $\sqrt{3y + 1} - 1 = \sqrt{y + 4}$
(c) $\sqrt{2x - 5} = 1 + \sqrt{x - 3}$
(d) $\sqrt{x - 3} + \sqrt{x + 5} = 4$
(e) $\sqrt{3x + 5} + \sqrt{2x + 3} + 1 = 0$
(f) $\sqrt{20 - x} = \sqrt{9 - x} + 3$
(g) $\sqrt{6y + 7} - \sqrt{3y + 3} = 1$
(h) $\sqrt{t + 2} + \sqrt{3t + 4} = 2$

B.3. Assume that all letters represent positive numbers and solve each equation for the required letter.

(a) $A = \sqrt{1 + \dfrac{a^2}{b^2}}$ for b (c) $K = \sqrt{1 - \dfrac{x^2}{u^2}}$ for u

(b) $T = 2\pi \sqrt{\dfrac{m}{g}}$ for g (d) $R = \sqrt{d^2 + k^2}$ for d

B.4. Solve these equations by making an appropriate substitution.

(a) $2x^{2/3} - x^{1/3} - 6 = 0$ (d) $x^{1/3} + x^{1/6} - 2 = 0$

(b) $x^{2/3} - 2x^{1/3} - 8 = 0$ (e) $x - 10\sqrt{x} + 9 = 0$

(c) $x^{1/2} - x^{1/4} - 2 = 0$ (f) $(1 + \sqrt{x})^2 + (1 + \sqrt{x}) = 6$

B.5. Solve these equations. [*Hint:* in part (a), use the substitution $u = x^{-1}$, then factor and solve. Use a similar technique for the other parts.]

(a) $x^{-2} - x^{-1} - 6 = 0$ (c) $x^{-1} + 2x^{-1/2} - 3 = 0$

(b) $2x^{-2} + x^{-1} - 1 = 0$ (d) $x^{-2/3} - x^{-1/3} - 6 = 0$

2. EXPONENTIAL FUNCTIONS

Exponential functions have numerous applications (some of which are discussed below) and will play an important role in calculus. In this section we carefully examine the graphs and general behavior of such functions.

As we saw in Section 1: for every positive real number a and every real number x,

$$a^x \text{ is a well-defined real number}$$

Consequently, for *each* positive real number a, there is a function (called an **exponential function**) whose rule is: $f(x) = a^x$. For example, there are the exponential functions

$$f(x) = 10^x, \qquad g(x) = 2^x, \qquad h(x) = (\tfrac{1}{2})^x, \qquad k(x) = (\tfrac{3}{7})^x, \qquad r(x) = \pi^x$$

GRAPH OF f(x) = aˣ WHEN a > 1

When $a > 1$, all the various exponential functions $f(x) = a^x$ have similar graphs, as the following examples illustrate.

EXAMPLE Let $a = 10$. In order to sketch the graph of $f(x) = 10^x$, we use a calculator or tables to compute the value of the function at various numbers:

$x =$	-3	-2	-1	$-.5$	0	$.5$	1	1.5	2	2.125	3
$10^x \approx$	.001	.01	.1	.316	1	3.16	10	31.6	100	133.4	1000

Next we plot the corresponding points and make a reasonable guess as to the shape of the graph (see Figure 5-1).

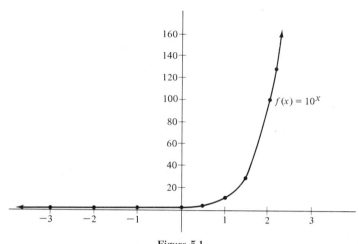

Figure 5-1

EXAMPLE   In order to sketch the graphs of the exponential functions

$$g(x) = (\tfrac{3}{2})^x, \qquad h(x) = 2^x, \qquad k(x) = 3^x$$

we first evaluate the functions at various numbers (using a calculator when necessary).

x	-3	-2	-1	$-.5$	0	.5	1	2	2.5	3	4	5
$g(x) = (\tfrac{3}{2})^x$	$\tfrac{8}{27}$	$\tfrac{4}{9}$	$\tfrac{2}{3}$	.82	1	1.22	1.5	2.25	2.76	3.38	5.06	7.59
$h(x) = 2^x$	$\tfrac{1}{8}$	$\tfrac{1}{4}$	$\tfrac{1}{2}$	.71	1	1.41	2	4	5.66	8	16	32
$k(x) = 3^x$	$\tfrac{1}{27}$	$\tfrac{1}{9}$	$\tfrac{1}{3}$	.58	1	1.73	3	9	15.59	27	81	243

Now we plot the points corresponding to the values on the table and sketch the graphs on the same set of coordinate axes (see Figure 5-2). For comparison purposes, the graph of $f(x) = 10^x$ from the previous example is also included.

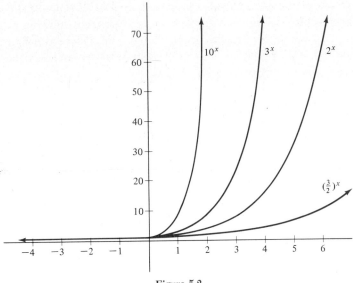

Figure 5-2

Notice that *all of the graphs* in the preceding examples have the same general shape. The principal difference between them is that

> *If* $b > a > 1$, *then the graph of* $f(x) = b^x$ *rises more steeply than the graph of* $g(x) = a^x$.

For instance, $10 > \frac{3}{2}$ and the graph of $f(x) = 10^x$ rises more steeply than the graph of $g(x) = (\frac{3}{2})^x$. Furthermore, observe that $\frac{3}{2} < 2 < 3$ and the graph of $h(x) = 2^x$ lies *between* the graph of $g(x) = (\frac{3}{2})^x$ and $k(x) = 3^x$. In general,

> *If* $1 < a < b < c$, *then the graph of* $h(x) = b^x$ *lies between the graph of* $g(x) = a^x$ *and* $k(x) = c^x$.

EXAMPLE   There is a certain irrational number that plays an important role in calculus. It is denoted e. The infinite decimal expansion of e begins $e = 2.71828 \cdots$.

One example of how the number e arises is given in the DO IT YOURSELF! segment on page 269. The exponential function $t(x) = e^x$ occurs frequently in various applications. Since $2 < e < 3$, we know that the graph of $t(x) = e^x$ lies between the graphs of $h(x) = 2^x$ and $k(x) = 3^x$. Using a calculator, we can sketch a reasonably accurate graph of the function t, as shown in Figure 5-3.

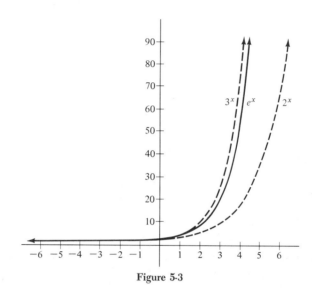

x	$t(x) = e^x$
-3	.05
-2	.14
-1	.37
0	1
1	2.72
2	7.39
3	20.09
4	54.6
4.5	90.02

Figure 5-3

GRAPH OF $f(x) = a^x$ WHEN $0 < a < 1$

When $0 < a < 1$, the graph of $f(x) = a^x$ has a different shape, as we now see.

EXAMPLE   In order to sketch the graph of $s(x) = (\frac{1}{2})^x$, we use a calculator to determine some values of the function.

x	-5	-4.5	-4	-3.5	-3	-2.5	-2	-1	$-.5$	0	.5	1	2	3
$(\frac{1}{2})^x$	32	22.63	16	11.31	8	5.66	4	2	1.41	1	.707	.5	.25	.125

By plotting the corresponding points, we see that the graph of $s(x) = (\frac{1}{2})^x$ is as shown in Figure 5-4. [The graph of $h(x) = 2^x$ is included for comparison purposes.]

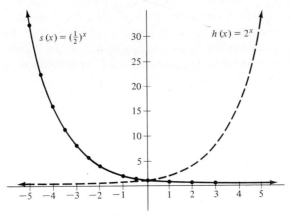

Figure 5-4

If you compare the graph of $h(x) = 2^x$ with the graph of $s(x) = (\frac{1}{2})^x = 1/2^x$, you will see that they are mirror images of each other (with the y-axis as mirror). In a similar manner, the graphs of the exponential functions

$$r(x) = (\tfrac{2}{3})^x = \left(\frac{1}{3/2}\right)^x, \qquad s(x) = (\tfrac{1}{3})^x, \qquad w(x) = (\tfrac{1}{10})^x$$

are just mirror images (in the y-axis) of the graphs of $g(x) = (\tfrac{3}{2})^x$, $k(x) = 3^x$, and $f(x) = 10^x$, respectively.

Here is a summary of some of the properties illustrated in the preceding examples.

BASIC PROPERTIES OF EXPONENTIAL FUNCTIONS

(*i*) $f(x) = a^x$ is always positive. *The entire graph of f lies above the x-axis and crosses the y-axis at* $y = 1$.

(*ii*)   *If* $a > 1$, *then* $f(x) = a^x$ *is an* increasing *function; that is, the graph of f is always rising from left to right (see p. 135).*

(*iii*)   *If* $0 < a < 1$, *then* $f(x) = a^x$ *is a* decreasing *function; that is, the graph of f is always falling from left to right (see p. 136).*

EXPONENTIAL GROWTH AND DECAY

We have seen that when $a > 1$ and x takes increasing positive values, the corresponding values of the function $f(x) = a^x$ increase sharply. In other words, the graph of $f(x) = a^x$ rises steeply to the right. In order to get an idea of just *how* greatly the values of the function increase, it is instructive to compare the graph of $g(x) = 2^x$, for instance, with the graph of the polynomial function $f(x) = x^4$. The graph of $f(x) = x^4$ on page 213 rises steeply as x takes larger and larger positive values. Indeed, for $0 \leq x \leq 16$ the graph of $f(x) = x^4$ lies considerably *above* the graph of $g(x) = 2^x$. But for larger values of x, the story is quite different, as we see from the following table.

x	16	17	18	19	20	21	22	35
$g(x) = 2^x$	65,536	131,072	262,144	524,288	1,048,576	2,097,152	4,194,304	34,359,738,368
$f(x) = x^4$	65,536	83,521	104,976	130,321	160,000	194,481	234,256	1,500,625

If we choose appropriate scales for the axes and sketch the graphs for $16 \leq x \leq 22$, the difference in steepness becomes dramatically apparent, as shown in Figure 5-5.

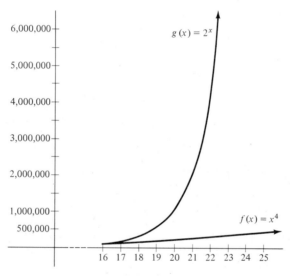

Figure 5-5

If we keep the same scale on the axes and try to extend the graph of $g(x) = 2^x$ just to $x = 40$, we would need a sheet of paper over 10 miles long!

The phenomenon of **exponential growth,** as demonstrated by the graph of $f(x) = a^x$ with $a > 1$, is explored further (via real-life examples) in the DO IT YOURSELF! segment at the end of this section. If the preceding discussion has left you wondering about how exponential functions arise in real life and what they *really* mean, you might find the discussion interesting.

When $a < 1$, then the function $f(x) = a^x$ dies out toward 0 as x increases. This phenomenon is sometimes described by the term **exponential decay.** To get an idea of the rate of such decay, it is instructive to compare the exponential function $s(x) = (\frac{1}{2})^x = 1/2^x$ and the rational function $f(x) = 1/x^2$ (with $x > 0$). As x takes larger and larger positive values, both graphs remain positive but die out toward zero. But $s(x) = 1/2^x$ dies out much faster. Using a calculator, we find that

$$\text{if } x = 10, \quad \text{then } \frac{1}{x^2} = .01, \quad \text{but} \quad \frac{1}{2^x} = \frac{1}{1024} < \frac{1}{1000} = .001$$

$$\text{if } x = 100, \quad \text{then } \frac{1}{x^2} = \frac{1}{10^4}, \quad \text{but} \quad \frac{1}{2^x} < \frac{1}{10^{30}}$$

EXERCISES

Note Unless advised otherwise, feel free to use a calculator for any of the computations involved in these problems.

A.1. Find a value of x (with $x > 1$) for which $2^x > x^{10}$.

A.2. Let h and k be the functions given by $h(x) = 2^x$ and $k(x) = 3^x$. Evaluate:
  (a) $(h \circ k)(2)$ (c) $(h \circ k)(0)$ (e) $(k \circ h)(-2)$
  (b) $(k \circ h)(1)$ (d) $(k \circ h)(-1)$ (f) $(h \circ k)(1)$

A.3. Graph these functions for the given values of x.
  (a) $f(x) = 4^x \ (-2 \le x \le 2)$ (d) $k(x) = (\frac{2}{5})^x \ (-3 \le x \le 3)$
  (b) $g(x) = 5^x \ (-2 \le x \le 2)$ (e) $f(x) = (1.2)^x \ (-6 \le x \le 12)$
  (c) $h(x) = (\frac{3}{4})^x \ (-8 \le x \le 4)$ (f) $g(x) = (.8)^x \ (-10 \le x \le 5)$

A.4. For each of the following functions, compute the difference quotient $\dfrac{f(x + \Delta x) - f(x)}{\Delta x}$. (See Exercises B.7 and B.8 on p. 106.)
  (a) $f(x) = 10^x$ (c) $h(x) = 3^{-x}$ (e) $f(x) = 2^x + 2^{-x}$
  (b) $g(x) = 2^x$ (d) $k(x) = 5^{x^2}$ (f) $g(x) = 3^x - 3^{-x}$

A.5. For each pair of functions f and g, compute the rule of the product function fg. For example, if $f(x) = e^{2x}$ and $g(x) = e^{-x+2}$, then $(fg)(x) = f(x)g(x) = e^{2x}e^{-x+2} = e^{2x-x+2} = e^{x+2}$.
  (a) $f(x) = e^x; g(x) = e^{2x+1}$

  (b) $f(x) = e^{x+1}; g(x) = e^{-2x-3}$

  (c) $f(x) = 5^x + 5^{-x}; g(x) = 5^x - 5^{-x}$

  (d) $f(x) = \dfrac{2^x + 2^{-x}}{2^x - 2^{-x}}; g(x) = \dfrac{2^x + 2^{-x}}{2}$

B.1. Observe that $a^{-x} = 1/a^x = (1/a)^x$. Use this fact to graph these functions.
  (a) $f(x) = 4^{-x}$ (c) $f(x) = (\frac{4}{3})^{-x}$ (e) $f(x) = (.8)^{-x}$
  (b) $f(x) = 5^{-x}$ (d) $f(x) = (\frac{5}{2})^{-x}$ (f) $f(x) = (1.5)^{-x}$

B.2. Carefully graph $h(x) = 2^x$. Then use this graph to graph each of the functions below *without* plotting points. (*Hint:* Section 6 of Chapter 3 should be helpful here.)
  (a) $f(x) = 2^x - 5$ (d) $g(x) = 2^{x-1}$ [*Hint: h(x - 1) = g(x)*.]
  (b) $g(x) = -(2^x)$ (e) $k(x) = 2^{x+2}$
  (c) $k(x) = 3(2^x)$ (f) $r(x) = 4(2^{x-5})$

B.3.   Graph each of these functions.

(a) $f(x) = 2^{3x}$ (d) $k(x) = 3^{-2x}$ (g) $f(x) = 2^{3-x^2}$

(b) $g(x) = 3^{x/2}$ (e) $k(x) = 2^{-3x}$ (h) $h(x) = 3^x - 2^x$

(c) $h(x) = 2^{x/3}$ (f) $q(x) = 2^{-x^2}$ (i) $p(x) = 2^x - 2^{-x}$

B.4.   Use your knowledge of the behavior of the function $t(x) = e^x$ (not a calculator) to answer these questions.

(a) When $|x|$ is large and x is positive, is $f(x) = e^x + e^{-x}$ positive or negative? Is $|e^x + e^{-x}|$ large or small?

(b) Answer (a) for the function $g(x) = e^x - e^{-x}$.

B.5.   Use Exercise B.4, together with a calculator to plot a few points, to graph these functions.

(a) $h(x) = \dfrac{e^x + e^{-x}}{2}$ (b) $k(x) = \dfrac{e^x - e^{-x}}{2}$

B.6.   Determine whether each of these functions is even, odd, or neither (see pp. 133 and 143 for definitions).

(a) $f(x) = 10^x$ (c) $h(x) = 3^x + 3^{-x}$ (e) $f(x) = \dfrac{e^x + e^{-x}}{2}$

(b) $g(x) = 2^{-x}$ (d) $k(x) = 5^x - 5^{-x}$ (f) $h(x) = \dfrac{e^x - e^{-x}}{2}$

B.7.   Show graphically that the equation $2^x = x$ has *no* solutions.

B.8.   Oil is piped into a storage tank that already contains some oil in such a way that the volume of the oil in the tank doubles every hour and it takes 15 hr to fill the tank. How long does it take for the tank to become one-fourth full?

B.9.   If you begin with k mg of radium, then the amount $M(t)$ of radium remaining after t years is given by $M(t) = k2^{-t/1600}$. If you begin with 100 mg of radium, how much is left after 800 years? after 1600 years? after 3200 years?

B.10.   Water and salt are continuously added to a tank in such a way that the number of kilograms of salt in the tank at time t min is $200 - 100e^{-t/20}$.

(a) How much salt is in the tank at the beginning (that is, when $t = 0$)?

(b) How much salt is in the tank after 10 min? after 20 min? after 40 min?

B.11.   The number of bacteria in a certain culture at time t is given by the function $B(t) = (5000)3^t$, where the time t is measured in *hours* after 4:00 P.M.

(a) What is the initial number of bacteria at 4:00 P.M. (that is, when $t = 0$)?

(b) What is the number of bacteria at 4:10 P.M.? at 4:30 P.M.? at 5 P.M.? at 5:15 P.M.? at 6:20 P.M.?

C.1.   (a) Find a function $f(x)$ with the property $f(r + s) = f(r)f(s)$ for all real numbers r and s.

(b) Find a function $g(x)$ with the property $g(2x) = (g(x))^2$ for every real number x. (*Hint:* think exponential.)

C.2. **(a)** Graph each of the following on the same set of axes for the indicated values of x.

    (i) $g(x) = 2^x$ (ii) $h(x) = (.1)2^x$ (iii) $k(x) = 2^{x/4}$
        $(-5 \le x \le 5)$ $(-8 \le x \le 8)$ $(-20 \le x \le 20)$

Based on your results in part (a), answer the following questions (in which $a > 1$ and x is positive).

  (b) If c is a fixed real number greater than 1, does $g(x) = a^{cx}$ increase faster or slower than $f(x) = a^x$?

  (c) If $c < 1$, does $g(x) = a^{cx}$ increase faster or slower than $f(x) = a^x$?

  (d) Under what circumstances does $h(x) = ka^x$ (k a fixed real number) increase *faster* than $f(x) = a^x$?

  (e) Suppose $k < 1$ and $c < 1$. How does the growth of $t(x) = ka^{cx}$ compare with that of $f(x) = a^x$?

C.3. Look back at Section 3 of Chapter 4 where the basic properties of graphs of polynomial functions were listed. Then review the basic properties of the graph of $f(x) = a^x$ discussed in this section. Using these various properties, give an argument to show that for any fixed positive number a ($\ne 1$), it is *not* possible to find a polynomial function $g(x) = c_n x^n + \cdots + c_1 x + c_0$ such that $a^x = g(x)$ for *all* numbers x. In other words, *no exponential function is a polynomial function*.

C.4. An eccentric billionaire offers you a job for the month of September. She says that she will pay you 2¢ on the first day, 4¢ on the second day, 8¢ on the third day, and so on, doubling your pay on each successive day.

  (a) Let $P(x)$ denote your salary in *dollars* on day x. Find the rule of the function P.

  (b) Would you be better off financially if instead you were paid $10,000 per day? [*Hint:* consider $P(30)$.]

DO IT YOURSELF!

EXPONENTIAL GROWTH

Thus far we have discussed only the function $f(x) = a^x$. There are many other functions, loosely labeled exponential, which behave similarly. For instance,

$$f(x) = 67a^x, \qquad h(x) = a^{x/10}, \qquad T(x) = (\tfrac{1}{50})3^{x-1}, \qquad S(x) = 15(4^{77x})$$

Such variations on the basic exponential theme occur quite often in real life situations, as the following examples illustrate. We begin with a subject close to all of us these days, *money*.

COMPOUND INTEREST

You are probably familiar with compound interest. You put some money in the bank. At the end of some time period (a year, quarter, month, or day, depending on the bank), the bank pays interest on the amount you deposited. At the end of the next time period, the bank pays interest on the original amount *and* on the interest earned in the first time period (this is what is meant by *compounding* the interest). And so it goes.

Suppose you deposit P dollars at $r\%$ interest. If you leave your money in the bank, then at the end of one time period you will have your original amount, P dollars, *plus* the interest it has earned. The interest earned by P dollars at $r\%$ per time period is just ($r\%$ of P) dollars. As is the usual custom, we think of r as a decimal and just write rP for $r\%$ of P. For instance,

$$5\% \text{ of } P \text{ is } .05P, \qquad 17\% \text{ of } P \text{ is } .17P, \qquad \text{and so on}$$

Therefore after one time period at $r\%$ interest, your P dollars grow to

$$P + rP \text{ dollars} = P(1 + r) \text{ dollars}$$

If you leave this money in the bank for a *second* period, you will then have $P(1 + r)$ dollars *plus* the interest this amount has earned. The interest on $P(1 + r)$ dollars for one period is

$$r\% \text{ of } P(1 + r), \text{ that is, } r[P(1 + r)]$$

Consequently, after two periods at $r\%$ interest, you will have

$$P(1 + r) + r[P(1 + r)] = P(1 + r)[1 + r] = P(1 + r)^2 \text{ dollars}$$

Similarly, at the end of the *third* time period, you will have the amount you started the period with, $P(1 + r)^2$, *plus* the interest on this amount, $r[P(1 + r)^2]$, for a total of

$$P(1 + r)^2 + r[P(1 + r)^2] = P(1 + r)^2[1 + r] = P(1 + r)^3$$

This pattern continues on for each time period, and leads to this conclusion:

THE BASIC FORMULA FOR COMPOUND INTEREST

If P *dollars are invested at* r% *interest per time period, then after* x *time periods, you will have*

$$P(1 + r)^x \text{ dollars}$$

In many banks interest is paid from day of deposit to day of withdrawal, regardless of the time period used for compounding interest, so that this formula is used even when x is not an integer. For instance, if the interest rate is $r\%$ per *year* and you leave your money there for 6 years and 5 months, then $x = 6\frac{5}{12} = \frac{77}{12}$ years, and your total amount is $P(1 + r)^{77/12}$. Hereafter we assume we are dealing with such a bank, so that in the basic formula x is allowed to be any positive real number.

If the interest rate r stays fixed, then $1 + r$ is a constant, as is the original amount P. If we denote $1 + r$ by a, then the basic formula for total amount becomes

$$T = P(1 + r)^x = Pa^x$$

In other words, the total amount T is an *exponential function* of the number of periods x you leave the money in the bank. In functional notation, $T(x) = Pa^x$.

Since $a = 1 + r > 1$, our experience with exponential functions tells us that as x gets larger, $T(x) = Pa^x$ becomes *enormous*. Translated into money, this means that if you leave your money in the bank *long enough* (x is large), you or your descendants will end up *rich* ($T(x) = Pa^x$ is enormous).

EXAMPLE° Suppose the interest rate is 10% per year, and a father invests a thousand dollars on the day his daughter is born. Then $a = 1 + r = 1 + .1 = 1.1$ and the total amount available after x years is given by the function $T(x) = Pa^x = 1000(1.1)^x$. Here are some possible values for x and $T(x)$:

x	10	30	50	65	75	100
$T(x) = 1000(1.1)^x$	$2594	$17,449	$117,391	$490,371	$1,271,895	$13,780,612

Thus the daughter could retire at age 65 with almost *half a million* dollars in the bank, or wait until age 75 and collect over a *million* dollars. If the daughter doesn't touch the money and lives to the ripe old age of 100, then *her* children (and the tax man) will inherit more than *13 million* dollars.

The most common method of paying interest by banks is to pay an *annual rate* of interest *compounded quarterly*, or monthly, or even daily. What this means is explained in the next example.

EXAMPLE Suppose you invest P dollars for 1 year at 5% per year, compounded quarterly. The annual interest rate is .05, so the interest for one quarter $(= \frac{1}{4}$ year) is $\frac{1}{4}$ of the interest for a full year, that is, .05/4. In the basic formula

$$T(x) = P(1 + r)^x$$

the time period is a quarter year, the interest rate r per quarter is .05/4, and the number x of time periods to total 1 year is 4. Hence at the end of a year you will have

$$P\left(1 + \frac{.05}{4}\right)^4 = P(1 + .0125)^4 = P(1.0125)^4 = P(1.05095)$$

whereas at a straight 5% per year you would have only

$$P + .05P = P(1.05)$$

Thus 5% compounded quarterly yields a slightly better return than 5% compounded annually.

As a general rule, the more often your interest is compounded, the better off you are. But there is, alas, a limit to how well you can do, as the following example illustrates.

° In this example and all the ones below, we have used either tables or a calculator to perform the necessary calculations.

EXAMPLE   You have \$1 to invest for 1 year. The Exponential Bank offers to pay you 100% interest° per year, compounded n times per year. "But what number is n?" you ask. The banker replies, "Oh, you can pick any value you want for n. Of course, all interest amounts are rounded off to the nearest penny."

> *Question:* can you choose n so large that at the end of the year, your dollar will have grown to some huge amount?

To answer this question we use the basic formula with $P = 1$:

$$T(x) = P(1 + r)^x = (1 + r)^x$$

Since the interest rate is 100% ($= 1.00$) compounded n times per year, then for a time period of $1/n$ year, the interest rate is $\frac{1}{n}(1.00) = \frac{1}{n}$ and the number of time periods to total 1 year is n. Thus $T(n) = \left(1 + \frac{1}{n}\right)^n$ is the amount \$1 will grow to if 100% interest is compounded n times per year. Let's see what happens for various values of n.

Interest is Compounded	$n =$	$\left(1 + \dfrac{1}{n}\right)^n = $ °°
Annually	1	$(1 + \frac{1}{1})^1 = 2$
Semiannually	2	$(1 + \frac{1}{2})^2 = 2.25$
Quarterly	4	$(1 + \frac{1}{4})^4 \approx 2.4414$
Monthly	12	$(1 + \frac{1}{12})^{12} \approx 2.6130$
Daily	365	$(1 + \frac{1}{365})^{365} \approx 2.71457$
Every 12 hr	$2 \cdot 365 = 730$	$(1 + \frac{1}{730})^{730} \approx 2.71642$
Every 2 hr	$12 \cdot 365 = 4380$	$(1 + \frac{1}{4380})^{4380} \approx 2.71797$
Hourly	$24 \cdot 365 = 8760$	$(1 + \frac{1}{8760})^{8760} \approx 2.718127$
Half-hourly	$2 \cdot 8{,}760 = 17{,}520$	$(1 + \frac{1}{17,520})^{17,520} \approx 2.718204$
Every 15 min	$2 \cdot 17{,}520 = 35{,}040$	$(1 + \frac{1}{35,040})^{35,040} \approx 2.718243$
Every minute	$15 \cdot 35{,}040 = 525{,}600$	$(1 + \frac{1}{525,600})^{525,600} \approx 2.7182792$
Every 30 sec	$2 \cdot 525{,}600 = 1{,}051{,}200$	$(1 + \frac{1}{1,051,200})^{1,051,200} \approx 2.7182805$
Every second	$30 \cdot 1{,}051{,}200 = 31{,}536{,}000$	$(1 + \frac{1}{31,536,000})^{31,536,000} \approx 2.7182818$

° The number 100 is chosen for computational convenience. Essentially the same point can be made with a more realistic interest rate.

°° The calculations in this table were made on a large computer, using double precision. The results are accurate, except possibly for rounding off in the last digit. A hand calculator is quite likely to compute $(1 + \frac{1}{n})^n$ inaccurately when n is large. (Can you figure out why?)

The bank rounds off all amounts to the nearest penny. Consequently, once n is 730 or larger, the number of times the bank compounds doesn't make any difference. Your investment is worth \$2.72 at the end of the year no matter how big n is

THE NUMBER e

The calculations in the preceding example suggest that as n takes larger and larger values, then the corresponding values of $(1 + 1/n)^n$ get closer and closer to a specific real number, whose decimal expansion begins $2.71828 \cdots$. This is indeed the case, as will be shown in calculus.

The real number $2.71828 \cdots$ is denoted e. It is an irrational number and appears (more or less naturally, as it did here) in several different mathematical contexts. When mathematicians or scientists speak of *the* exponential function, they mean the function $t(x) = e^x$. Its graph was given on page 260.

THE POPULATION EXPLOSION

If we neglect special inhibiting or stimulating factors, a population normally grows at a rate proportional to its size. This means that the ratio

$$\frac{\text{current rate of growth of population}}{\text{current size of population}}$$

is a fixed number at all times. Bacteria colonies grow this way, provided they have a normal environment. The same is true of human populations.

Let $S(t)$ denote the *size* of the population at time t and let k denote the constant ratio of growth rate to size of population. In calculus it is shown that

$$S(t) = ce^{kt}$$

where c is the original population [that is, $S(0)$] and e is the irrational number just introduced. If $k = 1$ and $c = 1$, then the population size at time t is given by the exponential function $S(t) = e^t$, whose graph is on page 260. It increases extremely rapidly, even more rapidly than the function $g(x) = 2^x$ discussed above.

If the constants c and k are small, then $S(t) = ce^{kt}$ grows more slowly than does e^t, as we saw in Exercise C.2 on page 265. But even in such a case, as time goes on (t gets larger), the value of $S(t)$ soon becomes *huge*. This is what the "population explosion" is all about.

RADIOACTIVE DECAY AND RADIOCARBON DATING

The **half-life** of a radioactive element is the time in which a given quantity decays to one-half of its original mass. Denote the mass of the element at time t by $M(t)$. It is shown in calculus that

$$M(t) = k2^{-t/h} = k(\tfrac{1}{2})^{t/h}$$

where h is the half-life and k is the original mass of the element [that is, $k = M(0)$, the mass at starting time $t = 0$].

The radioactive isotope carbon-14 is present in all living organisms. When the organism dies, its carbon-14 begins to decay. Since the half-life of carbon-14 is 5730 years, we have the function

$$M(t) = k2^{-t/5730} = k(\tfrac{1}{2})^{t/5730}$$

This function is used by archaeologists and paleontologists to determine the age of various artifacts and fossils up to 50,000 years old. Here is how they do it.

Suppose you have a sample of some organic matter whose age is to be determined. For instance, the sample might be part of a fossilized tree knocked over by a glacier, or some charcoal from the hearth of some prehistoric family, or an object immersed in lava during a volcanic eruption. In each case time t is measured from the death of the organism, when its carbon-14 began decaying. If the object is 7000 yr old, for example, then the present value of t is 7000. We don't know the present value of t for our sample, but we *can* find the present value of $M(t)$, by measuring the amount of carbon-14 now present in the sample. This is done via radioactive emissions and a Geiger counter.

It is also possible to determine the original amount k of carbon-14 in our sample. The details of this determination need not concern us here. We simply note for the record that they depend on measuring both the carbon-12 and the carbon-14 in a present-day sample of the same material and the fact that the ratio of carbon-14 to carbon-12 is essentially constant over a 50,000-yr period.

Consequently, in the functional relationship for our sample

$$M(t) = k2^{-t/5730}$$

we know both $M(t)$ and k. *If* we can solve the resulting exponential equation for t, then we have the age of the sample.

EXAMPLE The present mass of the carbon-14 in a sample is .5470. Its original mass is determined to have been 1.273. How old is the sample? Substituting $M(t) = .547$ and $k = 1.273$ into the basic decay equation for carbon-14 yields

$$M(t) = k2^{-t/5730}$$
$$.547 = (1.273)2^{-t/5730}$$

As we shall see later in this chapter, this equation can be easily solved. It turns out that $t = 6982.69$. Thus the sample is approximately 6983 years old.

EXERCISES

Note Use a calculator when necessary. Most of these exercises require the use of one of the formulas developed in the text, such as the basic formula for compound interest. In some cases, it may also be necessary to solve exponential equations, such as $(1.12)^x = 2$. The methods for doing this precisely are discussed later in this chapter. For now, just experiment and find an approximate answer. For example, $(1.12)^6 \approx 1.97$ and $(1.12)^{6.2} \approx 2.02$, so the solution of $(1.12)^x = 2$ is approximately $x = 6.1$.

A.1. The half-life of radium is approximately 1660 yr. If the original mass of a sample of radium is 1 gram, how much radium is left after 830 yr?

A.2. The half-life of a certain element is 1.4 days. If you begin with 2 grams, how much is left after 1 week? after 30 days?

A.3. If you put $500 in a savings account that pays interest at 5% per year, how much will you have at the end of 10 years if
  (a) interest is compounded annually?
  (b) interest is compounded quarterly?
  (c) interest is compounded twice a day?

B.1. Bankers have a rule of thumb that tells you *approximately* how many years it will take you to double your money if you leave it in the bank at a fixed interest rate, compounded annually. Find this rule of thumb by using the basic formula for compound interest and some experimentation as follows.
  (a) Determine how many years (rounded to the nearest year) it takes to double your money at *each* of these interest rates: 3%, 6%, 8%, 10%, 12%, 18%, 24%, and 36%.
  (b) Compare the answers in (a) to the numbers $\frac{72}{3}$, $\frac{72}{6}$, $\frac{72}{8}$, $\frac{72}{10}$, and so on.

B.2. At what annual rate of interest should $1000 be invested so that it will double in 10 yr if interest is compounded quarterly?

B.3. How long does it take $500 to triple if it is invested at 6% compounded annually? compounded quarterly? compounded daily?

B.4. Any quantity of uranium decays to two-thirds of its original mass in .26 billion yr. Find the half-life of uranium.

3. LOGARITHMS

The invention of logarithms in the seventeenth century was a major advance in the technique of numerical calculation. For over two centuries logarithms were the only effective tool for doing complicated computations in astronomy, chemistry, physics, and other fields. Today, of course, such calculations can be quickly and easily performed by computers or hand calculators.

Despite their decline as computational tools, logarithms still play an important role in mathematics. Several areas of calculus and the sciences require a good understanding of the concept and basic properties of logarithms and logarithmic functions.

COMMON LOGARITHMS

The graph of the function $f(x) = 10^x$, which is shown in Figure 5-6 on the next page, has a crucial geometric property:

> A horizontal straight line that lies above the x-axis
> intersects the graph of $f(x) = 10^x$ at exactly one point

Translated into algebraic terms, this geometric property is just the statement:

> For each positive number v, there is one
> and only one number u such that $10^u = v$

A typical example is shown in Figure 5-6.

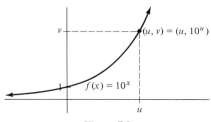

Figure 5-6

The unique u such that $10^u = v$ is called the **logarithm of v to the base 10** (or the **common logarithm** of v) and is denoted by the symbol log v. Thus

---

For each positive number v, *log* v *is the unique number such that*
$10^{\log v} = $ v.

In other words,

> **log v is the exponent to which 10 must be raised to produce v**

Keep this fact in mind. *A logarithm is just a particular kind of exponent,* so we won't really be saying anything new here. We'll just be restating known facts about exponents in the language and symbolism of logarithms.

EXAMPLE   What is log 100? Answer: log 100 is the exponent to which 10 must be raised to produce 100. Since $10^2 = 100$, we see that log 100 = 2.

EXAMPLE   What is log .001? Answer: log .001 is the exponent to which 10 must be raised to produce .001. But $.001 = \frac{1}{1000} = 1/10^3 = 10^{-3}$, so that log .001 = -3.

EXAMPLE   log $\sqrt{10}$ is the exponent to which 10 must be raised to produce $\sqrt{10}$. Since $\sqrt{10} = 10^{1/2}$, we have log $\sqrt{10} = \frac{1}{2}$.

Just as in the preceding examples, the logarithm of any power of 10 can be easily found. For any real number k, $\log 10^k$ is the exponent to which 10 must be raised to produce 10^k. Obviously, this exponent is just k itself. Therefore

$$\log 10^k = k \text{ for every real number } k.$$

This fact is true whether k has simple or complicated form. For instance,

$$\begin{array}{ll}
\log 1 = \log 10^0 = 0 & \text{(here } k = 0) \\
\log 10^{(3-\sqrt{2})} = 3 - \sqrt{2} & \text{(here } k = 3 - \sqrt{2}) \\
\log 10^{2x+1} = 2x + 1 & \text{(here } k = 2x + 1)
\end{array}$$

Computing $\log v$ when v is not an obvious power of 10 can be quite time-consuming. Fortunately, these computations have already been made for us and are available in various tables of logarithms. (In this context, you can think of a calculator as a pushbutton logarithm table.)

But even without a table, it is possible to make some rough estimates of the logarithms of various number by using our knowledge of the exponential function $f(x) = 10^x$. It's a good idea to be adept at such estimates, since they will warn you of any gross errors that you might make in reading a table or using a calculator.

EXAMPLE In order to estimate $\log 225$, we first note that $10^2 = 100$ and $10^3 = 1000$. Since 225 lies between 10^2 and 10^3, the exponent to which 10 must be raised to produce 225 (namely, $\log 225$) must be a number between 2 and 3; that is, $2 < \log 225 < 3$. Since 225 is quite a bit closer to $100 = 10^2$ than to $1000 = 10^3$, the exponent $\log 225$ probably lies closer to 2 than to 3. A rough estimate might be that $2 < \log 225 < 2.5$.

EXAMPLE To estimate $\log (-5)$, you must answer the question: to what exponent must 10 be raised to produce -5? But as we know, *every* power of 10 is *positive* (the graph of 10^x always lies above the x-axis). It is *impossible* to have some power of 10 equal to -5. Therefore $\log (-5)$, or the logarithm of any negative number, is *not defined*.

Since logarithms are just exponents, a table of common logarithms is nothing more than a table of powers of 10. In order to emphasize this fact, the following sample logarithm table is written in both logarithmic and exponential language. (Most logarithm tables assume that you know logarithms are exponents and don't mention exponents.) You can use this table to check the accuracy of our estimate of $\log 225$ in the preceding example.

Logarithmic Statement	Equivalent Exponential Statement	Logarithmic Statement	Equivalent Exponential Statement
$\log 0.1 = -1.$	$10^{-1} = .1$	$\log 45.1 = 1.6542$	$10^{1.6542} = 45.1$
$\log 0.2 = -.6990$	$10^{-.6990} = .2$	$\log 45.2 = 1.6551$	$10^{1.6551} = 45.2$
$\log 0.4 = -.3979$	$10^{-.3979} = .4$	$\log 45.3 = 1.6561$	$10^{1.6561} = 45.3$
$\log 0.6 = -.2218$	$10^{-.2218} = .6$	$\log 45.4 = 1.6571$	$10^{1.6571} = 45.4$
$\log 0.8 = -.0969$	$10^{-.0969} = .8$	$\log 45.5 = 1.6580$	$10^{1.6580} = 45.5$
$\log 1 = 0$	$10^0 = 1$	$\log 220 = 2.3424$	$10^{2.3424} = 220$
$\log 2 = .3010$	$10^{.3010} = 2$	$\log 222.5 = 2.3473$	$10^{2.3473} = 222.5$
$\log 3 = .4771$	$10^{.4771} = 3$	$\log 225 = 2.3522$	$10^{2.3522} = 225$
$\log 4 = .6021$	$10^{.6021} = 4$	$\log 227.5 = 2.3570$	$10^{2.3570} = 227.5$
$\log 5 = .6990$	$10^{.6990} = 5$	$\log 230 = 2.3617$	$10^{2.3617} = 230$

Since most logarithms are actually irrational numbers, most of the entries in this table (as in any logarithm table) are rational number *approximations*. For example, $10^{.6990}$ actually works out to $5.00034535 \cdots$ rather than 5. But for most purposes such approximations are adequate, so it is customary to write $\log 5 = .6990$ and $10^{.6990} = 5$ rather than the more accurate statements $\log 5 \approx .6990$ and $10^{.6990} \approx 5$.° Although a typical calculator computes logarithms to a greater degree of accuracy (for instance, $\log 5 \approx .698970004$), it too provides only rational approximations in most cases.

LOGARITHMS TO OTHER BASES

The definition and discussion of logarithms above depended only on the properties of the exponential function $f(x) = 10^x$. If b is any real number greater than 1, then the exponential function $g(x) = b^x$ has the same basic properties as does $f(x) = 10^x$. Consequently, it is possible to define logarithms in terms of the base b in the same way that logarithms to the base 10 were defined above. We shall do this now, since logarithms to bases other than 10 are often needed.

Let b be a fixed real number with $b > 1$. The graph of the function $g(x) = b^x$ has the same general shape as the graph of $f(x) = 10^x$, as shown in Figure 5-7. Furthermore, the function g has the same crucial property:

For each positive real number v, there is one and only one number u such that $b^u = v$

° $\approx$ means "approximately equal."

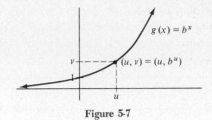

Figure 5-7

The unique number u such that $b^u = v$ is called the **logarithm of v to the base b** and is denoted by the symbol $\log_b v$.° Thus

> For each positive number v, $\log_b$v *is the unique number such that* $b^{\log_b v} = $ v.

In other words,

$\log_b v$ **is the exponent to which b must be raised to produce** v.

EXAMPLE $\log_2 16$ is the exponent to which 2 must be raised in order to produce 16. Since $2^4 = 16$, we see that $\log_2 16 = 4$. Similarly,

$$\log_2 1 = 0 \qquad \text{since } 2^0 = 1$$
$$\log_2 32 = 5 \qquad \text{since } 2^5 = 32$$
$$\log_2\left(\tfrac{1}{16}\right) = -4 \qquad \text{since } 2^{-4} = \tfrac{1}{16}$$
$$\log_2 \sqrt{2} = \tfrac{1}{2} \qquad \text{since } 2^{1/2} = \sqrt{2}$$

EXAMPLE Here are some examples of logarithms to various bases. Since logarithms are exponents, each of these logarithmic statements can be translated into a statement in exponential language.

Logarithmic Statement	Equivalent Exponential Statement
$\log_3 81 = 4$	$3^4 = 81$
$\log_4 64 = 3$	$4^3 = 64$
$\log_{125} 5 = \tfrac{1}{3}$	$125^{1/3} = 5$ (since $125^{1/3} = \sqrt[3]{125} = 5$)
$\log_4\left(\tfrac{1}{16}\right) = -2$	$4^{-2} = \tfrac{1}{16}$
$\log_8\left(\tfrac{1}{4}\right) = -\tfrac{2}{3}$	$8^{-2/3} = \tfrac{1}{4}$ (Verify!)

° If the base b happens to be the number 10, we shall often write $\log v$ as above instead of $\log_{10} v$.

As the preceding examples illustrate, the logarithm of any power of the base b is easily found. For any base b and any real number u, $\log_b(b^u)$ is the exponent to which b must be raised to produce the number b^u. This exponent is just u itself:

$$\boxed{\log_b(b^u) = u \text{ for every real number u.}}$$

NATURAL LOGARITHMS

The irrational number e whose decimal expansion begins $2.71828 \cdots$ was introduced on page 259. Logarithms to the base e are called **natural logarithms** and are frequently used in many scientific contexts. (The reason why this is so will become apparent in calculus.) In some books and on most calculators the natural logarithm of the number v is denoted by the symbol $\ln v$ instead of $\log_e v$. But no matter what notation is used, the basic fact is the same:

$\log_e v$ **(or $\ln v$)   is the exponent to which e must be raised to produce** v

Except for obvious cases, such as $\log_e(e^u) = u$, it is necessary to use either tables or a calculator to determine $\log_e v$.

EXAMPLE   Find $\log_e 81$. Now $\log_e 81$ is the exponent to which e must be raised to produce 81. Since $e \approx 2.72$ and since $3^4 = 81$, it seems likely that $\log_e 81$ is a number a bit bigger than 4. A calculator shows that $\log_e 81 \approx 4.3944$; that is, $e^{4.3944} \approx 81$.

BASIC PROPERTIES OF LOGARITHMS

Although the following discussion applies equally well to logarithms to any base, the most important cases occur when the base is either 10 or e. So if you want to keep the discussion concrete, just read "10" or "e" whenever you see "b."

Let b be a fixed real number with $b > 1$. The first basic property of logarithms to the base b is just the definition:

$$\boxed{b^{\log_b v} = v \text{ for every positive real number v.}}$$

The second basic property is the fact discussed above:

$$\boxed{\log_b(b^u) = u \text{ for every real number u.}}$$

The remaining properties of logarithms are essentially restatements in logarithmic language of well-known properties of exponents. For instance, the first law of exponents states that $b^m b^n = b^{m+n}$, or in words

The exponent of a product is the sum of the exponents of the factors

Since logarithms are just particular kinds of exponents, this statement translates as

The logarithm of a product is the sum of the logarithms of the factors

We now write this statement in formal symbolic language:

FIRST LAW OF LOGARITHMS

$log_b(vw) = log_b v + log_b w$ *for all positive real numbers* v, w.

PROOF The definition of logarithms (second box on the preceding page) tells us that

$$b^{\log_b v} = v \qquad \text{and} \qquad b^{\log_b w} = w$$

Therefore by the first law of exponents (with $m = \log_b v$ and $n = \log_b w$),

$$vw = (b^{\log_b v})(b^{\log_b w}) = b^{\log_b v + \log_b w}$$

Thus raising b to the exponent $(\log_b v + \log_b w)$ produces the number vw. But the exponent to which b must be raised to produce vw is precisely $\log_b(vw)$. Therefore

$$\log_b(vw) = \log_b v + \log_b w$$

This completes the proof.

EXAMPLE Suppose $b = 10$. From the table on page 274 we know that $\log_{10} 3 \approx .4771$ and $\log_{10} 4 \approx .6021$. Consequently, by the logarithm property just proved,

$$\log_{10} 12 = \log_{10}(3 \cdot 4) = \log_{10} 3 + \log_{10} 4 \approx .4771 + .6021 = 1.0792$$

Warning Be sure to use the first law of logarithms correctly. Avoid this well-known error: $\log_b 5 + \log_b 9 = \log_b(5 + 9) = \log_b 14$. The correct statement in this situation is $\log_b 5 + \log_b 9 = \log_b(5 \cdot 9) = \log_b 45$.

The second law of exponents—namely, $b^m / b^n = b^{m-n}$—can be roughly stated in words as:

The exponent of the quotient is the difference of the exponents.

If the exponents happen to be logarithms, this statement says:

The logarithm of a quotient is the difference of the logarithms.

In formal and more accurate terminology we have the

SECOND LAW OF LOGARITHMS

$$log_b\left(\frac{v}{w}\right) = log_b v - log_b w \qquad \textit{for all positive real numbers v,w.}$$

PROOF The definition of logarithms states that

$$v = b^{\log_b v} \qquad \text{and} \qquad w = b^{\log_b w}$$

Therefore by the second law of exponents (with $m = \log_b v$ and $n = \log_b w$),

$$\frac{v}{w} = \frac{b^{\log_b v}}{b^{\log_b w}} = b^{\log_b v - \log_b w}$$

Thus if b is raised to the exponent $(\log_b v - \log_b w)$, the result is the number v/w. But the exponent to which b must be raised to produce v/w is precisely $\log_b(v/w)$. Therefore

$$\log_b\left(\frac{v}{w}\right) = \log_b v - \log_b w$$

and the proof is complete.

EXAMPLE Using the table on page 274 we see that

$$\log_{10}\left(\frac{230}{45.3}\right) = \log_{10} 230 - \log_{10} 45.3 \approx 2.3617 - 1.6561 = .7056$$

EXAMPLE $\log_e(\frac{17}{44}) = \log_e 17 - \log_e 44.$

Warning Do not confuse $\log_b(v/w)$ with $\log_b v/\log_b w$. They are *different* numbers. For instance,

$$\log_{10}\left(\frac{1000}{100}\right) = \log_{10} 10 = 1 \qquad \text{but} \qquad \frac{\log_{10} 1000}{\log_{10} 100} = \frac{3}{2}$$

Consequently, statements such as these are *false:*

$$\frac{\log_2 32}{\log_2 4} = \log_2\left(\frac{32}{4}\right) \qquad \text{and} \qquad \frac{\log_2 32}{\log_2 4} = \log_2 32 - \log_2 4$$

The third law of exponents—namely, $(b^m)^n = b^{mn}$—is rather awkward to state in words, but it too can be translated into logarithmic language:

> ## THIRD LAW OF LOGARITHMS
>
> $log_b(v^k) = k(log_b v)$ *for all real numbers* k *and* v *with* v > 0.

PROOF Since $v = b^{\log_b v}$ by the definition of logarithms, the third law of exponents (with $m = \log_b v$ and $n = k$) shows that

$$v^k = (b^{\log_b v})^k = b^{(\log_b v)k} = b^{k(\log_b v)}$$

Thus raising b to the exponent $k(\log_b v)$ produces the number v^k. But the exponent to which b must be raised to produce v^k is precisely the number $\log_b(v^k)$. Therefore $\log_b(v^k) = k(\log_b v)$, and the proof is complete.

EXAMPLE Is 5^9 larger than 1 million? Using the table on page 274 we see that

$$\log_{10} 5^9 = 9(\log_{10} 5) \approx 9(.6990) = 6.291$$

So that $10^{6.291} \approx 5^9$. Since $10^{6.291} > 10^6$ and $10^6 = 1{,}000{,}000$, we see that 5^9 must be larger than 1,000,000.

EXAMPLE What is $\log_{10} \sqrt{5}$? Using the table on page 274 and the fact that $\sqrt{5} = 5^{1/2}$ we have

$$\log_{10} \sqrt{5} = \log_{10}(5^{1/2}) = \tfrac{1}{2}(\log_{10} 5) \approx \tfrac{1}{2}(.6990) = .3495$$

that is, $10^{.3495} \approx \sqrt{5}$. Observe that this answer is consistent with our intuition, since $\sqrt{5}$ is a number slightly larger than 2 and the table shows that $2 = 10^{.3010}$.

EXAMPLE For any base b and positive number v we have

$$\log_b \left(\frac{1}{v}\right) = \log_b(v^{-1}) = (-1)(\log_b v) = -\log_b v$$

For instance, we saw above that $\log_e 81 \approx 4.3944$. Therefore $\log_e(\tfrac{1}{81}) \approx -4.3944$.

EXAMPLE $\log \sqrt[5]{(\tfrac{7}{39})^4} = \log (\tfrac{7}{39})^{4/5} = \tfrac{4}{5}(\log \tfrac{7}{39})$
$$= \tfrac{4}{5}(\log 7 - \log 39) = \tfrac{4}{5}\log 7 - \tfrac{4}{5}\log 39.$$

THE RELATIONSHIP OF LOGARITHMS TO DIFFERENT BASES

A given positive number has many different logarithms, depending on the base that is used. For example,

$$\log_{10} 220, \quad \log_e 220, \quad \log_4 220, \quad \log_{75} 220$$

are four different numbers. We can use the table on page 274 to find the approximation $\log_{10} 220 \approx 2.3424$ and a calculator to find the approximation $\log_e 220 \approx 5.3936$. Just how are these two numbers related? And how do you find $\log_4 220$ and $\log_{75} 220$? Surprisingly enough, the answer to both these questions is quite easy. The logarithm to any base b of a given number w can be found directly from its base 10 logarithm by this formula:

$$\text{If } b > 1, \text{ then for any positive number } w,$$
$$\log_b w = \frac{\log_{10} w}{\log_{10} b} = \left(\frac{1}{\log_{10} b}\right) \log_{10} w$$

We shall prove this statement. But first, you should note carefully what it says. For a fixed base b, the number $1/\log_{10} b$ is a *constant*. The statement in the box says that $\log_b w$ *can be obtained from* $\log_{10} w$ *simply by multiplying by this constant.* In other words, logarithms to the base b are proportional to logarithms to base 10. Once you have an accurate table of logarithms to base 10, you can find logarithms to any base.

EXAMPLE In order to find logarithms to the base e (remember $e \approx 2.71828 \cdots$), we use the formula in the box above with $b = e$ and the fact that $\log_{10} e \approx .4343$. Then the constant multiplier is the number

$$\frac{1}{\log_{10} b} = \frac{1}{\log_{10} e} \approx \frac{1}{.4343} \approx 2.3026°$$

Using the table of common logarithms on page 274 we have

Number w	$\log_{10} w$	(Multiply by $\dfrac{1}{\log_{10} e} \approx 2.3026$) $\longrightarrow$ $\log_e w$
3	.4771	$(.4771)(2.3026) \approx 1.0986$
45.2	1.6551	$(1.6551)(2.3026) \approx 3.8110$
220	2.3424	$(2.3424)(2.3026) \approx 5.3936$

Now that its meaning is clear, here is a proof of the statement in the preceding box. For any base b and positive number w, we have $b^{\log_b w} = w$. If we rewrite the number b as a power of 10, namely, $b = 10^{\log_{10} b}$, then the preceding equation becomes

$$(10^{\log_{10} b})^{\log_b w} = w$$
$$10^{(\log_{10} b)(\log_b w)} = w$$

Thus if 10 is raised to the exponent $(\log_{10} b)(\log_b w)$, the result is the number w. But the exponent to which 10 must be raised to produce w is precisely the number $\log_{10} w$. Therefore

$$(\log_{10} b)(\log_b w) = \log_{10} w$$

° All logarithms and other numbers have been rounded off to four decimal places, so all results are approximate.

Dividing both sides of this equation by the nonzero number $\log_{10} b$ yields

$$\log_b w = \frac{\log_{10} w}{\log_{10} b} = \left(\frac{1}{\log_{10} b}\right)\log_{10} w$$

which is what we wanted to prove.

LOGARITHMIC EQUATIONS

The various properties of logarithms can be used to solve equations involving logarithms, as illustrated in the following examples.

EXAMPLE   Here is a step-by-step solution of the equation

$$\log (x - 15) = 2 - \log x$$

$$\log (x - 15) + \log x = 2 \qquad\qquad \text{(add } \log x \text{ to both sides)}$$

$$\log (x - 15)x = 2 \qquad\qquad \text{(first law of logarithms)}$$

$$\log (x^2 - 15x) = 2 \qquad\qquad \text{(multiply out left side)}$$

This last equation states that the exponent to which 10 must be raised in order to produce the number $x^2 - 15x$ is 2, that is,

$$x^2 - 15x = 10^2$$

$$x^2 - 15x - 100 = 0 \qquad\qquad \text{(subtract } 10^2 = 100 \text{ from both sides)}$$

$$(x - 20)(x + 5) = 0 \qquad\qquad \text{(factor left side)}$$

$$\begin{array}{ccc} x - 20 = 0 & \text{or} & x + 5 = 0 \\ x = 20 & & x = -5 \end{array}$$

Therefore the only *possible* solutions of the original equations are $x = 20$ and $x = -5$. We now must check to see if either of these numbers actually *is* a solution. To do this it is convenient to write the original equation in the equivalent form

$$\log (x - 15) + \log x = 2$$

Substituting $x = 20$ in the left side of this equation and using the first law of logarithms yields:

$$\log (20 - 15) + \log 20 = \log 5 + \log 20 = \log (5 \cdot 20) = \log 100 = 2$$

Therefore $x = 20$ is a solution. However, substitution of $x = -5$ in the equation yields

$$\log (-5 - 15) + \log (-5) = 2$$

This is impossible since logarithms of negative numbers are not defined. Therefore $x = -5$ is *not* a solution.

EXAMPLE   In order to solve the equation

$$\log_3(x + 3) - \log_3(x - 5) = 2$$

we apply the second law of logarithms to obtain:

$$\log_3\left(\frac{x + 3}{x - 5}\right) = 2$$

This equation states that the exponent to which 3 must be raised to produce the number $(x + 3)/(x - 5)$ is the number 2, that is,

$$\frac{x + 3}{x - 5} = 3^2$$

$$\frac{x + 3}{x - 5} - 9 = 0 \qquad \text{(subtract 9 from both sides)}$$

$$\frac{(x + 3) - 9(x - 5)}{x - 5} = 0 \qquad \text{(put left side over common denominator)}$$

$$\frac{-8x + 48}{x - 5} = 0 \qquad \text{(simplify left side)}$$

Since the only way that a fraction can be zero is to have its denominator nonzero and its numerator zero, we must have $x \neq 5$ and

$$-8x + 48 = 0$$
$$-8x = -48$$
$$x = 6$$

We can check whether or not $x = 6$ actually is a solution by substituting it in the original equation:

$$\log_3(6 + 3) - \log_3(6 - 5) = \log_3 9 - \log_3 1 = 2 - 0 = 2$$

Therefore $x = 6$ checks and is the solution of the equation.

EXERCISES

A.1.   Find the common (base 10) logarithm of:

(a)   10,000

(c)   1,000,000

(e) $\sqrt[3]{.01}$

(g) $\dfrac{10^5}{.01}$

(b)   .001

(d)   .01

(f) $\sqrt[5]{1000}$

(h) $\dfrac{\sqrt{10}}{1000}$

A.2.   Find the logarithm to base 2 of:

(a)   16

(c) $\dfrac{1}{2\sqrt{2}}$

(e) $\frac{1}{64}$

(g) $\frac{1}{256}$

(b) $\sqrt[3]{256}$

(d)   128

(f) $\dfrac{1}{(\sqrt[3]{2})^5}$

(h)   4096

A.3. Translate each of these exponential statements into an equivalent logarithmic statement:

(a) $10^{-2} = .01$ (d) $10^{.4771} \approx 3$ (g) $10^{7k} = r$

(b) $10^3 = 1000$ (e) $10^{1.4367} \approx 27.3$ (h) $10^{(a+b)} = c$

(c) $\sqrt[3]{10} = 10^{1/3}$ (f) $10^{3.9488} \approx 8888$ (i) $10^{x^2+2} = y$

A.4. Translate each of these exponential statements into an equivalent logarithmic statement:

(a) $5^4 = 625$ (d) $3^{-2} = \frac{1}{9}$ (g) $e^{-4} \approx .0183$

(b) $7^8 = 5{,}764{,}801$ (e) $b^{14} = 3379$ (h) $e^{12/7} \approx 5.553$

(c) $2^{-3} = \frac{1}{8}$ (f) $e^{3.25} \approx 25.79$ (i) $a^{-b} = c$

A.5. Translate each of these logarithmic statements into an equivalent exponential statement (remember $\log v$ means $\log_{10} v$):

(a) $\log 10{,}000 = 4$ (d) $\log 500 \approx 2.699$ (g) $\log a = b$

(b) $\log .001 = -3$ (e) $\log (.8) \approx -.097$ (h) $\log (a + c) = d$

(c) $\log 750 \approx 2.86$ (f) $\log (.005) \approx -2.3$ (i) $\log (x^2 + 2y) = z + w$

A.6. Translate each of these logarithmic statements into an equivalent exponential statement:

(a) $\log_2 \sqrt{2} = \frac{1}{2}$ (c) $\log_8(\frac{1}{4}) = -\frac{2}{3}$ (e) $\log_e 3 \approx 1.099$

(b) $\log_5 125 = 3$ (d) $\log_2(\frac{1}{4}) = -2$ (f) $\log_e 10 \approx 2.303$

A.7. Simplify each of these expressions. For example, $\log_e(e^{17,737}) = 17{,}737$.

(a) $\log 10^{\sqrt{43}}$ (d) $\log_{17}(17^{17})$ (g) $\log_{k+1}(k + 1)^{14}$

(b) $\log 10^{\sqrt{49}}$ (e) $\log 10^{\sqrt{x^2+y^2}}$ (h) $10^{\log 57.3}$

(c) $\log_5(5^{4.7})$ (f) $\log_{3.5}(3.5^{(x^2-1)})$ (i) $e^{\log_e 931}$

A.8. Evaluate each of the following. *Examples:* $\log_{\sqrt{2}} 8 = 6$ since $(\sqrt{2})^6 = 8$ and $\log_{27} 9 = \frac{2}{3}$ since $27^{2/3} = \sqrt[3]{27^2} = \sqrt[3]{3^6} = 3^2 = 9$.

(a) $\log (.0001)$ (d) $\log_{16} 4$ (g) $\log_{16} 32$

(b) $\log 100{,}000$ (e) $\log_e \left(\frac{1}{e}\right)$ (h) $\log_8 4$

(c) $\log_2 64$ (f) $\log_{\sqrt{3}}(27)$ (i) $\log_{\sqrt{3}}(\frac{1}{9})$

A.9. In each of these statements, replace u, b, or v by a number, so that the resulting statement is true.

(a) $\log_3 81 = u$ (c) $\log_{81} 27 = u$ (e) $\log (10 \sqrt{10}) = u$

(b) $\log_{27} v = \frac{1}{3}$ (d) $\log_5 v = -4$ (f) $\log_b(\frac{1}{9}) = -\frac{2}{3}$

A.10. Suppose b is a fixed number >1 and that $\log_b 2 = .13$, $\log_b 3 = .2$, and $\log_b 5 = .3$. Use the laws of logarithms to calculate each of the following. For example, $\log_b 6 = \log_b(2 \cdot 3) = \log_b 2 + \log_b 3 = .13 + .2 = .33$.

(a) $\log_b 10$ (d) $\log_b 27$ (g) $\log_b 18$ (*Hint:* $18 = 2 \cdot 3^2$)

(b) $\log_b 15$ (e) $\log_b(\frac{5}{3})$ (h) $\log_b 48$

(c) $\log_b 4$ (f) $\log_b(\frac{3}{2})$ (i) $\log_b 45$

A.11. Use the laws of logarithms to express each of the following as a single logarithm. For example, $\log x + 2(\log y) = \log x + \log y^2 = \log (xy^2)$.

(a) $\log_e x^2 + 3 \log_e y$ (d) $\log_e 3x - 2(\log_e x - \log_e(2 + y))$

(b) $\log 2x + 2(\log x) - \log 3y$ (e) $2(\log_e x) - 3(\log_e x^2 + \log_e x)$

(c) $\log_e(x^2 - 9) - \log_e(x + 3)$ (f) $\log_e \left(\dfrac{e}{\sqrt{x}}\right) - \log_e \sqrt{ex}$

A.12. If $\log_b 12 = 7.4$ and $\log_b 8.86 = 19.61$, then what is $\log_b 8.86/\log_b 12$?

B.1. Use the following approximations and the table on page 274 to find these logarithms.

$\log_{10}(\frac{14}{3}) \approx \frac{2}{3}$, $\log_{10}(18) \approx 1.25$, $\log_{10}(317) \approx 2.5$, $\log_{10}(465) \approx \frac{8}{3}$

(a) $\log_{317}(.8)$ (d) $\log_{465}(3)$ (g) $\log_{465}(220)$

(b) $\log_{18}(.2)$ (e) $\log_{18}(45.1)$ (h) $\log_{317}(5)$

(c) $\log_{14/3}(.4)$ (f) $\log_{14/3}(225)$ (i) $\log_{18}(230)$

B.2. Answer true or false, and give reasons for your answers. Assume all letters represent positive numbers.

(a) $\log_b \left(\dfrac{r}{5}\right) = \log_b r - \log_b 5$ (e) $\log_5(5x) = 5(\log_5 x)$

(b) $\dfrac{\log_b a}{\log_b c} = \log_b \left(\dfrac{a}{c}\right)$ (f) $\log_b(ab)^t = t \log_b a + t$

(c) $\dfrac{\log_b r}{t} = \log_b(r)^{1/t}$ (g) $\dfrac{\log_e 10}{\log_e 5} = \log_e 2$

(d) $\log_b(cd) = \log_b c + \log_b d$ (h) $\log_e(r^e) = r$

B.3. Suppose $\log_b x = 3$. What is $\log_{1/b} x$?

B.4. Solve these equations.

(a) $\log x + \log (x - 3) = 1$ (e) $\log (x + 9) - \log x = 1$

(b) $\log (x - 1) + \log (x + 2) = 1$ (f) $\log (2x + 1) = 1 + \log (x - 2)$

(c) $\log_5(x + 3) = 1 - \log_5(x - 1)$ (g) $\log (x + 1) + \log (x - 1) = -2$

(d) $\log_4(x - 5) = 2 - \log_4(x + 1)$ (h) $\log x = \log (x + 3) - 1$

B.5. Solve these equations.

(a) $\log \sqrt{x^2 - 1} = 2$

(b) $\log \sqrt[3]{x^2 + 21x} = \frac{2}{3}$

(c) $\log (x + 2) - \log (4x + 3) = \log \left(\dfrac{1}{x}\right)$

(d) $\log (x + 1) = \frac{1}{2} + \log x$

(e) $\log (x^2 + 1) - \log (x - 1) = 1 + \log (x + 1)$

(f) $\dfrac{\log (x + 1)}{\log (x - 1)} = 2$

B.6. Solve these equations.
  (a) $(\log x)^2 = \log x^2$ [*Hint:* let $u = \log x$; then $\log x^2 = 2(\log x) = 2u$, so that the equation becomes $u^2 = 2u$. Solve for u; then find x.]
  (b) $\log x^4 = (\log x)^3$ (d) $(\log x)^2 - \log x^5 = -6$
  (c) $(\log x)^2 - \log x^2 = 3$ (e) $(\log_5 x)^2 = 5 + \log_5 x^4$

B.7. Suppose $b > 1$ and that x and v are positive numbers such that $\log_b x = \frac{1}{2}\log_b v + 3$. Show that $x = (b^3)\sqrt{v}$.

B.8. In chemistry, the concentration of hydrogen ions in a given substance is denoted $[H^+]$ and is measured in moles per liter. The pH of the substance is defined to be the number

$$pH = -\log_{10}[H^+]$$

For example, for bananas $[H^+]$ is 3×10^{-5} moles/liter, so that $pH = -\log(3 \times 10^{-5}) = -(\log 3 + \log 10^{-5}) = -((\log 3) - 5) = -\log 3 + 5 \approx -.4771 + 5 = 4.5229$. For each substance listed below, the value of $[H^+]$ is given. Use the logarithm table on page 274 and the logarithm laws to determine the corresponding pH.
  (a) beer, $[H^+] = .8 \times 10^{-4}$ (d) beets, $[H^+] = .4 \times 10^{-5}$
  (b) wine, $[H^+] = 4 \times 10^{-4}$ (e) wheat flour, $[H^+] = 2 \times 10^{-6}$
  (c) hominy, $[H^+] = 5 \times 10^{-8}$

B.9. Which is larger: 97^{98} or 98^{97}? [*Hint:* $\log 97 \approx 1.9868$ and $\log 98 \approx 1.9912$ and $f(x) = 10^x$ is an increasing function.]

DO IT YOURSELF!

SCIENTIFIC NOTATION

Every positive real number can always be written as a product of some power of 10 and a number between 1 and 10. For example,

$$356 = 3.56 \times 100 = 3.56 \times 10^2$$
$$1{,}563{,}627 = 1.563627 \times 1{,}000{,}000 = 1.563627 \times 10^6$$
$$.072 = 7.2 \times \tfrac{1}{100} = 7.2 \times 10^{-2}$$
$$.000862 = 8.62 \times \tfrac{1}{10{,}000} = 8.62 \times .0001 = 8.62 \times 10^{-4}$$

When a number is written in this form, it is said to be written in **scientific notation.** Scientific notation is needed in order to use logarithm tables, as we shall see below.

Calculator Note Some calculators are equipped with scientific notation. Usually they have a key labeled EE (for enter exponent). If you enter the number 7.235, press the EE key, and then enter the number -12, the calculator display will read: $\boxed{7.235 \qquad -12.}$ This indicates the number $7.235 \times 10^{-12} = .000000000007235$. Notice that this number cannot even be entered on a 10-digit calculator which does not have scientific notation capability.

LOGARITHM TABLES

Throughout this discussion *we deal only with common logarithms (base 10)*. Obviously, there can't be a table listing the logarithm of *every* real number. A typical logarithm table (such as the one at the end of this book) lists the logarithms of all two-place decimals from 1 to 10 (that is, 1, 1.01, 1.02, 1.03, . . . , 9.97, 9.98. 9.99, 10). Since the logarithm of a decimal need not be a rational number, the entries in a logarithm table are decimal approximations of the actual logarithms. In the table in this book all logarithms are approximated to four decimal places.

LOGARITHMS OF NUMBERS BETWEEN 1 AND 10

For convenience, we reproduce here a portion of the logarithm table at the back of the book. It will be used in the next eleven examples.

x	0	1	2	3	4	5 . . . .
5.5	.7404	.7412	.7419	.7427	.7435	.7443 . . .
5.6	.7482	.7490	.7497	.7505	.7513	.7520 . . .
5.7	.7559	.7556	.7574	.7582	.7589	.7597 . . .
.						
.						
.						

EXAMPLE In order to find the logarithm of 5.63, we first look in the *left*-hand column for the first two digits of our number, namely, 5.6. Now on the same line as 5.6 we look at the entry in the column labeled 3; it is .7505. So the logarithm of 5.63 is approximately .7505.

EXAMPLE To find the logarithm of 5.7 = 5.70, look in the first column for 5.7 and then across the same line to the entry in the column labeled 0. It is .7559, so $\log 5.7 \approx .7559$.

INTERPOLATION

When the number whose logarithm is wanted does not appear in the table, we must do some approximating.

EXAMPLE We cannot find $\log 5.527$ immediately since our table does *not* include 5.527. However, the table does include 5.52 = 5.520 and 5.53 = 5.530 and

$$5.520 < 5.527 < 5.530$$

Since 27 lies $\frac{7}{10}$ of the way from 20 to 30, we see that 5.527 lies $\frac{7}{10}$ of the way from 5.520 to 5.530. Therefore it seems reasonable that log 5.527 lies approximately $\frac{7}{10}$ of the way from log 5.520 to log 5.530. Using the table, we find that

$$\log 5.530 = \log 5.53 \approx .7427$$
$$\log 5.520 = \log 5.52 \approx .7419$$

difference .0008

Therefore log 5.527 lies $\frac{7}{10}$ of the way between .7419 and .7427. The distance from .7419 to .7427 is .0008 and $\frac{7}{10}$ of this distance is $(.7)(.0008) = .00056$. Since we are using four-place logarithms, we round off this number to four places: $.00056 \approx .0006$. Then we have:

$$\log 5.527 \approx \log 5.52 + .0006 \approx .7419 + .0006 = .7425$$

The process used to find log 5.527 in the preceding example is called **linear interpolation**. Due to the approximating involved there may be a slight error in the results. But for most purposes this error is insignificant. The entire process will be justified (in a more general setting) in calculus and the size of the possible error will be discussed.

EXAMPLE   In order to find log 5.732 we first find log 5.73 and log 5.74 in the table:

$$\log 5.74 \approx .7589$$
$$\log 5.73 \approx .7582$$

difference .0007

Since 5.732 lies $\frac{2}{10}$ of the way between $5.73 = 5.730$ and $5.74 = 5.740$, it seems likely that log 5.732 lies $\frac{2}{10}$ of the way from $\log 5.73 = .7582$ to $\log 5.74 = .7589$. The distance from .7582 to .7589 is .0007 and $\frac{2}{10}(.0007) = .2(.0007) = .00014$. Rounding this number to four decimal places gives .0001, so that

$$\log 5.732 \approx \log 5.73 + .0001 \approx .7582 + .0001 = .7583$$

LOGARITHMS OF OTHER NUMBERS

Once you know how to find logarithms of numbers from 1 to 10, it is easy to find the logarithm of any positive real number. The key is to write the number in scientific notation.

EXAMPLE   In order to find the logarithm of 573.2, we note that

$$573.2 = 5.732 \times 10^2$$

Consequently,

$$\log 573.2 = \log (10^2 \cdot 5.732) = \log 10^2 + \log 5.732$$

But $\log 10^2 = 2$ and 5.732 is a number between 1 and 10. Use the tables to find log 5.732. We actually did this in the preceding example and found that $\log 5.732 \approx .7583$. Therefore

$$\log 573.2 = \log 10^2 + \log 5.732 \approx 2 + .7583 = 2.7583$$

EXAMPLE In order to find log .00563, we write .00563 in scientific notation:

$$.00563 = 5.63 \times .001 = 5.63 \times 10^{-3}$$

Consequently,

$$\log .00563 = \log(10^{-3} \cdot 5.63) = \log 10^{-3} + \log 5.63$$

We know that $\log 10^{-3} = -3$. Using the tables in the first example above, we found that $\log 5.63 \approx .7505$. Therefore,

$$\log .00563 = \log 10^{-3} + \log 5.63 \approx -3 + .7505$$

Warning Be careful here: $-3 + .7505$ is *not* the number -3.7505. If you do your arithmetic carefully, you see that $-3 + .7505 = -2.2495$. However, we shall soon see that for purposes of computation, it will be more convenient to write $\log .00563 \approx -3 + .7505$ rather than -2.2495.

The preceding examples show that

> **The logarithm of any positive number can always be approximated by the sum of an integer and a number between 0 and 1**

For instance, $\log 5.75 \approx .7597 = 0 + .7597$ and $\log 573.2 \approx 2.7583 = 2 + .7583$ and $\log .000563 \approx -3 + .7505$. A logarithm written in this way is said to be in **standard form**. The integer part of a logarithm in standard form is called the **characteristic**. The decimal fraction part of a logarithm in standard form (that is, the number between 0 and 1) is called the **mantissa**.

ANTILOGARITHMS

We now know how to find the logarithm of a given number. Equally important is the ability to reverse this process:

> Given a logarithm, find the number which has this logarithm

The number whose logarithm is u is called the **antilogarithm** of u, so what we are dealing with here is just the problem: given a number u, find the antilogarithm of u.

Finding antilogarithms is quite simple, in theory. You just use the basic property of logarithms discussed on page 273.

$$\log 10^u = u \text{ for every real number } u$$

This property is just another way of saying that the number whose logarithm is u is precisely 10^u, that is,

> The antilogarithm of u is 10^u

The practical problem of determining the antilogarithm of u (that is, computing 10^u) is easily handled if you have a calculator equipped with either a 10^x or a y^x key. Antilogarithms can be found without a calculator by using the logarithm tables "in reverse," as illustrated in the following examples.

EXAMPLE The antilogarithm of .7435 is the number y such that $\log y = .7435$. In order to find y, look through the logarithm table until you find the entry .7435. As shown on page 286, this entry lies on the same line as 5.5 (left column) and in the column labeled 4 at the top. This means that $\log 5.54 \approx .7435$, so that $y \approx 5.54$. Thus the antilogarithm of .7435 is 5.54. But we already know that the antilogarithm of any number u is 10^u. So what we have also shown here is that

$$5.54 \approx \text{antilogarithm of } .7435 = 10^{.7435}$$

EXAMPLE Suppose $\log y = .7495$. In order to find y (that is, the antilogarithm of .7495), we look through the logarithm table for the entry .7495. But it isn't there. The entries closest in value to .7495 are .7490 and .7497. Reading from the table on page 286 we see that

$$\log 5.62 \approx .7497$$
$$\log 5.61 \approx .7490$$

difference .0007

Since .7495 lies $\frac{5}{7}$ of the way from .7490 to .7497, it seems reasonable that the number y with $\log y = .7495$ lies $\frac{5}{7}$ of the way from 5.61 to 5.62. Since the distance from 5.61 to 5.62 is .01, and since $\frac{5}{7} \approx .7143$, the distance from 5.61 to the number y is approximately

$$\tfrac{5}{7}(.01) = (.7143)(.01) = .007143$$

Rounding this number to four decimal places gives .0071, so that

$$y \approx 5.61 + .0071 = 5.6171$$

Therefore 5.6171 is the antilogarithm of .7495. Since 10^u is known to be the antilogarithm of u, we can now conclude that $10^{.7495} \approx 5.6171$.

EXAMPLE As we have seen, all logarithms in the table are numbers between 0 and 1. So the antilogarithm of 2.7435 (that is, the number x with $\log x = 2.7435$) cannot be found directly from the tables. But we know that the antilogarithm of any number u is just 10^u. Therefore the antilogarithm of 2.7435 is just $10^{2.7435}$. Simple arithmetic and the laws of exponents show that

$$10^{2.7435} = 10^{2+.7435} = (10^2)(10^{.7435})$$

Now $10^{.7435}$ is just the antilogarithm of .7435. So once we know this number, we can multiply by $10^2 = 100$ to get the antilogarithm of 2.7435. But .7435 *is* a number between 0 and 1 and we have just seen how to use the tables to find antilogarithms of such numbers. In fact, in an example above we found that

$$10^{.7435} = \text{antilogarithm of } .7435 \approx 5.54$$

Therefore the antilogarithm of 2.7435 is just

$$10^2(\text{antilogarithm of } .7435) \approx 10^2(5.54) = 100(5.54) = 554$$

EXAMPLE Suppose $\log x = -3.2505$. In order to use the table to find x, it is first necessary to write $\log x = -3.2505$ in standard form (that is, as the sum of an integer and a number between 0 and 1). Be careful—it is *not* true that $-3.2505 = -3 + .2505$. It is true that $-3.2505 = -3 - .2505$, but $-.2505$ does not lie between 0 and 1. Here's how to write -3.2505 in standard form:

$$\log x = -3.2505 = (-4 + 4) - 3.2505 = -4 + (4 - 3.2505) = -4 + .7495$$

Now we can proceed as before. Since the antilogarithm of -3.2505 is known to be $10^{-3.2505}$ and since

$$10^{-3.2505} = 10^{-4+.7495} = (10^{-4})(10^{.7495})$$

we need only find $10^{.7495}$, the antilogarithm of $.7495$. For once we have this, we just multiply by $10^{-4} = .0001$ to obtain the antilogarithm of $-4 + .7495 = -3.2505$. Since $.7495$ lies between 0 and 1, this can be done via tables. In fact, it was done in the second example on page 289, where we found that

$$10^{.7495} = \text{antilogarithm of } .7495 \approx 5.6171$$

Therefore the antilogarithm of $-3.2505 = -4 + .7495$ is

$$10^{-4}(\text{antilogarithm of } .7495) \approx (10^{-4})(5.6171) = (.0001)(5.6171) = .00056171$$

Here is a summary of the procedure used in the preceding examples to find antilogarithms.

GIVEN *u*, FIND THE ANTILOGARITHM OF *u* AS FOLLOWS:

   (*i*)   *Write* u *in standard form, as the sum of an integer* k *and a number* v *between 0 and 1:* u = k + v. *(Note:* k *may be positive, negative, or zero.)*

  (*ii*) *Use the tables (and interpolation, if necessary) to find the antilogarithm of* v *(that is, the number* y *with* log y = v).

 (*iii*) *Then the antilogarithm of* u *is* (10^k)y.

COMPUTATIONS WITH LOGARITHMS

Now that you know how to use the tables to find both logarithms and antilogarithms, it is relatively easy to compute with logarithms.

EXAMPLE Find the product $(24.86)(.01392)(1.787)$. Let x denote this product. Using the First Law of Logarithms, we see that

$$\log x = \log\,[(24.86)(.01392)(1.787)]$$
$$= \log 24.86 + \log .01392 + \log 1.787$$

We now use the logarithm tables and simple addition to calculate the right-hand side of this equation. Note that all logarithms are written in standard form, as the sum of an integer (characteristic) and a number between 0 and 1 (mantissa). When the characteristic is negative, it is usually written after the mantissa:

$$\begin{array}{rl} \log 24.86 \approx & 1.3955 \\ \log .01392 \approx & .1436 - 2 \\ \underline{\log 1.787 \approx \ \ .2521} \\ \text{sum:} \approx & 1.7912 - 2 = .7912 - 1 \end{array}$$

Therefore the equation becomes:

$$\log x \approx .7912 - 1$$

According to the procedure for finding antilogarithms, we know that $x \approx (10^{-1})y$, where $\log y = .7912$. Using the logarithm tables and interpolation, we find that $y \approx 6.183$. Consequently,

$$x \approx (10^{-1})y \approx (.1)(6.183) = .6183, \text{ that is, } (24.86)(.01392)(1.787) \approx .6183$$

Note that our answer is only carried out to four decimal places, whereas the actual product involves ten decimal places. The missing places in our answer represent part of the relatively small error in our approximation.

EXAMPLE In order to compute $(2.4)^{37.8}$, let $x = (2.4)^{37.8}$. Then by the third law of logarithms

$$\log x = \log [(2.4)^{37.8}] = (37.8)(\log 2.4)$$

The logarithm tables show that $\log 2.4 \approx .3802$, so that

$$\log x = (37.8)(\log 2.4) \approx (37.8)(.3802) = 14.37156 \approx 14.3716$$

The antilogarithm procedure shows that $x \approx (10^{14})y$, where $\log y = .3716$. Using the logarithm tables and interpolation, we find that $y \approx 2.353$, so that

$$x \approx (10^{14})y \approx (10^{14})(2.353), \text{ that is, } (2.4)^{37.8} \approx (10^{14})(2.353)$$

It is worth noting that this last example, unlike the preceding one, cannot be worked out by hand or, for that matter, on many simple calculators. Here is another such example.

EXAMPLE In order to compute $\sqrt[7]{2.4/3780}$, let $x = \sqrt[7]{2.4/3780}$ and use the second and third laws of logarithms:

$$\log x = \log \sqrt[7]{\frac{2.4}{3780}} = \log \left(\frac{2.4}{3780}\right)^{1/7} = \frac{1}{7}\left(\log \frac{2.4}{3780}\right) = \frac{1}{7}(\log 2.4 - \log 3780)$$

The logarithm tables show that $\log 2.4 \approx .3802$ and $\log 3780 \approx 3.5775$ so that

$$\log x = \tfrac{1}{7}(\log 2.4 - \log 3780) \approx \tfrac{1}{7}(.3802 - 3.5775) = \tfrac{1}{7}(-3.1973) \approx -.4568$$

In order to find x, we must write $\log x \approx -.4568$ in standard form:

$$\log x \approx -.4568 = (1 - 1) - .4568 = (1 - .4568) - 1 = .5432 - 1$$

Now the antilogarithm procedure shows that $x \approx (10^{-1})y$, where $\log y = .5432$. Using the logarithm tables and interpolation we find that $y \approx 3.493$. Therefore

$$\sqrt[7]{\frac{2.4}{3780}} = x \approx (10^{-1})y \approx (.1)(3.493) = .3493$$

Note   It is customary to write an equal sign ($=$) where we have written an "approximately equal" sign ($\approx$). For instance, $\log 2.4 = .3802$ instead of $\log 2.4 \approx .3802$. Since everyone familiar with logarithms *knows* that the entries given in the tables or by calculators are rational approximations, this inaccuracy causes no difficulty in practice.

EXERCISES

A.1.   Do Exercise B.9 on page 253.

A.2.   Use the logarithm table at the end of this book to find the following logarithms. Write each logarithm in standard form and label the characteristic.
   (a) $\log (1.01)$
   (b) $\log (3.79)$
   (c) $\log (45.6)$
   (d) $\log (666)$
   (e) $\log (7880)$
   (f) $\log (.00842)$
   (g) $\log (.000915)$
   (h) $\log (.0000327)$
   (i) $\log (57,300,000)$

A.3.   Use the logarithm table and interpolation to find each of these logarithms. Write each one in standard form.
   (a) $\log (2.345)$
   (b) $\log (34.73)$
   (c) $\log (467.9)$
   (b) $\log (.07171)$
   (e) $\log (.0008463)$
   (f) $\log (.009752)$
   (g) $\log (.7631)$
   (h) $\log (492,700)$
   (i) $\log (5,324,000)$

A.4.   Use the logarithm tables to find the antilogarithms of these numbers. Remember to put negative numbers in standard form.
   (a) .3464
   (b) .6937
   (c) 1.9528
   (d) $-.6126$
   (e) -1.1791
   (f) -3.8729

A.5.   Use the tables and interpolation to find the antilogarithms of
   (a) .8776
   (b) 2.3797
   (c) 5.2955
   (d) -1.1247
   (e) -4.2608
   (f) -2.2705

A.6.   Use tables and interpolation, if necessary, to solve each equation.
   (a) $\log x = 2.6415$
   (b) $\log y = -2.0357$
   (c) $\log x = 1.4735$
   (d) $\log x = 3.9196$
   (e) $\log z = -1.6328$
   (f) $\log y = -.1293$

A.7.   Use the logarithm table to approximate the following numbers.
   (a) $(256)(.0123)$
   (b) $(493)(3.41)(.0412)$
   (c) $\frac{934}{727}$
   (d) $(3.94)^{12}$
   (e) $(7.25)^8(1.26)^5$
   (f) $\sqrt[3]{756}$
   (g) $\sqrt{93.7}$
   (h) $\dfrac{7.98}{(314)(6.12)}$

B.1. Use the logarithm tables to approximate the following.

(a) $(3.841)(710.2)$

(b) $\dfrac{.7812}{.01204}$

(c) $\dfrac{7.981}{(127.6)(8.054)}$

(d) $(32.7)^{14.1}$

(e) $\sqrt[9]{\dfrac{413}{(.072)^7}}$

4. LOGARITHMIC FUNCTIONS

The concept of logarithm leads to a new family of functions. Since logarithms are just exponents, it isn't surprising that these logarithmic functions are closely related to the exponential functions studied in Section 2. What may be surprising, however, is the fact that a variety of physical and other phenomena can be mathematically described by logarithmic functions. Consequently, logarithmic functions play an important role in calculus and certain other branches of mathematics.

Let b be a fixed real number, greater than 1. Each positive number x has a logarithm to the base b. Let f be the function whose domain is the set of all positive real numbers and whose rule is:

$$f(x) = \log_b x$$

The function f is called a **logarithmic function.**

For each base b there is a different logarithmic function. For instance, corresponding to the bases 10, 2, and e ($\approx 2.71828 \cdots$), we have the three functions whose rules are:

$$g(x) = \log x,^\circ \qquad h(x) = \log_2 x, \qquad f(x) = \log_e x$$

By using an appropriate table or a calculator if necessary, we can compute the value of each of these functions for various positive numbers:

$$g(4) = \log 4 \approx .6021, \qquad h(4) = \log_2 4 = 2,$$
$$g(1000) = \log 1000 = 3, \qquad h(1000) = \log_2 1000 \approx 9.9658,$$
$$g(1) = \log 1 = 0, \qquad h(1) = \log_2 1 = 0,$$

$$f(4) = \log_e 4 \approx 1.3863,$$
$$f(1000) = \log_e 1000 \approx 6.9078,$$
$$f(1) = \log_e 1 = 0$$

Although there is a different logarithm function for each base b, all of these functions are very closely related. As we saw in the last section, logarithms to the base b are just constant multiples of logarithms to the base 10. More specifically,

$$\log_b x = \left(\frac{1}{\log b}\right)(\log x)$$

° As usual for base 10, we write $\log x$ instead of $\log_{10} x$.

Consequently,

> *The logarithmic function* $f(x) = \log_b x$ *is a constant multiple of the common logarithm function* $g(x) = \log x$. *The constant multiplier is* $1/\log b$.

For example, if $f(x) = \log_e x$, then

$$f(x) = \left(\frac{1}{\log e}\right) g(x) \approx (2.3026) g(x)$$

since $1/\log e \approx 1/.4343 \approx 2.3026$.

As we saw in the last section, logarithms to any base b satisfy the laws of logarithms. If we translate these laws into functional notation, we see that the logarithmic function $f(x) = \log_b x$ has several interesting algebraic properties:

Logarithm Law	Equivalent Functional Statement
$\log_b(vw) = \log_b v + \log_b w$	$f(vw) = f(v) + f(w)$
$\log_b\left(\dfrac{v}{w}\right) = \log_b v - \log_b w$	$f\left(\dfrac{v}{w}\right) = f(v) - f(w)$
$\log_b(v^k) = k(\log_b v)$	$f(v^k) = k \cdot f(v)$

For example, $f(150) = f(3 \cdot 50) = f(3) + f(50)$ and $f(16) = f(2^4) = 4 \cdot f(2)$.

GRAPHS OF LOGARITHMIC FUNCTIONS

Once we have the graph of the common logarithm function $g(x) = \log x$, the graphs of other logarithmic functions can be easily determined.

EXAMPLE   Using the properties of logarithms and the table at the end of the book, we first plot some points on the graph of the function $g(x) = \log x$. It then seems that the graph of g looks like the one in Figure 5-8.

x	$g(x) = \log x$
10^{-300}	-300
.001	-3
.01	-2
.1	-1
1	0
3	.48
5	.70
10	1
15	1.18
50	1.70
100	2
1,000	3
100,000	5

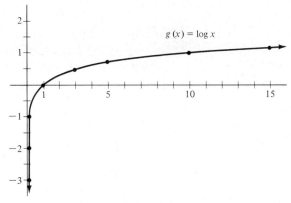

Figure 5-8

Observe that as x moves to the *left* from 1 to 0, the graph falls very sharply and gets closer and closer to the y-axis. The graph never touches the y-axis since the function is not defined when $x = 0$. Thus in the negative direction, the y-axis is a vertical asymptote of the graph. As x moves to the *right* from $x = 1$, the graph continually rises, but extremely slowly. At $x = 100,000$, the height of the graph over the x-axis is only 5 units.

EXAMPLE   One way to graph the logarithmic function $f(x) = \log_e x$ is to plot a number of points. A faster way to graph f is to use the relationship between f and the common logarithm function $g(x) = \log x$ given in the box on page 280:

$f(x)$ is a constant multiple of $g(x)$; specifically,

$$f(x) = \left(\frac{1}{\log e}\right)g(x) \approx (2.3026)g(x)$$

Therefore, as we saw on page 150,* the graph of f is just the graph of g stretched away from the x-axis by a factor of $1/(\log e) \approx 2.3026$ as shown in Figure 5-9.

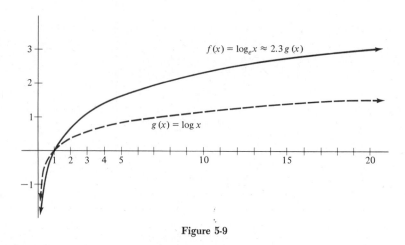

Figure 5-9

For any base $b > 1$, the graph of the logarithmic function $f(x) = \log_b x$ may be obtained from the graph of $g(x) = \log x$ as in the preceding example. It will have the same basic shape as the graph of g.

EXAMPLE   The term "logarithmic" is often applied to any function whose rule involves logarithms, such as $h(x) = \log(100/x^3\sqrt{x})$. It is sometimes possible to use the algebraic properties of logarithms to rewrite the rule of such a function and to find its graph.** We have:

$$h(x) = \log\frac{100}{x^3\sqrt{x}} = \log 100 - \log(x^3)\sqrt{x} = \log 100 - (\log x^3 + \log \sqrt{x})$$

$$= \log 100 - \log x^3 - \log x^{1/2} = 2 - 3(\log x) - \tfrac{1}{2}(\log x) = 2 - \tfrac{7}{2}(\log x)$$

Now let f be the function given by $f(x) = \log x$. As we saw in Section 6 of Chapter 3, the graph of the function h is just the graph of the function f stretched away from the x-axis by a factor of $\tfrac{7}{2}$, reflected in the x-axis, and moved 2 units upward, as shown in Figure 5-10.

---

* If you have not read this material before, you may prefer to obtain the graph of f by plotting points instead.
** If you have not read Section 6 of Chapter 3, you may wish to omit this example.

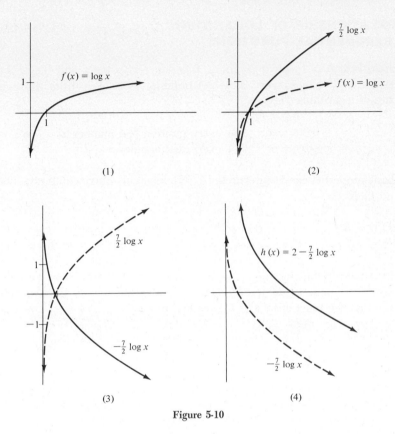

$f(x) = \log x$

$\frac{7}{2} \log x$

$f(x) = \log x$

(1)   (2)

$\frac{7}{2} \log x$

$-\frac{7}{2} \log x$

$h(x) = 2 - \frac{7}{2} \log x$

$-\frac{7}{2} \log x$

(3)   (4)

Figure 5-10

Warning   Since the logarithm laws deal only with positive numbers, you must be careful when applying them to logarithmic functions. There wasn't any difficulty in the preceding example since both of the functions $\log (100/x^3 \sqrt{x})$ and $2 - \frac{7}{2}(\log x)$ are defined only when x is positive. But this isn't always the case. For instance, the third law of logarithms guarantees that

$$\log x^2 = 2(\log x) \qquad \text{for every } positive \text{ real number } x$$

But the functions

$$g(x) = \log x^2 \qquad \text{and} \qquad h(x) = 2(\log x)$$

are *not* the same function since they have *different domains*. For whenever x is nonzero, the number x^2 is positive, so that $g(x) = \log x^2$ is defined. In particular, $g(x)$ is defined for every *negative* real number. But $h(x) = 2(\log x)$ is only defined when x is *positive*.

THE RELATIONSHIP OF LOGARITHMIC AND EXPONENTIAL FUNCTIONS

For a fixed base $b > 1$, the logarithmic function $f(x) = \log_b x$ and the exponential function $h(x) = b^x$ have a very close relationship, as is clear from the two basic properties of logarithms:

$$b^{\log_b x} = x \qquad \text{for every positive real number } x$$
$$\log_b(b^x) = x \qquad \text{for every real number } x$$

These basic properties can be interpreted as statements about the composite functions°
$h \circ f$ and $f \circ h$:

$$(h \circ f)(x) = h(f(x)) = h(\log_b x) = b^{\log_b x} = x \qquad \text{for every positive real number } x$$
$$(f \circ h)(x) = f(h(x)) = f(b^x) = \log_b(b^x) = x \qquad \text{for every real number } x$$

Since the domain of the function $f(x) = \log_b x$ is the set of all positive real numbers and the domain of the function $h(x) = b^x$ is the set of all real numbers, the two statements above say that the functions f and h have this property:

$$(h \circ f)(x) = x \qquad \text{for every number } x \text{ in the domain of } f$$
$$(f \circ h)(x) = x \qquad \text{for every number } x \text{ in the domain of } h$$

This property may be paraphrased by saying that "h undoes what f does" and "f undoes what h does." For this reason any two functions that have this property are said to be **inverses** of each other. The general theory of inverse functions is discussed in detail in Section 3 of Chapter 7.

A calculator equipped with a "log" key and a "10^x" key will provide a visual electronic demonstration of the inverse functions $f(x) = \log x$ and $h(x) = 10^x$. For instance, to evaluate the composite function $h \circ f$ at a given number,

enter the number	press the log key	press the 10^x key
x	$f(x)$	$h(f(x)) = (h \circ f)(x)$

The final display will be the number with which you started,°° thus demonstrating that h does indeed undo what f does, that is, that $(h \circ f)(x) = x$. Similar remarks apply to the function $f \circ h$. If the calculator has an "ln" key and an "e^x" key, you can provide the same demonstration for the inverse functions $f(x) = \log_e x$ and $h(x) = e^x$.

If the graph of $f(x) = \log x$ (p. 295) and the graph of $h(x) = 10^x$ are sketched on the same set of coordinate axes, the result is the picture shown in Figure 5-11.

° Composite functions were introduced in Section 7 of Chapter 3.
°° Since all calculators round off after a certain number of digits, there may be a slight error, so that the final display sometimes differs slightly form the original number.

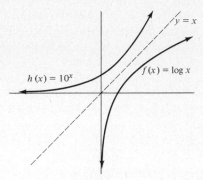

Figure 5-11

Observe that the graph of f is the mirror image of the graph of h (with the dotted line $y = x$ acting as the mirror). In more technical terms, the graph of f is the reflection of the graph of h in the line $y = x$. As we shall see in Chapter 7, the graphs of any two inverse functions are related in this way.

EXERCISES

A.1. In each case assume that the given point lies on the graph of the function $f(x) = \log_b x$. Then find b.
(a) $(100, 2)$ (b) $(8, 3)$ (c) $(\sqrt{8}, \frac{3}{2})$ (d) $(\sqrt[3]{(125)^2}, 2)$

A.2. Find the domain of each of the functions below (where, as usual, the domain is the largest set of real numbers for which the rule of the function produces well-defined real numbers).
(a) $f(x) = \log(x + 1)$
(b) $g(x) = \log_e(x + 2)$
(c) $h(x) = \log(-x)$
(d) $k(x) = \log_3(-x + 2)$
(e) $f(x) = (\log x) - 10$
(f) $r(x) = \log_e(-x - 3) - 10$
(g) $k(x) = \log(x^2 - 1)$
(h) $f(x) = \log_2 x - \log_2(x + 1)$
(i) $g(x) = \log_e x + \log_e(x - 2)$

A.3. Use the graph of $g(x) = \log x$ on page 295 and the fact that every logarithmic function is a constant multiple of g to sketch the graph of these functions, without plotting points. (The logarithm table on p. 274 may be helpful.)
(a) $f(x) = \log_4 x$ (b) $p(x) = \log_{225} x$ (c) $k(x) = \log_{45.2} x$

A.4. Use the graphs on page 259 and the fact that exponential and logarithmic functions are inverses of each other to sketch the graphs of
$$p(x) = \log_{(3/2)} x, \qquad g(x) = \log_2 x, \qquad t(x) = \log_3 x.$$

B.1. In Section 2 we saw that if $1 < a < b$, then the graph of the exponential function $f(x) = b^x$ rose much faster than the graph of $g(x) = a^x$. What can be said about the growth of the graphs of $h(x) = \log_a x$ and $k(x) = \log_b x$ when $1 < a < b$ and $x > 0$?

B.2. For each pair of functions f and g, determine the values of x for which $f(x) = g(x)$. *Example: if $f(x) = \log x^4$ and $g(x) = 4(\log x)$, then $f(x) = g(x)$ for all positive x.*

(a) $f(x) = \log x^4$; $g(x) = 2(\log x^2)$

(b) $f(x) = \log \sqrt{x + 1}$; $g(x) = \frac{1}{2}(\log(x + 1))$

(c) $f(x) = (\log x^3) + 1$; $g(x) = 3(\log x) + 1$

(d) $f(x) = \log x^4$; $g(x) = 2(\log x) + \log x^2$

(e) $f(x) = \log\left(\dfrac{x}{x - 1}\right)$; $g(x) = \log x - \log(x - 1)$

(f) $f(x) = \log(x\sqrt{x - 2})$; $g(x) = \log x + \frac{1}{2}(\log(x - 2))$

B.3. Use the logarithm laws and the techniques of Section 6 of Chapter 3 to obtain the graph of each of the following functions from the graph of $g(x) = \log x$ without plotting very many points.

(a) $f(x) = \log(5x)$ (c) $k(x) = \log(x - 4)$ (e) $h(x) = \log\left(\dfrac{1}{x^3}\right)$

(b) $h(x) = (\log x) - 7$ (d) $f(x) = \log x^3$ (f) $k(x) = \log x \sqrt{x}$

B.4. Graph the following functions. Whenever possible use algebraic techniques to determine the shape of the graph, rather than plotting points. If the domain of the function includes negative numbers, be careful.

(a) $g(x) = \log x^2$ (c) $f(x) = \log |x|$ (e) $p(x) = \log x + \log x^2$

(b) $h(x) = \log x^4$ (d) $k(x) = |\log x|$ (f) $h(x) = \log x - \log \sqrt{x}$

B.5. (a) Graph the function $f(x) = 1/\log x$.

(b) How does the graph of f compare to the graph of $h(x) = \log(1/x)$?

B.6. If $1 < b < 10$, does the graph of $f(x) = \log_b x$ rise at a faster or slower rate than that of $g(x) = \log x$? Why? What's the answer when $b > 10$?

B.7. Suppose $f(x) = A \log x + B$, where A and B are constants. If $f(1) = 10$ and $f(10) = 1$, then find A and B.

C.1. (a) By how much must x be increased in order that the graph of $g(x) = \log_2 x$ rise 1 unit?

(b) If you turn the volume of your HiFi up and down, the "loudness" of the music changes. The output of the HiFi can be exactly measured in watts. The law of Weber-Fechner states that the relationship between output x in watts and loudness L at output x is given by the function

$$L(x) = \log_2 x$$

You are accustomed to listening to your HiFi at the level $L = 4$. After some time you develop callous ears, and you turn up the volume to $L = 5$. This goes on and eventually your turn it up from $L = 19$ to $L = 20$. Now whenever you increase the loudness L, you use more watts of power and your power bill increases. By how much does your power bill increase when you go from $L = 4$ to $L = 5$? from $L = 8$ to $L = 9$? from $L = 12$ to $L = 13$? from $L = 19$ to $L = 20$? Can you state a general rule for the increase when L goes from k to $k + 1$? [*Hint:* part (a) may shed some light on this.]

(c) If power costs $\frac{1}{2}$¢ per watt per hr, by how many *dollars* does your bill for 1 hr of power increase when you go from $L = 4$ to $L = 5$? from $L = 8$ to $L = 9$? from $L = 12$ to $L = 13$? from $L = 19$ to $L = 20$?

DO IT YOURSELF!

LOGARITHMIC GROWTH

In order to have a better feeling for the rate at which the graph of a logarithmic function rises when x is large, it is useful to consider the common logarithm function $f(x) = \log x$. By the properties of logarithms, we know that for any positive number v,

$$f(10v) = \log (10v) = \log 10 + \log v = 1 + \log v = 1 + f(v)$$

Therefore the y-coordinate of the point $(10v, f(10v))$ on the graph of f is just the number $f(v) + 1$. This means that the vertical distance between the points

$$(v, f(v)) \quad \text{and} \quad (10v, f(10v)) = (10v, f(v) + 1)$$

is just 1 unit, as shown in Figure 5-12.

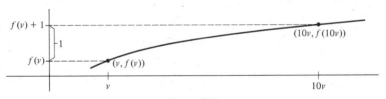

Figure 5-12

In other words, in order for the graph of $f(x) = \log x$ to rise just 1 unit, x must be increased *10 times*. Thus from $x = 5$ to $x = 10 \cdot 5 = 50$ (a horizontal distance of 45 units), the graph rises just 1 unit. Similarly, from $x = 750,000$ to $x = 10 \cdot 750,000 = 7,500,000$ (a horizontal distance of 6.75 *million units*), the graph rises only *1 unit*.

More generally, for any number k,

$$f(10^k v) = \log (10^k v) = \log 10^k + \log v = k + \log v = k + f(v)$$

Therefore

> *Changing the value of* x *by a factor of* 10^k *changes the value of*
> $f(x) = log \ x$ *by just* k *units.*

The implications of this fact can best be seen in an example from the real world, such as
the following one.

EXAMPLE   The magnitude of an earthquake is measured on the Richter scale. A mild
quake that does little damage might register 3.2 on the Richter scale, whereas the great
San Francisco earthquake of 1906 registered 8.4. As we shall now see, magnitude on
the Richter scale is determined by a certain logarithmic function.

    At the time the Richter scale was invented, the smallest earthquake recorded up to
that time was chosen as the "zero earthquake" to which all other quakes would be
compared. The magnitude $R(i)$ on the Richter scale of a given earthquake is
defined to be

$$R(i) = \log \frac{i}{i_0}$$

where i is the amplitude of the ground motion of the quake (as measured by a
seismograph at a fixed distance from the epicenter of the quake) and i_0 is the amplitude
of the ground motion of the zero earthquake (as measured by a seismograph at the same
fixed distance from the epicenter). Observe that the magnitude of the zero earthquake
is in fact 0 since

$$R(i_0) = \log \frac{i_0}{i_0} = \log 1 = 0$$

    Since the Richter magnitude $R(i)$ is given by a logarithmic function, the comments
above about logarithmic growth are applicable here. In particular,

> The value of i must be increased by a factor of 10^k in order
> for the magnitude $R(i)$ to increase by just k units

For example, the zero quake has magnitude 0 and a quake that has a thousand times as
much ground motion as the zero quake (that is, $i = 1000i_0 = 10^3 i_0$) has magnitude 3
since

$$R(10^3 i_0) = \log \frac{10^3 i_0}{i_0} = \log 10^3 = 3$$

    Similarly, if the Richter magnitude of two earthquakes differs by 5, then one quake is
100,000 ($= 10^5$) times stronger than the other. To see this, suppose one quake has

magnitude $R(k) = 8$ and the other has magnitude $R(j) = 3$. Then

$$5 = R(k) - R(j) = \log \frac{k}{i_0} - \log \frac{j}{i_0}$$
$$5 = (\log k - \log i_0) - (\log j - \log i_0)$$
$$5 = \log k - \log j$$
$$5 = \log \left(\frac{k}{j} \right)$$

Since $10^{\log x} = x$ for every positive number x, we have:

$$10^5 = 10^{\log (k/j)} = \frac{k}{j}$$

so that $k = 10^5 j$, as claimed.

Other applications of logarithmic functions include physical phenomena, such as sound, and psychological phenomena, such as the rate at which people forget learned information. See the exercises below.

EXERCISES

A.1. What is the magnitude on the Richter scale of an earthquake that is
    (a)   100 times stronger than the zero quake.
    (b) $10^{4.7}$ times stronger than the zero quake.
    (c)   350 times stronger than the zero quake.
    (d)   2500 times stronger than the zero quake.

A.2. The energy intensity i of a sound is related to the loudness of the sound by the function

$$L(i) = 10 \log \left(\frac{i}{i_0} \right)$$

where i_0 is the minimum intensity detectable by the human ear (the threshold of hearing) and $L(i)$ is measured in decibels. [If $L(i)$ is to be measured in bels (1 bel = 10 decibels), then the function is $L(i) = \log (i/i_0)$.]
    (a)   The intensity of the sound of a ticking watch is approximately 100 times the minimum intensity i_0. What is the decibel measure of the loudness of the watch?
    (b)   The intensity of soft music is 10,000 times greater than the minimum intensity i_0. How many decibels of loudness is this?
    (c)   The sound of the Victoria Falls in Africa is 10 billion times more intense than the minimum sound. How many decibels is this?

B.1. Students in a calculus class were given a final exam. Each month thereafter, they took an equivalent exam, to test how much they remembered. The class average on an exam taken after t months is given by the "forgetting function"

$$F(t) = 82 - 18 \log (t + 1)$$

Thus the average on the original exam $(t = 0)$ was $F(0) = 82 - 18 \log (0 + 1)$ $= 82 - 18 \log 1 = 82 - 0 = 82$. After 2 months, the average was $F(2) =$ $82 - 18 \log 3 \approx 82 - 8.5878 = 73.4122$.
 (a) What was the class average at the end of 3 months? 6 months? 9 months? 1 year?
 (b) How many months will it take until the class average falls below 60?

B.2. Third graders are taught the elements of Sanskrit. At the end of the instruction period they are tested, and then tested at weekly intervals thereafter. The average score on the exam given after t weeks is given by the function

$$G(t) = 77 - 24 \log (t + 1)$$

 (a) What was the average score on the original exam (that is, when $t = 0$)?
 (b) What was the average score at the end of 2 weeks? 5 weeks? 10 weeks?
 (c) How long will it take for the average score to drop below 50? below 40? below 30?

B.3. How much stronger was the San Francisco earthquake of 1906 (8.4 on the Richter scale) than the Seattle earthquake of 1964 (6.1 on the Richter scale)? How much stronger than the 1964 Alaska quake (7.5 on the Richter scale)?

B.4. (a) Using Exercises B.1 and B.2 as a model, construct a forgetting function $H(t)$ satisfying these conditions: (i) $H(t)$ is the class average on an exam given t months after the instruction period in a certain subject has ended. (ii) The class average immediately after the instruction period was 90. (iii) The class average after 9 months was 75.
 (b) Using the function $H(t)$ constructed in part (a), determine the class average after 5 months, 15 months, and 20 months.

B.5. Refer to Exercise A.2 for the loudness function $L(i)$. The sound of freeway traffic measures 85 decibels, while the sound of traffic on a quiet residential street measures 45 decibels. How many times more intense is the sound of the freeway traffic? [*Hint:* If j denotes the sound intensity of freeway traffic and k the intensity of residential traffic, then $L(j) = 85$ and $L(k) = 45$ so that $L(j) - L(k) = 40$.]

5. EXPONENTIAL EQUATIONS AND APPLICATIONS

We now use the information about logarithms and exponents developed earlier in this chapter to solve exponential equations, such as

$$3 \cdot 5^x + 4 = 10, \qquad 2^{x+1} = 3^{x-1}, \qquad 3e^{2x^2-1} - 7 = 17, \qquad 2^x - 3 - 14.2^{-x} = 0$$

The ability to solve such equations will enable us to deal effectively with a number of applications, including compound interest, population growth, atmospheric pressure, and radioactive decay.

EXPONENTIAL EQUATIONS

The key to solving equations such as those just given, in which the unknown appears as an exponent, is to use the third law of logarithms:

$$\log_b(v^k) = k(\log_b v) \qquad \text{for all real numbers } v, k \text{ with } v > 0$$

The following examples show how this fact is used to solve exponential equations.

EXAMPLE   Here is a step-by-step solution of the equation:

$$3 \cdot 5^x + 4 = 10$$

$$3 \cdot 5^x = 6 \qquad \text{(subtract 4 from both sides)}$$

$$5^x = 2 \qquad \text{(divide both sides by 3)}$$

Now take the logarithm of each side (to base 10). Since equal numbers have equal logarithms, we obtain:

$$\log 5^x = \log 2$$

$$x(\log 5) = \log 2 \qquad \text{(Third Law of Logarithms)}$$

$$x = \frac{\log 2}{\log 5} \qquad \text{(divide both sides by the number } \log 5)$$

For many purposes this answer is perfectly adequate. But if a numerical approximation is needed, we can use a calculator or the table on page 274 to find that

$$x = \frac{\log 2}{\log 5} \approx \frac{.3010}{.6990} \approx .4306$$

(Remember $\log 2/\log 5$ is *not* the same as $\log \frac{2}{5}$ or $\log 2 - \log 5$.)

In order to solve the equation $5^x = 2$ in the preceding example, we took logarithms to base 10. If we had used logarithms to some other base, such as e or 5, the solutions would have gone like this:

Base e	**Base 5**
$5^x = 2$	$5^x = 2$
$\log_e(5^x) = \log_e 2$	$\log_5(5^x) = \log_5 2$
$x(\log_e 5) = \log_e 2$	$x(\log_5 5) = \log_5 2$
$x = \dfrac{\log_e 2}{\log_e 2}$	$x = \dfrac{\log_5 2}{\log_5 5} = \dfrac{\log_5 2}{1} = \log_5 2$

At first glance, each of these answers seems quite different from the answer above, $x = \log 2/\log 5$. But actually *all three answers are exactly the same number.* As we saw on page 280, logarithms to the base e are just constant multiples of logarithms to base 10, with the constant multiplier being the nonzero number $1/\log e$. Therefore,

$$\frac{\log_e 2}{\log_e 5} = \frac{(1/\log e)(\log 2)}{(1/\log e)(\log 5)} = \frac{\log 2}{\log 5}$$

Similarly, $\log_5 v = (1/\log 5) \log v$, so that

$$\log_5 2 = \frac{\log_5 2}{1} = \frac{\log_5 2}{\log_5 5} = \frac{(1/\log 5)(\log 2)}{(1/\log 5)(\log 5)} = \frac{\log 2}{\log 5}$$

Consequently, *when solving exponential equations, you may use logarithms to any base.* If numerical approximations are needed, base 10 is usually the most convenient. But in some cases it may be more helpful to use a different base in order that the answer have a simpler form.

EXAMPLE   To solve $2^{x+1} = 3^{x-1}$, we take logarithms of both sides (base 10):

$$\log 2^{x+1} = \log 3^{x-1}$$

$(x + 1)(\log 2) = (x - 1)(\log 3)$ (third law of logarithms)

$x(\log 2) + \log 2 = x(\log 3) - \log 3$ (multiply out both sides)

$x(\log 2) - x(\log 3) = -\log 2 - \log 3$ (rearrange terms)

$x(\log 2 - \log 3) = -\log 2 - \log 3$ (factor left side)

$$x = \frac{-\log 2 - \log 3}{\log 2 - \log 3}$$ (divide both sides by $(\log 2 - \log 3)$).

A numerical approximation of this answer can be found by using the table on page 274:

$$x = \frac{-\log 2 - \log 3}{\log 2 - \log 3} \approx \frac{-.3010 - .4771}{.3010 - .4771} = \frac{-.7781}{-.1761} \approx 4.4185$$

EXAMPLE   To solve $2^x - 5 - 14 \cdot 2^{-x} = 0$ we begin by multiplying both sides of the equation by 2^x. Since $2^x > 0$ for every x, the resulting equation is equivalent to the original one:

$$2^x 2^x - 5 \cdot 2^x - 14 \cdot 2^{-x} 2^x = 0$$
$$(2^x)^2 - 5 \cdot 2^x - 14 = 0$$

Now let $u = 2^x$ so that the equation becomes

$$u^2 - 5u - 14 = 0$$
$$(u + 2)(u - 7) = 0$$

$$u + 2 = 0 \qquad \text{or} \qquad u - 7 = 0$$
$$u = -2 \qquad\qquad u = 7$$
$$2^x = -2 \qquad\qquad 2^x = 7$$

Since 2^x is always positive, the equation $2^x = -2$ has no solution. To solve $2^x = 7$ we use logarithms to base 10:

$$\log 2^x = \log 7$$
$$x(\log 2) = \log 7$$

$$x = \frac{\log 7}{\log 2} \approx \frac{.8451}{.3010} \approx 2.8076$$

This is the only solution of the original equation.

APPLICATIONS

The necessary background for understanding the following examples was presented on pages 265–270. If you have not already done so, you may wish to read those pages before going further in this section.

EXAMPLE Three thousand dollars is to be invested at an interest rate of 8% per year, compounded quarterly. How many years will it be until the investment is worth $8750?
    According to the basic formula for compound interest (p. 266),

$$T(x) = P(1 + r)^x$$

where P is the original amount invested, r is the interest rate per time period (expressed as a decimal), x is the number of time periods, and $T(x)$ is the value of the investment after x time periods In this case $P = 3000$. Since interest is compounded quarterly, the basic time period is one-fourth of a year. Since the interest rate for the full year is 8% $(= .08)$, the interest rate for a quarter-year is $\frac{1}{4}(.08) = .02$. Therefore after x quarter-years the value of the investment will be

$$T(x) = 3000(1 + .02)^x = 3000(1.02)^x$$

In order to find out when the investment has the value 8750, we must find the value of x for which $T(x) = 8750$. In other words, we must solve the equation

$$3000(1.02)^x = 8750$$

Dividing both sides by 3000 and taking logarithms yields

$$(1.02)^x = \frac{8750}{3000}$$

$$\log (1.02)^x = \log \frac{8750}{3000}$$

By the second and third laws of logarithms, this equation becomes

$$x(\log 1.02) = \log 8750 - \log 3000$$

Dividing both sides by the nonzero number log 1.02 and using a calculator or logarithm table, we find that

$$x = \frac{\log 8750 - \log 3000}{\log 1.02} \approx \frac{3.9420 - 3.4771}{.0086} = \frac{.4649}{.0086} \approx 54.0581$$

Since x is the number of quarter-years, we must divide by 4 to find the number of years: $54.0581/4 = 13.514525$.

EXAMPLE A biologist observes that a culture contains 1000 of a certain type of bacteria. Seven hours later there are 5000 bacteria in the culture. How many bacteria will there be in 12 hours? in 1 day?

As we saw on page 269, the normal population growth is given by the function $S(t) = ce^{kt}$, where c is the original population, $S(t)$ is the population at time t, and k is a constant. In this case $c = 1000$, so that

$$S(t) = 1000e^{kt}$$

In order to determine the constant k we use the fact that after 7 hr the bacteria population is 5000, that is,

$$S(7) = 5000, \quad \text{or equivalently,} \quad 1000e^{7k} = 5000$$

We need only solve the right-hand equation to find k:

$$1000e^{7k} = 5000$$
$$e^{7k} = 5$$
$$\log_e(e^{7k}) = \log_e 5$$
$$7k = \log_e 5$$

$$k = \frac{\log_e 5}{7} \approx \frac{1.609}{7} \approx .2299 \approx .23$$

Therefore the function describing the growth of this bacteria population is given by $S(t) \approx 1000e^{.23t}$. After 12 hours, the number of bacteria is

$$S(12) \approx 1000e^{(.23)12} = 1000e^{2.76} \approx 1000(15.8)° = 15,800$$

After 1 day ($t = 24$ hours) the number of bacteria is

$$S(24) \approx 1000e^{(.23)24} = 1000e^{5.52} \approx 1000(249.635)° = 249,635,000$$

EXAMPLE The skeleton of a mastodon has been found. Analysis shows that the bones have lost 58% of the carbon-14 that was present when the mastodon died. How old is the skeleton?

As we saw on page 269, the basic formula for radioactive decay is $M(t) = k2^{-t/h}$, where k is the original mass of the element, h is its half-life, and $M(t)$ is the mass of the element at time t. In this case the element is carbon-14, which has a half-life of 5730

° A calculator was used to evaluate $e^{2.76}$ and $e^{5.52}$.

years. If time is measured from the death of the mastodon, then we want to find the value of t corresponding to the present time. We know that the present mass of carbon-14 is .58 less than the original mass k. Therefore the present mass is $k - .58k = .42k$, and we have

$$M(t) = k2^{-t/h}$$
$$.42k = k2^{-t/5730}$$

The solution of this equation is the desired value of t, that is, the age of the skeleton. Dividing both sides by k and taking logarithms yields

$$.42 = 2^{-t/5730}$$
$$\log(.42) = \log(2^{-t/5730})$$
$$\log(.42) = \left(\frac{-t}{5730}\right)(\log 2) = -\frac{t(\log 2)}{5730}$$

Multiplying both sides by $-5730/\log 2$ shows that

$$t = -\frac{5730}{\log 2}(\log .42) \approx -\frac{5730(-.3768)}{.301} = \frac{2159.064}{.301} = 7172.9701$$

Therefore the skeleton is approximately 7173 years old.

EXERCISES

A.1. Solve these equations.
  (a) $3^x = 81$
  (b) $3^x + 3 = 30$
  (c) $2^{5-x} = 16$
  (d) $2^{(2x+1)} = \frac{1}{8}$
  (e) $3^{(x-1)} = 9^{5x}$
  (f) $4^{5x} = 16^{(2x-1)}$
  (g) $3^{5x} \cdot 9^{x^2} = 27$
  (h) $2^{(x^2+5x)} = \frac{1}{16}$
  (i) $9^{x^2} = 3^{(-5x-2)}$

A.2. Explain why none of the final answers in Exercise A.1 involves any logarithms.

In Exercises A.3–A.5, solve the given equations. Express your answers in terms of common logarithms [for instance, $x = (2 + \log 5)/(\log 3)$] unless directed otherwise. If you have a calculator, find a rational approximation for each answer [for instance, $x = (2 + \log 5)/\log 3 \approx 5.6568$].

A.3. (a) $3^x = 5$
  (b) $5^x = 4$
  (c) $10^x = 2$
  (d) $(\frac{1}{3})^x = 3000$
  (e) $4 \cdot 7^x + 1 = 21$
  (f) $5 \cdot 2^x - 3 = 27$

A.4. (a) $2^x = 3^{x-1}$
  (b) $4^{x+2} = 2^{x-1}$
  (c) $3^{1-2x} = 5^{x+5}$
  (d) $4^{3x-1} = 3^{x-2}$
  (e) $2^{1-3x} = 3^{x+1}$
  (f) $3^{z+3} = 2^z$

A.5. (a) $9^x - 4 \cdot 3^x + 3 = 0$ [*Hint:* note that $9^x = (3^2)^x = (3^x)^2$ and use the substitution $u = 3^x$.]
  (b) $4^x - 6 \cdot 2^x = -8$
  (c) $2^x - 2 + 2^{-x} = 0$
  (d) $7^x - 6 \cdot 7^{-x} = 1$
  (e) $4^x + 6 \cdot 4^{-x} = 5$
  (f) $5^x + 3 = 10 \cdot 5^{-x}$
  (g) $2^x + 5 = 6(\sqrt{2})^x$

B.1.   Use logarithms to base 5 to solve each of these equations for x.

(a) $\dfrac{5^x + 5^{-x}}{2} = 3$ (*Hint:* use an appropriate multiplication and substitution, as in Exercise A.5, to obtain a quadratic equation; solve this equation via the quadratic formula.)

(b) $\dfrac{5^x - 5^{-x}}{2} = 2$ (c) $\dfrac{5^x + 5^{-x}}{2} = t$

B.2.   Use natural logarithms to solve each of these equations for x. (Compare Exercise B.1.)

(a) $\dfrac{e^x + e^{-x}}{2} = 2$ (b) $\dfrac{e^x - e^{-x}}{2} = t$ (c) $\dfrac{e^x + e^{-x}}{e^x - e^{-x}} = t$

In Exercises B.3–B.12, set up each problem and find the answers in terms of either common or natural logarithms, as appropriate. If you have a calculator (or are willing to use logarithm tables), find a rational approximation for each answer.

B.3.   (a) How many years will it take for an investment of $2000 to double its value when interest is compounded annually at a rate of 7%?

(b) How long will it take for the investment in part (a) to double its value if the 7% interest is compounded monthly instead of annually?

(c) What is the "doubling time" in part (a) if 7% interest is compounded daily? (Use 365 days per year for every year, including leap year.)

B.4.   (a) Find a formula which tells how many years it will take for an investment of k dollars to double at an interest rate of 7% per year, compounded annually.

(b) Do part (a) with 7% replaced by r% (r constant).

B.5.   (a) How long will it take to triple your money if you invest $500 at a rate of 5% per year, compounded annually?

(b) How long will it take if the 5% interest is compounded quarterly?

B.6.   At what rate of interest (compounded annually) should you invest $500 if you want to have $1500 in 12 years?

B.7.   How much money should be invested in a bank that pays 5% interest (compounded quarterly) so that 9 years later the balance will be $5000?

B.8.   Under normal conditions, the atmospheric pressure (in millibars) at height h feet above sea level is given by $P(h) = 1015e^{-kh}$, where k is a positive constant.

(a) Assume that the pressure at 18,000 ft is half the pressure at sea level and find k.

(b) Using the information from part (a), determine the atmospheric pressure at these heights: 1000 ft, 5000 ft, 15,000 ft.

B.9.   How old is a piece of ivory that has lost 36% of its carbon-14?

B.10.   A certain radioactive substance loses one-third of its original mass in 5 days. Find its half-life.

B.11. The half-life of a certain substance is 3.6 days. How long will it take for 20 grams of the substance to decay to 3 grams?

B.12. In 1960 the population of the United States was 179 million, and in 1970 it was 203 million.
(a) Use the population growth function to predict the population in 1990.
(b) In what year will the population reach 416 million?
(c) If you have a reasonably reliable estimate of the U.S. population for the current year, compare it to the number predicted by the growth function. How far apart are the two figures? What could account for the difference?

Trigonometric Functions

Trigonometry was invented by the ancient Greeks and Babylonians in order to solve certain problems in astronomy and navigation involving angles and triangles. In fact, "trigonometry" means "triangle measurement." This kind of trigonometry remains useful even today, as any surveyor or civil engineer can tell you.

But with the invention of calculus in the seventeenth century and the subsequent explosion of knowledge in physics and other fields, a different viewpoint toward trigonometry arose. Whereas the ancients dealt only with *angles*, it became clear that much more could be accomplished by considering the classic trigonometric concepts of sine and cosine as *functions* with domain the set of *all real numbers*. The result of this switch in viewpoint is that the sine function and the cosine function are probably *the* most important functions for the description of the physical universe. Almost any phenomena involving rotation or vibration can be described in terms of the sine and cosine functions, including vibrating strings, pendulums, planetary orbits, electron orbitals, building sways, radio transmission and reception, light rays, sound waves, and many more.

The presentation of trigonometry in this chapter reflects this modern viewpoint. Nevertheless, angles still play an important role in defining the trigonometric functions, so the chapter begins with them.

1. ANGLES AND THEIR MEASUREMENT

An angle consists of two half-lines which begin at the same point. This point is called the **vertex** of the angle. One half-line is called the **initial side** of the angle; the other is

called the **terminal side** of the angle. One usually draws a curved arrow from the initial
side to the terminal side. For example, see Figure 6-1.

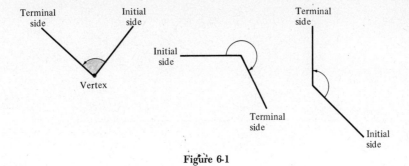

Figure 6-1

When drawing angles in the Cartesian plane, it is customary to put the vertex at the
origin and the initial side on the positive x-axis, and to move counterclockwise to the
terminal side. In this chapter we shall deal primarily with angles in this position, which
is called **standard position.** For example, the angles in Figure 6-2 are in standard
position.

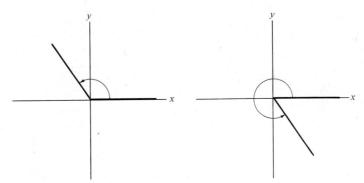

Figure 6-2

DEGREES

In order to *measure* angles, we must assign a number to each angle. Here is the classical
method for doing this: given an angle in standard position, construct a circle with center
at the origin and divide its circumference into 360 equal parts, called **degrees.** Then

note where the terminal side intersects the circle. For example, Figure 6-3 shows an angle θ with measure 25 degrees (in symbols, 25°).

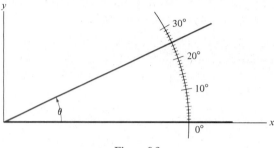

Figure 6-3

An angle that goes completely around the circle has measure 360°. An angle that goes halfway around the circle has measure $\frac{1}{2} \times 360° = 180°$. One that goes one-sixth of the way around has measure $\frac{1}{6} \times 360° = 60°$, and so on. You should be familiar with all of the angles shown in Figure 6-4.

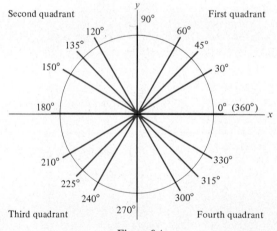

Figure 6-4

RADIANS

The measurement of angles by degrees can be compared with the English system of linear measurement. The English system has proved too cumbersome and inefficient for scientific work and has been replaced by the metric system. For similar reasons, degree measure has been replaced in many mathematical situations by **radian** measure.

Radian measure is based upon the unit circle. The radian measure of an angle in standard position is defined to be

The distance (measured along the unit circle)
from the point (1, 0) to the point where the terminal side
of the angle intersects the unit circle

For instance, the angle in Figure 6-5 has measure t radians, where the number t is the distance along the unit circle from $(1, 0)$ to P.

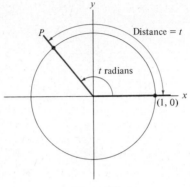

Figure 6-5

EXAMPLE   To find an angle of 3.5 radians, we measure a distance of 3.5 along the unit circle and draw the terminal side through that point, as shown in Figure 6-6.

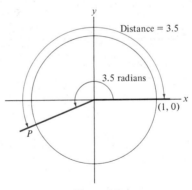

Figure 6-6

You should become reasonably adept at translating back and forth between radian measure and degree measure. One way to do this is to use the fact proved in the DO IT YOURSELF! segment below:

1 radian is approximately $57.3°$

But there is an even better way of directly determining the radian measure of many familiar angles. The key to doing this is the fact that the circumference of a circle of radius r is precisely $2\pi r$. Consequently, the distance all the way around the unit circle (that is, the circumference of a circle of radius $r = 1$) is precisely 2π. Thus a full circle angle (360°) has radian measure 2π. By considering a given angle as a fraction of a full circle angle, we can determine its radian measure quite easily. For example, an angle of 90° is $\frac{90}{360} = \frac{1}{4}$ of a full circle angle, as shown in Figure 6-7.

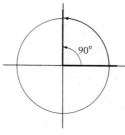

Figure 6-7

So its radian measure is $\frac{1}{4}$ of the measure of a full circle angle, namely, $\frac{1}{4}(2\pi) = \pi/2$. Figure 6-8 shows some more examples.

Degree Measure	Fraction of Full Circle		Radian Measure
180°	$\dfrac{180}{360} = \dfrac{1}{2}$ circle		$\dfrac{1}{2}(2\pi) = \pi$ radians
30°	$\dfrac{30}{360} = \dfrac{1}{12}$ circle		$\dfrac{1}{12}(2\pi) = \dfrac{\pi}{6}$ radians

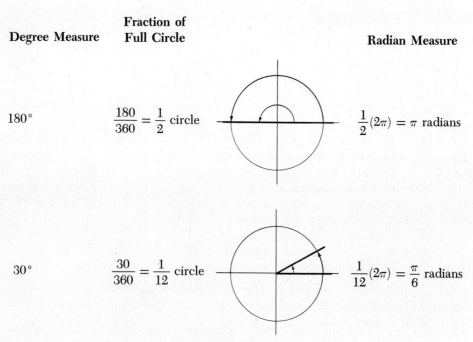

Figure 6-8

Degree Measure	Fraction of Full Circle		Radian Measure
270°	$\dfrac{270}{360} = \dfrac{3}{4}$ circle	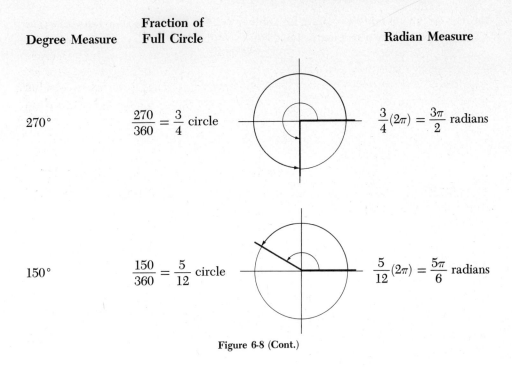	$\dfrac{3}{4}(2\pi) = \dfrac{3\pi}{2}$ radians
150°	$\dfrac{150}{360} = \dfrac{5}{12}$ circle		$\dfrac{5}{12}(2\pi) = \dfrac{5\pi}{6}$ radians

Figure 6-8 (Cont.)

By measuring angles in radians, we can assign an angle to every number between 0 and 2π (≈ 6.283). It will prove useful to assign an angle to every real number, including those less than 0 and greater than 2π. The following paragraphs explain how this is done, considering separately the two cases of numbers greater than 2π and less than 0.

Remember that the measure of an angle is the number of radians through which the terminal side must be rotated (beginning at the initial side) to arrive at its final position. We now adopt the first of two conventions:

> Angles of more than 2π radians ($= 360°$) occur when the terminal side is rotated counterclockwise through more than one full circle before coming to rest at its final position

EXAMPLE An angle in standard position of measure $5\pi/2$ radians is obtained as follows. Note that $5\pi/2 = 2\pi + (\pi/2)$. Rotate the terminal side counterclockwise through 2π radians *and then* rotate for another $\pi/2$ radians (that is, one full circle plus $\pi/2$ radians), as shown in Figure 6-9.

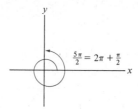

Figure 6-9

EXAMPLE   In order to construct an angle of $16\pi/3$ radians (≈ 16.76 radians), observe that

$$\frac{16\pi}{3} = \frac{6\pi}{3} + \frac{6\pi}{3} + \frac{4\pi}{3} = 2\pi + 2\pi + \frac{4\pi}{3}$$

An angle of $16\pi/3$ radians in standard position is obtained by first rotating the terminal side counterclockwise through 2π radians, then once again through 2π radians, and finally through $4\pi/3$ radians (that is, two full circles, plus $4\pi/3$ radians), as shown in Figure 6-10.

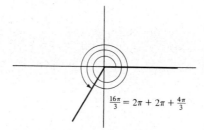

Figure 6-10

And finally, we adopt a second convention:

> **Angles of negative radian measure occur when the terminal side is rotated clockwise to its final position**

EXAMPLE   An angle of $-\pi/4$ is obtained by rotating the terminal side *clockwise* through $\pi/4 (= \frac{1}{8}$ circle$)$, as shown in Figure 6-11.

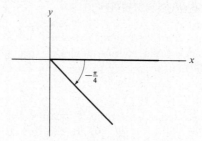

Figure 6-11

EXAMPLE An angle of $-.72$ radians is obtained by traveling *clockwise* along the unit circle from the point $(1, 0)$ for a distance of $.72$, and then drawing the terminal side of the angle through this point (Figure 6-12).

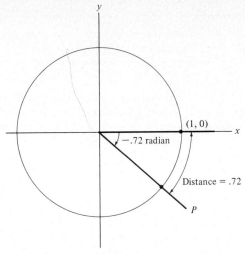

Figure 6-12

EXAMPLE An angle of $-17\pi/6$ radians (approximately -8.9 radians) is obtained by rotating the terminal side clockwise through $17\pi/6$ radians [that is, one full circle plus $5\pi/6$ radians since $17\pi/6 = 12\pi/6 + 5\pi/6 = 2\pi + 5\pi/6$], as shown in Figure 6-13.

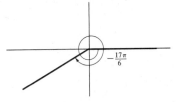

Figure 6-13

As the preceding examples illustrate, several different numbers may correspond to the same angle. For instance, since angles of radian measure $\pi/2$ and $5\pi/2(= \pi/2 + 2\pi)$ and $9\pi/2(= \pi/2 + 4\pi)$ all have the same terminal side, we shall consider them to be the same angle (Figure 6-14).

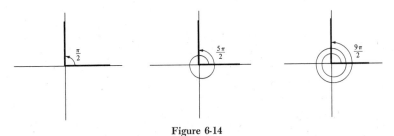

Figure 6-14

The same is true of any angle: whenever we increase *or decrease* the radian measure of an angle by 2π, 4π, 6π, and so on, the terminal side of the angle ends up in the same place. Since $4\pi = 2 \cdot 2\pi$, and $6\pi = 3 \cdot 2\pi$, and so on, we see that:

> *Increasing or decreasing the radian measure of an angle by an integer multiple of 2π results in the same angle.*

EXAMPLE   An angle of $5\pi/4$ radians is the same as an angle of $(5\pi/4) - 2\pi = -3\pi/4$ radians. An angle of $\pi/6$ radians is the same as an angle of $(\pi/6) + 2 \cdot 2\pi = 25\pi/6$ radians and also the same as an angle of $(\pi/6) - 7 \cdot 2\pi = -83\pi/6$ radians.

EXERCISES

A.1. **(a)** How many degrees is an angle in standard position formed by rotating the terminal side by $\frac{1}{9}$ of a full circle?

(b) Answer part (a) for rotations of $\frac{1}{24}$ of a circle and $\frac{1}{16}$ of a circle.

A.2. **(a)** The diagram in Figure 6-15 is the same one that appears on page 314. Fill in the *degree measure* of each of the indicated angles. You should be familiar enough with degree measurement to do this with reasonable accuracy.

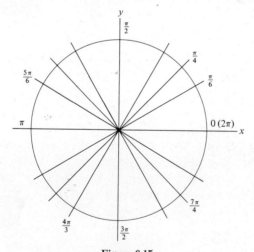

Figure 6-15

(b) State the *radian measure* of *every* angle in Figure 6-15 above by comparing the given angle to a full circle angle ($360° = 2\pi$ radians) as on pages 316–317.

A.3. Convert each of these radian measures to degrees by using the fact that π radians $= 180°$.

(a) $\dfrac{\pi}{2}$ (d) $\dfrac{\pi}{5}$ (g) $\dfrac{\pi}{45}$ (j) $\dfrac{7\pi}{15}$

(b) $\dfrac{\pi}{3}$ (e) $\dfrac{\pi}{6}$ (h) $\dfrac{\pi}{60}$ (k) $\dfrac{2\pi}{9}$

(c) $\dfrac{\pi}{4}$ (f) $\dfrac{\pi}{10}$ (i) $\dfrac{5\pi}{12}$ (l) $\dfrac{6\pi}{5}$

A.4. Find a number between 0 and 2π representing the same angle as the given numbers. (*Hint:* add or subtract 2π a sufficient number of times.)

(a) $-\dfrac{\pi}{3}$ (d) $\dfrac{16\pi}{3}$ (g) $\dfrac{-17\pi}{7}$ (j) 18.5

(b) $-\dfrac{3\pi}{4}$ (e) $-\dfrac{7\pi}{5}$ (h) $\dfrac{-32\pi}{9}$ (k) 27.43

(c) $\dfrac{19\pi}{4}$ (f) $\dfrac{45\pi}{8}$ (i) 7   (l) -10

A.5. State four other real numbers representing the same angle as the given number.

(a) $\dfrac{\pi}{4}$ (b) $\dfrac{7\pi}{5}$ (c) $-\dfrac{\pi}{6}$ (d) $\dfrac{-9\pi}{7}$

B.1. Through how many radians does the second hand of a clock move in

     (a) 40 sec   (c) 35 sec   (e) 3 min and 25 sec
     (b) 50 sec   (d) 2 min and 15 sec   (f) 1 min and 55 sec

B.2. Consider the situation in Figure 6-16 in which the circle has radius r, the angle θ has measure t radians, the circular arc from P to Q determined by θ has length s, and the area of the (shaded) circular segment determined by θ is A.

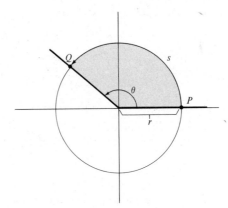

Figure 6-16

It is proved in plane geometry that these two equations always hold:

$$\frac{t}{2\pi} = \frac{s}{2\pi r} \quad \text{and} \quad \frac{t}{2\pi} = \frac{A}{\pi r^2}$$

Warning These equations are only valid when t is in *radians*. Use these equations to derive
(a) an equation that expresses s in terms of r and t.
(b) an equation that expresses A in terms of r and t.
(c) an equation that expresses A in terms of r and s.

Use the equations derived in Exercise B.2 to do Exercises B.3–B.6.

B.3. (a) Suppose the radius of the circle in Figure 6-17 is 20 cm and the length of the arc s is 25 cm. What is the radian measure of the angle θ?

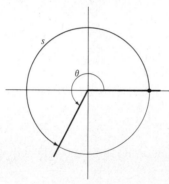

Figure 6-17

(b) Find the radian measure of θ if the circle has *diameter* 150 cm and s has length 360 cm.

B.4. A wheel on a car has a radius of 36 cm. Through what angle (in radians) does the wheel turn while the car travels
(a) 2 m ($= 200$ cm)   (c) 720 m
(b) 5 m ($= 500$ cm)   (d) 1 km ($= 1000$ m)

B.5. On a circle of *diameter* 10, how long is the circular arc determined by a central angle of radian measure
(a) 1 radian   (c) 1.75 radians
(b) 2 radians   (d) 2.2 radians

B.6. The latitude of Springfield, Illinois, is 40° north. Assuming that the earth is a perfect sphere with radius 4000 mi, what is the distance from Springfield to the North Pole?

C.1. Consider a wheel that is rotating around its axle at a rate of 1 revolution per min (rpm). Through what angle (in radians) does the wheel move
(a) in 1 min?   (c) in 3.5 min?   (e) in $\frac{9}{8}$ min?
(b) in 2 min?   (d) in 4.25 min?   (f) in t min ($t \geq 0$)?

C.2. (a) Do Exercise C.1 for a wheel that rotates at a rate of 2 rpm.

(b) Do Exercise C.1 for a wheel rotating at 5 rpm.

(c) Do Exercise C.1 for a wheel rotating at k rpm (where k is a fixed positive integer).

DO IT YOURSELF!

CONVERSION BETWEEN DEGREES AND RADIANS

In the case of angle measurement, the basic relation is

$$360° = 2\pi \text{ radians}$$

Therefore 1 radian is $360/2\pi$ degrees. So the way to convert from radians to degrees is to multiply by $360/2\pi = 180/\pi$:[°]

$$degrees = \frac{180}{\pi} \times radians$$

For example, $11\pi/12$ radians $= (180/\pi) \cdot (11\pi/12)° = 165°$ and 1 radian $= (180/\pi) \cdot (1)° \approx 57.3°$.[°°]

The equation $360° = 2\pi$ radians also shows that $1°$ is $2\pi/360$ radians (that is, $\pi/180$ radians). Thus

$$radians = \frac{\pi}{180} \times degrees$$

For instance, $300° = (\pi/180) \cdot (300)$ radians $= 5\pi/3$ radians and $37° = (\pi/180) \cdot (37)$ radians $\approx .645$ radians.

Since $180/\pi \approx 57.3$ and $\pi/180 \approx .0175$, we have these (approximate) conversion formulas:

$$degrees \approx 57.3 \times radians$$
$$radians \approx .0175 \times degrees$$

For example, $170° \approx .0175 \times 170 = 2.975$ radians.

[°] This is analogous to converting feet to inches by multiplying by 12, since 1 ft = 12 in.
[°°] $\approx$ means "approximately equal to." We must use $\approx$ here since we used the decimal approximation 3.14 of π in order to compute the quotient.

If you have a calculator which handles trigonometric functions, it can most likely convert from degrees to radians and vice versa. If this is the case, you should learn how to use the calculator for these conversions.

MINUTES AND SECONDS

When measuring angles, decimal notation is frequently used. For instance, one speaks of an angle of 6.327 radians or an angle of 159.483°. In ancient times, however, decimal notation was unknown and other units were used for measuring angles of less than one degree. These units, called minutes and seconds, were widely used until relatively recently:

> A *minute* is $\frac{1}{60}$ of a degree.
> A *second* is $\frac{1}{60}$ of a minute ($= \frac{1}{3600}$ of a degree).

The usual notation for minutes is ′ and for seconds ″. For example, 215°56′34″ denotes an angle of 215 degrees, 56 minutes, 34 seconds. Using the facts that 60 minutes make a degree and 60 seconds make a minute, we see that

$$215°56′34″ = 215 + \tfrac{56}{60} + \tfrac{34}{3600} \text{ degrees} \approx 215.943°$$

EXERCISES

A.1. Change the following angles from degrees to radians.
    (a) 17°   (b) 266°   (c) $-72°$

A.2. Change the following angles from radians to degrees.
    (a) $\dfrac{9\pi}{5}$ radians   (b) $\dfrac{12\pi}{7}$ radians   (c) 6.5 radians

A.3. Convert the following angle measurements to decimal notation (in degrees).
    (a) 27°30′45″   (b) 106°50′24″   (c) 13°42′18″

B.1. Locate the following angles given in radians by first converting to degrees.
    (a) 2 radians   (c) 4 radians   (e) .29 radians
    (b) 3 radians   (d) 5 radians   (f) 17.6 radians

2. THE SINE AND COSINE FUNCTIONS

A thorough understanding of the sine and cosine functions is absolutely necessary in order to deal with the applications of calculus to various physical situations.

> **WARNING TO PEOPLE WHO HAVE HAD TRIGONOMETRY BEFORE.** *There are many different ways to define the sine and cosine functions. The one presented here is geared toward the eventual use of these functions in calculus and in physical applications. At first, it may appear to be quite different from the definition you learned previously (especially if the earlier definition involved triangles). For now, just concentrate on the definition presented here, and don't worry about relating it to any definitions you remember from the past. In Section 4, we shall discuss other ways of describing the sine and cosine functions. At that time any confusion you may have should get cleared up, since it will turn out that all the definitions lead to the very same functions.*

THE SINE FUNCTION

The rule of the sine function is not given by an algebraic formula but is based instead on a geometric construction. Consequently, it can sometimes be difficult to calculate the values of the sine function for particular numbers. On the other hand, many useful properties of the sine function can be easily derived from this geometric definition. It turns out that these properties of the sine function are the important things, rather than the details of computing particular values of the function.

The *domain of the sine function* is the set of all real numbers. For any real number t, the value of the sine function at t will be written $\sin(t)$. The *rule for determining* $\sin(t)$ is given by the following three-step process:

(i) Construct an angle of t radians in standard position.

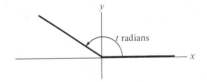

(ii) Find the point P where the terminal side of this angle intersects the unit circle $x^2 + y^2 = 1$. That is, find the coordinates of P.

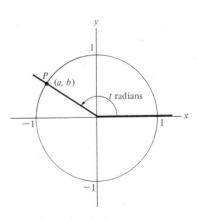

(iii) The value of the sine function at t is the *y-coordinate* of the point P. Thus if the coordinates of P are (a, b), then $\sin(t) = b$.

For the present, *don't worry about evaluating the sine function,* except in the routine cases illustrated in the following examples and exercises. It is more important at this stage to become comfortable with the definition, the notation, and the basic properties of the sine function. Various ways of evaluating the sine function will be discussed in Section 4. All that you need to know for now is that all the necessary calculations are available either from a suitable calculator or from trigonometric tables. For example, a calculator° shows that

$$\sin{(2)} \approx .91, \qquad \sin{(12)} \approx -.54, \qquad \sin{(-4.1)} \approx .82$$

EXAMPLE We shall find the value of the sine function for $t = \pi/2$; that is, we shall find $\sin{(\pi/2)}$. Construct an angle of $\pi/2$ radians and note that its terminal side intersects the unit circle at the point $P = (0, 1)$, as shown in Figure 6-18.

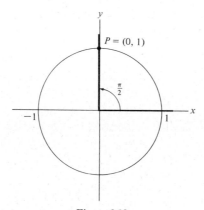

Figure 6-18

Since the y-coordinate of P is 1, we have:

$$\sin{\left(\frac{\pi}{2}\right)} = 1$$

EXAMPLE The value of the sine function at $t = 0$ is easily computed since an angle of 0 radians has its terminal side on the positive x-axis, so that the point P has coordinates $(1, 0)$. See Figure 6-19.

° As is often the case with calculators, these numbers are only rational approximations to the correct values. When using a calculator, be sure it is operating in **radian mode,** not degree mode. For a discussion of the sine function in terms of degrees, see Section 9.

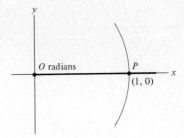

Figure 6-19

Therefore

$$\sin(0) = y\text{-coordinate of } P = 0$$

EXAMPLE   Let t be a real number such that the terminal side of an angle of t radians in standard position intersects the unit circle at the point P with coordinates $(2/\sqrt{13}, -3/\sqrt{13})$, as shown in Figure 6-20.

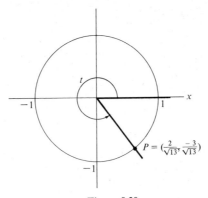

Figure 6-20

Then

$$\sin(t) = y\text{-coordinate of } P = \frac{-3}{\sqrt{13}}$$

THE COSINE FUNCTION

As the name suggests, the cosine function is closely related to the sine function. Its definition, like that of the sine function, is based on a geometric construction. *The domain of the cosine function is the set of all real numbers.* For any real number t, the value of the cosine function at t will be written cos (t). The *rule for determining cos (t)* is given by the following three-step process:

(i)   Construct an angle of t radians in standard position.

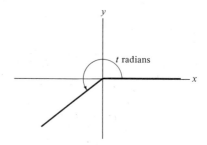

(ii)   Find the point P where the terminal side of this angle intersects the unit circle. That is, find the coordinates of P.

(iii)   The value of the cosine function at t is the *x-coordinate* of the point P. Thus if the coordinates of P are (a, b), then $\cos(t) = a$.

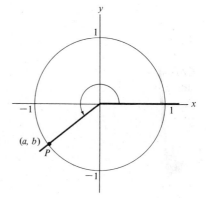

The remarks on page 326 about evaluating the sine function apply equally well to the cosine function. If you have a calculator, you can provide yourself with examples, such as

$$\cos(5) = .28, \qquad \cos(8.72) = -.76, \qquad \cos(2.5) = -.8$$

But other than the simple examples and exercises below, you should concentrate primarily on learning the definition, notation, and basic properties of the cosine function.

EXAMPLE   We shall find the value of the cosine function for $t = \pi$; that is, we shall find $\cos(\pi)$. Construct an angle of π radians, and note that it intersects the unit circle at the point $(-1, 0)$, as shown in Figure 6-21.

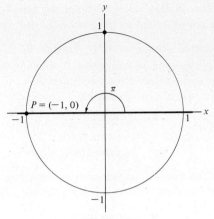

Figure 6-21

Since the x-coordinate of P is -1, we have $\cos(\pi) = -1$.

EXAMPLE $\text{Cos}(\pi/2)$ is easily computed since an angle of $\pi/2$ radians has its terminal side along the positive y-axis, which intersects the unit circle at $(0, 1)$, as shown in Figure 6-22. Therefore $\cos(\pi/2) = 0$.

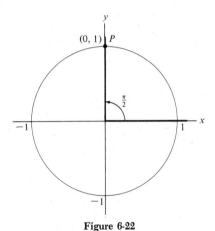

Figure 6-22

ALGEBRA WITH TRIGONOMETRIC FUNCTIONS

As we saw in Section 7 of Chapter 3, new functions can be constructed from given ones via addition, subtraction, multiplication, division, and composition. In particular, these algebraic techniques can be used to construct new trigonometric functions from the sine and cosine functions. When dealing with such functions, you should be aware of certain notational customs that are observed in the mathematical world:

(i) **Parentheses are omitted whenever it is possible to do so without confusion.** Hereafter we shall write $\sin t$ instead of $\sin (t)$. But the meaning remains exactly the same: $\sin t$ denotes the value of the sine function at the number t; and similarly for cosine. Here are some more examples:

$$-(\sin t) \text{ is written } -\sin t;$$
$$4(\cos (t)) \text{ is written } 4 \cos t;$$
$$(\sin (4t^5) - \pi) \text{ is written } \sin 4t^5 - \pi$$

(ii) **When dealing with powers of trigonometric functions, exponents are written inside** between the function symbol and the variable. For instance,

$$(\cos t)^3 \text{ is written } \cos^3 t;$$
$$(\sin t)^4 (\cos t)^2 \text{ is written } \sin^4 t \cos^2 t;$$
$$(\sin 7t)^5 \text{ is written } \sin^5 7t$$

But be careful: $\cos t^3$ means $\cos (t^3)$, *not* $(\cos t)^3$ and *not* $\cos^3 t$.

Except for these conventions, the algebra of trigonometric functions is just like the algebra of any other functions.

EXAMPLE   If $f(t) = \sin^2 t + \cos t$ and $g(t) = \cos^3 t + 5$, then the product function fg is given by the rule:

$$(fg)(t) = f(t)g(t) = (\sin^2 t + \cos t)(\cos^3 t + 5)$$
$$= \sin^2 t \cos^3 t + 5 \sin^2 t + \cos^4 t + 5 \cos t$$

EXAMPLE   If $f(t) = \cos^2 t - 9$ and $g(t) = \cos t + 3$, then the quotient function f/g is given by the rule:

$$\left(\frac{f}{g}\right)(t) = \frac{f(t)}{g(t)} = \frac{\cos^2 t - 9}{\cos t + 3}$$

This rule can be simplified by factoring the numerator according to the usual rules of algebra:

$$\left(\frac{f}{g}\right)(t) = \frac{\cos^2 t - 9}{\cos t + 3} = \frac{(\cos t + 3)(\cos t - 3)}{\cos t + 3} = \cos t - 3$$

provided that $\cos t + 3 \neq 0$. As we shall see in Section 3, this is always the case.

Warning   To avoid mistakes, remember that you are dealing with *functional notation* here. For one thing, the symbol $\sin t$ is a *single entity*, as is the symbol $\cos t$. Don't get caught performing some nonsensical "canceling" operation such as $\sin t/\cos t = \sin/\cos$ or $\cos t^2/\cos t = \cos t/\cos = t$. Parentheses are another potential source of error, as we now see.

EXAMPLE Let f be the sine function, $f(t) = \sin t$, and let g be the linear function given by $g(t) = t - \pi$. Then the composite function $g \circ f$ is given by the rule:

$$(g \circ f)(t) = g(f(t)) = g(\sin t) = (\sin t) - \pi = \sin t - \pi$$

On the other hand, the composite function $f \circ g$ has the rule:

$$(f \circ g)(t) = f(g(t)) = f(t - \pi) = \sin(t - \pi)$$

In this case the parentheses are absolutely necessary and cannot be removed, for the functional notation $\sin(t - \pi)$ indicates that we first substract π from the number t, and then apply the sine function to the result. But the expression $\sin t - \pi$ indicates that we first evaluate the sine function at the number t, and then subtract π from the result. To see the difference, just evaluate each of these composite functions at the number $t = \pi$:

$$(f \circ g)(\pi) = \sin(\pi - \pi) = \sin 0 = 0 \quad \text{but} \quad (g \circ f)(\pi) = \sin \pi - \pi = 0 - \pi = -\pi$$

EXERCISES

A.1. Use the definition of the sine and cosine functions to determine

(a) $\sin\left(\dfrac{3\pi}{2}\right)$ (d) $\cos(0)$ (g) $\cos\left(-\dfrac{3\pi}{2}\right)$

(b) $\sin(\pi)$ (e) $\sin(4\pi)$ (h) $\sin\left(\dfrac{9\pi}{2}\right)$

(c) $\cos\left(\dfrac{3\pi}{2}\right)$ (f) $\cos\left(-\dfrac{\pi}{2}\right)$ (i) $\cos\left(-\dfrac{11\pi}{2}\right)$

A.2. (a) Show that the point $(-2/\sqrt{5}, 1/\sqrt{5})$ lies on the unit circle.
(b) Suppose the terminal side of an angle of t radians passes through the point $(-2/\sqrt{5}, 1/\sqrt{5})$. Find $\sin t$ and $\cos t$.

A.3. Do both parts of Exercise A.2 for each of the following points:

(a) $\left(\dfrac{1}{\sqrt{10}}, -\dfrac{3}{\sqrt{10}}\right)$ (b) $\left(-\dfrac{3}{5}, -\dfrac{4}{5}\right)$ (c) $(.6, -.8)$

A.4. Find the rule of the product function of the two given functions. For example, if $f(t) = \sin t + \cos t$ and $g(t) = \cos^2 t$, then $(fg)(t) = (\sin t + \cos t)\cos^2 t = \sin t \cos^2 t + \cos^3 t$.
(a) $f(t) = 3\sin t; g(t) = \sin t + 2\cos t$
(b) $f(t) = 5\cos t; g(t) = \cos^3 t - 1$
(c) $f(t) = 3 + \sin^2 t; g(t) = \sin t + \cos t$
(d) $f(t) = \sin(2t) + \cos^4 t; g(t) = \cos(2t) + \cos^2 t$

A.5. Factor each of these expressions. For example, $\sin^2 t - 9 = (\sin t + 3)(\sin t - 3)$.
(a) $\cos^2 t - 4$ (d) $\sin^3 t - \sin t$ (g) $6\sin^2 t - \sin t - 1$
(b) $25 - \sin^2 t$ (e) $\sin^2 t + 6\sin t + 9$ (h) $\sin t \cos t + \cos^2 t$
(c) $\sin^2 t - \cos^2 t$ (f) $\cos^2 t - \cos t - 2$ (i) $\cos^4 t + 4\cos^2 t - 5$

A.6. For each pair of functions, find the rule of both composite functions $f \circ g$ and $g \circ f$. For example, if $f(t) = \sin t$ and $g(t) = t^2 + 3$, then $(f \circ g)(t) = f(g(t)) = \sin(t^2 + 3)$ and $(g \circ f)(t) = g(f(t)) = \sin^2 t + 3$.

(a) $f(t) = \cos t$; $g(t) = 2t + 4$ (c) $f(t) = \sin(t + 3)$; $g(t) = t^2 - 1$

(b) $f(t) = \sin t + 2$; $g(t) = t^2$ (d) $f(t) = \cos^2(t - 2)$; $g(t) = 5t + 2$

B.1. (a) Use the examples in the text and Exercise A.1 to show that for certain numbers t and u, $\sin(t + u)$ need *not* be the same number as $\sin t + \sin u$.

(b) Show that $\cos(t + u)$ need not be the same number as $\cos t + \cos u$.

B.2. Draw a rough sketch to determine which of these numbers are positive.

(a) $\sin 1$ (c) $\sin 3$ (e) $\cos(1.7)$

(b) $\cos 2$ (d) $(\cos 2)(\sin 2)$ (f) $\cos 3 + \sin 3$

B.3. (a) Find all numbers t such that $0 \le t < 2\pi$ and $\sin t < 1$.

(b) Find all numbers t such that $0 \le t < 2\pi$ and $\cos t > 0$.

B.4. Which number is larger: $\sin(\cos 0)$ or $\cos(\sin 0)$?

B.5. Find all solutions to the given equation.

(a) $\sin t = 1$ (c) $\cos t = 0$ (e) $|\sin t| = 1$

(b) $\cos t = -1$ (d) $\sin t = -1$ (f) $|\cos t| = 1$

C.1. Figure 6-23 is a picture of the unit circle. The full circle angle (2π radians) is divided into 36 equal angles, each of measure $\pi/18$ radians ($\approx .175$ radians). For clarity, the scale on the vertical axis is noted to the right of the circle and the scale on the horizontal axis is noted below the circle.

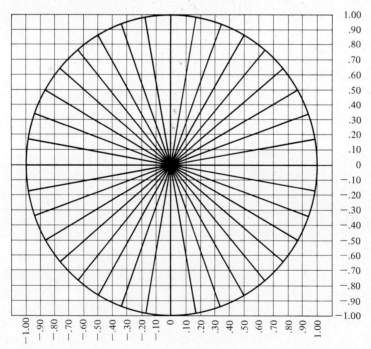

Figure 6-23

Use Figure 6-23 and careful measurement to estimate the functional values needed to fill the blanks in the following table. (*Note:* a decimal approximation of each value of t is given in parentheses.)

t	0	$\dfrac{\pi}{9}$	$\dfrac{2\pi}{9}$	$\dfrac{\pi}{3}$	$\dfrac{4\pi}{9}$	$\dfrac{5\pi}{9}$	$\dfrac{2\pi}{3}$	$\dfrac{7\pi}{9}$	$\dfrac{8\pi}{9}$	π
		(.35)	(.7)	(1.05)	(1.4)	(1.75)	(2.1)	(2.45)	(2.8)	(3.14)
$\sin t$	0	.34						.64		
$\cos t$	1		.77			−.17				

t	π	$\dfrac{19\pi}{18}$	$\dfrac{7\pi}{6}$	$\dfrac{23\pi}{18}$	$\dfrac{25\pi}{18}$	$\dfrac{3\pi}{2}$	$\dfrac{29\pi}{18}$	$\dfrac{31\pi}{18}$	$\dfrac{11\pi}{6}$	$\dfrac{35\pi}{18}$	2π
	(3.14)	(3.32)	(3.67)	(4.01)	(4.36)	(4.71)	(5.06)	(5.41)	(5.76)	(6.11)	(6.28)
$\sin t$			−.5				−.94				0
$\cos t$					−.34					.98	1

C.2. (a) Use the table constructed in Exercise C.1 to plot some points and sketch a graph of the function $f(t) = \sin t$ with $0 \le t \le 2\pi$.

     (b) Sketch the graph of $g(t) = \cos t$ with $0 \le t \le 2\pi$.

C.3. Figure 6-24 is a diagram of a merry-go-round that includes horses A, B, C, D, E, F.

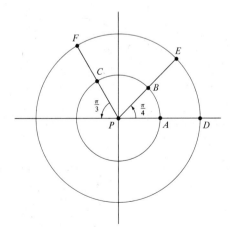

Figure 6-24

The distance from the center P to A is 1 unit, and the distance from P to D is 5 units. Define six functions as follows:

$A(t)$ = vertical distance from horse A to the x-axis at time t;
$B(t)$ = vertical distance from horse B to the x-axis at time t;

and similarly for $C(t)$, $D(t)$, $E(t)$, $F(t)$. The time t is measured in minutes. Assume that the merry-go-round rotates counterclockwise at a rate of 1 revolution per minute and that the horses are in the positions shown in the diagram at the starting time $t = 0$. As the merry-go-round rotates, the horses move around the circles indicated in the diagram.

(a) Show that $B(t) = A(t + \frac{1}{8})$ for every t.
(b) In a similar manner, express $C(t)$ in terms of the function $A(t)$.
(c) Express $E(t)$ and $F(t)$ in terms of the function $D(t)$.
(d) Explain why Figure 6-25 is valid, and use it and similar triangles to express $D(t)$ in terms of $A(t)$.

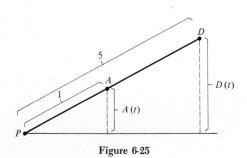

Figure 6-25

(e) In a similar manner, express $E(t)$ and $F(t)$ in terms of $A(t)$.
(f) Show that $A(t) = \sin(2\pi t)$ for every t.

(*Hint:* Exercise C.1 in Section 1 may be helpful for relating the time t and the angle through which the merry-go-round has moved at time t.)

(g) Use parts (a), (b), and (f) to express $B(t)$ and $C(t)$ in terms of the sine function.
(h) Use parts (d), (e), and (f) to express $D(t)$, $E(t)$, and $F(t)$ in terms of the sine function.

3. BASIC PROPERTIES OF THE SINE AND COSINE FUNCTIONS

In order to define $\sin t$ and $\cos t$ for a given real number t, we first construct an angle of t radians and then find the point P where the terminal side of the angle intersects the unit circle, as shown in Figure 6-26. Then $\cos t$ is defined to be the x-coordinate of P and $\sin t$ to be the y-coordinate of P.

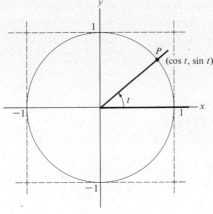

Figure 6-26

Now *both* coordinates of *every* point on the unit circle lie between -1 and $+1$ (as shown by the dotted lines in Figure 6-26), and they satisfy the equation $x^2 + y^2 = 1$. Since the point P with coordinates $(\cos t, \sin t)$ lies on the unit circle, we conclude that

$$-1 \leq \sin t \leq 1 \qquad and \qquad -1 \leq \cos t \leq 1$$
$$for\ every\ real\ number\ t$$

$$\cos^2 t + \sin^2 t = 1 \quad for\ every\ real\ number\ t$$

An equation that is valid for all numbers, such as this last one, is an example of an **identity.** This particular identity is called the **Pythagorean Identity** and is often rewritten as

$$\sin^2 t + \cos^2 t = 1 \ \text{ or } \ \sin^2 t = 1 - \cos^2 t \ \text{ or } \ \cos^2 t = 1 - \sin^2 t$$

EXAMPLE   If $\sin t = \frac{2}{3}$, find $\cos t$.   Since $\sin^2 t + \cos^2 t = 1$, we have $(\frac{2}{3})^2 + \cos^2 t = 1$, so that $\cos^2 t = 1 - (\frac{2}{3})^2 = 1 - (\frac{4}{9}) = \frac{5}{9}$. Taking square roots shows that there are two possibilities:

$$\cos t = \sqrt{\frac{5}{9}} = \frac{\sqrt{5}}{3} \qquad \text{or} \qquad \cos t = -\sqrt{\frac{5}{9}} = \frac{-\sqrt{5}}{3}$$

Without more information, we cannot answer the question further. But if we knew, for example, that the terminal side of an angle of t radians lies in the second quadrant, then we would know that $\cos t$, being the x-coordinate of a point in the second quadrant, is necessarily negative. In that case we could then conclude that $\cos t = -\sqrt{5}/3$.

EXAMPLE An important feature of an identity such as the Pythagorean Identity $\sin^2 t + \cos^2 t = 1$ is that *any* number or expression may be substituted for t and the equation remains true. For instance, we can use the number $t = 5x$ or $t = 3k + 7$:

$$\sin^2(5x) + \cos^2(5x) = 1 \qquad \text{and} \qquad \sin^2(3k + 7) + \cos^2(3k + 7) = 1$$

PERIODICITY

Let t be a real number. Consider what happens when we apply the rule of the sine function to each of the three numbers t, $t + 2\pi$, and $t - 2\pi$. As we saw in Section 1, the terminal side of an angle of t radians is the same as the terminal side of an angle of $t + 2\pi$ radians, and the same as the terminal side of an angle of $t - 2\pi$ radians. Therefore the point P where the terminal side meets the unit circle is the *same* in all three cases, as shown in Figure 6-27.

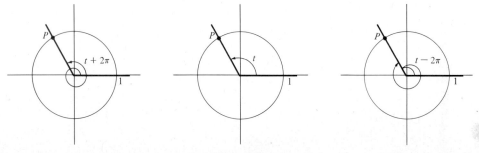

Figure 6-27

Since the second coordinate of P is the value of the sine function in each case, we see that $\sin(t + 2\pi) = \sin t = \sin(t - 2\pi)$. Since the first coordinate of P is the value of the cosine function in each case, this argument also shows that $\cos(t + 2\pi) = \cos t = \cos(t - 2\pi)$. It can also be shown (Exercises C.1 and C.2) that there is no positive number k smaller than 2π with the property that $\sin(t + k) = \sin t$ for *every* number t (and similarly for cosine). As you may recall from Section 5 of Chapter 3, there is a special name for functions with these properties:

---

*The sine and cosine functions are **periodic with period 2π:***   *for every real number t,*

$$\sin(t \pm 2\pi) = \sin t \qquad \text{and} \qquad \cos(t \pm 2\pi) = \cos t.$$

EXAMPLE Since $\pi/2 + 2\pi = 5\pi/2$ and $5\pi/2 + 2\pi = 9\pi/2$ and $9\pi/2 + 2\pi = 13\pi/2$, and so on, we see that

$$1 = \sin\frac{\pi}{2} = \sin\frac{5\pi}{2} = \sin\frac{9\pi}{2} = \sin\frac{13\pi}{2} = \sin\frac{17\pi}{2} = \cdots$$

Likewise, since $\pi/2 - 2\pi = -3\pi/2$ and $-3\pi/2 - 2\pi = -7\pi/2$, and so on, we have

$$1 = \sin\frac{\pi}{2} = \sin\left(-\frac{3\pi}{2}\right) = \sin\left(-\frac{7\pi}{2}\right) = \sin\left(-\frac{11\pi}{2}\right) = \cdots$$

SIGNS

For a given function f, it is often useful to know which values of t make $f(t)$ positive and which make $f(t)$ negative. In order to determine this information for the sine and cosine functions, just remember that for any number t, the numbers $\cos t$ and $\sin t$ are the coordinates of a point P on the terminal side of an angle of t radians. The quadrant that $P = (\cos t, \sin t)$ lies in determines immediately which of its coordinates are positive and which are negative, as shown in Figure 6-28.

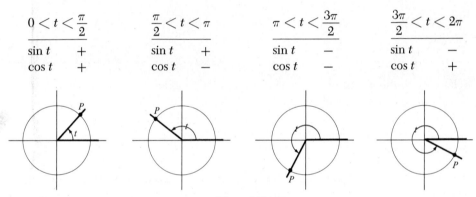

Figure 6-28

BASIC IDENTITIES

In addition to the Pythagorean Identity, $\sin^2 t + \cos^2 t = 1$, which relates the sine function to the cosine function, there are a number of useful identities involving only a single function. The periodicity of the sine and cosine function provides examples of such identities:

$$\sin(t + 2\pi) = \sin t \quad \text{and} \quad \cos(t + 2\pi) = \cos t$$

We shall now use some elementary geometry to establish several more single-function identities.

We begin with any real number t and its negative $-t$. We want to find the relationship between the numbers $\sin t$ and $\sin(-t)$, as well as the relationship between $\cos t$ and $\cos(-t)$. In order to evaluate the sine and cosine functions at both t and $-t$

we construct two angles in standard position, one of measure t radians and the other of measure $-t$ radians, as shown in Figure 6-29.

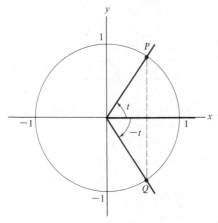

Figure 6-29

According to the definition of the sine and cosine functions, the coordinates of P are $(\cos t, \sin t)$ and the coordinates of Q are $(\cos (-t), \sin (-t))$. It is evident from the picture that the points Q and P lie on the same vertical line.° Therefore Q and P have the same x-coordinate; that is, $\cos (-t) = \cos t$. Since this argument works for any real number t, we have:

$$cos\,(-t) = cos\,t \quad for\ every\ real\ number\ t$$

This identity can be described in the language of Section 5 of Chapter 3 by saying that *cosine is an even function*. The picture above also shows that P and Q lie at equal distances from the x-axis.° This means that the y-coordinate of Q must be the negative of the y-coordinate of P, that is,

$$sin\,(-t) = -sin\,t \quad for\ every\ real\ number\ t$$

In other words, *sine is an odd function* (see Section 5 of Chapter 3).

° This fact can be proved by considering basic geometric facts about triangles. Although the picture above is drawn for a number t between 0 and $\pi/2$, similar pictures for other values of t show that P and Q still have these same properties.

EXAMPLE   As we shall see in the next section, $\sin(\pi/6) = \frac{1}{2}$. Therefore $\sin(-\pi/6) = -\frac{1}{2}$. We shall also see that $\cos(3\pi/4) = -\sqrt{2}/2$, so that $\cos(-3\pi/4) = \cos(3\pi/4) = -\sqrt{2}/2$.

EXAMPLE   Let v be a real number. Since the identities above are valid for *any* number, they must hold for $t = v - (\pi/2)$ and $-t = -[v - (\pi/2)] = \pi/2 - v$. Thus since $\cos(-t) = \cos t$, we have:

$$\cos\left(\frac{\pi}{2} - v\right) = \cos\left(v - \frac{\pi}{2}\right)$$

Similarly, since $\sin(-t) = -\sin t$ we have:

$$\sin\left(\frac{\pi}{2} - v\right) = -\sin\left(v - \frac{\pi}{2}\right)$$

Since both of these equations are valid for any real number v, they are further examples of identities.

Several more identities are the result of this fact about points on the unit circle:

If P and Q are on opposite ends of the same diameter and P has coordinates (a, b), then Q has coordinates $(-a, -b)$ (see Figure 6-30)

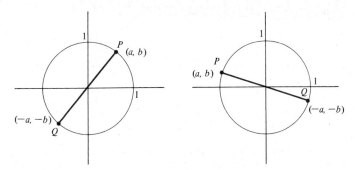

Figure 6-30

We shall assume this fact° and shall apply it as follows. Let t be any real number and construct two angles in standard position, one of measure t radians and the other of measure $t + \pi$ radians, as shown in Figure 6-31.

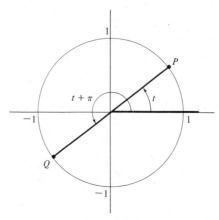

Figure 6-31

According to the definition of the sine and cosine functions, the coordinates of P are $(\cos t, \sin t)$, and the coordinates of Q are $(\cos (t + \pi)), \sin (t + \pi))$. Since π radians corresponds to a half-circle, it is clear that P and Q are on opposite ends of the same diameter of the circle. Therefore as we saw above,

The x-coordinate of Q is the negative of the x-coordinate of P.
The y-coordinate of Q is the negative of the y-coordinate of P.

In other words,

$$\cos (t + \pi) = -\cos t \quad \text{and} \quad \sin (t + \pi) = -\sin t$$
$$\textit{for every real number } t$$

It is now possible to establish a number of other identities either by using geometric arguments, as those just given, or by using known identities and some algebra, as in the following example.

EXAMPLE   Let v be any real number. We shall prove that $\sin (\pi - v) = \sin v$. By elementary arithmetic we have

$$\sin (\pi - v) = \sin ((-v) + \pi)$$

_____

° It can be proved by considering the equation $x^2 + y^2 = 1$ of the circle and the equation of the straight line through P and the origin.

Now apply the identity $\sin(t + \pi) = -\sin t$ with $t = -v$:

$$\sin((-v) + \pi) = -\sin(-v)$$

But we know that $\sin(-v) = -\sin v$. Therefore

$$-\sin(-v) = -(-\sin v) = \sin v$$

By putting all of these equations together we see that

$$\sin(\pi - v) = \sin v$$

This argument works for any real number. It doesn't matter whether we denote this number by v, as above, or by some other letter. In particular, we can express exactly the same information by saying

$$\boxed{\sin(\pi - t) = \sin t \qquad \textit{for every real number t}}$$

A similar argument for cosine (see Exercise B.7 below) shows that

$$\boxed{\cos(\pi - t) = -\cos t \qquad \textit{for every real number t}}$$

THE ALGEBRA OF IDENTITIES

The identities introduced above are just a few of many trigonometric identities. Others will be presented in the exercises and in later sections of this chapter. Because there are so many identities, it is usually possible to express the rule of a given trigonometric function in several different ways. Consequently, you need to become reasonably adept at manipulating identities in order to deal effectively with trigonometric functions.

EXAMPLE In order to find $(\sin t + \cos t)^2$, multiply out the expression and apply the Pythagorean Identity:

$$(\sin t + \cos t)^2 = \sin^2 t + 2\sin t \cos t + \cos^2 t$$
$$= (\sin^2 t + \cos^2 t) + 2\sin t \cos t = 1 + 2\sin t \cos t$$

EXAMPLE Prove that $\sin^2(\pi - t)\cos^2(\pi - t) = \sin^2 t - \sin^4 t$. The best way to deal with a problem such as this is to begin on one side of the equation and apply various identities to change it into the other side of the equation. Deciding just which identities to apply is partly trial and error and partly common sense. Whenever you see an equation involving sine squared or cosine squared such as this one, it's a good bet that the Pythagorean Identity is involved. In this case it seems reasonable to use the Pythagorean Identity on the number $\pi - t$ instead of t:

$$\sin^2(\pi - t) + \cos^2(\pi - t) = 1$$

Now use this fact on the left side of the equation above:

$$\sin^2(\pi - t)\cos^2(\pi - t) = \sin^2(\pi - t)(1 - \sin^2(\pi - t)) = \sin^2(\pi - t) - \sin^4(\pi - t)$$

Now the last expression on the right involves only powers of $\sin(\pi - t)$. But we saw on p. 341 that $\sin(\pi - t) = \sin t$. Using this identity on the right side of the last equation yields the conclusion we wanted to prove:

$$\sin^2(\pi - t)\cos^2(\pi - t) = \sin^2(\pi - t) - \sin^4(\pi - t) = \sin^2 t - \sin^4 t$$

Warning   A good deal of trial and error is usually involved in proving identities. This fact is sometimes obscured by examples such as the preceding ones in which it appears that everything follows very simply and logically. But remember, you are seeing only the final result, not all the scratch paper, wrong directions, and mistakes that were involved in producing the finished product. So don't get discouraged if you can't instantly work out all the identities in the Exercises.

EXERCISES

A.1.   Which of the following are possible pairs of values for $\sin t$ and $\cos t$?
   (a) $\sin t = \frac{5}{13}$, $\cos t = \frac{12}{13}$
   (b) $\sin t = -2$, $\cos t = 1$

   (c) $\sin t = -1$, $\cos t = 1$

   (d) $\sin t = \frac{1}{2}$, $\cos t = \frac{1}{2}$
   (e) $\sin t = \frac{8}{17}$, $\cos t = \frac{15}{17}$

   (f) $\sin t = -\dfrac{1}{\sqrt{2}}$, $\cos t = \sqrt{2}$

A.2.   (a)   In what intervals between -2π and 2π is $\sin t$ negative?

   (b)   In what intervals between $-\dfrac{\pi}{2}$ and $\dfrac{5\pi}{2}$ is $\cos t$ positive?

B.1.   (a)   If $\sin t = -\frac{4}{5}$ and an angle of t radians lies in the third quadrant, what is $\cos t$? (*Hint:* use the Pythagorean Identity.)
   (b)   If $\cos t = -3/\sqrt{10}$ and an angle of t radians lies in the second quadrant, what is $\sin t$?
   (c)   If $\sin t = \frac{1}{2}$ and an angle of t radians lies in the first quadrant, what is $\cos t$?
   (d)   If $\cos t = 2/\sqrt{5}$ and an angle of t radians lies in the fourth quadrant, what is $\sin t$?
   (e)   If $\sin t = -1$, what is $\cos t$?

B.2.   Assume that $\sin(\pi/4) = \sqrt{2}/2$ and $\sin(\pi/3) = \sqrt{3}/2$. Use the identities and other facts in the text to find:

   (a) $\cos\dfrac{\pi}{4}$ and $\cos\dfrac{\pi}{3}$ (think Pythagorean)

   (b) $\sin\left(\dfrac{5\pi}{4}\right)$ (*Hint:* $\dfrac{5\pi}{4} = \dfrac{\pi}{4} + \pi$; use an

      identity presented in the text.)

   (c) $\sin\left(\dfrac{-4\pi}{3}\right)$

   (d) $\cos\left(\dfrac{5\pi}{3}\right)$

   (e) $\cos\left(\dfrac{3\pi}{4}\right)$

B.3. Use your knowledge of the values of the sine function and cosine function for 0, $\pi/2$, π, and $3\pi/2$ and the concept of periodicity to find the following numbers. *Example:* $\sin(-\pi/2) = \sin((-\pi/2) + 2\pi) = \sin(3\pi/2) = -1$.

(a) $\cos\dfrac{9\pi}{2}$ (b) $\sin(-11\pi)$ (c) $\cos\dfrac{45\pi}{2}$ (d) $\cos\left(-\dfrac{17\pi}{2}\right)$

B.4. Show that the given function is periodic with period *less than* 2π. That is, show that in each case there is a fixed number k, with $k < 2\pi$, such that $f(t + k) = f(t)$ for every number t in the domain of the function.

(a) $f(t) = \sin 2t$

(b) $f(t) = \cos 3t$

(c) $f(t) = \sin^2 t$

(d) $k(t) = \cos\left(\dfrac{3\pi t}{2}\right)$

(e) $f(t) = \dfrac{\sin t}{\cos t}$

(f) $g(t) = \dfrac{\sin(t + \pi)}{\cos(t + \pi)}$

B.5. Multiply and simplify your answer (via algebra and suitable identities). Assume all denominators are nonzero.

(a) $(\sin t + \cos t)(\sin t - \cos t)$

(b) $(\sin t - \cos t)^2$

(c) $\sqrt{\sin^3 t \cos t}\,\sqrt{\cos t}$

(d) $(\sin(t + \pi) + \cos(t + \pi))^2$

(e) $\left(\dfrac{4\cos^3 t}{\sin^2 t}\right)\left(\dfrac{\sin t}{4\cos t}\right)^2$

(f) $\dfrac{5\cos t}{\sin^2 t} \cdot \dfrac{\sin^2 t - \sin t \cos t}{\sin^2 t - \cos^2 t}$

(g) $\dfrac{9\cos^2 t - 25}{2\cos t - 2} \cdot \dfrac{\cos^2 t - 1}{6\cos t - 10}$

B.6. Simplify these expressions. Assume all denominators are nonzero.

(a) $\dfrac{4\cos t \sin^3 t}{22\cos^2 t \sin t}$

(b) $\dfrac{\cos^2 t \sin t}{\sin^2 t \cos t}$

(c) $\dfrac{\sin^2 t - 2\sin t + 1}{\sin t - 1}$

(d) $\dfrac{\cos^2 t + 4\cos t + 4}{\cos t + 2}$

(e) $\dfrac{\sin^2 t - 1}{\sin(\pi - t) + 1}$

(f) $\dfrac{30\cos^3(t + \pi)\sin t}{6\sin^2(t + \pi)\cos t}$

B.7. Use algebra and the identities proved in the text to verify that each of these equations is an identity, that is, that each is valid for every number t. (*Note:* the hints suggest *one* way of doing this, but there are several *other* equally good ways.)

(a) $\sin(t - \pi) = -\sin t$ [*Hint:* $t - \pi = -(\pi - t)$.]
(b) $\cos(t - \pi) = -\cos t$ [*Hint:* $t - \pi = (t - 2\pi) + \pi$.]
(c) $\cos(\pi - t) = -\cos t$
(d) $\sin(2\pi - t) = -\sin t$
(e) $\cos(2\pi - t) = \cos t$

B.8. Verify that each of these equations is an identity, that is, that each is valid for every number t for which both sides of the equation are defined.

(a) $\sin^4 t - \cos^4 t = 2 \sin^2 t - 1$ (d) $\dfrac{\sin^2 t}{\cos^2 t} + 1 = \dfrac{1}{\cos^2 t}$

(b) $1 - 2 \cos^2 t + \cos^4 t = \sin^4 t$ (e) $\dfrac{\cos^2(\pi + t)}{\sin^2(\pi + t)} + 1 = \dfrac{1}{\sin^2(\pi - t)}$

(c) $\dfrac{\sin t}{1 - \cos t} = \dfrac{1 + \cos t}{\sin t}$

C.1. Here is a proof that the cosine function has period 2π. We saw in the text that $\cos (t + 2\pi) = \cos t$ for every t. We must show that there is no positive number smaller than 2π with this property. Do this as follows:
(a) Find all numbers k such that $0 < k < 2\pi$ and $\cos k = 1$. (*Hint:* draw a picture and use the definition of the cosine function.)
(b) Suppose k is a number such that $\cos (t + k) = \cos t$ for every number t. Show that $\cos k = 1$. (*Hint:* consider $t = 0$.)
(c) Use (a) and (b) to show that there is no positive number k less than 2π with the property that $\cos (t + k) = \cos t$ for *every* number t. Therefore $k = 2\pi$ is the smallest such number and the cosine function has period 2π.

C.2. Here is proof that the sine function has period 2π. We saw in the text that $\sin (t + 2\pi) = \sin t$ for every t. We must show that there is no positive number smaller than 2π with this property. Do this as follows:
(a) Find a number t such that $\sin (t + \pi) \neq \sin t$.
(b) Find all numbers k such that $0 < k < 2\pi$ and $\sin k = 0$. (*Hint:* draw a picture and use the definition of the sine function.)
(c) Suppose k is a number such that $\sin (t + k) = \sin t$ for every number t. Show that $\sin k = 0$. (*Hint:* consider $t = 0$.)
(d) Use parts (a)–(c) to show that there is no positive number k less than 2π with the property that $\sin (t + k) = \sin t$ for *every* number t. Therefore $k = 2\pi$ is the smallest such number and the sine function has period 2π.

4. EVALUATION OF THE SINE AND COSINE FUNCTIONS

Both sine and cosine are functions whose domain is the set of all real numbers. There are several different ways of evaluating these functions at a particular number t (that is, of calculating the numbers $\sin t$ and $\cos t$). All the usual methods are presented here. Since any one of them is liable to be used later (depending on the situation), you must become familiar with all of them.

CALCULATORS AND TABLES

If you have a calculator equipped to handle trigonometric functions, you should definitely learn how to use it.° It would be a good idea to reread the *Note on Calculators* on page 244. If you don't own a calculator, you can evaluate the sine and cosine functions (with a bit more work) by using the trigonometric tables at the end of this book. In fact, even people with calculators should know how to read such tables. The use of these tables is explained in the DO IT YOURSELF! segment at the end of this section.

In any case, you should be aware that both calculators and tables have definite shortcomings. Strictly speaking they can be used to compute $\sin t$ and $\cos t$ only when t is a rational number with a finite decimal expansion, say, 8 to 12 places for a typical calculator or 4 places for a typical table. Furthermore, the value given by a calculator or table for $\sin t$ or $\cos t$ will usually be a decimal *approximation* of the actual value, which might well be an irrational number.

In most practical situations, these decimal approximations are sufficiently accurate. But it is occasionally disconcerting to discover that your calculator doesn't "believe in" certain well-known identities. For example, since $\sin (\pi - t) = \sin t$ for every t, you would expect a calculator to show that for $t = 200$, $\sin (\pi - 200) - \sin 200 = 0$. Some calculators will, but others will produce an answer such as .000000000289.

As we shall presently see, the exact values of the trigonometric functions are readily available for certain numbers. In such cases a calculator or table may actually be more of a hindrance than a help. When working with irrational numbers such as π or $\sqrt{2}$, for instance, it is usually best to leave them in this form rather than to use decimal approximations.

EXAMPLE As we shall see later, $\sin (\pi/4)$ is the irrational number $\sqrt{2}/2$. For many purposes this is a very convenient number to work with. For instance, if you square it you obtain $\sin^2(\pi/4) = (\sqrt{2}/2)^2 = \frac{2}{4} = \frac{1}{2}$. But the table at the end of the book shows that $\sin (\pi/4) \approx \sin (.7854) \approx .7071$. This number is a 4-place decimal approximation of $\sqrt{2}/2$, which is sometimes rather inconvenient to work with. For example, its square is .49999041.

GEOMETRIC METHOD

As shown in the examples below, the *exact* value of $\sin t$ and $\cos t$ can easily be found whenever the number t is an integer multiple of either $\pi/4$ or $\pi/6$. Examples of such numbers are $-3\pi/4 = (-3) \times (\pi/4)$ and $8\pi/3 = 16 \times (\pi/6)$. The method of calcu-

---

° Most calculators operate in either the radian mode or the degree mode for trigonometric functions. Everything we have done so far is in terms of radians, so *use the radian mode for now*. Trigonometric functions in terms of degrees are explained in Section 9.

lation depends on three geometric facts. The first of these facts is that the sum of the measures of all three angles of a triangle is π radians (180°). Here are the other two:

(i) **Sides opposite equal angles in a triangle have the same length.** Therefore in a right triangle with two angles of $\pi/4$ radians (45°) and hypotenuse 1, the other two sides each have length $\sqrt{2}/2$, as shown in Figure 6-32 (see Geometry Review, p. 582, for details).

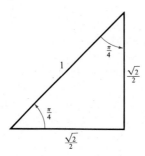

Figure 6-32

(ii) **In a right triangle with an angle of $\pi/6$ radians (30°), the length of the side opposite this angle is half the length of the hypotenuse.** Therefore if the hypotenuse has length 1, the other sides have lengths $\frac{1}{2}$ and $\sqrt{3}/2$, as shown in Figure 6-33 (see pp. 583–584 for details).

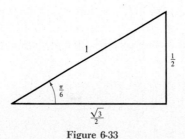

Figure 6-33

EXAMPLE   In order to find $\sin (\pi/4)$ and $\cos (\pi/4)$, construct an angle of $\pi/4$ radians in standard position. Let P be the point where the terminal side of this angle meets the unit circle. Now construct a right triangle by drawing a vertical line from P to the x-axis, as shown in Figure 6-34. This triangle has an angle of $\pi/4$ radians and a right angle of $\pi/2$ radians.

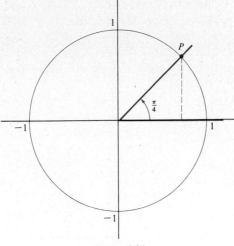

Figure 6-34

Since the sum of all the angles of a triangle is π radians, the third angle must have measure $\pi/4$ radians also (since $(\pi/4) + (\pi/4) + (\pi/2) = \pi$). The hypotenuse of the triangle is a radius of the unit circle, so it has length 1. Consequently, by statement (i) above, the other sides of the triangle each have length $\sqrt{2}/2$, as shown in Figure 6-35.

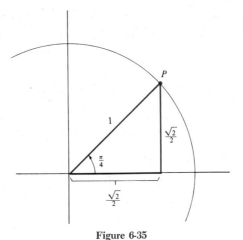

Figure 6-35

Therefore the coordinates of P are $(\sqrt{2}/2), \sqrt{2}/2)$. Since the y-coordinate of P is by definition $\sin(\pi/4)$, we conclude that $\sin(\pi/4) = \sqrt{2}/2$. Similarly, $\cos(\pi/4) =$ x-coordinate of $P = \sqrt{2}/2$. In this case both the sine and cosine functions have the same value at $t = \pi/4$, but this doesn't happen for most values of t.

EXAMPLE In order to find $\sin(11\pi/3)$ and $\cos(11\pi/3)$, construct an angle of $11\pi/3$ radians in standard position. Since $11\pi/3 = (6\pi/3) + (5\pi/3) = 2\pi + (5\pi/3)$, the terminal side of this angle is the same as the terminal side of an angle of $5\pi/3$ radians. Let P be the point where this terminal side meets the unit circle. Draw a vertical line from P to the x-axis, thus forming a right triangle with hypotenuse 1 and an angle of $\pi/3$ radians, as shown in Figure 6-36.

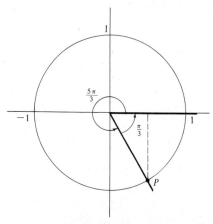

Figure 6-36

Since the sum of all the angles of a triangle is π radians, the third angle must have measure $\pi/6$ (for $(\pi/6) + (\pi/3) + (\pi/2) = \pi$). By statement (ii) above, the sides of the triangle must have lengths $\frac{1}{2}$ and $\sqrt{3}/2$, as shown in Figure 6-37.

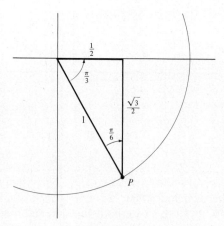

Figure 6-37

Since P is in the fourth quadrant, its x-coordinate is positive and its y-coordinate is negative. Hence the coordinates of P must be $(\frac{1}{2}, -\sqrt{3}/2)$. Therefore

$$\cos\frac{11\pi}{3} = x\text{-coordinate of } P = \tfrac{1}{2} \quad \text{and} \quad \sin\frac{11\pi}{3} = y\text{-coordinate of } P = \frac{-\sqrt{3}}{2}$$

EXAMPLE The same procedure can be used to find $\sin(-5\pi/4)$ and $\cos(-5\pi/4)$. Construct an angle of $-5\pi/4$ radians in standard position. Let P be the point where the terminal side of this angle meets the unit circle. Draw a vertical line from P to the x-axis, thus forming a right triangle with hypotenuse 1 and two angles of $\pi/4$ radians, as shown in Figure 6-38.

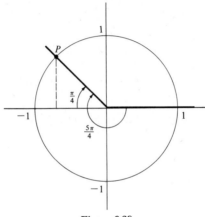

Figure 6-38

By statement (i) above, the other sides of the triangle must each have length $\sqrt{2}/2$, as shown in Figure 6-39.

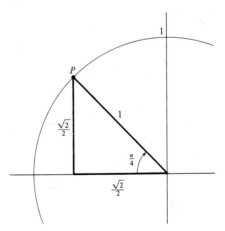

Figure 6-39

Since P lies in the second quadrant, its x-coordinate is negative and its y-coordinate is positive. Thus P must have coordinates $(-\sqrt{2}/2, \sqrt{2}/2)$. Therefore

$$\cos\left(\frac{-5\pi}{4}\right) = x\text{-coordinate of } P = \frac{-\sqrt{2}}{2}$$

and

$$\sin\left(\frac{-5\pi}{4}\right) = y\text{-coordinate of } P = \frac{\sqrt{2}}{2}$$

USING IDENTITIES *

Once you know the values of the sine and cosine functions at certain numbers, whether by geometry as above or by other means, you can then evaluate these functions at various numbers by using identities such as

$$\begin{array}{ll} \sin(-t) = -\sin t, & \sin(t+\pi) = -\sin t, \\ \cos(-t) = \cos t, & \cos(t+\pi) = -\cos t \end{array}$$

EXAMPLE On page 347 we saw that $\sin(\pi/4) = \sqrt{2}/2$ and $\cos(\pi/4) = \sqrt{2}/2$. Consequently,

$$\sin\left(-\frac{\pi}{4}\right) = -\sin\frac{\pi}{4} = -\frac{\sqrt{2}}{2} \quad \text{and} \quad \cos\left(-\frac{\pi}{4}\right) = \cos\frac{\pi}{4} = \frac{\sqrt{2}}{2}$$

POINT-IN-THE-PLANE-METHOD

One possible disadvantage of most of the calculations presented so far is their dependence on the unit circle. Fortunately, there is another method of describing $\sin t$ and $\cos t$, which does not use the unit circle:

> Let t be any real number. Let (x, y) be any *point (except the origin)* on the terminal side of an angle of t radians in standard position. Then
> $$\sin t = \frac{y}{r} \quad \text{and} \quad \cos t = \frac{x}{r}$$
> where r = $\sqrt{x^2 + y^2}$ is the distance from (x, y) to the origin.

Before proving this statement, we shall illustrate its meaning and usefulness by some examples.

---

* If you have not yet read Section 3, skip this subsection and go to the next one, *Point in the Plane Method.*

EXAMPLE Find $\sin t$ and $\cos t$ when the point $(-5, 7)$ lies on the terminal side of an angle of t radians in standard position (see Figure 6-40).

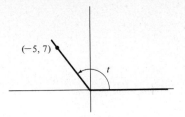

Figure 6-40

In this case we have $(x, y) = (-5, 7)$ so that $r = \sqrt{x^2 + y^2} = \sqrt{(-5)^2 + 7^2} = \sqrt{74}$. Therefore

$$\sin t = \frac{y}{r} = \frac{7}{\sqrt{74}} \qquad \text{and} \qquad \cos t = \frac{x}{r} = \frac{-5}{\sqrt{74}}$$

EXAMPLE Suppose the terminal side of an angle of t radians lies in the first quadrant and has slope $\frac{2}{3}$ (see Figure 6-41). Find $\sin t$ and $\cos t$.

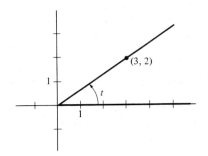

Figure 6-41

A line that has slope $\frac{2}{3}$ and passes through $(0, 0)$ must also contain the point $(3, 2)$. In order to find $\sin t$ and $\cos t$, we use the point $(3, 2)$ and $r = \sqrt{x^2 + y^2} = \sqrt{3^2 + 2^2} = \sqrt{13}$:

$$\sin t = \frac{y}{r} = \frac{2}{\sqrt{13}} \qquad \text{and} \qquad \cos t = \frac{x}{r} = \frac{3}{\sqrt{13}}$$

PROOF OF THE STATEMENT IN THE BOX ABOVE Given a real number t, construct an angle of t radians in standard position. Let $Q = (x, y)$ be *any* point (except the origin) on the terminal side of this angle.° Let P be the point where the terminal

---

° The picture on the next page shows Q lying outside the unit circle. But the discussion applies equally well when Q is inside the unit circle.

side meets the unit circle. Draw vertical lines from Q to the x-axis and from P to the x axis, as shown in Figure 6-42.

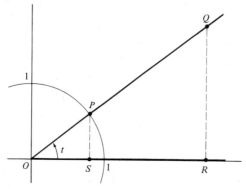

Figure 6-42

Both triangle QOR and triangle POS are right triangles containing an angle of t radians. Therefore these triangles are *similar.*° Consequently, the ratios of corresponding sides are equal:

$$\frac{\text{length } OP}{\text{length } OQ} = \frac{\text{length } PS}{\text{length } QR} \quad \text{and} \quad \frac{\text{length } OP}{\text{length } OQ} = \frac{\text{length } OS}{\text{length } OR}$$

The next step is to determine these various lengths. Since OP is a radius of the unit circle, length $OP = 1$. Since length OQ is the distance from $O = (0, 0)$ to $Q = (x, y)$, the distance formula shows that

$$\text{length } OQ = \sqrt{(x - 0)^2 + (y - 0)^2} = \sqrt{x^2 + y^2}$$

Denote the number $\sqrt{x^2 + y^2}$ by r, so that length $OQ = r$. Finally, the definition of sine and cosine shows that the coordinates of P are $(\cos t, \sin t)$. Thus since $Q = (x, y)$, the other lengths must be as shown in Figure 6-43.

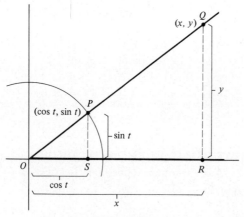

Figure 6-43

° Similar triangles are discussed in the *Geometry Review* at the end of the book.

Using this information, we can rewrite the equations

$$\frac{\text{length } OP}{\text{length } OQ} = \frac{\text{length } PS}{\text{length } QR} \quad \text{and} \quad \frac{\text{length } OP}{\text{length } OQ} = \frac{\text{length } OS}{\text{length } OR}$$

as follows:

$$\frac{1}{r} = \frac{\sin t}{y} \quad \text{and} \quad \frac{1}{r} = \frac{\cos t}{x}$$

$$r \sin t = y \quad \text{and} \quad r \cos t = x$$

$$\sin t = \frac{y}{r} \qquad\qquad \cos t = \frac{x}{r}$$

This completes the proof in the case when the line OP lies in the first quadrant. When it lies in other quadrants, the proof is essentially the same, except that care must be taken about signs (for instance, in the second quadrant x is negative so that length $OR = |x|$ instead of x and length $OS = |\cos t|$ instead of $\cos t$). These details are left as an exercise for those interested.

RIGHT-TRIANGLE METHOD

For numbers between 0 and $\pi/2$, the preceding discussion leads to still another way of evaluating the sine and cosine functions. Suppose t is a number between 0 and $\pi/2$ and we have a right triangle that has an angle of t radians. By sliding, rotating, and possibly flipping over this triangle, we can move the angle of t radians into standard position in such a way that the hypotenuse lies in the first quadrant and is the terminal side of the angle, as shown in Figure 6-44. Denote the length of the *hypotenuse* AC by r. Denote the length of the side AB *adjacent* to the angle of t radians by u. Denote the length of the side BC *opposite* the angle of t radians by v. Then the coordinates of the vertex C are (u, v) (see Figure 6-44).

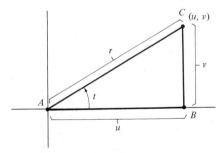

Figure 6-44

Since $C = (u, v)$ lies on the terminal side of an angle of t radians, we know from the discussion above that

$$\sin t = \frac{v}{r} = \frac{\text{length of opposite side}}{\text{length of hypotenuse}} \quad \text{and} \quad \cos t = \frac{u}{r} = \frac{\text{length of adjacent side}}{\text{length of hypotenuse}}$$

These facts can be succinctly summarized as follows:

> Let t be a number between 0 and $\pi/2$. Consider any right triangle with an angle of t radians. Then
>
> $$\sin t = \frac{opposite}{hypotenuse} \quad and \quad \cos t = \frac{adjacent}{hypotenuse}$$

This is the description of the sine and cosine functions that is usually presented first when trigonometry is studied from the point of view of angles and triangles (see Section 9 for more details). The primary advantage of this description is that it does not involve either the unit circle or the coordinate system in the plane. This description is valid no matter what position the right triangle is in.

EXAMPLE   Consider a right triangle whose other two angles measure $\pi/6$ radians (30°) and $\pi/3$ radians (60°) and whose hypotenuse has length 2. As explained in the *Geometry Review* (p. 583), the side opposite the angle of $\pi/6$ must have length 1 (half the hypotenuse) and the side adjacent to the angle of $\pi/6$ must have length $\sqrt{3}$ (see Figure 6-45).

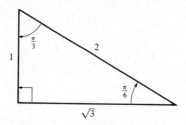

Figure 6-45

Now apply the facts in the box above to the angle of $\pi/6$ radians:

$$\sin\frac{\pi}{6} = \frac{opposite}{hypotenuse} = \frac{1}{2} \quad and \quad \cos\frac{\pi}{6} = \frac{adjacent}{hypotenuse} = \frac{\sqrt{3}}{2}$$

But we can apply the same facts to the angle of $\pi/3$ radians. As Figure 6-45 shows, the side opposite the angle of $\pi/3$ radians has length $\sqrt{3}$ and the side adjacent to it has the length 1. Therefore

$$\sin\frac{\pi}{3} = \frac{opposite}{hypotenuse} = \frac{\sqrt{3}}{2} \quad and \quad \cos\frac{\pi}{3} = \frac{adjacent}{hypotenuse} = \frac{1}{2}$$

EXAMPLE Find the number t, when a right triangle with sides of lengths 3, 4, and 5 has an angle of measure t radians, as shown in Figure 6-46.

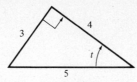

Figure 6-46

Here the hypotenuse has length 5 and the side adjacent to the angle of t radians has length 4. Therefore

$$\cos t = \frac{\text{adjacent}}{\text{hypotenuse}} = \frac{4}{5} = .8$$

According to the table at the end of the book (whose use is explained in the DO IT YOURSELF! segment below), the number t with $0 < t < \pi/2$ and $\cos t = .8$ is approximately .6429. In particular, the angle in question measures approximately .6429 radians ($\approx 36.84°$).

EXERCISES

A.1. Fill in the missing entries in the table below. Write each entry as a fraction with denominator 2 and with a radical in the numerator. For example, $\sin (\pi/2) = 1 = \sqrt{4}/2$. Observe the pattern that results. Some students find this scheme to be an easy way to remember these values of sine and cosine.

t	0	$\dfrac{\pi}{6}$	$\dfrac{\pi}{4}$	$\dfrac{\pi}{3}$	$\dfrac{\pi}{2}$
$\sin t$					
$\cos t$					

A.2. Find $\sin t$ and $\cos t$ when $t =$
 (a) $5\pi/6$ (c) $5\pi/4$ (e) 3π (g) $19\pi/6$
 (b) $11\pi/6$ (d) $7\pi/3$ (f) $3\pi/2$ (h) $-11\pi/4$

A.3. Assume that the terminal side of an angle of t radians in standard position passes through the given point. Find $\sin t$ and $\cos t$.
 (a) $(2, 7)$ (c) $(\sqrt{3}, -10)$ (e) $(\frac{1}{2}, 2)$
 (b) $(-3, 2)$ (d) $(-2, -\sqrt{2})$ (f) $(-5, \frac{1}{5})$

A.4. Referring to Figure 6-47, find $\sin \theta$ and $\cos \theta$, where θ is the radian measure of the indicated angle.

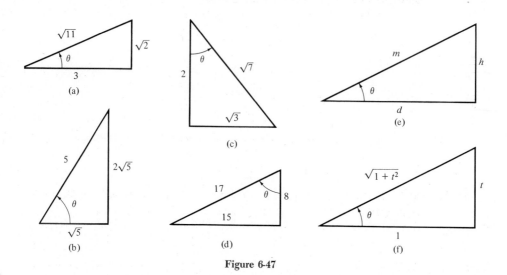

Figure 6-47

B.1. Express each of these expressions as a single real number:
   (a) $\sin (\pi/3) \cos \pi + \sin \pi \cos (\pi/3)$
   (b) $\sin (\pi/6) \cos (\pi/2) - \cos (\pi/6) \sin (\pi/2)$
   (c) $\cos (\pi/2) \cos (\pi/4) - \sin (\pi/2) \sin (\pi/4)$
   (d) $\cos (2\pi/3) \cos \pi + \sin (2\pi/3) \sin \pi$
   (e) $\sin (3\pi/4) \cos (5\pi/6) - \cos (3\pi/4) \sin (5\pi/6)$
   (f) $\sin (-7\pi/3) \cos (5\pi/4) + \cos (-7\pi/3) \sin (5\pi/4)$

B.2. Assume that the terminal side of an angle of t radians in standard position lies in the given quadrant and on the given straight line. Find $\sin t$ and $\cos t$. (*Hint:* use the equation of the line to find a point on the terminal side of the angle.)
   (a) quadrant IV; line with equation $y = -3x$
   (b) quadrant III; line with equation $2y - 4x = 0$
   (c) quadrant IV; line through $(-3, 5)$ and $(-9, 15)$
   (d) quadrant III; line through origin parallel to $7x - 2y + 6 = 0$

In Exercises B.3–B.6, use 3.14 for π and the table below. See Exercise B.3 for hints. *All angles are measured in radians.*

t	.5	.55	.6	.65	.7	.75	.8	.85	.9
$\sin t$	.49	.52	.56	.61	.64	.68	.72	.75	.78
$\cos t$	.88	.85	.83	.80	.76	.73	.70	.66	.62

B.3. In the right triangle in Figure 6-48,

    (a) find the radian measure of ∢B. [*Hint:* the sum of the radian measures of all three angles is π radians (3.14 radians).]

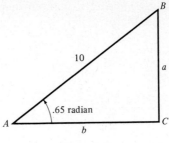

Figure 6-48

    (b) find a. (*Hint:* we have sin .65 = opposite/hypotenuse = $a/10$ and from the table above, sin .65 = .61. Hence .61 = $a/10$; solve for a.)

    (c) find b. (*Hint:* use the Pythagorean Theorem, or solve the equation .8 = cos .65 = adjacent/hypotenuse = $b/10$ for b.)

B.4. In the right triangle in Figure 6-49,   **(a)** find the radian measure of ∢A;   **(b)** find c;   **(c)** find a.

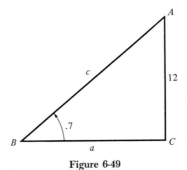

Figure 6-49

B.5. In the right triangle in Figure 6-50,   **(a)** find the radian measure of ∢B;   **(b)** find c;   **(c)** find a.

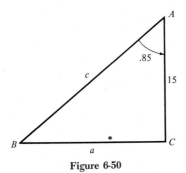

Figure 6-50

B.6.   Find the (approximate) radian measure t of the indicated angle of the given right triangle in Figure 6-51.

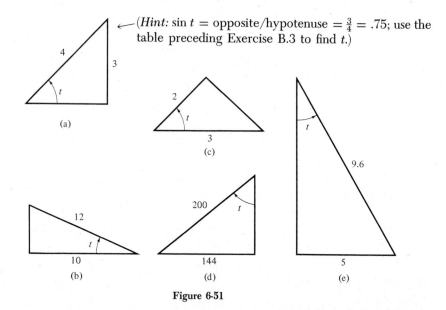

$\leftarrow$ (*Hint:* sin t = opposite/hypotenuse = $\frac{3}{4}$ = .75; use the table preceding Exercise B.3 to find t.)

Figure 6-51

B.7.   Find the length h of the indicated side in each of the right triangles in Figure 6-52. [*Hint:* see Exercise B.3(b).]

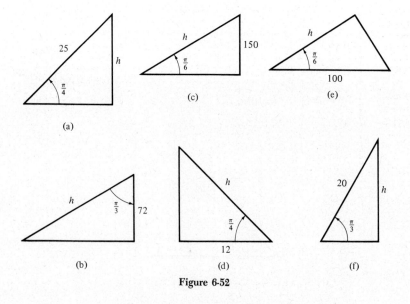

Figure 6-52

C.1. Imagine a 16-ft-long drawbridge on a medieval castle (see Figure 6-53).

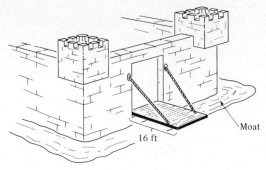

Figure 6-53

The Royal Army is engaged in an ignominious retreat. The King would like to raise the end of the drawbridge 8 ft off the ground so that Sir Rodney can jump onto the drawbridge and scramble into the castle, yet the enemy's cavalry will be held at bay. Through how much of an angle must the drawbridge be raised in order that the end of it be 8 ft off the ground?

DO IT YOURSELF!

TRIGONOMETRIC TABLES

A typical table of trigonometric functions can be found at the end of this book. A portion of it is reproduced below. On the far left and the far right of the complete table are three columns labeled DEG, MIN, and RADIANS. The first two refer to the measure of an angle in degrees and minutes, and are not needed until Section 9, so they are omitted from the portion of the table shown here.

RADIANS	SIN	COS	TAN	COT	SEC	CSC	
.0000	.0000	1.0000	.0000	—	1.0000	—	1.5709
.0029	.0029	1.0000	.0029	343.773	1.0000	343.774	1.5680
.0058	.0058	1.0000	.0058	171.885	1.0000	171.888	1.5651
⋮	⋮	⋮	⋮	⋮	⋮	⋮	⋮
.1513	.1507	.9886	.1524	6.561	1.012	6.636	1.4196
.1542	.1536	.9881	.1554	6.435	1.012	6.512	1.4167
.1571	.1564	.9877	.1584	6.314	1.012	6.392	1.4138
⋮	⋮	⋮	⋮	⋮	⋮	⋮	⋮
.2269	.2250	.9744	.2303	4.331	1.026	4.445	1.3440
.2298	.2278	.9737	.2339	4.275	1.027	4.390	1.3411
.2327	.2306	.9730	.2370	4.219	1.028	4.336	1.3382
	COS	SIN	COT	TAN	CSC	SEC	RADIANS

The six middle columns contain the values of the six trigonometric functions—sine, cosine, and four others to be defined later. The columns are labeled on the top and bottom, but note that the labels do *not* match up. Furthermore, the column labeled RADIANS on the right lists different numbers than does the column on the left.

The table gives values of the trigonometric functions only for real numbers between 0 and $\pi/2$. In fact, they can be used to obtain values of the trigonometric functions for any real number, as we shall see below.

USING TABLES WHEN t IS BETWEEN 0 AND $\pi/2$

For the reader's convenience, all of the following examples will involve only the portion of the tables reprinted above. However, the complete tables in the back of the book will be needed for the exercises.

EXAMPLE In order to find cos .2298, look through the RADIANS column of the table until you find .2298. It appears in the column on the *left* side of the page, so use the function labels at the *top* of the page. Reading across the line beginning .2298 radians, you will find that the entry under the COS column is the number .9737. Therefore cos .2298 $\approx$ .9737.

EXAMPLE In order to find sin 1.4167, look through the RADIANS column until you find 1.4167. This time it appears in the column on the *right* side of the page, so use the function labels at the *bottom* of the page. Reading from right to left across the line 1.4167 radians on the right, you will find that the entry in the SIN column is .9881. Therefore sin 1.4167 $\approx$ .9881.

REFERENCE NUMBERS

In order to use tables to find the values of the trigonometric functions for a number t not between 0 and $\pi/2$, we introduce the concept of a **reference number:**

If $\pi/2 < t \leq \pi$, then the reference number of t is the number $\pi - t$.
If $\pi < t \leq 3\pi/2$, then the reference number of t is the number $t - \pi$.
If $3\pi/2 < t \leq 2\pi$, then the reference number of t is the number $2\pi - t$.

To get a picture of the reference number of a given number t, think of t as the radian measure of an angle in standard position with terminal side L. The reference number of t is then the measure of the smallest positive angle (not necessarily in standard position) made by L and the x-axis, as shown in Figure 6-54.

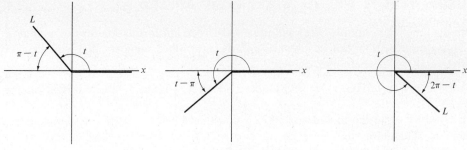

Figure 6-54

There are two basic facts about reference numbers. First,

> *The reference number of* t *is a number between* 0 *and* $\pi/2$.

This can be seen by looking at either the algebraic definition or the geometric picture of reference numbers given in Figure 6-54. Here is the second fact:

> *If a number* t *has reference number* k, *then*
>
> $\sin t = \pm \sin k$ *and* $\cos t = \pm \cos k$

This statement can be proved by using various identities from Section 3 (in particular, Exercise B.7). For example, if $\pi/2 < t \le \pi$, then the reference number of t is the number $k = \pi - t$. Using the identities proved on page 341, we see that

$$\sin k = \sin (\pi - t) = \sin t \qquad \text{and} \qquad \cos k = \cos (\pi - t) = -\cos t$$

The proof of the other cases is similar.

The *general procedure for using the reference number* of t to find $\sin t$ and $\cos t$ is as follows.

(i)   Find the reference number k of the number t.
(ii) Since k is between 0 and $\pi/2$ (first basic fact), the table can be used to find $\sin k$ and $\cos k$ (as in the examples above).
(iii) By the second basic fact, $\sin t = \pm \sin k$ and $\cos t = \pm \cos k$. The proper sign can be determined by considering in which quadrant the terminal side of an angle of t radians lies (see p. 337 for details).

The following numerical examples will make this procedure clear.

USING TABLES WHEN t IS BETWEEN $\pi/2$ AND 2π

In the following examples, we shall use 3.1416 as an approximation for the number π. This approximation can be used to approximate other numbers that will be used frequently in the examples, namely:

$$\pi/2 \approx 1.5708, \qquad \pi \approx 3.1416, \qquad 3\pi/2 \approx 4.7124, \qquad 2\pi \approx 6.2832$$

EXAMPLE In order to find sin 2.9874, we first note that $t = 2.9874$ lies between $\pi/2$ and π. Therefore the reference number of t is $\pi - t = \pi - 2.9874 \approx 3.1416 - 2.9874 = .1542$. By the second basic fact about reference numbers, $\sin 2.9874 \approx \pm\sin .1542$. Since .1542 lies between 0 and $\pi/2$, we can use the tables to find sin .1542. The number .1542 appears in the *left*-hand RADIANS column. The entry opposite .1542 in the SIN column (labels at *top* of table) is .1536. Thus

$$\sin 2.9874 \approx \pm\sin .1542 \approx \pm.1536$$

Since the sine function is positive for all numbers between $\pi/2$ and π (see p. 337), we conclude that $\sin 2.9874 \approx .1536$.

EXAMPLE In order to find sin 4.5612, we note that $t = 4.5612$ lies between π and $3\pi/2$. Hence the reference number of t is $t - \pi = 4.5612 - \pi \approx 4.5612 - 3.1416 = 1.4196$. By the second basic fact, $\sin 4.5612 \approx \pm\sin 1.4196$. Since 1.4196 lies between 0 and $\pi/2$, we can use the tables to find sin 1.4196. The number 1.4196 appears in the *right*-hand RADIAN column. The entry opposite 1.4196 in the SIN column (labels at *bottom* of table) is .9886. Therefore

$$\sin 4.5612 \approx \pm\sin 1.4196 \approx \pm.9886$$

Since the sine function is negative for all numbers between π and $3\pi/2$ (see p. 337), we conclude that $\sin 4.5612 \approx -.9886$.

USING TABLES WHEN t IS NOT BETWEEN 0 AND 2π

In order to find sin t and cos t when t is a number larger than 2π, we use the techniques above and the fact that both the sine and cosine functions have period 2π. This fact implies that for any number t,

$$\sin t = \sin (t - 2\pi) = \sin (t - 4\pi) = \cdots$$

and

$$\cos t = \cos (t - 2\pi) = \cos (t - 4\pi) = \cdots$$

EXAMPLE In order to find cos 17.5085, we first note that 17.5085 is larger than 2π. In fact, a little experimentation and arithmetic shows that $4\pi < 17.5085 < 6\pi$. Therefore $17.5085 - 4\pi$ is a number between 0 and 2π. By periodicity, we know that $\cos 17.5085 = \cos (17.5085 - 4\pi) \approx \cos (17.5085 - 12.5664) = \cos 4.9421$. Since 4.9421 lies between $3\pi/2$ and 2π, we can use reference numbers and tables to evaluate it. The reference number of 4.9421 is $2\pi - 4.9421 \approx 6.2832 - 4.9421 = 1.3411$. By the second basic fact about reference numbers, $\cos 4.9421 \approx \pm\cos 1.3411$. Using the tables, we find that $\cos 1.3411 \approx .2278$. Since cosine is positive for all numbers between $3\pi/2$ and 2π, we conclude that $\cos 4.9421 \approx .2278$. Therefore $\cos 17.5085 \approx \cos 4.9421 \approx .2278$.

The preceding examples show how to evaluate $\sin t$ and $\cos t$ for any positive number t. If t is negative, then we use the techniques above together with the identities developed in Section 3:

$$\sin(-t) = -\sin t \quad \text{and} \quad \cos(-t) = \cos t$$

EXAMPLE In the examples above, we found that $\cos 17.5085 \approx .2278$ and that $\sin 2.9874 \approx .1536$. Therefore

$$\cos(-17.5085) = \cos 17.5085 \approx .2278$$
$$\sin(-2.9874) = -\sin 2.9874 \approx -.1536$$

INTERPOLATION

In all the preceding examples, the number t or its reference number always appeared as an entry in the RADIANS column of the tables. In order to find $\sin t$ or $\cos t$ when t (or its reference number) do not appear in the RADIANS column, you must **interpolate** as illustrated in this example.

EXAMPLE Find $\cos .228$. The number $.228$ does not appear in the RADIANS column of the tables, but two numbers near it do appear: $.2269$ and $.2298$. So we have:

$$\text{difference } .0029 \left\{ \begin{array}{l} .2269 \\ .228 \\ .2298 \end{array} \right. \left. \begin{array}{l} \\ \end{array} \right\} \text{difference } .0011$$

Thus $.228$ lies $\frac{11}{29}$ of the way from $.2269$ to $.2298$. Consequently, it seems plausible that $\cos .228$ lies $\frac{11}{29}$ of the way from $\cos .2269$ to $\cos .2298$. Using the tables, we see that

$$\left. \begin{array}{l} \cos .2269 \approx .9744 \\ \cos .2298 \approx .9737 \end{array} \right\} \text{difference } .0007$$

Now $\frac{11}{29}$ of $.0007$ is the number $11(.0007)/29 \approx .0003$. Therefore the number that is $\frac{11}{29}$ of the way from $.9744$ to $.9737$ is $.9744 - .0003 = .9741$. We conclude that $\cos .228 \approx .9741$.

Note It is customary to write an equal sign $(=)$ instead of the "approximately equal" sign $(\approx)$ that has often been used above. For instance, one writes $\cos .2298 = .9737$ instead of $\cos .2298 \approx .9737$. Since everyone familiar with the subject knows that the values given for trigonometric functions in tables or by calculators are approximations, this inaccuracy causes no difficulty in actual practice.

EXERCISES

A.1. Use the tables at the end of the book to evaluate:
    (a) sin .2153   (c) sin 1.2887   (e) cos .7447
    (b) cos .5149   (d) cos 1.1927   (f) sin .9106

A.2. Use appropriate reference numbers and tables to find:
    (a) cos 2.7838   (c) sin 4.8723   (e) cos 6.1436
    (b) sin 3.5634   (d) cos 4.599

A.3. Use periodicity, reference numbers (if needed), and tables to find:
    (a) sin 7.6796   (c) sin 15.3415   (e) sin 22.6748
    (b) cos 6.833   (d) cos 10.0764

A.4. Use reference numbers (if needed), tables, and appropriate identities to find:
    (a) $\cos(-.2473)$ (c) $\cos(-4.2965)$
    (b) $\sin(-2.0739)$ (d) $\cos(-5.9225)$

B.1. Use tables to evaluate:
    (a) cos 18.9805   (c) $\sin(-3.5488)$ (e) $\sin(-14.9953)$
    (b) sin 1.7394   (d) cos 5.233   (f) $\cos(-24.3124)$

B.2. Use tables and interpolation (if necessary) to find:
    (a) sin .8854   (c) cos 1.1935   (e) sin .34
    (b) cos .5489   (d) sin 1.362   (f) cos 1

B.3. Evaluate:
    (a) sin (sin 2.62)   (b) sin (cos .751)   (c) cos (cos 1.44)

C.1. Compare t, $\sin t$, and $\cos t$ for small values of t by examining the tables. Can you draw any conclusions? Can you explain these conclusions, using the definition of $\sin t$ and $\cos t$?

C.2. (a) Verify that for any line in the trig tables at the end of the book, the sum of the entry in the left-hand RADIANS column and the entry in the right-hand RADIANS column is (approximately) $\pi/2$.
    (b) Use part (a) and the tables to find an identity involving $\sin t$ and $\cos((\pi/2) - t)$.
    (c) Do part (b) for $\cos t$ and $\sin((\pi/2) - t)$.

5. GRAPHS OF THE SINE AND COSINE FUNCTIONS

In this section we first obtain the graphs of the sine and cosine functions. The rest of the section deals with functions that are variations on the sine or cosine functions. Such graphs are often useful since they provide a schematic, visual picture of many periodic phenomena.

THE SINE FUNCTION

The table in Figure 6-55 provides 17 points on the graph of the sine function between $t = 0$, and $t = 2\pi$. All of these calculations can be done by the methods of Section 4, without using tables or calculators.

t	0	$\dfrac{\pi}{6}$	$\dfrac{\pi}{4}$	$\dfrac{\pi}{3}$	$\dfrac{\pi}{2}$	$\dfrac{2\pi}{3}$	$\dfrac{3\pi}{4}$	$\dfrac{5\pi}{6}$	π	$\dfrac{7\pi}{6}$	$\dfrac{5\pi}{4}$	$\dfrac{4\pi}{3}$	$\dfrac{3\pi}{2}$	$\dfrac{5\pi}{3}$	$\dfrac{7\pi}{4}$	$\dfrac{11\pi}{6}$	2π
$\sin t$	0	$\dfrac{1}{2}$	$\dfrac{\sqrt{2}}{2}$	$\dfrac{\sqrt{3}}{2}$	1	$\dfrac{\sqrt{3}}{2}$	$\dfrac{\sqrt{2}}{2}$	$\dfrac{1}{2}$	0	$-\dfrac{1}{2}$	$-\dfrac{\sqrt{2}}{2}$	$-\dfrac{\sqrt{3}}{2}$	-1	$-\dfrac{\sqrt{3}}{2}$	$-\dfrac{\sqrt{2}}{2}$	$-\dfrac{1}{2}$	0

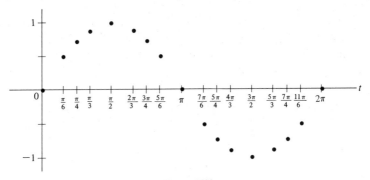

Figure 6-55

These points suggest that between $t = 0$ and $t = 2\pi$ the graph looks like the one in Figure 6-56.

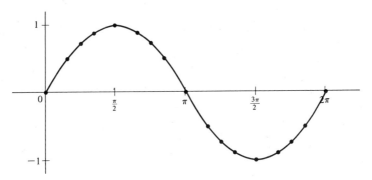

Figure 6-56

To obtain the rest of the graph, we recall that the function $f(t) = \sin t$ has period 2π, that is,

$$f(t + 2\pi) = f(t) \qquad \text{for every number } t$$

(see Section 3 for details). In graphical terms this fact says that for any number t, the points with first coordinates t and $t + 2\pi$ have the *same* second coordinate, namely, $f(t) = f(t + 2\pi)$. In other words, these two points are the same distance from the horizontal axis. For example, see Figure 6-57.

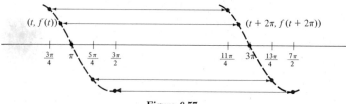

Figure 6-57

Consequently, the shape of the graph above or below a number t on the horizontal axis will be the *same* as the shape of the graph above or below the number $t + 2\pi$. In particular, the portion of the graph between $t = 2\pi$ and $t = 4\pi$ will look exactly the same as the portion between $t = 0$ and $t = 2\pi$. Similarly, the graph repeats the same pattern at intervals of 2π along the entire horizontal axis. Therefore the complete graph of $f(t) = \sin t$ looks like the one shown in Figure 6-58.

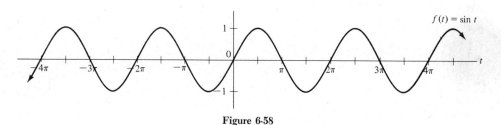

Figure 6-58

Thus the graph of the sine function is a wave continuing in both directions, not becoming extremely large (as do polynomial and exponential functions), but repeating forever. This repetition is a geometric way of visualizing the periodicity identity $\sin (t + 2\pi) = \sin t$.

THE COSINE FUNCTION

To graph the cosine function $g(t) = \cos t$, we begin by plotting some points between $t = 0$ and $t = 2\pi$, and then connecting them by a reasonable curve, as shown in Figure 6-59.

t	0	$\dfrac{\pi}{6}$	$\dfrac{\pi}{4}$	$\dfrac{\pi}{3}$	$\dfrac{\pi}{2}$	$\dfrac{2\pi}{3}$	$\dfrac{3\pi}{4}$	$\dfrac{5\pi}{6}$	π	$\dfrac{7\pi}{6}$	$\dfrac{5\pi}{4}$	$\dfrac{4\pi}{3}$	$\dfrac{3\pi}{2}$	$\dfrac{5\pi}{3}$	$\dfrac{7\pi}{4}$	$\dfrac{11\pi}{6}$	2π
$\cos t$	1	$\dfrac{\sqrt{3}}{2}$	$\dfrac{\sqrt{2}}{2}$	$\dfrac{1}{2}$	0	$-\dfrac{1}{2}$	$-\dfrac{\sqrt{2}}{2}$	$-\dfrac{\sqrt{3}}{2}$	-1	$-\dfrac{\sqrt{3}}{2}$	$-\dfrac{\sqrt{2}}{2}$	$-\dfrac{1}{2}$	0	$\dfrac{1}{2}$	$\dfrac{\sqrt{2}}{2}$	$\dfrac{\sqrt{3}}{2}$	1

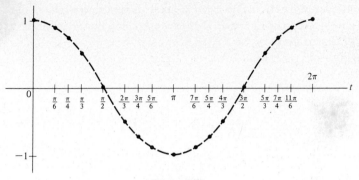

Figure 6-59

The cosine function also has period 2π; that is,

$$\cos(t + 2\pi) = \cos t \qquad \text{for every number } t$$

(see Section 3 for details). Just as with the sine function, the graphical meaning of this fact is that the portion of the graph between $t = 0$ and $t = 2\pi$ repeats at intervals of 2π along the entire horizontal axis. Thus the graph of the cosine function looks like the one shown in Figure 6-60.

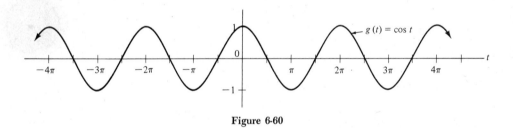

Figure 6-60

THE FUNCTIONS sin *bt* AND cos *bt*

Once you know the graphs of the sine and cosine functions, the graphs of many related functions can easily be obtained, as in the following examples. The underlying theme of these examples is to obtain the required graph with *little or no plotting of points*.

EXAMPLE In order to graph the function $h(t) = \sin 2t$, we first introduce some terminology. We say that the graph of the sine function $f(t) = \sin t$ makes one complete **cycle** from 0 to 2π. This means that as t takes all the values from 0 to 2π, the graph of $f(t) = \sin t$ begins at height 0, rises to its maximum 1, drops to its minimum -1, and returns to its starting height 0 (see the picture on p. 365). Now consider the function $h(t) = \sin 2t$. Note that

As t takes all values from 0 to π, $2t$ takes all values from 0 to 2π

Consequently, as t moves from 0 to π, sin $2t$ takes all the values of the sine function from 0 to 2π. Thus the graph of $h(t) = \sin 2t$ makes one complete cycle from $t = 0$ to $t = \pi$, as shown in Figure 6-61.

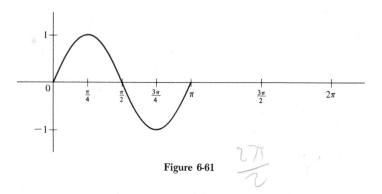

Figure 6-61 $\dfrac{2\pi}{2}$

Now when t runs from π to 2π, then $2t$ runs from 2π to 4π. Hence as t moves from π to 2π, sin $2t$ takes all values of the sine function from 2π to 4π and the graph of $h(t) = \sin 2t$ makes another complete cycle, as shown in Figure 6-62.

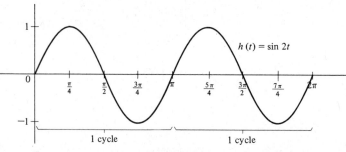

Figure 6-62

Thus the graph of sin $2t$ between 0 and 2π contains two "copies" of the graph of sin t between 0 and 2π. Another way to describe this situation is to say that the graph of sin $2t$ moves up and down twice as fast as the graph of sin t. The entire graph of $h(t) = \sin 2t$ continues this same pattern in both directions along the horizontal axis. The graph shows that the function $h(t) = \sin 2t$ is periodic. Its period is π (the length of one cycle).

EXAMPLE The graph of $k(t) = \cos 3t$ may be obtained by similar means. As t takes all values from 0 to $2\pi/3$, then $3t$ takes all values from 0 to $3(2\pi/3) = 2\pi$. Consequently, as t moves from 0 to $2\pi/3$, $k(t) = \cos 3t$ takes all the values of the cosine function from 0 to 2π. The graph of $k(t) = \cos 3t$ makes one complete cycle from 0 to $2\pi/3$ (just as the cosine function does from 0 to 2π), as shown in Figure 6-63.

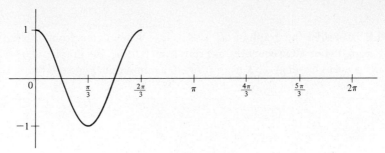

Figure 6-63

Similarly, when t moves from $2\pi/3$ to $4\pi/3$, then $3t$ moves from 2π to 4π and the graph of $k(t) = \cos 3t$ makes another complete cycle, as shown in Figure 6-64.

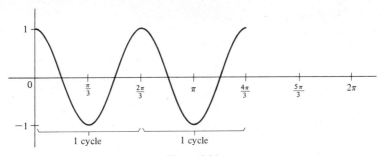

Figure 6-64

Finally, when t runs from $4\pi/3$ to 2π, then $3t$ moves from 4π to 6π and the graph of $k(t) = \cos 3t$ makes another complete cycle, as shown in Figure 6-65.

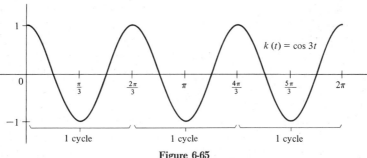

Figure 6-65

Thus the graph of $k(t) = \cos 3t$ between 0 and 2π contains 3 "copies" of the graph of $\cos t$ from 0 to 2π. The entire graph of the function k continues the same pattern in both directions. It shows that $k(t) = \cos 3t$ is a periodic function with period $2\pi/3$ (the length of one cycle).

EXAMPLE   Consider the graph of $h(t) = \sin\left(\frac{1}{2}t\right)$. When t runs from 0 to 2π, then $\frac{1}{2}t$ runs only from 0 to π. Consequently, as t runs from 0 to 2π, the graph of $h(t) = \sin\left(\frac{1}{2}t\right)$ makes *half* a cycle (as does $\sin t$ from 0 to π), as shown in Figure 6-66.

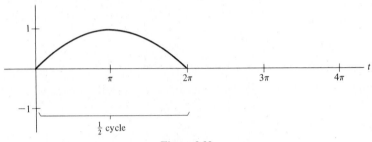

Figure 6-66

As t runs from 2π to 4π, then $\frac{1}{2}t$ runs from π to 2π. Hence as t runs from 2π to 4π, the graph of $h(t) = \sin\frac{1}{2}t$ behaves like the graph of the sine function from π to 2π, as shown in   Figure 6-67.

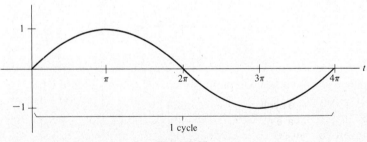

Figure 6-67

The entire graph repeats this pattern in both directions. The graph of $h(t) = \sin\frac{1}{2}t$ contains only *half* a "copy" of the graph of the sine function from 0 to 2π. In other words, the graph of $h(t) = \sin\frac{1}{2}t$ moves up and down only half as fast as that of $f(t) = \sin t$. The graph shows that $h(t) = \sin\frac{1}{2}t$ is a periodic function, with period 4π (the length of one cycle).

THE FUNCTIONS *A* sin *bt* AND *A* cos *bt*

The remainder of this section makes use of Section 6 of Chapter 3 (*New Graphs from Old*).

EXAMPLE   In order to graph $g(t) = 7 \cos 3t$, we first observe that this function is just a constant multiple of the function $k(t) = \cos 3t$:

$$g(t) = 7 \cos 3t = 7(\cos 3t) = 7k(t)$$

We know what the graph of $k(t) = \cos 3t$ looks like (it's on p. 369 above). As we saw on page 150, the fact that $g(t) = 7k(t)$ means that the graph of $g(t)$ is just the graph of $k(t)$, stretched away from the horizontal axis by a factor of 7, as shown in Figure 6-68.

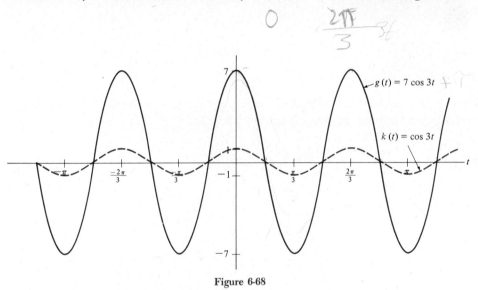

Figure 6-68

Observe that a complete cycle of either the graph of g or the graph of k has length $2\pi/3$ and that both functions are periodic with period $2\pi/3$. What is the principal difference between the two graphs? The graph of $k(t) = \cos 3t$ oscillates between 1 and -1. The maximum (vertical) distance from the graph of k to the horizontal axis is 1 unit. But the graph of $g(t) = 7 \cos 3t$ oscillates between 7 and -7. The maximum distance from the graph of g to the horizontal axis is 7 units. We describe this situation by saying that the function g (or its graph) has **amplitude** 7. Similarly, the amplitude of $k(t) = \cos 3t$ is 1.

EXAMPLE   The graph of $f(t) = -\frac{3}{2} \sin 2t$ may be obtained from the graph of $h(t) = \sin 2t$ (on p. 368) in two steps. As we saw on page 150, the graph of $\frac{3}{2} \sin 2t$ is just the graph of $h(t) = \sin 2t$ stretched away from the horizontal axis by a factor of $\frac{3}{2}$, as shown in Figure 6-69.

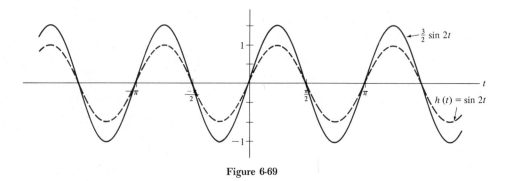

Figure 6-69

As we saw on page 151, the graph of $f(t) = -\frac{3}{2}\sin 2t$ is just the graph of $\frac{3}{2}\sin 2t$ reflected in the horizontal axis (Figure 6-70).

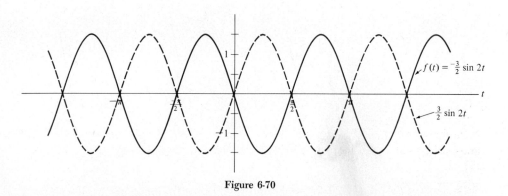

Figure 6-70

Both the function $h(t) = \sin 2t$ and $f(t) = -\frac{3}{2}\sin 2t$ have period π (the length of one cycle). The graph f oscillates between $\frac{3}{2}$ and $-\frac{3}{2}$. So the function $f(t) = -\frac{3}{2}\sin 2t$ has amplitude $\frac{3}{2}$.

EXAMPLE In order to graph $f(t) = \frac{1}{3}\sin 4t$, we first graph the function $g(t) = \sin 4t$, using the techniques introduced earlier. As t runs from 0 to $\pi/2$, then $4t$ runs from 0 to 2π. Thus the graph of $g(t) = \sin 4t$ makes one complete cycle from 0 to $\pi/2$ and four cycles from $-\pi$ to π, as shown in Figure 6-71.

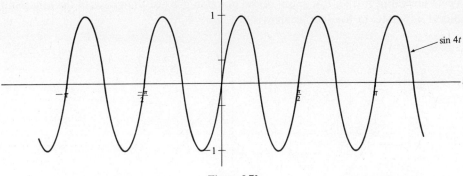

Figure 6-71

As we saw on page 150, the graph of $f(t) = \frac{1}{3}\sin 4t = \frac{1}{3}g(t)$ is just the graph of $g(t) = \sin 4t$ shrunk toward the horizontal axis by a factor of $\frac{1}{3}$ (Figure 6-72).

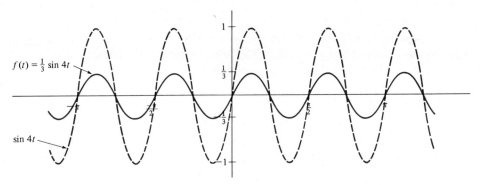

Figure 6-72

Note that the function $f(t) = \frac{1}{3}\sin 4t$ has period $\pi/2$ (the length of one cycle). The graph of f oscillates between $\frac{1}{3}$ and $-\frac{1}{3}$ and thus has amplitude $\frac{1}{3}$.

THE FUNCTIONS $A \sin (bt + c)$ AND $A \cos (bt + c)$

In order to understand the following examples, you should review the meaning of functional notation such as $g(t + 5)$ or $f(t - (\pi/4))$. For example, if $f(t) = t^2 + \sin t$, then $f(t - (\pi/4)) = (t - (\pi/4))^2 + \sin (t - (\pi/4))$. Other examples and a discussion of the graphs of $f(t)$, $f(t - (\pi/4))$, and so on, are in Section 6 of Chapter 3.

EXAMPLE   In order to graph the function $g(t) = \sin (t + (\pi/2))$, we first compare it to the sine function $f(t) = \sin t$. We see that

$$g(t) = \sin\left(t + \frac{\pi}{2}\right) = f\left(t + \frac{\pi}{2}\right)$$

As we saw on page 153, the graph of the function g is just the graph of the function f shifted $\pi/2$ units to the left, as shown in Figure 6-73.

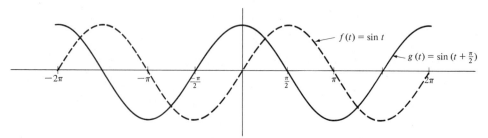

Figure 6-73

The graph shows that the function g has the same amplitude and period as the sine function f (namely, amplitude 1 and period 2π). The fact that the graph of g was obtained from the graph of f by shifting $\pi/2$ units to the left (the negative direction) is usually expressed by saying that the **phase shift** of the function g is $-\pi/2$.

EXAMPLE   Graph the function given by $h(t) = \sin(4t + (\pi/2))$. Based on the preceding example you might correctly expect that the graph of h is obtained by horizontally shifting the graph of $f(t) = \sin 4t$. But be careful. Many students have a tendency in this situation to say: we can get the graph of $h(t) = \sin(4t + (\pi/2))$ by shifting the graph of $f(t) = \sin 4t$ to the left by $\pi/2$ units. But **this is wrong!** To see why it's wrong, remember that the function whose graph is the graph of f shifted $\pi/2$ units to the left is the function $f(t + (\pi/2))$, as explained on page 153. Now the rule of the function $f(t + (\pi/2))$ is obtained by replacing t by $t + (\pi/2)$ in the rule of the function f:

$$f(t) = \sin 4t$$

$$\text{replace } t \text{ by } t + \frac{\pi}{2}$$

$$f\left(t + \frac{\pi}{2}\right) = \sin 4\left(t + \frac{\pi}{2}\right).$$

Next evaluate both the function $f(t + (\pi/2))$ and the function h at $t = 0$:

$$f\left(0 + \frac{\pi}{2}\right) = \sin 4\left(0 + \frac{\pi}{2}\right) = \sin 4\left(\frac{\pi}{2}\right) = \sin 2\pi = 0$$

$$h(0) = \sin\left(4 \cdot 0 + \frac{\pi}{2}\right) = \sin \frac{\pi}{2} = 1$$

Therefore the functions $f(t + (\pi/2))$ and h are different functions and cannot possibly have the same graph.

The *correct way* to deal with this problem is to use some simple factoring and write $4t + (\pi/2)$ as $4(t + (\pi/8))$. Now consider the function $f(t + (\pi/8))$:

$$f(t) = \sin 4t$$

replace t by $t + \dfrac{\pi}{8}$

$$f\left(t + \frac{\pi}{8}\right) = \sin 4\left(t + \frac{\pi}{8}\right)$$

Multiplying out the rule for the function $f(t + (\pi/8))$ shows that

$$f\left(t + \frac{\pi}{8}\right) = \sin 4\left(t + \frac{\pi}{8}\right) = \sin\left(4t + \frac{\pi}{2}\right) = h(t)$$

Therefore the function h is the same as the function $f(t + (\pi/8))$. As we saw on page 153, this means that the graph of h is just the graph of $f(t) = \sin 4t$ shifted $\pi/8$ units to the left. We know what the graph of f looks like (it's on p. 373), so the graph of h looks like the one in Figure 6-74.

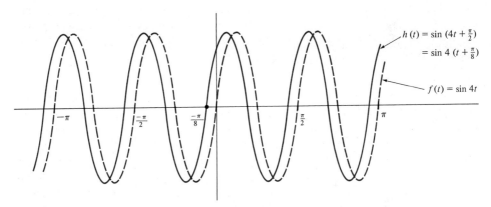

Figure 6-74

The graph shows that the function h has the same amplitude (1) and period $(\pi/2)$ as the graph of f. Since the graph of h is obtained from the graph of f by shifting $\pi/8$ units to the left (negative direction), the phase shift of h is $-\pi/8$.

EXAMPLE In order to graph $f(t) = 7\cos(3t - \pi)$, we use the same factorization technique as above and write $f(t)$ as

$$f(t) = 7\cos(3t - \pi) = 7\cos 3\left(t - \frac{\pi}{3}\right)$$

Let g be the function given by $g(t) = 7\cos 3t$. Then

$$f(t) = 7\cos 3\left(t - \frac{\pi}{3}\right) = g\left(t - \frac{\pi}{3}\right)$$

Therefore, as we saw on page 154, the graph of f is the graph of g shifted $\pi/3$ units to the *right*. So we can use the graph of g (see Figure 6-68 on p. 371) to obtain the graph of f, as shown in Figure 6-75.

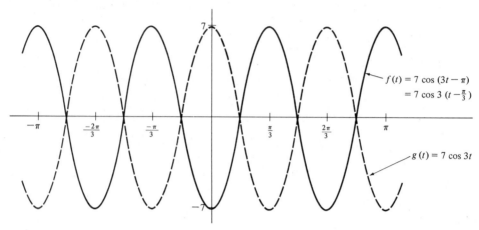

Figure 6-75

The graph shows that the function f has the same amplitude (7) and the same period $(2\pi/3)$ as the function g. Since the graph of g was shifted $\pi/3$ units to the *right* (positive direction) to obtain the graph of f, the phase shift is $+\pi/3$.

A SUMMARY AND A SHORTCUT

All of the functions discussed in this section have rules of one of these two forms:

$$f(t) = A \sin{(bt + c)} \qquad \text{or} \qquad f(t) = A \cos{(bt + c)}$$

where A, b, c are constants with A nonzero and b positive. For example, the function $f(t) = 7 \cos{(3t - \pi)}$ has this form with $A = 7$, $b = 3$, and $c = -\pi$. Similarly, the function $h(t) = \sin 2t$ has this form with $A = 1$, $b = 2$, and $c = 0$.

The graphs of all such functions have the same general "wave" shape as do the graphs of the sine and cosine function. The factors that distinguish the graph of one such function from another are the amplitude, period, and phase shift. You may have noticed that these three factors are closely related to the constants A, b, and c in the rule of the function. For instance, the example on page 375 shows that the function $f(t) = 7 \cos{(3t - \pi)}$ has amplitude 7, period $2\pi/3$, and phase shift $\pi/3$. All of this information can be seen in the rule of the function:

$$f(t) = 7 \cos{(3t - \pi)} = 7 \cos 3 \left(t - \frac{\pi}{3} \right)$$

amplitude 7

period $\dfrac{2\pi}{3}$

phase shift $+\dfrac{\pi}{3} = -\left(-\dfrac{\pi}{3} \right)$

Here is a similar analysis of some of the other functions discussed above:

function: $h(t) = \sin\left(4t + \dfrac{\pi}{2}\right) = 1\,\sin 4\left(t + \dfrac{\pi}{8}\right)$ (see pp. 374–375)

amplitude 1

period $\dfrac{\pi}{2} = \dfrac{2\pi}{4}$

phase shift $-\dfrac{\pi}{8}$

function: $f(t) = -\tfrac{3}{2}\sin 2t = -\tfrac{3}{2}\sin 2(t + 0)$ (see pp. 371–372)

amplitude $\tfrac{3}{2} = \left|-\tfrac{3}{2}\right|$

period $\pi = \dfrac{2\pi}{2}$

phase shift 0

These examples show how to find the amplitude, period, and phase shift in the general case:

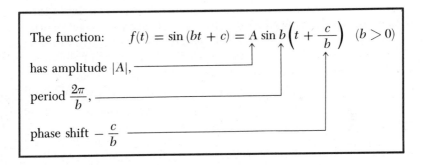

The function: $f(t) = \sin(bt + c) = A\,\sin b\left(t + \dfrac{c}{b}\right)$ $(b > 0)$

has amplitude $|A|$,

period $\dfrac{2\pi}{b}$,

phase shift $-\dfrac{c}{b}$

A similar analysis applies to $f(t) = A\cos(bt + c)$. Once you know the amplitude, period, and phase shift, you can immediately visualize the general shape of the graph, without plotting any points.

EXAMPLE If f is the function given by $f(t) = 4\sin(2t - (\pi/3))$, then $A = 4$, $b = 2$, and $c = -\pi/3$, and $f(t) = 4\sin(2t - (\pi/3)) = 4\sin 2(t - (\pi/6))$. Therefore f has

amplitude $|A| = |4| = 4$, period $\dfrac{2\pi}{b} = \dfrac{2\pi}{2} = \pi$, phase shift $-\dfrac{c}{b} = -\dfrac{(-\pi/3)}{2} = \dfrac{\pi}{6}$

Consequently, the graph of f oscillates between 4 and -4 (amplitude 4), completes one cycle every π units (period π), and begins a cycle at $t = \pi/6$ (phase shift $\pi/6$).

APPLICATIONS

The sine and cosine functions, or variations on them, can be used to describe many different phenomena. Here is a very simplified example to give you some idea how this is done.

EXAMPLE   Suppose that a weight hanging from a spring is set in motion by an upward push, as shown in Figure 6-76.

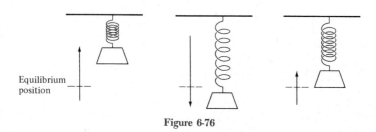

Equilibrium
position

Figure 6-76

For now we shall consider only an idealized situation in which the spring has perfect elasticity, and friction, air resistance, etc. are ignored. Thus the motion of the weight is uniform and regular. Suppose that the maximum distance the weight moves in either direction from its equilibrium point is 8 cm. Suppose that the length of time it takes for the weight to move upward from equilibrium to its maximum height, then downward to its maximum depth below equilibrium and upward to equilibrium again, is 5 sec. Under these circumstances, the motion of the weight can be described mathematically as follows.

Let $h(t)$ denote the height (or depth) of the weight above (or below) its equilibrium position at time t. Measure height in positive numbers and depth in negative numbers. Clearly, the height $h(t)$ is a function of the time t. The value of $h(t)$ is 0 when $t = 0$. As t takes increasing values from 0 to 5, $h(t)$ increases to 8, then decreases to 0, becomes negative and decreases to -8, then increases to 0 again. Consequently, the graph of h must look like the one shown in Figure 6-77.

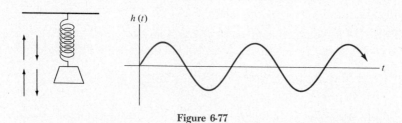

$h(t)$

Figure 6-77

In this graph the time is graphed on the horizontal axis and the height on the vertical axis. The up-and-down motion of the *spring* is fixed in one place, but the up-and-down motion of the *graph* moves to the right as t increases. Since the weight moves up and down uniformly, the "waves" in the graph of $h(t)$ are all the same size.

Thus the graph of h has the same general shape as the graphs presented earlier in this section. The amplitude of the function h is 8, its period is 5, and it has 0 phase shift. But the examples above show how to find a function with exactly these properties. We know that the function $A \sin (bt + c)$ has amplitude $|A|$, period $2\pi/b$, and phase shift $-c/b$. We need only choose the constants A, b, c so that

$$A = 8, \qquad \frac{2\pi}{b} = 5, \qquad -\frac{c}{b} = 0$$

or equivalently,

$$A = 8, \qquad b = \frac{2\pi}{5}, \qquad c = 0$$

Therefore the motion of the moving spring can be described by the function

$$h(t) = A \sin (bt + c) = 8 \sin \left(\frac{2\pi}{5}t + 0\right) = 8 \sin \frac{2\pi t}{5}$$

Motion that can be described by a function of the form $f(t) = A \sin (bt + c)$ or $f(t) = A \cos (bt + c)$ is called **simple harmonic motion.** Many kinds of physical motion are simple harmonic motions. Other periodic phenomena, such as sound waves, are more complicated to describe. Their graphs consist of waves of varying amplitude. Such graphs are discussed in Exercises C.1 and C.2.

EXERCISES

Directions When asked to graph a function, do so with a minimum of point plotting (preferably none), by using the graphs of the sine and cosine functions and the techniques discussed in the section.

A.1. Use the graph of the sine function in Figure 6-58 on page 366 to answer these questions. For *how many* values of t (with $0 \le t \le 2\pi$) is
(a) $\sin t = \frac{1}{2}$?   (c) $\sin t = 1$?   (e) $\sin t = 0$?
(b) $\sin t = .8$?   (d) $\sin t = .6$?   (f) $\sin t = -1$?

A.2. Sketch the graphs of these functions between $t = 0$ and $t = 4\pi$.
(a) $f(t) = \cos 2t$ (c) $h(t) = \sin 3t$ (e) $p(t) = \cos (4t/3)$
(b) $g(t) = \sin 5t$ (d) $k(t) = \sin (t/4)$ (f) $q(t) = \sin (3t/4)$

A.3. Sketch the graphs of these functions.
(a) $f(t) = 5 \sin t$ (c) $k(t) = -3 \sin t$ (e) $p(t) = -\frac{1}{2} \sin 2t$
(b) $h(t) = 4 \cos 2t$ (d) $y(t) = -2 \cos 3t$ (f) $q(t) = \frac{2}{3} \cos \frac{3}{2}t$

B.1. Use the identities $\sin(-t) = -\sin t$ and $\cos(-t) = \cos t$, together with the methods discussed in the text, to graph each of these functions:
(a) $f(t) = 3 \sin(-t)$ (c) $h(t) = \sin(-t/2)$
(b) $g(t) = \cos(-2t)$ (d) $k(t) = \frac{1}{3} \cos(-3t)$

B.2. (a) What is the period of the function $f(t) = \sin 2\pi t$?
(b) For what values of t (with $0 \le t \le 2\pi$) is $f(t) = 0$?
(c) For what values of t (with $0 \le t \le 2\pi$) is $f(t) = 1$? or $f(t) = -1$?
(d) Graph the function f between 0 and 2π.

B.3. In addition to the ones presented in this section, there are several other techniques for obtaining new graphs from old ones (see pp. 146, 147, 151). Use these techniques and the graphs of the sine and cosine functions to graph each of these functions:
(a) $h(t) = \sin t + 2$ [*not* $\sin(t + 2)$]   (c) $q(t) = -\cos t$
(b) $k(t) = \cos t - 3$ (d) $k(t) = 2 - \sin t$

B.4. Sketch the graphs of these functions:
(a) $g(t) = \cos(t + (\pi/3))$ (c) $p(t) = -3 \sin(3t - \pi)$
(b) $k(t) = \sin(t - (2\pi/3))$ (d) $h(t) = -\frac{1}{2} \cos(2t - (3\pi/2))$

B.5. State the values of t (with $0 \le t \le 2\pi$) at which the given function reaches its maximum value. For example, $f(t) = 5 \sin t$ reaches its maximum value of 5 at $t = \pi/2$.
(a) $f(t) = 3 \cos 2t$ (c) $k(t) = 3 \cos(\frac{1}{2}t)$
(b) $h(t) = -5 \sin t$ (d) $f(t) = 2 \sin(2t - \pi)$

B.6. State the amplitude, period, and phase shift of each of these functions. Do not graph the functions.
(a) $g(t) = 3 \sin(2t - \pi)$ (e) $f(t) = \sin 2\pi t$
(b) $h(t) = -4 \cos(3t - (\pi/6))$ (f) $k(t) = \cos(2\pi t/3)$
(c) $q(t) = -7 \sin(7t + \frac{1}{7})$ (g) $p(t) = 6 \cos(3\pi t + 1)$
(d) $g(t) = 97 \cos(14t + 5)$

B.7. Give the rule of a periodic function with the given numbers as amplitude, period, and phase shift (in this order):
(a) 3, $\pi/4$, $\pi/5$ (c) $\frac{2}{3}$, 1, 0   (e) 7, $\frac{5}{3}$, $-\pi/2$
(b) 2, 3, 0   (d) $\frac{4}{5}$, 2, 3   (f) 19, 4, -5

B.8. For each graph in Figure 6-78 state the rule of a function which has the given graph. For example, (*) in Figure 6-78 is the graph of $f(t) = 3 \sin 2t$.

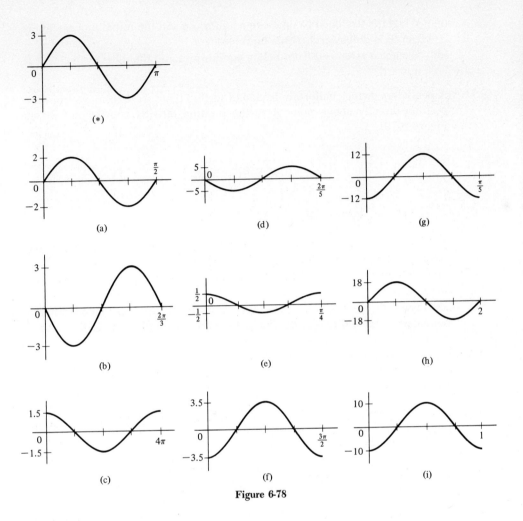

Figure 6-78

B.9. Sketch the graphs of $f(t) = |\sin t|$ and $g(t) = |\cos t|$. (*Hint:* see Exercise C.1 on p. 156).

B.10. Suppose there is a weight hanging from a spring (under the same idealized conditions described in the example on p. 378). The weight is given a push to start it moving. At any time t, let $h(t)$ be the height (or depth) of the weight above (or below) its equilibrium point. Assume that the maximum distance the weight moves in either direction from the equilibrium point is 6 cm and that it moves through a complete cycle every 4 sec. Express $h(t)$ in terms of the sine or cosine function under the stated conditions.

(a) Initial push is upward from the equilibrium.
(b) Initial push is *downward* from the equilibrium point. (*Hint:* what does the graph of $A \sin bt$ look like when $A < 0$?)

(c)   Weight is stretched 6 cm above equilibrium and the initial movement (at $t = 0$) is downward. (*Hint:* think cosine.)

(d)   Weight is stretched 6 cm below equilibrium and the initial movement is upward.

B.11.   A pendulum swings uniformly back and forth, taking 2 sec to move from the position directly above point A to the position directly above point B (see Figure 6-79).

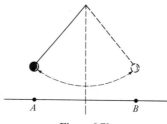

Figure 6-79

The distance from A to B is 20 cm. Let $d(t)$ be the horizontal distance from the pendulum to the (dotted) center line at time t sec (with distances to the right of the line measured by positive numbers and distances to the left by negative ones). Assume that the pendulum is on the centerline at time $t = 0$ and moving to the right. Proceed as in the example on p. 378 and find the rule of the function $d(t)$ and sketch its graph for the first 8 sec.

C.1. $f(t) = \sin t + \sin 2t$ is a periodic function whose graph consists of "waves" of different sizes. One way to graph f and similar functions is to use the method of **addition of ordinates** as follows:

(a)   Verify that $f(t) = f(t + 2\pi)$ for every t. Hence we need only graph f from $t = 0$ to $t = 2\pi$, then repeat this pattern to the right and left.

(b)   Graph the functions $\sin t$ and $\sin 2t$ on the same coordinate axes from 0 to 2π. Use dotted lines for the graphs.

(c)   On the same coordinate axes, plot some points on the graph of f by *visually inspecting* the graphs of $\sin t$ and $\sin 2t$. For example, see Figure 6-80.

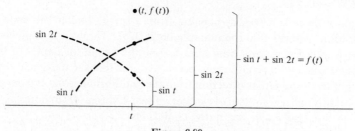

Figure 6-80

Some care must be used when one of $\sin t$ or $\sin 2t$ is negative.

(d)   Use the points plotted in part (c) to sketch the graph of f from 0 to 2π.

C.2. Use addition of ordinates (explained in Exercise C.1) to sketch the graphs of these functions.

(a) $f(t) = 2 \cos t - \sin 2t$

(b) $f(t) = \cos t + \sin 2t$

(c) $g(t) = 2 \sin t + \sin 2t$

(d) $h(t) = 2 \cos t + 3 \sin t$

(e) $h(t) = 3 \cos t - \sin 2t$

C.3. Figure 6-81 is a diagram of a merry-go-round which is turning counterclockwise at a constant rate, making 2 revolutions in 1 min. On the merry-go-round are horses A, B, C, and D at 4 m from the center and horses E, F, and G at 8 m from the center. There is a function $a(t)$, which gives the distance the horse A is from the y-axis (this is the x-coordinate of the position A is in) as a function of time t (measured in minutes). Similarly, $b(t)$ gives the x-coordinate for B as a function of time, and so on. Assume the diagram shows the situation at time $t = 0$.

(a) Which of the following functions does $a(t)$ equal: $4 \cos t$, $4 \cos \pi t$, $4 \cos 2t$, $4 \cos 2\pi t$, $4 \cos (\frac{1}{2}t)$, $4 \cos ((\pi/2)t)$, $4 \cos 4\pi t$? Explain.

(b) Describe the functions $b(t)$, $c(t)$, $d(t)$, and so on using the cosine function:

$$b(t) = \underline{\hspace{1cm}}, \qquad c(t) = \underline{\hspace{1cm}}, \qquad d(t) = \underline{\hspace{1cm}},$$

$$e(t) = \underline{\hspace{1cm}}, \qquad f(t) = \underline{\hspace{1cm}}, \qquad g(t) = \underline{\hspace{1cm}}.$$

(c) Suppose the x-coordinate of a horse S is given by the function $4 \cos (4\pi t - (5\pi/6))$ and the x-coordinate of another horse T is given by $8 \cos (4\pi t + (\pi/3))$. Where are these horses located in relation to the rest of the horses? Mark the positions of T and S at $t = 0$ into Figure 6-81.

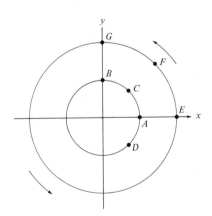

Figure 6-81

C.4. (a) Let f be the function given by $f(t) = 2^{-t} \sin t = (1/2^t) \sin t$. Use the basic properties of the sine and exponential functions to show that for every number t

$$-\frac{1}{2^t} \le f(t) \le \frac{1}{2^t}$$

(b)   Graph the function f from $t = -2\pi$ to $t = 2\pi$. [*Hint:* by part (a) the entire graph lies between the graphs of $g(t) = -1/2^t$ and $h(t) = 1/2^t$; these graphs are in Section 2 of Chapter 5. Subject to this condition, the graph of f moves up and down, making waves of varying size.]

6. THE TANGENT FUNCTION

In addition to the sine and cosine functions, there are four other trigonometric functions. The most important of these remaining four is the tangent function, whose basic properties are presented in this section.

The value of the tangent function at a number t is denoted by the symbol $\tan t$. The tangent function is defined to be the quotient function of the sine function by the cosine function. Thus

---

The tangent function is given by the rule:
$$\tan t = \frac{\sin t}{\cos t}$$

The domain of the tangent function consists of all real numbers t for which $\tan t = \sin t / \cos t$ is a well-defined real number. The number $\tan t$ will be defined whenever t is a number with $\cos t \neq 0$. The graph of the cosine function in Figure 6-60 on p. 367 shows that the only values of t for which $\cos t = 0$ are the odd multiples of $\pi/2$. Hence

---

The domain of the tangent function consists of all real numbers
except $\pm\dfrac{\pi}{2}, \pm\dfrac{3\pi}{2}, \pm\dfrac{5\pi}{2}, \ldots$

Since the tangent function is defined in terms of sine and cosine, its values are frequently difficult to compute directly. However, accurate approximations are readily available either from calculators* or tables. (The use of trigonometric tables for the tangent function is explained in the DO IT YOURSELF! segment at the end of this section.) Using a calculator we see, for example, that

$$\tan(-11) \approx 225.95, \qquad \tan 1.6 \approx -34.23, \qquad \tan 36 \approx 7.75$$

---

* When using a calculator to evaluate $\tan t$, be sure it is operating in the *radian mode* for the present. The tangent function in terms of degrees is discussed in Section 9.

When t is a multiple of either $\pi/4$ or $\pi/6$, then the number $\tan t$ can be computed exactly and directly. Just use one of the methods presented in Section 4 to find $\sin t$ and $\cos t$, then take the quotient.

You should memorize the values of the sine, cosine, and tangent functions for frequently used values of t. The following chart summarizes these values. All the computations for sine and cosine were done in Section 4. A dash indicates that the tangent function is not defined at that number.

t	0	$\dfrac{\pi}{6}$	$\dfrac{\pi}{4}$	$\dfrac{\pi}{3}$	$\dfrac{\pi}{2}$	π	$\dfrac{3\pi}{2}$	2π
$\sin t$	0	$\dfrac{1}{2}$	$\dfrac{\sqrt{2}}{2}$	$\dfrac{\sqrt{3}}{2}$	1	0	-1	0
$\cos t$	1	$\dfrac{\sqrt{3}}{2}$	$\dfrac{\sqrt{2}}{2}$	$\dfrac{1}{2}$	0	-1	0	1
$\tan t$	0	$\dfrac{1}{\sqrt{3}} = \dfrac{\sqrt{3}}{3}$	1	$\sqrt{3}$	$-$	0	$-$	0

ALTERNATE DESCRIPTIONS OF THE TANGENT FUNCTION

Since the point-in-the-plane method can be used to find the values of both the sine and cosine function (as in Section 4), the same is true for the tangent function:

> Let t be any real number in the domain of the tangent function. Let (x, y) be any point (except the origin) on the terminal side of an angle of t radians in standard position. Then
> $$\tan t = \frac{y}{x}$$

The proof of this statement is immediate. For as we saw on page 350, $\sin t = y/r$ and $\cos t = x/r$, where $r = \sqrt{x^2 + y^2}$. Therefore

$$\tan t = \frac{\sin t}{\cos t} = \frac{y/r}{x/r} = \frac{y}{x}$$

EXAMPLE Suppose the point $(-6, 15)$ lies on the terminal side of an angle of t radians in standard position, as shown in Figure 6-82. Then we use $(-6, 15)$ as the point (x, y) and obtain:

$$\tan t = \frac{y}{x} = \frac{15}{-6} = -\frac{5}{2} = -2.5$$

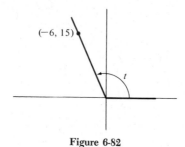

Figure 6-82

When t lies strictly between 0 and $\pi/2$, the number $\tan t$ can be described in terms of right triangles:

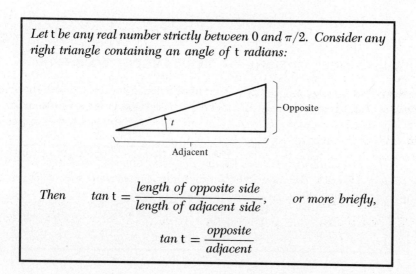

Let t be any real number strictly between 0 and $\pi/2$. Consider any right triangle containing an angle of t radians:

Then $\tan t = \dfrac{\text{length of opposite side}}{\text{length of adjacent side}}$, *or more briefly,*

$$\tan t = \frac{\text{opposite}}{\text{adjacent}}$$

The proof of this fact follows directly from the right triangle description of sine and cosine. As we saw on page 354,

$$\sin t = \frac{\text{opposite}}{\text{hypotenuse}} \quad \text{and} \quad \cos t = \frac{\text{adjacent}}{\text{hypotenuse}}$$

Therefore

$$\tan t = \frac{\sin t}{\cos t} = \frac{\dfrac{\text{opposite}}{\text{hypotenuse}}}{\dfrac{\text{adjacent}}{\text{hypotenuse}}} = \frac{\text{opposite}}{\text{adjacent}}$$

EXAMPLE   The right-triangle method can be used to find $\tan (\pi/3)$ and $\tan (\pi/6)$ without direct reference to sine and cosine. Consider a right triangle whose other angles measure $\pi/3$ radians (60°) and $\pi/6$ radians (30°) and whose hypotenuse has length 2. As explained in the *Geometry Review* (p. 583), the other sides have lengths 1 and $\sqrt{3}$, as shown in Figure 6-83.

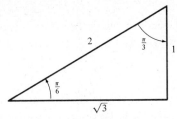

Figure 6-83

Using this picture we see that

$$\tan \frac{\pi}{3} = \frac{\text{opposite}}{\text{adjacent}} = \frac{\sqrt{3}}{1} = \sqrt{3} \quad \text{and} \quad \tan \frac{\pi}{6} = \frac{\text{opposite}}{\text{adjacent}} = \frac{1}{\sqrt{3}} = \frac{\sqrt{3}}{3}$$

Finally, there is a description of the tangent function that has no analogue for sine and cosine. To obtain it, let t be any real number in the domain of the tangent function and construct an angle of t radians in standard position. Since t is not an odd multiple of $\pm\pi/2$, the terminal side of this angle lies on a nonvertical straight line through the origin. We shall compute the *slope* of this line.

As we saw in Chapter 2, the slope of the line containing the points (x_1, y_1) and (x_2, y_2) is the number $\dfrac{y_2 - y_1}{x_2 - x_1}$. To apply this formula here we need two points on the terminal side of the angle. Use the origin $(0, 0)$ for one of them. Let the second point be the point P where the terminal side meets the unit circle, as shown in Figure 6-84.

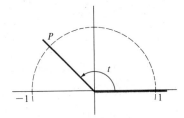

Figure 6-84

This point P was used to define sine and cosine and its coordinates are $(\cos t, \sin t)$. Using the points $(0, 0)$ and $(\cos t, \sin t)$ in the slope formula shows that

$$\text{slope} = \frac{\sin t - 0}{\cos t - 0} = \frac{\sin t}{\cos t} = \tan t$$

In summary,

> For any number t in the domain of the tangent function, the number tan t is the slope of the terminal side of an angle of t radians in standard position.

EXAMPLE Find the first-quadrant angle between the positive x-axis and the line $y = 4x$. Let t denote the radian measure of this angle (Figure 6-85).

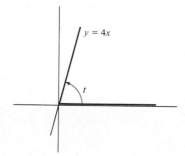

Figure 6-85

Recall that the slope of the line $y = 4x$ is 4 (see p. 61). Therefore $\tan t = 4$. Using the table at the end of the book, we find that $\tan t \approx 4$ when $t \approx 1.3265$ radians ($\approx 76°$).

BASIC PROPERTIES OF THE TANGENT FUNCTION

Since both sine and cosine are periodic functions, we would expect the tangent function to be periodic as well. But the *period* of the tangent function (the shortest interval over which it repeats its values) is actually smaller than 2π:

> The tangent function is periodic with period π; in particular,
>
> $$\tan (t + \pi) = \tan t$$
>
> for every number t in the domain of the tangent function.

The proof that $\tan (t + \pi) = \tan t$ rests on two identities proved in Section 3, namely, $\sin (t + \pi) = -\sin t$ and $\cos (t + \pi) = -\cos t$. Using these identities and the definition of the tangent function yields:

$$\tan(t + \pi) = \frac{\sin(t + \pi)}{\cos(t + \pi)} = \frac{-\sin t}{-\cos t} = \frac{\sin t}{\cos t} = \tan t$$

To complete the proof we must verify that π is the smallest positive number with this property. See Exercise C.1 for details.

EXAMPLE $\tan(4\pi/3) = \tan((\pi + 3\pi)/3) = \tan((\pi/3) + \pi) = \tan(\pi/3) = \sqrt{3}$. Similarly, $\tan 7\pi/6 = \tan((\pi/6) + \pi) = \tan(\pi/6) = \sqrt{3}/3$.

EXAMPLE Repeated application of the periodicity identity shows that $\tan t = \tan(t + \pi) = \tan((t + \pi) + \pi) = \tan(t + 2\pi)$, and similarly, $\tan(t + 2\pi) = \tan(t + 3\pi) = \tan(t + 4\pi)$, and so on. Thus $\tan t = \tan(t + k\pi)$ for $k = 1, 2, 3, \ldots$.

It is sometimes helpful to know for what values of t the function $h(t) = \tan t$ is positive or negative. The corresponding information for sine and cosine was compiled in Section 3. So it's easy to check the signs of $\tan t = \sin t/\cos t$. Here is a summary of the results for all three functions:

SIGNS OF TRIGONOMETRIC FUNCTIONS

$\frac{\pi}{2} < t < \pi$	$0 < t < \frac{\pi}{2}$
sin t +	sin t +
cos t −	cos t +
tan t −	tan t +
$\pi < t < \frac{3\pi}{2}$	$\frac{3\pi}{2} < t < 2\pi$
sin t −	sin t −
cos t −	cos t +
tan t +	tan t −

The various identities developed for the sine and cosine functions in Section 3 lead to analogous identities for the tangent function. We have already seen one example of this, the periodicity identity $\tan(t + \pi) = \tan t$. Here is another:

$$\tan(-t) = -\tan t$$

for every number t in the domain of the tangent function.

The proof of this fact uses the corresponding identities for sine and cosine:

$$\tan\left(-t\right) = \frac{\sin\left(-t\right)}{\cos\left(-t\right)} = \frac{-\sin t}{\cos t} = -\tan t$$

EXAMPLES $\tan\left(-\pi/4\right) = -\tan\left(\pi/4\right) = -1$. Combining this identity with the periodicity identity, we have $\tan\left(2\pi/3\right) = \tan\left(-(\pi/3) + \pi\right) = \tan\left(-\pi/3\right) = -\tan\left(\pi/3\right) = -\sqrt{3}$.

GRAPH OF THE TANGENT FUNCTION

Since the tangent function $h(t) = \tan t$ has period π, we need only determine its graph over a single interval of length π. The rest of the graph will repeat this pattern in both directions. So we shall begin by finding the graph of $h(t) = \tan t$ between $t = -\pi/2$ and $t = \pi/2$ (an interval of length π). By using facts developed above, we can construct a table of values for the tangent function and plot the corresponding points, as shown in Figure 6-86.

t	$h(t) = \tan t$
$-\pi/3$	$-\sqrt{3} \approx -1.7$
$-\pi/4$	-1
$-\pi/6$	$-\sqrt{3}/3 \approx -.58$
0	0
$\pi/6$	$\sqrt{3}/3 \approx .58$
$\pi/4$	1
$\pi/3$	$\sqrt{3} \approx 1.7$

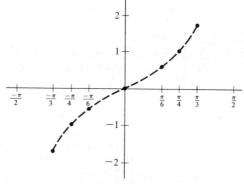

Figure 6-86

These points suggest the general shape of the graph. But a bit more analysis is needed to determine the behavior of the graph near $t = -\pi/2$ and $t = \pi/2$, values for which the function h is not defined.

Suppose t is a number very close to $\pi/2$ and a bit smaller than $\pi/2$. Then the graphs of the sine and cosine functions (p. 366 and 367) show that $\sin t$ is a number very close to 1 and $\cos t$ is a very small positive number. As t takes values closer and closer to $\pi/2$, $\sin t$ stays close to 1 and $\cos t$ gets smaller and smaller. Consequently, the number $\tan t = \sin t/\cos t$ takes larger and larger positive values.° Thus when t is close to $\pi/2$ and smaller than $\pi/2$, the graph of $h(t) = \tan t$ rises extremely sharply, as shown below.

° If this statement is not clear to you, reread the discussion of the Big-Little Principles on page 216.

A similar analysis works near $t = -\pi/2$ where $\tan t$ is negative and shows that the graph must be as shown in Figure 6-87.

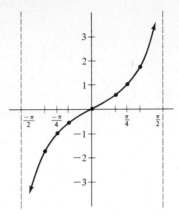

Figure 6-87

Although the graph gets closer and closer to the vertical lines through $t = \pi/2$ and $t = -\pi/2$, it never crosses them since the function $h(t) = \tan t$ is not defined at $\pm\pi/2$. These vertical lines are called **vertical asymptotes** of the graph.

Now that we have the graph between $t = -\pi/2$ and $t = \pi/2$, we can use the periodicity of the function $h(t) = \tan t$ to obtain the entire graph, as shown in Figure 6-88. Note that it has vertical asymptotes at $t = \pm\pi/2, \pm3\pi/2, \pm5\pi/2, \ldots$.

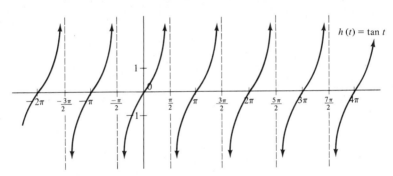

Figure 6-88

EXERCISES

A.1. Use the identities proved in the text and the table of values on page 385 to determine:

(a) $\tan(-\pi/6)$ (c) $\tan(5\pi/6)$ (e) $\tan(11\pi/4)$

(b) $\tan(-\pi/4)$ (d) $\tan(-5\pi/4)$ (f) $\tan(-17\pi/6)$

A.2. If the terminal side of an angle of t radians in standard position contains the given point, find $\tan t$.

(a) $\left(-\dfrac{3}{5}, -\dfrac{4}{5}\right)$

(c) $\left(\dfrac{2}{\sqrt{5}}, \dfrac{1}{\sqrt{5}}\right)$

(b) $\left(\dfrac{3}{\sqrt{10}}, \dfrac{-1}{\sqrt{10}}\right)$

(d) $\left(\dfrac{-3}{\sqrt{70}}, \dfrac{7}{\sqrt{70}}\right)$

A.3. Referring to Figure 6-89, find $\sin t$, $\cos t$, and $\tan t$, when t is the radian measure of angle A and the sides of the right triangle have the indicated lengths.

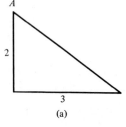
(a)

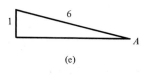
(e)

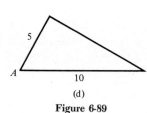
(c)

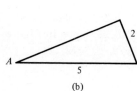
(b)

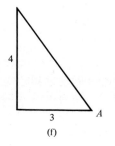

(d)

(f)

Figure 6-89

A.4. Use the graph of the tangent function to determine for *how many* values of t (with $-\pi/2 < t < \pi/2$) is

(a) $\tan t = 5$?

(c) $\tan t = c$ (c a fixed real number)?

(b) $\tan t = -57$?

B.1. Use the identities proved in this section and Section 3 to verify that each of these identities is valid.

(a) $\tan(\pi - t) = -\tan t$

(c) $1 + \tan^2 t = 1/\cos^2 t$

(b) $\tan(2\pi - t) = -\tan t$

B.2. Find all real numbers t such that $\tan t = 1$.

B.3. Find the length of side h in each of the right triangles in Figure 6-90 (all angles measured in radians).

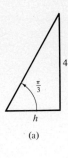

(a)

$$\left(\text{Hint: } \sqrt{3} = \tan\frac{\pi}{3} = \frac{\text{opposite}}{\text{adjacent}} = \frac{4}{h}. \text{ Solve } \sqrt{3} = \frac{4}{h} \text{ for } h.\right)$$

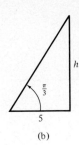

(b)

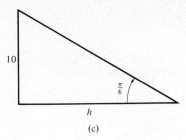

(c)

Figure 6-90

B.4. Use 3.14 for π and the table below to find the length of side h in each of the triangles in Figure 6-91 (all angles measured in radians). See Exercise B.3(a) for hints.

t	.5	.55	.6	.65	.7	.75	.8	.85	.9
$\tan t$	.55	.61	.68	.76	.84	.93	1.03	1.14	1.26

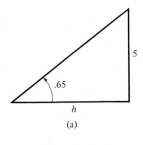

(a)

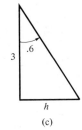

(c)

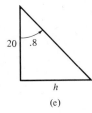

(e)

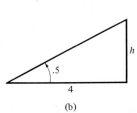

(b)

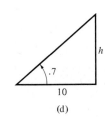

(d)

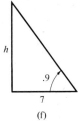

(f)

Figure 6-91

B.5. Use the table in Exercise B.4 to find the radian measure t of the indicated angle of the right triangle in Figure 6-92.

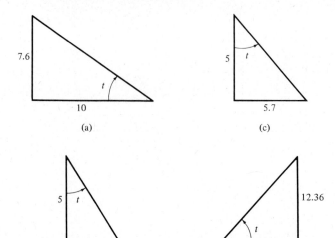

(a)

(c)

(b)

(d)

Figure 6-92

B.6. Find the equation of the straight line containing the terminal side of an angle in standard position whose measure is:
(a) $\pi/6$ radians   (c) $-\pi/3$ radians   (e) $7\pi/3$ radians
(b) $\pi/4$ radians   (d) $-3\pi/4$ radians   (f) $-19\pi/4$ radians

B.7. Use the table at the end of the book and your knowledge of the tangent function to find the approximate radian measure of the first-quadrant angle made by the positive x-axis and the line with equation
(a) $y = 11x$ (b) $y = 1.5x$ (c) $y = 1.4x$

B.8. Prove that each of these equations is an identity, that is, that it is valid for every t for which both sides are defined.

(a) $\tan t \cos t = \sin t$

(c) $\dfrac{\cos t}{1 - \sin t} = \dfrac{1}{\cos t} + \tan t$

(b) $\dfrac{1}{\cos t} - \cos t = \sin t \tan t$

(d) $\dfrac{1 - \tan^2 t}{1 + \tan^2 t} = 1 - 2 \sin^2 t$

C.1. Let k be a positive number and consider the property:

$$\tan (t + k) = \tan t \text{ for every } t \text{ in the domain of the tangent function}$$

(a) If k is any number with this property, show that $\tan k = 0$.
(b) Verify that there is no positive number t with $t < \pi$ and $\tan t = 0$.
(c) Use (a) and (b) to show that $k = \pi$ is the smallest positive number with the property. Hence the tangent function has period π.

DO IT YOURSELF!

TRIGONOMETRIC TABLES FOR THE TANGENT FUNCTION

The procedure for using tables to find $\tan t$ is essentially the same as the one given on pages 359–363 for the sine and cosine functions. But it can be simplified a bit since the tangent function has period π instead of 2π.

(i) If $0 \leq t < \pi/2$, then use the tables directly, as with sine and cosine. Use interpolation if necessary.

(ii) If $\pi/2 < t \leq \pi$, then find the reference number for t, namely, the number $k = \pi - t$. It will lie between 0 and $\pi/2$. Use the tables (with interpolation if necessary) to find the value of $\tan k$. As shown in Exercise B.1(a) above, $\tan t = -\tan k$.

(iii) If $t > \pi$, then subtract π from t as many times as necessary to obtain a number u between 0 and π. Since the tangent function has period π, $\tan t = \tan u$. Evaluate $\tan u$ as in step (i) or (ii) above.

(iv) If $t < 0$, then $-t$ is positive and $\tan(-t) = -\tan t$, or equivalently, $\tan t = -\tan(-t)$. Evaluate $\tan(-t)$ as in steps (i), (ii), or (iii) above.

EXAMPLE Find $\tan 2.5773$. Since $t = 2.5773$ lies between $\pi/2$ and π, we proceed as in step (ii) above and find the reference number $k = \pi - t \approx 3.1416 - 2.5773 = .5643$. Since .5643 lies between 0 and $\pi/2$, we can use the tables to find that $\tan .5643 \approx .6330$.   Therefore $\tan 2.5773 = \tan t = -\tan k = -\tan .5643 \approx -.6330$.

EXAMPLE Find $\tan 18.2853$. Since $18.2853 > \pi$, we begin subtracting π from 18.2853 as explained in step (iii) above. We find that $18.2853 - 5\pi \approx 18.2853 - 15.708 = 2.5773$, a number between 0 and π. Now using periodicity and the preceding example we have $\tan 18.2853 = \tan(18.2853 - 5\pi) \approx \tan 2.5773 \approx -.6330$.

EXAMPLE Find $\tan(-18.2853)$. Here $t = -18.2853$ so that $-t = 18.2853$. Proceeding as in step (iv) and using the preceding example, we have $\tan(-18.2853) = \tan t = -\tan(-t) = -\tan 18.2853 \approx -(-.6330) = .6330$.

7. OTHER TRIGONOMETRIC FUNCTIONS

The basic properties of the three remaining trigonometric functions are presented here. The discussion is rather brief since most of it parallels the presentation of the sine, cosine, and tangent functions above.

The three other trigonometric functions are defined in terms of sine and cosine as follows:

Name of Function	Value of Function at t Is Denoted	Rule of Function
cotangent	cot t	$\cot t = \dfrac{\cos t}{\sin t}$
secant	sec t	$\sec t = \dfrac{1}{\cos t}$
cosecant	csc t	$\csc t = \dfrac{1}{\sin t}$

Each of these functions is the quotient of functions that are defined for all real numbers. Consequently, the domain of each consists of all real numbers for which the denominator is nonzero. The graphs of sine and cosine in Section 5 show that $\sin t = 0$ only when t is an integer multiple of π and $\cos t = 0$ only when t is an odd integer multiple of $\pi/2$. Consequently,

> The domain of both the cotangent and cosecant functions consists of all real numbers except $0, \pm\pi, \pm2\pi, \pm3\pi, \ldots$.
>
> The domain of the secant function consists of all real numbers except $\pm\pi/2, \pm3\pi/2, \pm5\pi/2, \pm7\pi/2, \ldots$.

As you might expect, direct calculation of the values of these functions for a specific number t is often difficult. The cases where it is not (namely, when t is a multiple of $\pi/4$ or $\pi/6$) are discussed below. In other cases, you must use approximations obtained from a calculator° or trigonometric tables. For example, a calculator shows that

$$\cot(-5) \approx .2958, \qquad \sec 7 \approx 1.3264, \qquad \csc 18.5 \approx -2.9199,$$
$$\cot 15.7 \approx -125.57, \qquad \sec(-8) \approx -6.8729, \qquad \csc(-22) \approx 112.98$$

The use of trigonometric tables for the secant and cosecant functions is exactly the same as with sine and cosine, as explained on pages 359–363. The procedure for the cotangent function is the same as for the tangent function, as explained on page 395.

° Most calculators do not have keys for cotangent, secant, and cosecant. You have to use the sine, cosine, or tangent keys and find the appropriate quotient. Remember to use your calculator in *radian mode* for the present.

ALTERNATE DESCRIPTIONS

The various descriptions of the sine and cosine have analogues for these three new functions:

> *Let t be a real number and* (x, y) *any point (except the origin) on the terminal side of an angle of* t *radians in standard position. Let* $r = \sqrt{x^2 + y^2}$. *Then*
>
> $$\cot t = \frac{x}{y}, \qquad \sec t = \frac{r}{x}, \qquad \csc t = \frac{r}{y}$$
>
> *for each number* t *in the domain of the given function.*

These statements are proved by using the similar descriptions of sine and cosine (given on p. 350). For instance,

$$\cot t = \frac{\cos t}{\sin t} = \frac{x/r}{y/r} = \frac{x}{y}$$

The proofs of the other statements are similar.

EXAMPLE Evaluate all six trigonometric functions at $t = 3\pi/4$. The terminal side of an angle of $3\pi/4$ radians in standard position lies on the line $y = -x$, as shown in Figure 6-93.

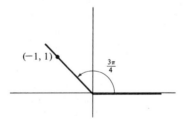

Figure 6-93

We shall use the point $(-1, 1)$ on this line to compute the function values. In this case $r = \sqrt{x^2 + y^2} = \sqrt{(-1)^2 + 1^2} = \sqrt{2}$. Therefore

$$\sin \frac{3\pi}{4} = \frac{y}{r} = \frac{1}{\sqrt{2}} = \frac{\sqrt{2}}{2}, \qquad \cos \frac{3\pi}{4} = \frac{x}{r} = \frac{-1}{\sqrt{2}} = \frac{-\sqrt{2}}{2},$$

$$\tan \frac{3\pi}{4} = \frac{y}{x} = \frac{1}{-1} = -1, \qquad \csc \frac{3\pi}{4} = \frac{r}{y} = \frac{\sqrt{2}}{1} = \sqrt{2},$$

$$\sec \frac{3\pi}{4} = \frac{r}{x} = \frac{\sqrt{2}}{-1} = -\sqrt{2}, \qquad \cot \frac{3\pi}{4} = \frac{x}{y} = \frac{-1}{1} = -1$$

Right triangles can be used to find cot t, sec t, and csc t when $0 < t < \pi/2$.

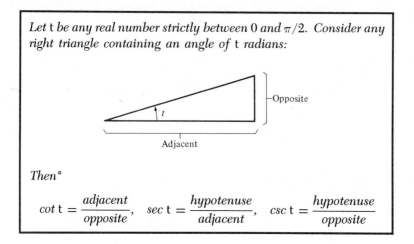

Let t *be any real number strictly between* 0 *and* $\pi/2$. *Consider any right triangle containing an angle of* t *radians:*

Then°

$$\cot t = \frac{adjacent}{opposite}, \quad \sec t = \frac{hypotenuse}{adjacent}, \quad \csc t = \frac{hypotenuse}{opposite}$$

All three of these statements are consequences of the right triangle description of sine and cosine. For instance,

$$\sec t = \frac{1}{\cos t} = \frac{1}{\dfrac{adjacent}{hypotenuse}} = \frac{hypotenuse}{adjacent}$$

EXAMPLE In order to evaluate all six trigonometric functions at $t = \pi/3$, we construct a right triangle containing an angle of $\pi/3$ radians and having hypotenuse of length 2. As explained in the *Geometry Review* (p. 583), the other sides of this triangle have lengths 1 and $\sqrt{3}$ as shown in Figure 6-94.

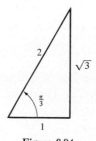

Figure 6-94

° Here, as above, "opposite" is short for "length of the side opposite the angle of t radians," and similarly for "adjacent."

We can now read off

$$\sin \frac{\pi}{3} = \frac{\text{opposite}}{\text{hypotenuse}} = \frac{\sqrt{3}}{2}, \qquad \cos \frac{\pi}{3} = \frac{\text{adjacent}}{\text{hypotenuse}} = \frac{1}{2},$$

$$\tan \frac{\pi}{3} = \frac{\text{opposite}}{\text{adjacent}} = \frac{\sqrt{3}}{1} = \sqrt{3}, \qquad \cot \frac{\pi}{3} = \frac{\text{adjacent}}{\text{opposite}} = \frac{1}{\sqrt{3}} = \frac{\sqrt{3}}{3},$$

$$\sec \frac{\pi}{3} = \frac{\text{hypotenuse}}{\text{adjacent}} = \frac{2}{1} = 2, \qquad \csc \frac{\pi}{3} = \frac{\text{hypotenuse}}{\text{opposite}} = \frac{2}{\sqrt{3}} = \frac{2\sqrt{3}}{3}.$$

BASIC PROPERTIES

We begin by noting that the relationship between the cotangent and tangent functions is essentially the same as that between secant and cosine or between cosecant and sine:

> *The cotangent and tangent functions are reciprocals; that is,*
>
> $$\cot t = \frac{1}{\tan t}, \qquad \tan t = \frac{1}{\cot t}$$
>
> *for every number t in the domain of both functions.*

The proof of these facts comes directly from the various definitions; for instance,

$$\cot t = \frac{\cos t}{\sin t} = \frac{1}{\dfrac{\sin t}{\cos t}} = \frac{1}{\tan t}$$

Like the trigonometric functions introduced earlier, these new functions are also periodic:

> *The secant and cosecant functions are periodic with period 2π and the cotangent function is periodic with period π. In particular,*
>
> $$\sec(t + 2\pi) = \sec t, \qquad \csc(t + 2\pi) = \csc t,$$
> $$\cot(t + \pi) = \cot t$$
>
> *for every number t in the domain of the given function.*

The proof of these statements uses the fact that each of these functions is the reciprocal of a function whose period is known. For instance,

$$\csc(t + 2\pi) = \frac{1}{\sin(t + 2\pi)} = \frac{1}{\sin t} = \csc t$$

$$\cot(t + \pi) = \frac{1}{\tan(t + \pi)} = \frac{1}{\tan t} = \cot t$$

The other details are left as an exercise.

Many of the identities proved earlier in this chapter lead to identities involving the secant, cosecant, and cotangent functions. Here are the two most important ones:

$$1 + tan^2\, t = sec^2\, t \qquad and \qquad 1 + cot^2\, t = csc^2\, t$$

for every number t *in the domain of both functions.*

The proof of these identities uses the definitions of the functions and the Pythagorean Identity $\sin^2 t + \cos^2 t = 1$:

$$1 + \tan^2 t = 1 + \frac{\sin^2 t}{\cos^2 t} = \frac{\cos^2 t + \sin^2 t}{\cos^2 t} = \frac{1}{\cos^2 t} = \left(\frac{1}{\cos t}\right)^2 = \sec^2 t$$

The second identity is proved similarly. These two identities are also called **Pythagorean Identities.** These and the other identities proved so far have a number of applications.

EXAMPLE   Simplify the expression $\dfrac{30 \cos^3 t \sin t}{6 \sin^2 t \cos t}$, assuming $\sin t \neq 0$, $\cos t \neq 0$:

$$\frac{30 \cos^3 t \sin t}{6 \sin^2 t \cos t} = \frac{5 \cos^3 \sin t}{\cos t \sin^2 t} = \frac{5 \cos^2 t}{\sin t} = 5\,\frac{\cos t}{\sin t}\,\cos t = 5 \cot t \cos t$$

EXAMPLE   Assume $\cos t \neq 0$ and simplify $\cos^2 t + \cos^2 t \tan^2 t$:

$$\cos^2 t + \cos^2 t \tan^2 t = \cos^2 t\,(1 + \tan^2 t) = \cos^2 t \sec^2 t = \cos^2 t \cdot \frac{1}{\cos^2 t} = 1$$

EXAMPLE   If $\tan t = \frac{3}{4}$ and $\sin t < 0$, find $\cot t$, $\cos t$, $\sin t$, $\sec t$, $\csc t$. First we have $\cot t = 1/\tan t = 1/\frac{3}{4} = \frac{4}{3}$. Next we use the Pythagorean Identity to obtain:

$$\sec^2 t = 1 + \tan^2 t = 1 + \left(\frac{3}{4}\right)^2 = 1 + \frac{9}{16} = \frac{25}{16}$$

$$\sec t = \pm \sqrt{\frac{25}{16}} = \pm \frac{5}{4}$$

$$\frac{1}{\cos t} = \pm \frac{5}{4} \qquad \text{or equivalently} \qquad \cos t = \pm \frac{4}{5}$$

Since sin t is given as negative and tan $t = \sin t/\cos t$ is positive, cos t must be negative. Hence $\cos t = -\frac{4}{5}$. Consequently,

$$\frac{3}{4} = \tan t = \frac{\sin t}{\cos t} = \frac{\sin t}{(-4/5)}$$

so that

$$\sin t = \left(-\frac{4}{5}\right)\left(\frac{3}{4}\right) = -\frac{3}{5}$$

Therefore

$$\sec t = \frac{1}{\cos t} = \frac{1}{(-4/5)} = -\frac{5}{4} \quad \text{and} \quad \csc t = \frac{1}{\sin t} = \frac{1}{(-3/5)} = -\frac{5}{3}$$

GRAPHS

Cotangent Function   Since the cotangent function $f(t) = \cot t$ has period π, we need only determine the graph between $t = 0$ and $t = \pi$. The rest of the graph will just repeat this pattern. Using facts developed above, we can construct a table of values and plot the corresponding points, as shown in Figure 6-95.

t	$f(t) = \cot t$
$\pi/6$	$\sqrt{3} \approx 1.7$
$\pi/4$	1
$\pi/3$	$\sqrt{3}/3 \approx .58$
$\pi/2$	0
$2\pi/3$	$-\sqrt{3}/3 \approx -.58$
$3\pi/4$	-1
$5\pi/6$	$-\sqrt{3} \approx -1.7$

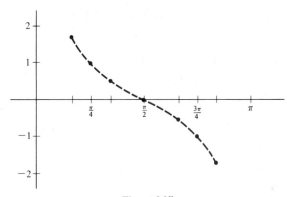

Figure 6-95

These points suggest the general shape of the graph, except near $t = 0$ and $t = \pi$ where cot t is not defined. An analysis very similar to the one used when graphing the tangent function shows that the vertical lines through $t = 0$ and $t = \pi$ are vertical asymptotes of the graph. Using this fact and periodicity, we see that the entire graph of $f(t) = \cot t$ looks like the one shown in Figure 6-96.

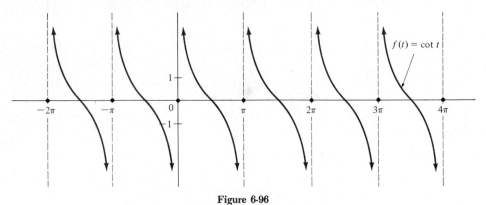

Figure 6-96

The graph gets closer and closer to, but never crosses, the vertical asymptotes through $t = 0, \pm\pi, \pm2\pi, \pm3\pi.\ldots$

Secant Function Since $g(t) = \sec t$ has period 2π, we begin by plotting some points between $t = 0$ and $t = 2\pi$, as shown in Figure 6-97. Note that the function is not defined at $t = \pi/2$ and $t = 3\pi/2$.

t	0	$\dfrac{\pi}{6}$	$\dfrac{\pi}{4}$	$\dfrac{\pi}{3}$	$\dfrac{\pi}{2}$	$\dfrac{2\pi}{3}$	$\dfrac{3\pi}{4}$	$\dfrac{5\pi}{6}$	π	$\dfrac{7\pi}{6}$	$\dfrac{5\pi}{4}$	$\dfrac{4\pi}{3}$	$\dfrac{3\pi}{2}$	$\dfrac{7\pi}{4}$	2π
$\sec t$	1	$\dfrac{2\sqrt{3}}{3}$	$\sqrt{2}$	2	—	-2	$-\sqrt{2}$	$-\dfrac{2\sqrt{3}}{3}$	-1	$-\dfrac{2\sqrt{3}}{3}$	$-\sqrt{2}$	-2	—	$\sqrt{2}$	1

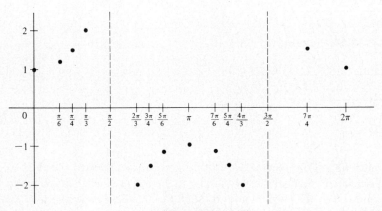

Figure 6-97

In order to determine the behavior of the graph near $t = \pi/2$ and $t = 3\pi/2$, we recall that $\sec t = 1/\cos t$. The graph of the cosine function on page 367 shows that:

When t is very close to $\dfrac{\pi}{2}$ or $\dfrac{3\pi}{2}$,   then $\cos t$ is very close to 0

Consequently,

When t is very close to $\dfrac{\pi}{2}$ or $\dfrac{3\pi}{2}$,   then

$\sec t = \dfrac{1}{\cos t}$ is extremely large in absolute value°

It follows that the graph near $\pi/2$ and $3\pi/2$ moves very quickly away from the x-axis, as shown in Figure 6-98.

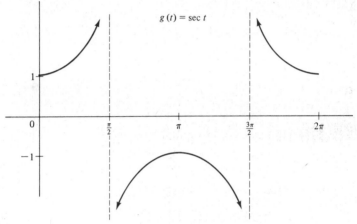

Figure 6-98

Since the secant function has period 2π, the graph repeats this pattern both left and right. The entire graph looks like the one in Figure 6-99.

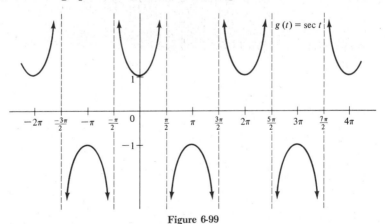

Figure 6-99

° It's the Big-Little Principle again; see page 216.

The vertical lines through $\pm\pi/2$, $\pm3\pi/2$, $\pm5\pi/2$, ... are vertical asymptotes of the graph. Note that the connected pieces of this graph (for instance, the piece between $-\pi/2$ and $\pi/2$) are *not* parabolas. For each such piece is completely contained between two vertical asymptotes, whereas parabolas extend indefinitely to the left and right.

Cosecant Function Using techniques similar to those used for the secant function (but with sine in place of cosine), we obtain the graph of the cosecant function, as shown in Figure 6-100. Note that it has vertical asymptotes at $t = 0$, $\pm\pi$, $\pm2\pi$, $\pm3\pi$, ... and that the connected pieces of the graph are not parabolas.

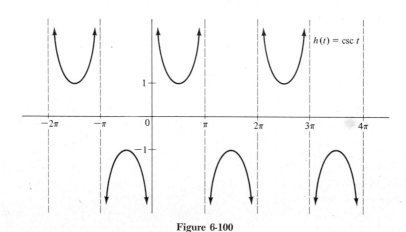

Figure 6-100

EXERCISES

A.1. In what quadrant does the terminal side of an angle of t radians (in standard position) lie if

    (a) $\cos t > 0$ and $\sin t < 0$ (d) $\csc t < 0$ and $\sec t > 0$
    (b) $\sin t < 0$ and $\tan t > 0$ (e) $\sec t > 0$ and $\cot t < 0$
    (c) $\sec t < 0$ and $\cot t < 0$ (f) $\sin t > 0$ and $\sec t < 0$

A.2. Evaluate all six trigonometric functions at t, where the given point lies on the terminal side of an angle of t radians in standard position:

    (a) $(3, 4)$ (d) $(-2, -3)$ (g) $(\sqrt{2},\ \sqrt{3})$
    (b) $(0, 6)$ (e) $(-\frac{1}{5}, 1)$ (h) $(1 + \sqrt{2}, 3)$
    (c) $(-5, 12)$ (f) $(\frac{4}{5}, -\frac{3}{5})$ (i) $(-2\sqrt{3},\ \sqrt{3})$

A.3. Evaluate all six trigonometric functions at the number t which is the radian measure of the indicated angle of the given right triangle, as shown in Figure 6-101.

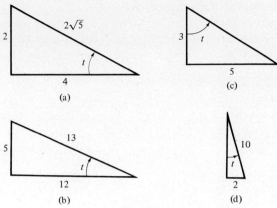

Figure 6-101

A.4. Examine the graph of the given function and state the values of t for which the function is positive and those for which it is negative when $0 \le t \le 2\pi$.
(a) $f(t) = \cot t$ (b) $f(t) = \sec t$ (c) $f(t) = \csc t$

A.5. Perform the indicated operations, then simplify your answers by using appropriate definitions and identities.
(a) $\tan t\,(\cos t - \csc t)$ (d) $(\sin t - \csc t)^2$
(b) $\cos t \sin t\,(\csc t + \sec t)$ (e) $(\cot t - \tan t)(\cot^2 t + 1 + \tan^2 t)$
(c) $(1 + \cot t)^2$ (f) $(\sin t + \csc t)(\sin^2 t + \csc^2 t - 1)$

B.1. Evaluate all six trigonometric functions at $t =$
(a) $\dfrac{4\pi}{3}$ (b) $\dfrac{7\pi}{4}$ (c) $-\dfrac{7\pi}{6}$

B.2. Fill in the missing entries of this table:

t	0	$\pi/6$	$\pi/4$	$\pi/3$	$\pi/2$	$2\pi/3$	$3\pi/4$	$5\pi/6$	π	$3\pi/2$
$\sin t$										
$\cos t$										
$\tan t$					—					—
$\cot t$	—								—	
$\sec t$					—					—
$\csc t$	—								—	

B.3. Factor and simplify these expressions:

(a) $\sec t \csc t - \csc^2 t$ (d) $4 \sec^2 t + 8 \sec t + 4$

(b) $\tan^2 t - \cot^2 t$ (e) $\cos^3 t - \sec^3 t$

(c) $\tan^4 t - \sec^4 t$ (f) $\csc^4 t + 4 \csc^2 t - 5$

B.4. Simplify these expressions. Assume all denominators are nonzero.

(a) $\dfrac{\cos^2 t \sin t}{\sin^2 t \cos t}$ (d) $\dfrac{\sec^2 t \csc t}{\csc^2 t \sec t}$

(b) $\dfrac{\sec^2 t + 2 \sec t + 1}{\sec t}$ (e) $(2 + \sqrt{\tan t})(2 - \sqrt{\tan t})$

(c) $\dfrac{4 \tan t \sec t + 2 \sec t}{6 \sin t \sec t + 2 \sec t}$ (f) $\dfrac{6 \tan t \sin t - 3 \sin t}{9 \sin^2 t + 3 \sin t}$

B.5. Use 3.14 for π and the table below to find the length of side h in each of the given triangles in Figure 6-102 (all angles measured in radians).

t	.2	.3	.5	.7	.9	1	1.1	1.3	1.5
cot t	4.93	3.23	1.83	1.19	.79	.64	.51	.28	.07
sec t	1.02	1.05	1.14	1.31	1.61	1.85	2.2	3.74	14.14
csc t	5.03	3.38	2.09	1.55	1.28	1.19	1.12	1.04	1.01

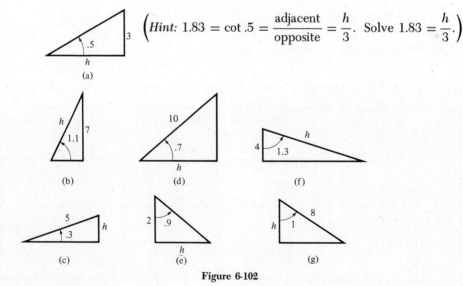

$\left(Hint:\ 1.83 = \cot .5 = \dfrac{\text{adjacent}}{\text{opposite}} = \dfrac{h}{3}.\ \text{Solve } 1.83 = \dfrac{h}{3}.\right)$

Figure 6-102

B.6. Use the table in Exercise B.5 to find the approximate radian measure t of the indicated angle of the right triangle, as shown in Figure 6-103.

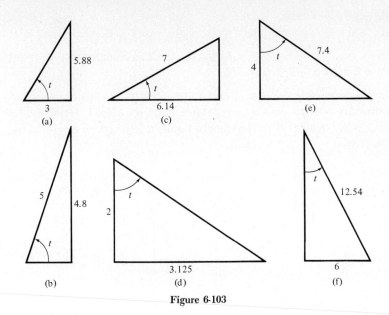

Figure 6-103

B.7. Prove the identity $1 + \cot^2 t = \csc^2 t$.

B.8. Prove each of these identities.

(a) $\cot(-t) = -\cot t$ (b) $\sec(-t) = \sec t$ (c) $\csc(-t) = -\csc t$

B.9. Use identities proved in Section 3 to prove these identities.

(a) $\sec(\pi - t) = -\sec t$ (d) $\csc(t + \pi) = -\csc t$
(b) $\sec(t + \pi) = -\sec t$ (e) $\csc(\pi - t) = \csc t$
(c) $\cot(\pi - t) = -\cot t$

B.10. Use the definitions and identities proved in this section to prove each of these identities.

(a) $\cos t \csc t = \cot t$ (d) $\dfrac{\tan^2 t}{\sec t + 1} = \sec t - 1$

(b) $\sin t (\csc t - \sin t) = \cos^2 t$ (e) $\tan t \sin t + \cos t = \sec t$

(c) $\dfrac{\sec t}{\csc t} = \tan t$ (f) $\dfrac{\sec t}{\tan t} = \csc t$

B.11. Prove each of these identities.

(a) $\sec^4 t - \sec^2 t = \dfrac{\sin^2 t}{\cos^4 t}$ (c) $\cot t + \tan t = \sec t \csc t$

(b) $(-\sin^2 t)(1 - \csc^2 t) = \cos^2 t$ (d) $1 - \tan^4 t = 2 \sec^2 t - \sec^4 t$

B.12. Find the values of all six trigonometric functions at t if the given conditions are true.

(a) $\cos t = -\frac{1}{2}$ and $\sin t > 0$ (*Hint:* $\sin^2 t + \cos^2 t = 1$)

(b) $\cos t = 0$ and $\sin t = 1$ (d) $\sec t = -\frac{13}{5}$ and $\tan t < 0$

(c) $\sin t = -\frac{2}{3}$ and $\sec t > 0$ (e) $\csc t = 8$ and $\cos t < 0$

B.13. Show graphically that the equation $\sec t = t$ has infinitely many solutions, but none between $-\pi/2$ and $\pi/2$.

C.1. In the diagram of the unit circle in Figure 6-104. find six line segments whose respective lengths are $\sin t$, $\cos t$, $\tan t$, $\cot t$, $\sec t$, $\csc t$. (*Hint:* $\sin t =$ length CA. Why? Note that OC has length 1 and various right triangles in the picture are similar.)

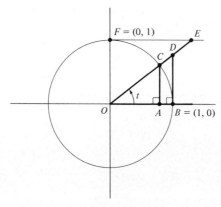

Figure 6-104

8. TRIGONOMETRIC IDENTITIES

A number of trigonometric identities have been introduced earlier in this chapter. They are reviewed here and several important new identities are discussed. Trigonometric identities are used in a variety of ways. They can be used to:

(i) Perform numerical calculations.

(ii) Simplify complicated expressions.

(iii) Express the rule of a function in a different form, not necessarily simpler, but more convenient for the purpose at hand.

(iv) Prove other widely used identities.

(v) Prove identities needed only in a particular proof or calculation.

All of these uses are illustrated in this section.

Since there are so many ways of using identities, it shouldn't be surprising that there are no hard-and-fast rules or set procedures for handling every possible situation. Some general guidelines and suggestions are presented below. But the only way to become comfortable with identities and proficient in using them is to read many examples carefully, making sure you understand them, and to work as many exercises as you possibly can. You may find it helpful to reread the subsection "The Algebra of Identities" on page 341, especially the warning at the end of that section.

NEW NOTATION

The functions $f(t) = \sin t$ and $g(t) = \cos t$ are defined in terms of the y-coordinate and x-coordinate of a certain point. Until now, the variable t has been used for trigonometric functions in order to avoid confusion with the x's and y's that occur in their definition. But by this time you should be used to these functions. Consequently, the letter used to denote the variable of a trigonometric function shouldn't make any difference. So from now on we shall follow the custom used in most calculus texts and often use the letter x (or sometimes y) to denote the variable in various trigonometric functions. Once again, we emphasize that this is just a *notational* change, not a mathematical one.

$$f(t) = \sin t, \qquad f(x) = \sin x, \qquad f(y) = \sin y$$

all denote *exactly the same* function. Analogous statements apply to the other trigonometric functions.

IDENTITIES

An equation involving a variable is said to be an **identity** if it is valid for every value of the variable for which every term of the equation is a well-defined real number. For example, both of the equations

$$(x - 3)^2 = x^2 - 6x + 9 \qquad \text{and} \qquad \sin^2 x + \cos^2 x = 1$$

are true for every real number x, and hence are identities. In the equations

$$1 + \cot^2 x = \csc^2 x \qquad \text{and} \qquad \cot x = \frac{1}{\tan x}$$

there are many values of x for which some of the terms are not defined. For instance, if $x = \pi/2$, then $\tan x$ is not defined. When $x = \pi$, neither $\cot x$ nor $\csc x$ is defined and $\tan x = 0$, so that $1/\tan x$ is not defined. But as we saw in the last section, for all numbers x for which $\cot x$, $\csc x$, and $\tan x$ *are* defined, these equations are true. Thus these equations are also examples of identities.

The phrases "prove the identity" and "verify the identity" are often used. Both mean "prove that the given equation is an identity."

REVIEW OF BASIC IDENTITIES

Here is a summary of the basic identities proved earlier in this chapter. Those involving sine and cosine were proved in Section 3, those involving tangent in Section 6, and all others in Section 7. *You should memorize all the basic identities in the first five boxes below.* To begin with, certain definitions can be considered as identities:

BASIC DEFINITIONS

$$tan \ x = \frac{sin \ x}{cos \ x}, \qquad cot \ x = \frac{cos \ x}{sin \ x},$$

$$sec \ x = \frac{1}{cos \ x}, \qquad csc \ x = \frac{1}{sin \ x}$$

The basic definitions lead to some identities for tangent and cotangent:

$$cot \ x = \frac{1}{tan \ x}, \qquad tan \ x = \frac{1}{cot \ x}$$

Since all six trigonometric functions are periodic, we have the:

PERIODICITY IDENTITIES

$$sin \ (x + 2\pi) = sin \ x, \qquad cos \ (x + 2\pi) = cos \ x, \qquad tan \ (x + 2\pi) = tan \ x,$$
$$csc \ (x + 2\pi) = csc \ x, \qquad sec \ (x + 2\pi) = sec \ x, \qquad cot \ (x + 2\pi) = cot \ x$$

Furthermore, the tangent and cotangent functions satisfy:

$$tan \ (x + \pi) = tan \ x, \qquad cot \ (x + \pi) = cot \ x$$

Some important two-function identities are the following:

PYTHAGOREAN IDENTITIES

$$sin^2 \ x + cos^2 \ x = 1, \qquad 1 + tan^2 \ x = sec^2 \ x, \qquad 1 + cot^2 \ x = csc^2 x$$

By comparing the values of the three basic trigonometric functions at the numbers x and $-x$, we obtain the so-called negative angle identities:

NEGATIVE ANGLE IDENTITIES

$$\sin(-x) = -\sin x, \qquad \cos(-x) = \cos x, \qquad \tan(-x) = -\tan x$$

Several other identities were proved in Section 3, such as $\sin(x + \pi) = -\sin x$. These identities won't be listed here since they are all special cases of some identities proved later in this section.

SIMPLIFYING EXPRESSIONS AND PROVING IDENTITIES

There are usually many ways to simplify a trigonometric expression. Likewise, a given identity can often be proved in several different ways. Here are some examples, involving only the basic identities, that illustrate the usual methods for simplifying expressions and proving identities. The suggestions printed in boldface type apply to all identities, not just the ones discussed here.

EXAMPLE Simplify the expression $(\csc x + \cot x)(1 - \cos x)$. One way of dealing with problems such as this is to

Express everything in terms of sine and cosine, then simplify.

$$(\csc x + \cot x)(1 - \cos x)$$
$$= \left(\frac{1}{\sin x} + \frac{\cos x}{\sin x}\right)(1 - \cos x) \qquad \text{(basic definitions)}$$
$$= \left(\frac{1 + \cos x}{\sin x}\right)(1 - \cos x) = \frac{(1 + \cos x)(1 - \cos x)}{\sin x}$$
$$= \frac{1 - \cos^2 x}{\sin x} = \frac{\sin^2 x}{\sin x} = \sin x \qquad \text{(Pythagorean Identity)}$$

EXAMPLE Prove that $\cos^2 y + \cos^2 y \tan^2 y = 1$. This is a typical situation in which the expression on one side of the equal sign is more complicated than the expression on the other side. The best way to handle such cases is to

Begin on the side with the more complicated expression; use algebra and identities to transform it into the expression on the other side.

This could be done, as in the preceding example, by first expressing everything in terms of sine and cosine. But it can be done more quickly this way:

$$\cos^2 y + \cos^2 y \tan^2 y = \cos^2 y(1 + \tan^2 y)$$
$$= \cos^2 y(\sec^2 y) \qquad \text{(Pythagorean identity)}$$
$$= \cos^2 y \left(\frac{1}{\cos^2 y}\right) = 1 \qquad \text{(basic definition)}$$

EXAMPLE Verify that $\dfrac{\tan^2 x}{1 + \sec x} = \dfrac{1 - \cos x}{\cos x}$. When dealing with fractions such as this, one strategy is to

Eliminate the fractions by cross multiplying.

For as you know, when $B \neq 0$ and $D \neq 0$,

$$\frac{A}{B} = \frac{C}{D} \qquad \text{exactly when} \qquad AD = BC$$

Applying this fact in the present situation, we see that

$$\frac{\tan^2 x}{1 + \sec x} = \frac{1 - \cos x}{\cos x} \quad \text{exactly when} \quad \tan^2 x \cos x = (1 + \sec x)(1 - \cos x)$$

So we need only prove the identity on the right. In cases such as this, when the expressions on each side of the equal sign seem equally complicated, a good approach is to

**Deal separately with each side: use identities and algebra to transform
one side into a certain expression; then use (possibly different) identities
and algebra to transform the other side into the same expression.**

If this is done in such a way that each step is *reversible*, then the identity will be proved. We begin with the right side:

$$(1 + \sec x)(1 - \cos x) = 1 + \sec x - \cos x - \sec x \cos x$$

$$= 1 + \sec x - \cos x - \left(\frac{1}{\cos x}\right)\cos x \qquad \text{(basic definition)}$$

$$= 1 + \sec x - \cos x - 1$$
$$= \sec x - \cos x$$

This is a reasonably simple expression and seems a good place to stop. Now we try to transform the left side into this same expression:

$$\tan^2 x \cos x = \frac{\sin^2 x}{\cos^2 x}\cos x = \frac{\sin^2 x}{\cos x} \qquad \text{(basic definition)}$$

$$= \frac{1 - \cos^2 x}{\cos x} = \frac{1}{\cos x} - \frac{\cos^2 x}{\cos x} \qquad \text{(Pythagorean Identity)}$$

$$= \sec x - \cos x \qquad \text{(basic definition)}$$

We have now transformed both the right and left sides of the original equation into the expression: $\sec x - \cos x$. Since each step in our calculations is reversible, we can proceed from the right side of the original identity to the expression $\sec x - \cos x$, and then to the left side of the original identity. Therefore the identity is verified.

EXAMPLE Prove that $\dfrac{1 + \cos z}{\sin z} = 2 \csc z - \dfrac{\sin z}{1 + \cos z}$. Since both sides seem equally complicated, we shall work separately with each side. A common strategy in situations such as this is to

Express each side in terms of a common denominator.

The obvious common denominator to choose here is $\sin z(1 + \cos z)$. Keeping this in mind, we begin with the left side:

$$\frac{1 + \cos z}{\sin z} = \frac{1 + \cos z}{\sin z} \cdot \frac{1 + \cos z}{1 + \cos z}$$

$$= \frac{(1 + \cos z)^2}{\sin z(1 + \cos z)}$$

Next we express the right side in terms of the same common denominator and attempt to transform it into this same expression.

$$2 \csc z - \frac{\sin z}{1 + \cos z}$$

$$= \frac{2 \csc z(1 + \cos z) - \sin z}{1 + \cos z} \qquad \text{(express as fraction)}$$

$$= \frac{2 \csc z(1 + \cos z) - \sin z}{1 + \cos z} \cdot \frac{\sin z}{\sin z} \qquad \begin{array}{l}\text{(express in terms of} \\ \text{common denominator)}\end{array}$$

$$= \frac{2 \csc z \sin z(1 + \cos z) - \sin^2 z}{\sin z(1 + \cos z)}$$

$$= \frac{2(1/\sin z) \sin z(1 + \cos z) - \sin^2 z}{\sin z(1 + \cos z)} \qquad \text{(basic definition)}$$

$$= \frac{2(1 + \cos z) - \sin^2 z}{\sin z(1 + \cos z)}$$

$$= \frac{2(1 + \cos z) - (1 - \cos^2 z)}{\sin z(1 + \cos z)} \qquad \text{(Pythagorean Identity)}$$

$$= \frac{2 + 2 \cos z - 1 + \cos^2 z}{\sin z(1 + \cos z)} = \frac{\cos^2 z + 2 \cos z + 1}{\sin z(1 + \cos z)} \qquad \text{(simplify)}$$

$$= \frac{(\cos z + 1)^2}{\sin z(1 + \cos z)}$$

Since both sides reduce to the same expression and each step is reversible, the identity is proved.

ADDITION AND SUBTRACTION IDENTITIES

How are the numbers $\sin 3$, $\sin 5$, and $\sin 8$ related to each other? Even though $3 + 5 = 8$, it is *not* true that $\sin 3 + \sin 5 = \sin 8$. In fact $\sin(x + y)$ is almost *never* equal to $\sin x + \sin y$. To what, then, is $\sin(x + y)$ equal? Similar questions could be asked about $\cos(x + y)$, as well as $\sin(x - y)$ and $\cos(x - y)$. The answers to all of these questions are contained in the addition and subtraction identities.

ADDITION AND SUBTRACTION IDENTITIES

$$\cos(x - y) = \cos x \cos y + \sin x \sin y$$
$$\cos(x + y) = \cos x \cos y - \sin x \sin y$$
$$\sin(x - y) = \sin x \cos y - \cos x \sin y$$
$$\sin(x + y) = \sin x \cos y + \cos x \sin y$$

These are probably the most important of all the trigonometric identities, in the sense that all the identities proved in this chapter (except the basic definitions and the Pythagorean Identities) are just special cases of the addition and subtraction identities. The proof of these identities is given at the end of this section. Before reading it, you should first become familiar with the examples and special cases below. They show *how* these identities are used in practice, and it is essential that you be able to use them efficiently.

EXAMPLE   These identities can be used to evaluate precisely $\sin t$ and $\cos t$ when t is an integer multiple of $\pi/12$, without calculators or tables. For instance, to find $\sin(5\pi/12)$, observe that $5\pi/12 = (\pi/6) + (\pi/4)$. Applying the identity for $\sin(x + y)$ with $x = \pi/6$ and $y = \pi/4$ shows that

$$\sin\frac{5\pi}{12} = \sin\left(\frac{\pi}{6} + \frac{\pi}{4}\right)$$

$$= \sin\frac{\pi}{6}\cos\frac{\pi}{4} + \cos\frac{\pi}{6}\sin\frac{\pi}{4} = \frac{1}{2}\cdot\frac{\sqrt{2}}{2} + \frac{\sqrt{3}}{2}\cdot\frac{\sqrt{2}}{2} = \frac{\sqrt{2}(\sqrt{3}+1)}{4}$$

Similarly, applying the identity for $\cos(x + y)$ with $x = \pi/6$ and $y = \pi/4$ yields:

$$\cos\frac{5\pi}{12} = \cos\left(\frac{\pi}{6} + \frac{\pi}{4}\right)$$

$$= \cos\frac{\pi}{6}\cos\frac{\pi}{4} - \sin\frac{\pi}{6}\sin\frac{\pi}{4} = \frac{\sqrt{3}}{2}\cdot\frac{\sqrt{2}}{2} - \frac{1}{2}\cdot\frac{\sqrt{2}}{2} = \frac{\sqrt{2}(\sqrt{3}-1)}{4}$$

These results can now be used to evaluate the four other trigonometric functions at $5\pi/12$. For example,

$$\cot\frac{5\pi}{12} = \frac{\cos\dfrac{5\pi}{12}}{\sin\dfrac{5\pi}{12}} = \frac{\dfrac{\sqrt{2}(\sqrt{3}-1)}{4}}{\dfrac{\sqrt{2}(\sqrt{3}+1)}{4}} = \frac{\sqrt{3}-1}{\sqrt{3}+1}$$

Rationalizing this last denominator yields:

$$\cot \frac{5\pi}{12} = \frac{\sqrt{3}-1}{\sqrt{3}+1} = \frac{(\sqrt{3}-1)(\sqrt{3}-1)}{(\sqrt{3}+1)(\sqrt{3}-1)} = \frac{3-2\sqrt{3}+1}{3-1} = \frac{4-2\sqrt{3}}{2} = 2 - \sqrt{3}$$

EXAMPLE   To determine $\sin(x - \pi)$, just apply the identity for $\sin(x - y)$ with $y = \pi$:

$$\sin(x - \pi) = \sin x \cos \pi - \cos x \sin \pi = (\sin x)(-1) - (\cos x)(0) = -\sin x$$

Similarly, applying the identity for $\cos(x - y)$ with $x = 2\pi$ yields:

$$\cos(2\pi - y) = \cos 2\pi \cos y + \sin 2\pi \sin y = (1)(\cos y) + (0)(\sin y) = \cos y$$

EXAMPLE   In calculus, it will be necessary to show that for the function $f(x) = \sin x$ and any number $h \neq 0$,

$$\frac{f(x+h) - f(x)}{h} = \sin x \left(\frac{\cos h - 1}{h}\right) + \cos x \left(\frac{\sin h}{h}\right)$$

This can be done by using the addition identity for $\sin(x + y)$ with $y = h$:

$$\frac{f(x+h) - f(x)}{h} = \frac{\sin(x+h) - \sin x}{h}$$

$$= \frac{\sin x \cos h + \cos x \sin h - \sin x}{h}$$

$$= \frac{\sin x (\cos h - 1) + \cos x \sin h}{h}$$

$$= \sin x \left(\frac{\cos h - 1}{h}\right) + \cos x \left(\frac{\sin h}{h}\right)$$

EXAMPLE   Prove that $\cos x \cos y = \frac{1}{2}(\cos(x+y) + \cos(x-y))$. We begin with the more complicated right side and use the addition and subtraction identities for cosine to transform it into the left side:

$$\frac{1}{2}(\cos(x+y) + \cos(x-y)) = \frac{1}{2}((\cos x \cos y - \sin x \sin y) + (\cos x \cos y + \sin x \sin y))$$
$$= \frac{1}{2}(\cos x \cos y + \cos x \cos y)$$
$$= \frac{1}{2}(2 \cos x \cos y) = \cos x \cos y$$

The addition and subtraction identities for sine and cosine can be used to obtain

---

ADDITION AND SUBTRACTION IDENTITIES FOR TANGENT

$$\tan(x + y) = \frac{\tan x + \tan y}{1 - \tan x \tan y}, \qquad \tan(x - y) = \frac{\tan x - \tan y}{1 + \tan x \tan y}$$

A proof of these identities is outlined in Exercise B.2.

EXAMPLE If $\tan 8 \approx -6.8$ and $\tan 16 \approx .3$, find an approximation for $\tan 24$:

$$\tan 24 = \tan(8 + 16) = \frac{\tan 8 + \tan 16}{1 - (\tan 8)(\tan 16)} \approx \frac{-6.8 + .3}{1 - (-6.8)(.3)} = \frac{-6.5}{3.04} \approx -2.138$$

DOUBLE ANGLE IDENTITIES

The following identities relate the values of a trigonometric function at a number x with the value of the same function at $2x$.

> ### DOUBLE ANGLE IDENTITIES
>
> $$\sin 2x = 2 \sin x \cos x,$$
>
> $$\cos 2x = \cos^2 x - \sin^2 x,$$
>
> $$\tan 2x = \frac{2 \tan x}{1 - \tan^2 x}$$

To prove these identities, just let $x = y$ in the addition identities:

$$\sin 2x = \sin(x + x) = \sin x \cos x + \cos x \sin x = 2 \sin x \cos x,$$
$$\cos 2x = \cos(x + x) = \cos x \cos x - \sin x \sin x = \cos^2 x - \sin^2 x,$$

$$\tan 2x = \tan(x + x) = \frac{\tan x + \tan x}{1 - \tan x \tan x} = \frac{2 \tan x}{1 - \tan^2 x}$$

EXAMPLE If $\pi/2 < x < \pi$ and $\cos x = -8/17$, find $\sin 2x$, $\cos 2x$, $\tan 2x$, and show that $\pi < 2x < 3\pi/2$. In order to use the double angle identities, we first must determine $\sin x$ and $\tan x$. Now $\sin x$ can be found by using the Pythagorean Identity:

$$\sin^2 x = 1 - \cos^2 x = 1 - (-\tfrac{8}{17})^2 = 1 - \tfrac{64}{289} = \tfrac{225}{289}$$

Since $\pi/2 < x < \pi$, we know $\sin x$ is positive. Therefore

$$\sin x = +\sqrt{\tfrac{225}{289}} = \tfrac{15}{17}$$

Consequently, $\tan x = \sin x / \cos x = \tfrac{15}{17} / (-\tfrac{8}{17}) = -\tfrac{15}{8}$. We now substitute these values in the double angle identities:

$$\sin 2x = 2 \sin x \cos x = 2(\tfrac{15}{17})(-\tfrac{8}{17}) = -\tfrac{240}{289} \approx -.83,$$
$$\cos 2x = \cos^2 x - \sin^2 x = (-\tfrac{8}{17})^2 - (\tfrac{15}{17})^2 = \tfrac{64}{289} - \tfrac{225}{289} = -\tfrac{161}{289} \approx -.56,$$

$$\tan 2x = \frac{2 \tan x}{1 - \tan^2 x} = \frac{2(-\tfrac{15}{8})}{1 - (-\tfrac{15}{8})^2} = \frac{-\tfrac{15}{4}}{1 - (\tfrac{225}{64})} = \frac{-\tfrac{15}{4}}{-\tfrac{161}{64}} = \tfrac{240}{161} \approx 1.49$$

Since $\pi/2 < x < \pi$, we know that $\pi < 2x < 2\pi$. The calculations above show that at $2x$ both sine and cosine are negative and tangent is positive. This can only occur if $2x$ lies between π and $3\pi/2$ (see the chart of signs on p. 389).

EXAMPLE   Express the rule of the function $f(x) = \sin 3x$ in terms of $\sin x$ and constants. We first use the addition identity for $\sin(x + y)$ with $y = 2x$:

$$f(x) = \sin 3x = \sin(x + 2x) = \sin x \cos 2x + \cos x \sin 2x$$

Next apply the double angle identities for $\cos 2x$ and $\sin 2x$:

$$\begin{aligned}
f(x) = \sin 3x &= \sin x \cos 2x + \cos x \sin 2x \\
&= \sin x(\cos^2 x - \sin^2 x) + \cos x(2 \sin x \cos x) \\
&= \sin x \cos^2 x - \sin^3 x + 2 \sin x \cos^2 x \\
&= 3 \sin x \cos^2 x - \sin^3 x
\end{aligned}$$

Finally, use the Pythagorean Identity:

$$\begin{aligned}
f(x) = \sin 3x &= 3 \sin x \cos^2 x - \sin^3 x = 3 \sin x(1 - \sin^2 x) - \sin^3 x \\
&= 3 \sin x - 3 \sin^3 x - \sin^3 x = 3 \sin x - 4 \sin^3 x
\end{aligned}$$

The double angle identity for $\cos 2x$ can be rewritten in several useful ways. For instance, we can use the Pythagorean Identity in the form $\cos^2 x = 1 - \sin^2 x$ to obtain:

$$\cos 2x = \cos^2 x - \sin^2 x = (1 - \sin^2 x) - \sin^2 x = 1 - 2 \sin^2 x$$

Similarly, using the Pythagorean identity in the form $\sin^2 x = 1 - \cos^2 x$, we have:

$$\cos 2x = \cos^2 x - \sin^2 x = \cos^2 x - (1 - \cos^2 x) = 2 \cos^2 x - 1$$

In summary:

MORE DOUBLE ANGLE IDENTITIES

$$\cos 2x = 1 - 2 \sin^2 x,$$
$$\cos 2x = 2 \cos^2 x - 1$$

If we solve the first equation in this box for $\sin^2 x$ and the second one for $\cos^2 x$ we obtain the following:

POWER-REDUCING IDENTITIES

$$\sin^2 x = \frac{1 - \cos 2x}{2} \quad and \quad \cos^2 x = \frac{1 + \cos 2x}{2}$$

EXAMPLE   Express the rule of the function $f(x) = \sin^4 x$ in terms of constants and first powers of the cosine function. We begin by applying the power-reducing identity for sine:

$$f(x) = \sin^4 x = \sin^2 x \sin^2 x = \frac{1 - \cos 2x}{2} \cdot \frac{1 - \cos 2x}{2} = \frac{1 - 2 \cos 2x + \cos^2 2x}{4}$$

Next we apply the power-reducing identity for cosine to $\cos^2 2x$. Note that this means using $2x$ in place of x in the identity:

$$\cos^2 2x = \frac{1 + \cos 2(2x)}{2} = \frac{1 + \cos 4x}{2}$$

Finally, we substitute this last result in the expression for $\sin^4 x$ above:

$$f(x) = \sin^4 x = \frac{1 - 2\cos 2x + \cos^2 2x}{4} = \frac{1 - 2\cos 2x + \dfrac{1 + \cos 4x}{2}}{4}$$

$$= \tfrac{1}{4} - \tfrac{1}{2}\cos 2x + \tfrac{1}{8}(1 + \cos 4x)$$
$$= \tfrac{3}{8} - \tfrac{1}{2}\cos 2x + \tfrac{1}{8}\cos 4x$$

HALF-ANGLE IDENTITIES

If we use the power-reducing identity with $x/2$ in place of x, we obtain:

$$\sin^2\left(\frac{x}{2}\right) = \frac{1 - \cos 2(x/2)}{2} = \frac{1 - \cos x}{2}$$

Consequently, we must have:

$$\sin\frac{x}{2} = \pm\sqrt{\frac{1 - \cos x}{2}}$$

This proves the first of the half-angle identities.

> ### HALF-ANGLE IDENTITIES
>
> $$sin\,\frac{x}{2} = \pm\sqrt{\frac{1 - cos\,x}{2}},$$
>
> $$cos\,\frac{x}{2} = \pm\sqrt{\frac{1 + cos\,x}{2}}$$

The second of these identities is derived from the second power-reducing identity, in the same way the first one was obtained. The sign in front of the radical depends on the quadrant in which the terminal side of an angle of $x/2$ radians lies.

EXAMPLE Find $\sin(\pi/8)$. Since $\pi/8$ lies between 0 and $\pi/2$, we know that $\sin(\pi/8)$ is positive. Observe that $\pi/8 = (1/2)(\pi/4) = (\pi/4)/2$ and apply the half-angle identity with $x = \pi/4$:

$$\sin\frac{\pi}{8} = \sin\frac{\pi/4}{2} = \sqrt{\frac{1 - \cos(\pi/4)}{2}}$$

$$= \sqrt{\frac{1 - (\sqrt{2}/2)}{2}} = \sqrt{\frac{(2 - \sqrt{2})/2}{2}} = \sqrt{\frac{2 - \sqrt{2}}{4}} = \frac{\sqrt{2 - \sqrt{2}}}{2}$$

EXAMPLE Prove that $1 - 2\sin^2(x/2) = \cos x$. We begin with the more complicated side (the left side) and use various identities to transform it into the other side:

$$1 - 2\sin^2\left(\frac{x}{2}\right) = 1 - 2\left(\sin\frac{x}{2}\right)^2 = 1 - 2\left(\pm\sqrt{\frac{1-\cos x}{2}}\right)^2 \quad \text{(half-angle identity)}$$

$$= 1 - 2\left(\frac{1-\cos x}{2}\right) = 1 - (1 - \cos x) = \cos x$$

The half-angle identities for sine and cosine can be used in the obvious way to obtain an identity for $\tan x/2$. But with a little more work, we can obtain two more useful identities (whose proofs are outlined in Exercises C.1 and C.2):

HALF-ANGLE IDENTITIES FOR TANGENT

$$\tan\frac{x}{2} = \frac{\sin x}{1 + \cos x} \quad and \quad \tan\frac{x}{2} = \frac{1 - \cos x}{\sin x}$$

EXAMPLE It is known that $\sin 26 \approx .76$ and $\cos 26 \approx .65$. Therefore

$$\tan 13 = \tan\frac{26}{2} = \frac{\sin 26}{1 + \cos 26} = \frac{.76}{1 + .65} = \frac{.76}{1.65} \approx .46$$

COFUNCTION IDENTITIES

Another special case of the addition and subtraction identities are the cofunction identities.

COFUNCTION IDENTITIES

$$\sin x = \cos\left(\frac{\pi}{2} - x\right), \quad \cos x = \sin\left(\frac{\pi}{2} - x\right)$$

$$\tan x = \cot\left(\frac{\pi}{2} - x\right), \quad \cot x = \tan\left(\frac{\pi}{2} - x\right)$$

$$\sec x = \csc\left(\frac{\pi}{2} - x\right), \quad \csc x = \sec\left(\frac{\pi}{2} - x\right)$$

The first cofunction identity is proved by using the identity for $\cos(x - y)$ with $\pi/2$ in place of x and x in place of y (be careful, this can be a bit confusing):

$$\cos\left(\frac{\pi}{2} - x\right) = \cos\frac{\pi}{2}\cos x + \sin\frac{\pi}{2}\sin x = (0)(\cos x) + (1)(\sin x) = \sin x$$

Since the first cofunction identity is valid for *every* number x, it is also valid with the number $x - (\pi/2)$ in place of x:

$$\sin\left(\frac{\pi}{2} - x\right) = \cos\left(\frac{\pi}{2} - \left(\frac{\pi}{2} - x\right)\right) = \cos x$$

Thus we have proved the second cofunction identity. The others now follow from these two. For instance,

$$\tan\left(\frac{\pi}{2} - x\right) = \frac{\sin\left((\pi/2) - x\right)}{\cos\left((\pi/2) - x\right)} = \frac{\cos x}{\sin x} = \cot x$$

EXAMPLE Verify that $\dfrac{\cos\left(x - (\pi/2)\right)}{\cos x} = \tan x$. Beginning on the left side, we see that the term $\cos\left(x - (\pi/2)\right)$ looks almost, but not quite, like the term $\cos\left((\pi/2) - x\right)$ in the cofunction identity. But note that $-(x - (\pi/2)) = (\pi/2) - x$. Therefore

$$\frac{\cos\left(x - \dfrac{\pi}{2}\right)}{\cos x} = \frac{\cos\left(-\left(x - \dfrac{\pi}{2}\right)\right)}{\cos x} = \frac{\cos\left(\dfrac{\pi}{2} - x\right)}{\cos x} \qquad \text{(negative angle identity}$$
$$\text{with } x - \frac{\pi}{2} \text{ in place of } x)$$

$$= \frac{\sin x}{\cos x} \qquad\qquad\qquad \text{(cofunction identity)}$$

$$= \tan x \qquad\qquad\qquad \text{(basic definition)}$$

PROOF OF THE ADDITION AND SUBTRACTION IDENTITIES

We first prove that $\cos(x - y) = \cos x \cos y + \sin x \sin y$ when $x > y$. Let P be the point where the terminal side of an angle of x radians in standard position meets the unit circle and let Q be the point where the terminal side of an angle of y radians in standard position meets the unit circle, as shown in Figure 6-105. According to the definitions of sine and cosine, P has coordinates $(\cos x, \sin x)$ and Q has coordinates $(\cos y, \sin y)$, as shown.

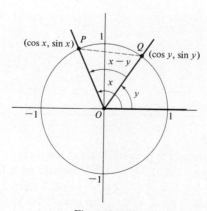

Figure 6-105

Observe that the angle QOP formed by the two terminal sides has radian measure $x - y$. Rotate this angle clockwise until side OQ lies on the horizontal axis, as shown in Figure 6-106.

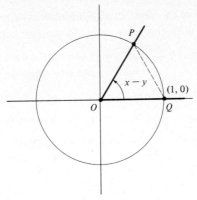

Figure 6-106

Angle QOP is now in standard position and its terminal side meets the unit circle at P. Since angle QOP has radian measure $x - y$, the definitions of sine and cosine show that P, in this new location, has coordinates $(\cos(x - y), \sin(x - y))$. Q now has coordinates $(1, 0)$.

The distance from P to Q (the length of the straight line segment PQ) is the same both before and after the rotation of the angle. The next step is to compute this distance. Using the coordinates of P and Q *before* the angle was rotated and the distance formula, we have:

distance from P to Q

$$= \sqrt{(\cos x - \cos y)^2 + (\sin x - \sin y)^2}$$
$$= \sqrt{\cos^2 x - 2\cos x \cos y + \cos^2 y + \sin^2 x - 2\sin x \sin y + \sin^2 y}$$
$$= \sqrt{(\cos^2 x + \sin^2 x) + (\cos^2 y + \sin^2 y) - 2\cos x \cos y - 2\sin x \sin y}$$
$$= \sqrt{1 + 1 - 2\cos x \cos y - 2\sin x \sin y}$$
$$= \sqrt{2 - 2\cos x \cos y - 2\sin x \sin y}$$

But using the coordinates of P and Q *after* the angle is rotated shows that

$$\text{distance from } P \text{ to } Q = \sqrt{(\cos(x - y) - 1)^2 + (\sin(x - y) - 0)^2}$$
$$= \sqrt{\cos^2(x - y) - 2\cos(x - y) + 1 + \sin^2(x - y)}$$
$$= \sqrt{\cos^2(x - y) + \sin^2(x - y) - 2\cos(x - y) + 1}$$
$$= \sqrt{1 - 2\cos(x - y) + 1} \qquad \text{(Pythagorean Identity)}$$
$$= \sqrt{2 - 2\cos(x - y)}$$

The two expressions for the distance from P to Q must be equal. Hence

$$\sqrt{2 - 2\cos(x - y)} = \sqrt{2 - 2\cos x \cos y - 2\sin x \sin y}$$

Squaring both sides of this equation and simplifying the result yields:

$$2 - 2\cos(x - y) = 2 - 2\cos x \cos y - 2\sin x \sin y$$
$$-2\cos(x - y) = -2(\cos x \cos y + \sin x \sin y)$$
$$\cos(x - y) = \cos x \cos y + \sin x \sin y$$

This completes the proof of the first addition identity when $x > y$. If $y > x$, then the proof just given is valid with the roles of x and y interchanged; it shows that

$$\cos(y - x) = \cos y \cos x + \sin y \sin x$$
$$= \cos x \cos y + \sin x \sin y$$

The negative angle identity with $x - y$ in place of x shows that

$$\cos(x - y) = \cos(-(x - y)) = \cos(y - x)$$

Combining this fact with the previous one shows that

$$\cos(x - y) = \cos x \cos y + \sin x \sin y$$

in this case also. Therefore the first addition identity is proved.

The identity for $\cos(x + y)$ now follows from the one for $\cos(x - y)$ by using the negative angle identities for sine and cosine:

$$\cos(x + y) = \cos(x - (-y)) = \cos x \cos(-y) + \sin x \sin(-y)$$
$$= \cos x \cos y + (\sin x)(-\sin y) = \cos x \cos y - \sin x \sin y$$

Next we note that the proof of the first two cofunction identities on page 419 depended only on the addition identity for $\cos(x - y)$. Since that has been proved, we can validly use the first two cofunction identities in the remainder of the proof. In particular:

$$\sin(x - y) = \cos\left(\frac{\pi}{2} - (x - y)\right) = \cos\left(\left(\frac{\pi}{2} - x\right) + y\right)$$

Applying the proven identity for $\cos(x + y)$ with $(\pi/2) - x$ in place of x and the two cofunction identities now yields

$$\sin(x - y) = \cos\left(\left(\frac{\pi}{2} - x\right) + y\right)$$
$$= \cos\left(\frac{\pi}{2} - x\right)\cos y - \sin\left(\frac{\pi}{2} - x\right)\sin y$$
$$= \sin x \cos y - \cos x \sin y$$

This proves the third of the addition and subtraction identities. The fourth is obtained from the third in the same way the second was obtained from the first.

EXERCISES

A.1. Evaluate all six trigonometric functions at

(a) $t = \dfrac{\pi}{12}$ (b) $t = \dfrac{5\pi}{12}$ (c) $t = \dfrac{7\pi}{12}$ (d) $t = \dfrac{11\pi}{12}$.

A.2. Simplify:
 (a) $\cos(x + y)\cos y + \sin(x + y)\sin y$
 (b) $\sin(x - y)\cos y + \cos(x - y)\sin y$
 (c) $\cos(x + y) - \cos(x - y)$
 (d) $\sin(x + y) - \sin(x - y)$

A.3. Evaluate all six trigonometric functions at:
 (a) $t = \dfrac{\pi}{8}$ (b) $\dfrac{3\pi}{8}$ $\left(Note: \dfrac{3\pi}{8} = \dfrac{\pi}{2} - \dfrac{\pi}{8}\right)$ (c) $\dfrac{5\pi}{8}$ (d) $\dfrac{7\pi}{8}$

A.4. Express each of the following in terms of $\sin x$ and $\cos x$.
 (a) $\sin\left(\dfrac{\pi}{2} + x\right)$ (c) $\cos\left(x - \dfrac{3\pi}{2}\right)$ (e) $\csc\left(x + \dfrac{\pi}{2}\right)$
 (b) $\cos\left(x + \dfrac{\pi}{2}\right)$ (d) $\cot(x + \pi)$ (f) $\sec(x - \pi)$

B.1. Prove these identities.
 (a) $\sec^2 x - \csc^2 x = \tan^2 x - \cot^2 x$
 (b) $\sec^2 x + \csc^2 x = \sec^2 x \csc^2 x$
 (c) $\sec^4 x - \tan^4 x = 1 + 2\tan^2 x$
 (d) $\sin^2 x \cot^2 x + \cos^2 x \tan^2 x = 1$

 (e) $\dfrac{1 + \cos x}{\sin x} + \dfrac{\sin x}{1 + \cos x} = 2\csc x$

 (f) $\dfrac{\sin x}{\csc x + \cot x} = 1 - \cos x$

 (g) $\dfrac{\sec x + \csc x}{1 + \tan x} = \csc x$

 (h) $\dfrac{\cot x - 1}{1 - \tan x} = \dfrac{\csc x}{\sec x}$

B.2. Prove the addition and subtraction identities for the tangent function (p. 415). [*Hint:* use the addition identities for $\sin(x + y)$ and $\cos(x + y)$ to express $\tan(x + y)$. Then divide both numerator and denominator of the resulting expression by $\cos x \cos y$.]

B.3. Prove these identities.

 (a) $\dfrac{\cos x}{1 - \sin x} = \sec x + \tan x$

 (b) $\csc x + \cot x = \dfrac{\sin x}{1 - \cos x}$

 (c) $\dfrac{1}{\sin x \cos x} - \cot x = \dfrac{\sin x \cos x}{1 - \sin^2 x}$

 (d) $\dfrac{\cot x + \cos x}{\cos x \cot x} = \dfrac{\cos x \cot x}{\cot x - \cos x}$

 (e) $(\csc x - \cot x)^4(\csc x + \cot x)^4 = 1$

(f) $1 + \sin x \cos x = \dfrac{\cos^3 x - \sin^3 x}{\cos x - \sin x}$

(g) $\dfrac{1 + \sec x}{\sin x + \tan x} = \csc x$

(h) $\sin^6 x + \cos^6 x = 1 - 3\sin^2 x \cos^2 x$

B.4. Express $\cos 3x$ in terms of $\cos x$.

B.5. Prove $\dfrac{\sin(x + y)}{\sin(x - y)} = \dfrac{\tan x + \tan y}{\tan x - \tan y}$.

B.6. If $f(x) = \cos x$ and h is a fixed nonzero number, prove that $\dfrac{f(x + h) - f(x)}{h} =$
$\cos x \left(\dfrac{\cos h - 1}{h}\right) - \sin x \left(\dfrac{\sin h}{h}\right)$.

B.7. Assume $\sin x = .8$ and $\sin y = \sqrt{.75} \approx .87$, and that x, y lie between 0 and $\pi/2$. Find
(a) $\cos(x + y)$ (c) $\cos(x - y)$ (e) $\tan(x + y)$
(b) $\sin(x + y)$ (d) $\sin(x - y)$ (f) $\tan(x - y)$

B.8. Find $\sin 2x$, $\cos 2x$, and $\tan 2x$ and determine in which quadrant the terminal side of an angle of $2x$ radians in standard position lies.

(a) $\sin x = \dfrac{5}{13} \left(0 < x < \dfrac{\pi}{2}\right)$ (c) $\cos x = \dfrac{-3}{5}\left(\pi < x < \dfrac{3\pi}{2}\right)$

(b) $\sin x = \dfrac{-4}{5} \left(\pi < x < \dfrac{3\pi}{2}\right)$ (d) $\tan x = \dfrac{3}{4}\left(\pi < x < \dfrac{3\pi}{2}\right)$

B.9. Prove these identities:

(a) $\dfrac{1 + \cos 2x}{\sin 2x} = \cot x$ (c) $\sin 2x = \dfrac{2 \cot x}{\csc^2 x}$

(b) $\cos^4 x - \sin^4 x = \cos 2x$

B.10. **(a)** Express the rule of the function $f(x) = \cos^3 x$ in terms of constants and first powers of the cosine function as in the example on p. 417.
(b) Do the same for $f(x) = \cos^4 x$.

B.11. If $\sin x = .6$ and $0 < x < \pi/2$, then find
(a) $\sin 2x$ (d) $\sin 4x$

(b) $\cos 4x$ (e) $\sin \dfrac{x}{2}$

(c) $\cos 2x$ (f) $\cos \dfrac{x}{2}$

B.12. Find $\sin (x/2)$ and $\cos (x/2)$ when

  (a) $\sin x = \dfrac{-3}{5}$ and $\dfrac{3\pi}{2} < x < 2\pi$

  (b) $\cot x = 1$ and $-\pi < x < \dfrac{-\pi}{2}$

B.13. Simplify

  (a) $\dfrac{\sin 2x}{2 \sin x}$ (d) $\cos^2 \left(\dfrac{x}{2}\right) - \sin^2 \left(\dfrac{x}{2}\right)$

  (b) $1 - 2 \sin^2 \left(\dfrac{x}{2}\right)$ (e) $(\sin x + \cos x)^2 - \sin 2x$

  (c) $2 \cos 2y \sin 2y$ (Think!)   (f) $2 \sin x \cos^3 x - 2 \sin^3 x \cos x$

C.1. (a) Prove that $\tan \dfrac{x}{2} = \pm \sqrt{\dfrac{1 - \cos x}{1 + \cos x}}$. $\left(Hint: \tan (x/2) = \dfrac{\sin (x/2)}{\cos (x/2)}.\right)$

  (b) Verify that $\left|\tan \dfrac{x}{2}\right| = \sqrt{\dfrac{1 - \cos x}{1 + \cos x}}$. (*Hint:* see the first box on p. 15.)

  (c) Prove that $\left|\tan \dfrac{x}{2}\right| = \dfrac{|\sin x|}{|1 + \cos x|}$. $\Big(Hint:$ multiply the right-hand side of

    (b) by $1 = \dfrac{1 + \cos x}{1 + \cos x}.\Big)$

  (d) Verify that $\left|\tan \dfrac{x}{2}\right| = \dfrac{|\sin x|}{1 + \cos x}$. (*Hint:* the absolute value of a nonnega-
    tive number is the number itself.)

  (e) Verify that $\tan (x/2)$ and $\sin x$ always have the same sign. (*Hint:* both
    functions have period 2π; hence it suffices to show that both are positive
    when $0 < x < \pi$ and both negative when $\pi < x < 2\pi$.)

  (f) Use (d) and (e) to verify that $\tan \dfrac{x}{2} = \dfrac{\sin x}{1 + \cos x}$.

C.2. Prove that $\tan \dfrac{x}{2} = \dfrac{1 - \cos x}{\sin x}$. (*Hint:* adapt the proof that $\tan \dfrac{x}{2} = \dfrac{\sin x}{1 + \cos x}$
  outlined in Exercise C.1.)

9. RIGHT-TRIANGLE TRIGONOMETRY — START

In this section we shall present the basic concepts of classical trigonometry. This
approach, which was developed by the ancient Greeks and Babylonians, differs slightly
from the viewpoint we have taken up to now. But it is often the best way to deal with
certain problems involving triangles.

TRIGONOMETRIC FUNCTIONS OF ANGLES

The definitions and the methods of evaluating the six basic trigonometric functions all involve angles in one way or another. A typical example is the sine function. The process of evaluating the sine function may be summarized as follows:

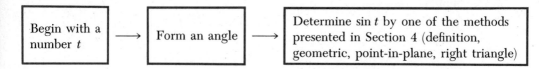

The classical approach to the sine function is somewhat different. It begins at the second step of the evaluation process, with *angles* rather than numbers. To each angle θ there is associated a number called its sine. Although the ancients would not have used these terms, their approach amounts to defining the sine function as a function whose domain consists of all *angles* instead of all real numbers. Analogous remarks apply to the other trigonometric functions.

Even today there are situations in which the classical approach is the most useful. Consequently, we shall take this approach whenever convenient. This means that instead of beginning with numbers and *then* passing to angles, we shall just begin with the angles. This is the only point at which there's any significant difference from our past procedures. It amounts to considering the domains of the trigonometric functions to consist of *angles* instead of numbers. From there on, everything is essentially the same. The *values* of the various trigonometric functions are still *numbers* and are obtained just as before. For example, the point-in-the-plane method yields this:

> Let θ be an angle in standard position and let (x, y) be any point (except the origin) on the terminal side of θ. Let $r = \sqrt{x^2 + y^2}$. Then the values of the six trigonometric functions of the angle θ are given by
>
> $$\sin \theta = \frac{y}{r}, \quad \cos \theta = \frac{x}{r}, \quad \tan \theta = \frac{y}{x},$$
>
> $$\csc \theta = \frac{r}{y}, \quad \sec \theta = \frac{r}{x}, \quad \cot \theta = \frac{x}{y}.$$

EXAMPLE Evaluate the six trigonometric functions at the angle θ shown in Figure 6-107.

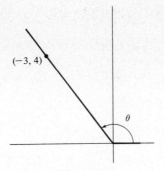

Figure 6-107

We use $(-3, 4)$ as the point (x, y), so that $r = \sqrt{x^2 + y^2} = \sqrt{9 + 16} = \sqrt{25} = 5$. Thus

$$\sin \theta = \frac{y}{r} = \frac{4}{5}, \qquad \cos \theta = \frac{x}{r} = \frac{-3}{5}, \qquad \tan \theta = \frac{y}{x} = \frac{4}{-3},$$

$$\csc \theta = \frac{r}{y} = \frac{5}{4}, \qquad \sec \theta = \frac{r}{x} = \frac{5}{-3}, \qquad \cot \theta = \frac{x}{y} = \frac{-3}{4}$$

When θ is an angle between 0 and 90°, we can use the right-triangle method:

Consider a right triangle containing an angle θ:

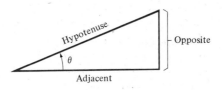

The values of the six trigonometric functions of the angle θ are given by:

$$\sin \theta = \frac{opposite}{hypotenuse}, \qquad \cos \theta = \frac{adjacent}{hypotenuse}, \qquad \tan \theta = \frac{opposite}{adjacent},$$

$$\csc \theta = \frac{hypotenuse}{opposite}, \qquad \sec \theta = \frac{hypotenuse}{adjacent}, \qquad \cot \theta = \frac{adjacent}{opposite}$$

From our work earlier in this chapter we know that both this method and the previous one result in the same functional values for a given angle θ.

EXAMPLE   Evaluate the six trigonometric functions at the angle θ shown in Figure 6-108.

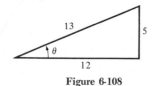

Figure 6-108

$$\sin \theta = \frac{\text{opposite}}{\text{hypotenuse}} = \frac{5}{13}, \qquad \csc \theta = \frac{\text{hypotenuse}}{\text{opposite}} = \frac{13}{5},$$

$$\cos \theta = \frac{\text{adjacent}}{\text{hypotensue}} = \frac{12}{13}, \qquad \sec \theta = \frac{\text{hypotenuse}}{\text{adjacent}} = \frac{13}{12},$$

$$\tan \theta = \frac{\text{opposite}}{\text{adjacent}} = \frac{5}{12}, \qquad \cot \theta = \frac{\text{adjacent}}{\text{opposite}} = \frac{12}{5}$$

DEGREES AND RADIANS

When dealing with the trigonometric functions as functions of angles, you must remember that angles can be measured in either radians or degrees. If radian measure is used (as was always the case earlier in this chapter), then everything is the same as before. For example, sin 30 denotes the value of the sine function at an *angle* of 30 radians. As the definition of the sine function shows, this is exactly the same as the value of the sine function at the number 30. Thus the same notation used before works equally well for numbers or for angles measured in radians.

But when angles are measured in degrees (as will be the case in the rest of this section), we must adopt some new notation. In order to indicate the value of the sine function at an angle of 30 *degrees*, we write

$$\sin 30° \quad \text{(note the degree symbol)}$$

Similarly, cot 115° denotes the value of the cotangent function at an angle of 115 degrees. The degree symbol here is not just a decoration. It is absolutely essential in order to avoid errors, as we now see.

EXAMPLE   Find sin 30 and find sin 30° (note degree symbol). Since an angle of 30 degrees is the *same* as one of $\pi/6$ radians, sin 30° is the same number as $\sin(\pi/6)$. But we know that $\sin(\pi/6) = \frac{1}{2}$. Therefore $\sin 30° = \frac{1}{2}$. On the other hand, 30 degrees is not the same as 30 radians, so sin 30° is *not* the same number as sin 30. As we saw above, sin 30 denotes the value of the sine function at an angle of 30 *radians*. A calculator shows that $\sin 30 \approx -.988$.

The values of the six trigonometric functions at angles of 0°, 30°, 45°, 60°, and 90° are frequently needed. Since these angles are just the angles of 0, $\pi/6$, $\pi/4$, $\pi/3$, and

$\pi/2$ radians, respectively, the functional values are easily found. For example, 45° is the same as $\pi/4$, radians, so that

$$\sin 45° = \sin \frac{\pi}{4} = \frac{\sqrt{2}}{2}, \qquad \cos 45° = \cos \frac{\pi}{4} = \frac{\sqrt{2}}{2}, \qquad \tan 45° = \tan \frac{\pi}{4} = 1$$

Similarly, since 60° is the same as $\pi/3$ radians,

$$\sin 60° = \sin \frac{\pi}{3} = \frac{\sqrt{3}}{2}, \qquad \cos 60° = \cos \frac{\pi}{3} = \frac{1}{2}, \qquad \tan 60° = \tan \frac{\pi}{3} = \sqrt{3}$$

Here is a summary of the values of the six trigonometric functions at frequently used angles. A dash indicates that the function is not defined; all denominators have been rationalized. You should memorize the values for sine, cosine, and tangent.

θ	$\sin \theta$	$\cos \theta$	$\tan \theta$	$\cot \theta$	$\sec \theta$	$\csc \theta$
0°	0	1	0	—	1	—
30°	$\frac{1}{2}$	$\frac{\sqrt{3}}{2}$	$\frac{\sqrt{3}}{3}$	$\sqrt{3}$	$\frac{2\sqrt{3}}{3}$	2
45°	$\frac{\sqrt{2}}{2}$	$\frac{\sqrt{2}}{2}$	1	1	$\sqrt{2}$	$\sqrt{2}$
60°	$\frac{\sqrt{3}}{2}$	$\frac{1}{2}$	$\sqrt{3}$	$\frac{\sqrt{3}}{3}$	2	$\frac{2\sqrt{3}}{3}$
90°	1	0	—	0	—	1

Except for special angles, such as those in the table above, it's usually necessary to use a calculator or tables to evaluate trigonometric functions of angles. Most calculators and tables contain the functional values in terms of both radians and degrees. For example, a calculator operated in *degree* mode shows that

$$\sin 346° \approx -.2419, \qquad \cos 127° \approx -.6018, \qquad \tan 253° \approx 3.2709$$

Trigonometric tables can be used directly to approximate functional values for angles between 0 and 90°. For other angles, you should use the reference number, as explained in the DO IT YOURSELF! segments at the end of Sections 4 and 6. The only changes that need to be made in order to apply that discussion to angles expressed in degrees are to replace π by 180°, replace 2π by 360°, and use the term "**reference angle**" instead of "reference number."

SOLVING RIGHT TRIANGLES

The principal application of classical trigonometry is the solution of triangles. "Solving a triangle" means finding the lengths of all three sides and the measure of all three angles when only three of these six quantities are known in advance. Solving right triangles depends on this fact:

> The right-triangle description of a trigonometric function (for instance, sin θ = opposite/hypotenuse) relates three quantities: the angle θ and two sides of a right triangle

If two of these three quantities are known, then the third one can always be found, as we now see.

EXAMPLE   Find the lengths of sides b and c in the right triangle shown in Figure 6-109.

Figure 6-109

Since the side c is opposite the 75° angle and the hypotenuse is 17, we know two of the three quantities related by the sine function:

$$\sin 75° = \frac{\text{opposite}}{\text{hypotenuse}} = \frac{c}{17}$$

Solving the equation sin 75° = c/17 for c yields:

$$c = 17(\sin 75°)$$

A calculator or table shows that sin 75° ≈ .9659. Therefore $c = 17(\sin 75°) \approx$ 17(.9659) = 16.4203. Since side b is adjacent to the 75° angle and the hypotenuse is known, we use the cosine function to find b:

$$\cos 75° = \frac{\text{adjacent}}{\text{hypotenuse}} = \frac{b}{17}, \quad \text{or equivalently,} \quad b = 17(\cos 75°)$$

A calculator or table shows that $b = 17(\cos 75°) \approx 17(.2588) = 4.3996$.

EXAMPLE Find the measures of angles α and β in the right triangle shown in Figure 6-110.

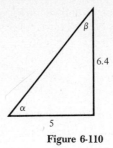

Figure 6-110

Since the lengths of the side opposite and the side adjacent to the angle α are known, it seems natural to use the tangent function here:

$$\tan \alpha = \frac{\text{opposite}}{\text{adjacent}} = \frac{6.4}{5} = 1.28$$

We need only look through the tables to find the angle between $0°$ and $90°$ whose tangent is 1.28. We find that $\tan 52° \approx 1.28$. Therefore $\alpha \approx 52°$. Since the known angles (namely, α and the right angle) have measures $52°$ and $90°$ (a total of $142°$) and the sum of all three angles is $180°$, angle β must have measure $180° - 142° = 38°$.

APPLICATIONS

The solution of many practical problems reduces to solving certain right triangles. Some typical examples are given below.

EXAMPLE A straight road leads from an ocean beach into the nearby hills. The road has a constant upward grade of $3°$. After going from the beach for 1 mi along this road, how high are you above sea level? Figure 6-111 is a picture of the situation (remember that 1 mi = 5280 ft).

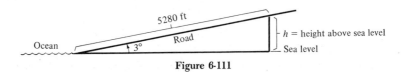

Figure 6-111

We know the $3°$ angle and the hypotenuse of this triangle and must find the side opposite the $3°$ angle. So we use the sine function:

$$\sin 3° = \frac{\text{opposite}}{\text{hypotenuse}} = \frac{h}{5280}, \quad \text{or equivalently,} \quad h = 5280(\sin 3°)$$

A calculator or table shows that $h = 5280(\sin 3°) \approx 5280(.0523) = 276.144$ ft.

In many practical applications, one uses the angle between the horizontal and some other line (for instance, the line of sight from an observer to a distant object). This angle is called the **angle of elevation** or the **angle of depression,** depending on whether the line is above or below the horizontal, as shown in Figure 6-112.

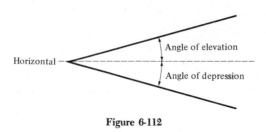

Figure 6-112

EXAMPLE   A wire is to be stretched from the top of a 10-meter-high building to a point on the ground. From the top of the building the angle of depression to the ground point is 22°. How long must the wire be? The picture is in Figure 6-113.

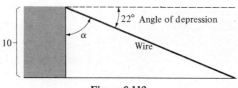

Figure 6-113

Since the sum of the angle of depression and the angle α is obviously 90°, angle α must measure 90° − 22° = 68°. We know the length of the side of the triangle adjacent to the angle α and we want to find the hypotenuse w (the length of the wire). Using the cosine function we see that

$$\cos 68° = \frac{\text{adjacent}}{\text{hypotenuse}} = \frac{10}{w}$$

Solving the equation $\cos 68° = 10/w$ for w yields

$$w = \frac{10}{\cos 68°} \approx \frac{10}{.3746} \approx 26.7$$

Hence the wire must be approximately 26.7 meters long.

EXAMPLE   A person standing on the edge of one bank of a canal observes a lamp post on the edge of the other bank of the canal. The person's eye level is 152 cm above the ground (approximately 5 ft). The angle of elevation from eye level to the top of the lamp post is 12° and the angle of depression from eye level to the bottom of the lamp post is 7°, as shown in Figure 6-114.

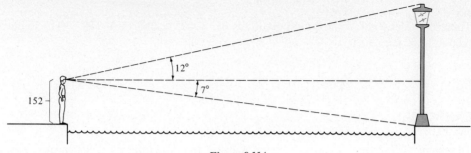

Figure 6-114

How high is the lamp post? How wide is the canal? Abstracting the essential information, we obtain the diagram shown in Figure 6-115.

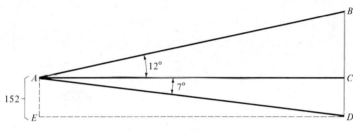

Figure 6-115

We must find the height of the lamp post BD and the width of the canal AC (or ED). The eye level height AE of the observer is 152 cm. Since AC and ED are parallel, CD also has length 152 cm. In right triangle ACD we know the angle of 7° and the side CD opposite it. We must find the adjacent side AC. The tangent function is needed:

$$\tan 7° = \frac{\text{opposite}}{\text{adjacent}} = \frac{152}{AC}, \qquad \text{or equivalently,} \qquad AC = \frac{152}{\tan 7°}$$

A calculator or table shows that $\tan 7° \approx .1228$, whence

$$AC = \frac{152}{\tan 7°} \approx \frac{152}{.1228} \approx 1237.79 \text{ cm}$$

So the canal is approximately 12.3779 meters° wide (about 40.6 ft). Now using right triangle ACB we see that

$$\tan 12° = \frac{\text{opposite}}{\text{adjacent}} = \frac{BC}{AC} \approx \frac{BC}{1237.79}$$

° Remember that 1 m = 100 cm.

or equivalently,

$$BC \approx 1237.79(\tan 12°) \approx 1237.79(.2126) \approx 263.15 \text{ cm}$$

Therefore the height of the lamp post BD is $BC + CD \approx 263.15 + 152 = 415.15$ cm (about 13.62 ft).

EXERCISES

Directions All of these exercises, except A.1–A.4 require either a calculator or tables. Round off all answers to two decimal places.

A.1. Referring to Figure 6-116, evaluate all six trigonometric functions at the given angle θ.

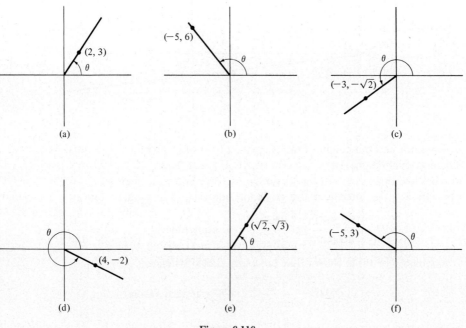

Figure 6-116

A.2. Referring to Figure 6-117, evaluate all six trigonometric functions at the given angle θ.

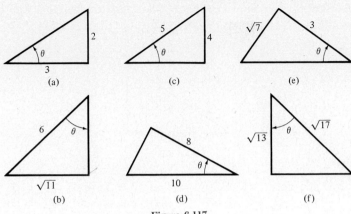

Figure 6-117

A.3. Without using a calculator, find
(a) sin 120°   (c) tan 300°   (e) sec 225°   (g) sin 150°
(b) cos 240°   (d) cot 135°   (f) csc 315°   (h) cos 210°

A.4. Consider the right triangle shown in Figure 6-118 and find c under the given conditions.

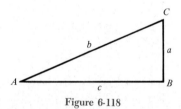

Figure 6-118

(a) $\cos A = \frac{12}{13}$ and $b = 39$ (d) $\sec A = 2$ and $b = 8$
(b) $\sin C = \frac{3}{4}$ and $b = 12$ (e) $\cot A = 6$ and $a = 1.4$
(c) $\tan A = \frac{5}{12}$ and $a = 15$ (f) $\csc C = 1.5$ and $b = 4.5$

A.5. Use the tables at the end of the book to find
(a) sin 27°   (c) tan 49°   (e) sec 65°
(b) cos 31°   (d) cot 24°   (f) csc 82°

A.6. Use appropriate reference angles and the tables at the end of the book to find these functional values.
(a) sin 100°   (c) tan 275°   (e) sec 327°
(b) cos 200°   (d) cot 157°   (f) csc 212°

B.1.   In the right triangle shown in Figure 6-119, find the lengths of all three sides under the given conditions.

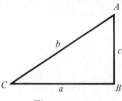

Figure 6-119

(a) $b = 10$ and $\angle C = 50°$ (e) $c = 5$ and $\angle A = 65°$
(b) $c = 12$ and $\angle C = 37°$ (f) $c = 4$ and $\angle C = 28°$
(c) $a = 6$ and $\angle A = 14°$ (g) $b = 3.5$ and $\angle A = 72°$
(d) $a = 8$ and $\angle A = 40°$ (h) $a = 4.2$ and $\angle C = 33°$

B.2.   Find angles A and C of the right triangle shown in Figure 6-120 under the given conditions.

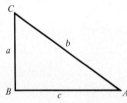

Figure 6-120

(a) $a = 4$ and $c = 6$ (e) $b = 18$ and $c = 12$
(b) $b = 14$ and $c = 5$ (f) $a = 4$ and $b = 9$
(c) $a = 7$ and $b = 10$ (g) $a = 2.5$ and $c = 1.4$
(d) $a = 5$ and $c = 3$ (h) $b = 3.7$ and $c = 2.2$

B.3.   Solve the right triangle shown in Figure 6-121 under the given conditions.

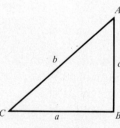

Figure 6-121

(a) $a = 5$, $b = 10$ (c) $b = 7.5$, $c = 2.5$ (e) $c = 10$, $\angle A = 35°$
(b) $a = 3$, $c = 4$ (d) $a = 4$, $\angle A = 22°$ (f) $b = 6.5$, $\angle C = 57°$

B.4. For maximum safety, the distance from the base of a ladder to the building wall should be one-fourth of the length of the ladder. If a ladder is in this position, what angle does it make with the ground?

B.5. The side of a hill makes a 20° angle with the horizontal. If you walk 500 ft up the hillside, how far have you risen vertically?

B.6. It is claimed that the Ohio Turnpike never has an uphill grade of more than 3°. How long must an uphill segment of the road be in order to allow a vertical rise of 450 ft?

B.7. A wire from the top of a TV tower to the ground makes an angle of 49.5° with the ground and touches ground 225 ft from the base of the tower. How high is the tower?

B.8. A buoy in the ocean is observed from the top of a 40-meter-high radar tower on shore. The angle of depression from the top of the tower to the base of the buoy is 6.5°. How far is the buoy from the base of the radar tower?

B.9. A swimming pool is 3 ft deep at the shallow end. The bottom of the pool has a steady downward drop of 12°. If the pool is 50 ft long, how deep is it at the deep end?

B.10. Suppose you are 5 ft tall at the shoulder and have an arm reach of 27 in. If you are standing straight up on a mountain whose side makes a 62° angle with the horizontal, can you reach your hand straight out and touch the mountain without bending over?

B.11. A plane passes directly over your head at an altitude of 500 ft. Two seconds later you observe that its angle of elevation is 42°. How far did the plane travel during those 2 sec?

B.12. A man stands 12 ft from a statue. The angle of elevation from eye level to the top of the statue is 30°, and the angle of depression to the base of the statue is 15°. How tall is the statue?

B.13. In aerial navigation, directions are given in degrees clockwise from north. Thus east is 90°, south is 180°, and so on, as shown in Figure 6-122.

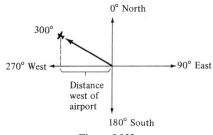

Figure 6-122

A plane travels from an airport for 200 mi in the direction 300°. How far west of the airport is the plane then?

B.14. A plane travels at a constant 300 mph in the direction 65° (see Exercise B.13).
(a) How far east of its starting point is the plane after half an hour?
(b) How far north of its starting point is the plane after 2 hr and 24 min?

B.15. Two boats lie on a straight line with the base of a lighthouse. From the top of the lighthouse (21 meters above water level) it is observed that the angle of depression of the nearest boat is 53° and the angle of depression of the farthest boat is 27°. How far apart are the boats?

B.16. A rocket shoots straight up from the launchpad. Five seconds after lift-off an observer 2 mi away notes that the rocket's angle of elevation is 3.5°. Four seconds later the angle of elevation is 41°. How far did the rocket rise during those 4 sec?

B.17. From a 35 meter high window, the angle of depression to a nearby streetlight is 55°. The angle of depression to the base of the streetlight is 57.8°. How high is the streetlight?

C.1. A 50-ft-high flagpole stands on top of a building. From a point on the ground the angle of elevation of the top of the pole is 43° and the angle of elevation of the bottom of the pole is 40°. How high is the building?

C.2. Two points on level ground are 500 meters apart. The angles of elevation from these points to the top of a nearby hill are 52° and 67°, respectively. How high is the hill?

Topics in Trigonometry

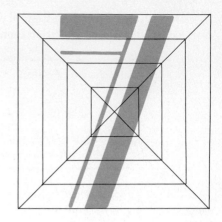

Further properties of trigonometric functions are presented in this chapter. These properties depend heavily on the fundamentals of trigonometry presented in the last chapter. The first part of Section 1 (*Basic Trigonometric Equations*) is needed to understand the other sections. Except for this, the three sections are independent of one another and can be read in any order.

1. TRIGONOMETRIC EQUATIONS

Equations containing trigonometric expressions with a variable, such as

$$\cos x = -.515, \qquad 10 \sin^2 x - 3 \sin x - 1 = 0, \qquad 12 \cot x \cos^2 x = 3 \cot x$$

are called **trigonometric equations.** Solutions of trigonometric equations may be expressed either as real numbers or as angles (usually in degree measure), depending on whether the trigonometric expressions involved are considered as functions of real numbers or as functions of angles.

BASIC TRIGONOMETRIC EQUATIONS

A **basic trigonometric equation** is one that involves a single trigonometric function and constant, such as $\cos x = -.515$ or $\tan x = 3$. The first thing needed to solve such equations is the fact that all of the trigonometric functions are periodic. For example,

$$\sin x = \sin (x \pm 2\pi) = \sin (x \pm 2(2\pi)) = \sin (x \pm 3(2\pi)) = \cdots$$

439

In short, $\sin x = \sin (x + k(2\pi))$ for $k = 0, \pm 1, \pm 2, \pm 3, \ldots$. Similar statements are true for the other trigonometric functions, so that we have:

If u *is a solution of a basic trigonometric equation, then so are the numbers*

$$u + 2k\pi \qquad (k = 0, \pm 1, \pm 2, \pm 3, \ldots)$$

Consequently, we need only find the solutions between 0 *and* 2π *in order to obtain all solutions.*

Although we are primarily interested in numerical solutions (rather than angles), the language of angles is frequently convenient. Recall that if angles in standard position are measured in radians, then angles between 0 and $\pi/2$ have their terminal sides in the first quadrant, angles between $\pi/2$ and π have their terminal sides in the second quadrant, and so on. Consequently, we shall sometimes refer to a **number in the first quadrant,** meaning a number between 0 and $\pi/2$ or a **number in the second quadrant,** meaning a number between $\pi/2$ and π, and so on.

Using this language, we can express the fact that $\cos x < 0$ when $\pi/2 < x < 3\pi/2$ by saying that the cosine function is negative in the second and third quadrants. Similarly, we say that the tangent function is positive in the first and third quadrants since $\tan x > 0$ when $0 < x < \pi/2$ or when $\pi < x < 3\pi/2$ (see the chart on p. 389).

It is now easy to determine the *number* of solutions of a given basic equation between 0 and 2π, and the *quadrants* in which these solutions lie.

EXAMPLE The equation $\sin x = 7$ has no solutions at all since $\sin x$ lies between -1 and 1 for every number x. The equation $\sec x = .42$ has no solutions since $|\sec x| \geq 1$ for every x (see the graph on p. 403).

EXAMPLE The only possibilities for solutions of either of the equations

$$\cos x = -.515 \qquad \text{or} \qquad \cos x = -1$$

are numbers in the second and third quadrants, since these are the only quadrants in which the cosine function is negative. The solutions of $\cos x -.515$ correspond to the points with second coordinate $-.515$ on the graph of the cosine function.

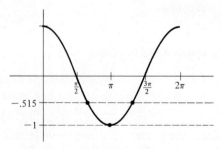

Figure 7-1

As shown in Figure 7-1, there are two such points, so there are exactly two solutions of $\cos x = -.515$ between 0 and 2π, one in the second and one in the third quadrant. Figure 7-1 also shows that there is just one point on the graph with second coordinate -1. Therefore $\cos x = -1$ has just one solution between 0 and 2π.

Similar arguments for the other trigonometric functions show that

> *A basic trigonometric equation has one, two, or no solutions in the interval* $[0, 2\pi]$. *The quadrants in which the solutions lie (if there are any) can be determined by finding the quadrants in which the given trigonometric function is positive or negative.*

If a basic equation with constant c is given, then its **reference equation** is the basic equation obtained by replacing c by $|c|$. For example,

The reference equation of $\tan x = 3$ is $\tan x = |3|$, that is, $\tan x = 3$;
the reference equation of $\sin x = -.47$ is $\sin x = |-.47|$, that is, $\sin x = .47$

Thus the reference equation of a basic equation with a nonnegative constant is just the equation itself. The reference equation of a basic equation with a negative constant is a basic equation with a nonnegative constant.

As we shall see in the examples below, any basic equation with nonnegative constant (in particular, any reference equation) that has a solution actually has a first-quadrant solution that can easily be found. Once we have a first-quadrant solution for the reference equation, we can find all solutions of the original basic equation by using this fact:

> *Given a basic trigonometric equation, let* r *be a first-quadrant solution of its reference equation. Then*
>
> | **If the given basic equation has a solution in this quadrant:** | **That solution is the number:** |
> |---|:---:|
> | *First quadrant* | r |
> | *Second quadrant* | $\pi - $ r |
> | *Third quadrant* | $\pi + $ r |
> | *Fourth quadrant* | $2\pi - $ r |

Here is a proof of this fact. A basic equation involving the sine function must look like either

$$\sin x = b \qquad \text{or} \qquad \sin x = -b$$

for some nonnegative constant b. The first of these equations has solutions in the first and second quadrants. The second one has solutions in the third and fourth quadrants. In either case, the reference equation is $\sin x = b$. Suppose r is a solution, so that

$\sin r = b$. According to the definition of the sine function, the number b is the second coordinate of the point K where the terminal side of an angle of r radians in standard position meets the unit circle. Let a denote the first coordinate of this point K, as shown in Figure 7-2. By using the symmetry of the unit circle or congruent triangles, you can verify that the points L, M, and N have the coordinates indicated in Figure 7-2.

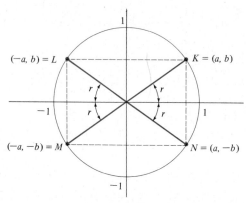

Figure 7-2

The picture shows not only that $\sin r = b$ but also that

$$\sin (\pi - r) = \text{second coordinate of } L = b$$

Therefore r and $\pi - r$ are the two solutions of $\sin x = b$. Similarly,

$$\sin (\pi + r) = \text{second coordinate of } M = -b,$$
$$\sin (2\pi - r) = \text{second coordinate of } N = -b$$

Therefore $\pi + r$ and $2\pi - r$ are the two solutions of $\sin x = -b$. The same picture and analogous arguments can be used in the case of the other trigonometric functions.

If a basic equation with nonnegative constant (such as a reference equation) has any solutions at all, then it must have one in the first quadrant. For all the trigonometric functions are positive in the first quadrant. This first-quadrant solution can be found (or at least approximated) by using one or more of these methods:

(i)   Your knowledge of special angles, such as 30°, 45°, 60°.
(ii)   A calculator.
(iii)   Trigonometric tables, such as those at the end of the book.

EXAMPLE   The equation $\cos x = -.515$ has solutions in the second and third quadrants, since those are the two quadrants where the cosine function is negative. The reference equation is $\cos x = .515$. To find a first-quadrant solution of the reference equation, we look through the cosine columns of the tables at the end of the book until we find the entry .515. We see that $\cos 1.0298 \approx .515$ (or equivalently, $\cos 59° \approx .515$). Therefore the number $r = 1.0298$ is an (approximate) first-quadrant solution of the reference equation. Therefore the solutions of $\cos x = -.515$ in $[0, 2\pi)$ are:

second quadrant: $\pi - r \approx \pi - 1.0298 \approx 2.1118$ (equivalently, $180° - 59° = 121°$)
third quadrant: $\pi + r \approx \pi + 1.0298 \approx 4.1714$ (equivalently, $180° + 59° = 239°$)

Consequently, all solutions of $\cos x = -.515$ are given by:

$$x \approx 2.1118 + 2k\pi \quad \text{and} \quad x \approx 4.1714 + 2k\pi \quad (k = 0, \pm 1, \pm 2, \ldots)$$

EXAMPLE $\sin x = \frac{1}{2}$ is its own reference equation and our knowledge of special angles tells us that a first-quadrant solution is the number $r = \pi/6$ (since $\sin(\pi/6) = \frac{1}{2}$). The sine function is also positive in the second quadrant, so another solution is given by $\pi - r = \pi - (\pi/6) = 5\pi/6$. In terms of degrees, these two solutions are $30°$ and $150°$. All solutions of $\sin x = \frac{1}{2}$ are given by:

$$x = \frac{\pi}{6} + 2k\pi \quad \text{and} \quad x = \frac{5\pi}{6} + 2k\pi \quad (k = 0, \pm 1, \pm 2, \ldots)$$

EXAMPLE Since 3 is positive, the equation $\tan x = 3$ is its own reference equation. To find a first-quadrant solution via calculator, we enter the number 3 and press the button labeled "$\tan^{-1}$" (or on some calculators "inv tan" or "arc tan"). The answer given by the calculator (in radian mode), namely, 1.249, is a number whose tangent is (approximately) 3. Just press the "tan" button to verify this. Therefore the number $r = 1.249$ is a first-quadrant (approximate) solution of $\tan x = 3$. Since the tangent function is positive in the first and third quadrants, there is another solution in the third quadrant:

$$\pi + r \approx \pi + 1.249 \approx 4.3906$$

All solutions of $\tan x = 3$ are given by:

$$x \approx 1.249 + 2k\pi \quad \text{and} \quad x \approx 4.3906 + 2k\pi \quad (k = 0, \pm 1, \pm 2, \ldots)$$

EXAMPLE The equation $\sin 2x = -\sqrt{2}/2$ is *not* a basic equation, but it can be solved by essentially the same method. We first make the substitution $u = 2x$ and then solve the basic equation $\sin u = -\sqrt{2}/2$ for u. The reference equation of $\sin u = -\sqrt{2}/2$ is the equation $\sin u = \sqrt{2}/2$. Our knowledge of special angles tells us that $\sin(\pi/4) = \sqrt{2}/2$. Therefore $r = \pi/4$ is a first-quadrant solution of the reference equation. Since the sine function is negative only in the third and fourth quadrants, the equation $\sin u = -\sqrt{2}/2$ has these solutions between 0 and 2π:

$$\text{third quadrant:} \quad \pi + r = \pi + \frac{\pi}{4} = \frac{5\pi}{4}$$

$$\text{fourth quadrant:} \quad 2\pi - r = 2\pi - \frac{\pi}{4} = \frac{7\pi}{4}$$

Consequently, all solutions of $\sin u = -\sqrt{2}/2$ are given by:

$$u = \frac{5\pi}{4} + 2k\pi \quad \text{and} \quad u = \frac{7\pi}{4} + 2k\pi \quad (k = 0, \pm 1, \pm 2, \ldots)$$

Using the fact that $u = 2x$, each of these solutions produces a solution of the original equation $\sin 2x = -\sqrt{2}/2$:

$$2x = u = \frac{5\pi}{4} + 2k\pi, \quad \text{or equivalently,} \quad x = \frac{1}{2}\left(\frac{5\pi}{4} + 2k\pi\right) = \frac{5\pi}{8} + k\pi,$$

$$2x = u = \frac{7\pi}{4} + 2k\pi, \quad \text{or equivalently,} \quad x = \frac{1}{2}\left(\frac{7\pi}{4} + 2\pi\right) = \frac{7\pi}{8} + k\pi$$

Therefore all solutions of $\sin 2x = -\sqrt{2}/2$ are given by:

$$x = \frac{5\pi}{8} + k\pi \quad \text{and} \quad x = \frac{7\pi}{8} + k\pi \quad (k = 0, \pm 1, \pm 2, \ldots)$$

In particular, this equation has *four* solutions in the interval $[0, 2\pi)$, obtained by letting $k = 0$ and $k = 1$ above: $5\pi/8$, $7\pi/8$, $13\pi/8$, and $15\pi/8$.

OTHER TRIGONOMETRIC EQUATIONS

The usual method for solving more complicated trigonometric equations is to use algebra to reduce the given problem to an equivalent one involving only basic equations. *In the rest of this section, we shall find only the solutions in the interval* $[0, 2\pi)$. Other solutions can then be obtained by adding suitable multiples of 2π.

EXAMPLE   To solve $10 \sin^2 x - 3 \sin x - 1 = 0$, we first factor:°

$$10 \sin^2 x - 3 \sin x - 1 = (2 \sin x - 1)(5 \sin x + 1)$$

Since a product is zero only when at least one of the factors is zero, the original equation reduces to:

$$
\begin{array}{ccc}
2 \sin x - 1 = 0 & \text{or} & 5 \sin x + 1 = 0 \\
2 \sin x = 1 & & 5 \sin x = -1 \\
\sin x = \tfrac{1}{2} & & \sin x = -\tfrac{1}{5} = -.2
\end{array}
$$

As we saw on page 443 the solutions of $\sin x = \frac{1}{2}$ are $\pi/6$ and $5\pi/6$. Using the methods for solving basic equations described above, we find that $\sin x = -.2$ has these (approximate) solutions in the interval $[0, 2\pi)$:

third quadrant: 3.343,   fourth quadrant: 6.0818

Therefore the solutions of the original equation are: $x = \pi/6$, $x = 5\pi/6$, $x \approx 3.343$, and $x \approx 6.0818$.

EXAMPLE   To solve $\tan^2 x + 5 \tan x + 3 = 0$, we first use the substitution $u = \tan x$, so that the equation becomes $u^2 + 5u + 3 = 0$. Since this equation does not readily factor, we use the quadratic formula to solve it:

$$u = \frac{-5 \pm \sqrt{5^2 - 4 \cdot 1 \cdot 3}}{2} = \frac{-5 \pm \sqrt{13}}{2}$$

° In order to see how to factor equations such as this, you may find it helpful to substitute u for $\sin x$, so that $10 \sin^2 x - 3 \sin x - 1$ becomes $10u^2 - 3u - 1 = (2u - 1)(5u + 1)$.

Since $u = \tan x$, the original equation is equivalent to

$$\tan x = \frac{-5 + \sqrt{13}}{2} \approx -.6972 \qquad \text{or} \qquad \tan x = \frac{-5 - \sqrt{13}}{2} \approx -4.3028$$

Solving these basic equations as above, we find that the (approximate) solutions of $\tan x = -.6972$ are $x = 2.5328$ and $x = 5.6744$. Similarly, the (approximate) solutions of $\tan x = -4.3028$ are $x = 1.7992$ and $x = 4.9408$. Thus $\tan^2 x + 5 \tan x + 3 = 0$ has four solutions in $[0, 2\pi)$.

SOLUTIONS VIA TRIGONOMETRIC IDENTITIES

If a trigonometric equation involves more than one trigonometric function, identities can often be used to obtain an equivalent equation involving just one function. There are usually several different ways of doing this, depending on which identities are used.

EXAMPLE To solve $\sec^2 x + 5 \tan x = -2$, we use the Pythagorean Identity $\sec^2 x = 1 + \tan^2 x$ to obtain an equivalent equation:

$$\sec^2 x + 5 \tan x = -2$$
$$\sec^2 x + 5 \tan x + 2 = 0$$
$$(1 + \tan^2 x) + 5 \tan x + 2 = 0$$
$$\tan^2 x + 5 \tan x + 3 = 0$$

As we saw in the preceding example, the solutions in $[0, 2\pi)$ of this equation are $x \approx 1.7992$, $x \approx 2.5328$, $x \approx 4.9408$, and $x \approx 5.6744$.

The solution of an equation involving terms such as $\sin 2x$ or $\cos 2x$ usually involves the double angle identities (p. 416 and p. 417).

EXAMPLE To solve $5 \cos x + 3 \cos 2x = 3$, we use the double angle identity: $\cos 2x = 2 \cos^2 x - 1$ as follows:

$$5 \cos x + 3 \cos 2x = 3$$

$$5 \cos x + 3(2 \cos^2 x - 1) = 3 \qquad \text{(double angle identity)}$$

$$5 \cos x + 6 \cos^2 x - 3 = 3 \qquad \text{(multiply out left side)}$$

$$6 \cos^2 x + 5 \cos x - 6 = 0 \qquad \text{(rearrange terms)}$$

$$(2 \cos x + 3)(3 \cos x - 2) = 0 \qquad \text{(factor left side)}$$

$$\begin{array}{ccc} 2 \cos x + 3 = 0 & \text{or} & 3 \cos x - 2 = 0 \\ 2 \cos x = -3 & & 3 \cos x = 2 \\ \cos x = -\tfrac{3}{2} & & \cos x = \tfrac{2}{3} \end{array}$$

The equation $\cos x = -\frac{3}{2}$ has no solutions since $\cos x$ always lies between -1 and 1. A calculator shows that $.841$ and $2\pi - .841 \approx 5.4422$ are the solutions in $[0, 2\pi)$ of $\cos x = \frac{2}{3}$, and hence of the original equation as well.

EXERCISES

Directions In questions where equations are to be solved, unless stated otherwise,

(i) Find only the solutions in $[0, 2\pi)$.
(ii) Express your answers as numbers; but if angles are called for, express them in degrees.
(iii) Find exact solutions when possible; otherwise, use a calculator or tables and approximate your answers to four decimal places at most.

A.1. Use the graph of the given function to find the *number* of solutions in $[0, 2\pi)$ of each of these basic equations.
(a) $\sin x = .7$ (c) $\sin x = -3$ (e) $\sin x = 1$
(b) $\cos x = \frac{1}{3}$ (d) $\cos x = -.01$ (f) $\csc x = 1.1$

A.2. For each equation below, find the *number* of solutions in $[0, 2\pi)$ and the *quadrants* in which they lie.
(a) $\sin x = -.2$ (c) $\tan x = -6$ (e) $\cos x = \pi/6$
(b) $\cos x = \frac{3}{4}$ (d) $\cos x = -.4$ (f) $\sin x = \pi/4$

A.3. The solutions of $\sin x = 0$ are precisely the points where the graph of the sine function crosses the x-axis. Use this fact (and the similar ones for other functions) to find *all* solutions of:
(a) $\sin x = 0$ (c) $\tan x = 0$ (e) $\sec x = 0$
(b) $\cos x = 0$ (d) $\cot x = 0$ (f) $\csc x = 0$

B.1. Use your knowledge of special angles to find all solutions of these equations.
(a) $\sin x = \sqrt{3}/2$ (d) $\tan x = 1$ (g) $2 \sin x + 1 = 0$
(b) $2 \cos x = \sqrt{2}$ (e) $2 \cos x = -\sqrt{3}$ (h) $\csc x = \sqrt{2}$
(c) $\tan x = -\sqrt{3}$ (f) $\sin x = 0$ (i) $\cot x = 1$

B.2. Use the tables at the end of the book to approximate all solutions in $[0, 2\pi)$ of the given equation.
(a) $\sin x = .119$ (c) $\tan x = -.237$ (e) $\sec x = -2.65$
(b) $\cos x = .958$ (d) $\cot x = 2.3$ (f) $\cos x = -.564$

B.3. Use a calculator to find all *angles* θ with $0° \le \theta < 360°$ such that:
(a) $\sin \theta = .39$ (c) $\tan \theta = 7.95$ (e) $\cos \theta = -.42$
(b) $\cos \theta = .3$ (d) $\tan \theta = 69.4$ (f) $\sin \theta = -.2$

B.4. Find all solutions in $[0, 2\pi)$ of each of these equations (there may be more than two). Describe *all* solutions of each equation.
(a) $\sin 2x = -\sqrt{3}/2$ (c) $\tan 3x = -\sqrt{3}$ (e) $\cos 3x = -.6$
(b) $\cos 2x = \sqrt{2}/2$ (d) $\sin 2x = .4$ (f) $\tan 4x = 8$

B.5. Use factoring and your knowledge of special angles to find all *angles* $0 \le \theta < 360°$ such that:
(a) $2 \sin^2 \theta + 3 \sin \theta + 1 = 0$ (c) $\tan^2 \theta - 3 = 0$
(b) $4 \cos^2 \theta + 4 \cos \theta - 3 = 0$ (d) $2 \sin^2 \theta = 1$

(e) $\cos^2 \theta - 2 \cos \theta + 1 = 0$ (g) $9 \tan^4 \theta - 1 = 0$
(f) $4 \cos^2 \theta + 4 \cos \theta + 1 = 0$ (h) $\sin^2 \theta - 3 \sin \theta = 10$

B.6. Solve these equations. [*Hint:* factor.]
(a) $3 \sin^2 x - 8 \sin x - 3 = 0$ (d) $3 \sin^2 x + 2 \sin x = 5$
(b) $5 \cos^2 x + 6 \cos x = 8$ (e) $32 \cos^2 x - 4 \cos x = 3$
(c) $2 \tan^2 x + 5 \tan x + 3 = 0$ (f) $2 \tan^2 x - 3 \tan x = 20$

B.7. Use factoring to solve these equations (and read Exercise C.4).
(a) $\cot x \cos x = \cos x$ (d) $\tan x \sec x + 3 \tan x = 0$
(b) $\tan x \sin x = \sin x$ (e) $4 \cos^2 x \tan x = \tan x$
(c) $\cos x \csc x = 2 \cos x$ (f) $\csc x \tan^2 x - 5 \csc x = 0$

B.8. Solve each of these equations. [*Hint:* first collect all terms on one side and factor; for instance, $2 \sin x \cos x - \cos x - 2 \sin x + 1 = 0$ factors as $(\cos x - 1)(2 \sin x - 1) = 0$.]
(a) $4 \sin x \tan x - 3 \tan x + 20 \sin x - 15 = 0$
(b) $25 \sin x \cos x - 5 \sin x + 20 \cos x = 4$
(c) $8 \cos x \tan x + 2 \tan x = 28 \cos x + 7$
(d) $2 \cos x \csc^2 x - 2 \cos x = \csc^2 x - 1$

B.9. Use the quadratic formula (as in the Example on p. 444) to solve these equations.
(a) $\sin^2 x + 2 \sin x - 2 = 0$ (d) $4 \cos^2 x - 2 \cos x = 1$
(b) $\cos^2 x + 5 \cos x = 1$ (e) $2 \tan^2 x - 1 = 3 \tan x$
(c) $\tan^2 x + 1 = 3 \tan x$ (f) $6 \sin^2 x + 4 \sin x = 1$

B.10. Use appropriate identities and your knowledge of special angles to solve:
(a) $\sin^2 x + 3 \cos^2 x = 0$ (d) $\cos 2x - \sin x = 1$
(b) $\sec^2 x - 2 \tan^2 x = 0$ (e) $\sin 4x + 2 \sin 2x = 0$
(c) $\sin 2x + \cos x = 0$ (f) $\sin 2x + \sin x + 2 \cos x + 1 = 0$

B.11. Solve these equations.
(a) $9 - 12 \sin x = 4 \cos^2 x$ (d) $2 \tan^2 x + \tan x = 5 - \sec^2 x$
(b) $\sec^2 x + \tan x = 3$ (e) $4 \sin 2x + 3 \sin x = 0$
(c) $\cos^2 x - \sin^2 x + \sin x = 0$ (f) $\sin x + \cos x = \sqrt{2}/2$

B.12. Use the half-angle, cofunction, and other identities to solve:
(a) $\sin (x/2) = 1 - \cos x$ (d) $\sin(x - \pi/2) - \cos x = 1$
(b) $4 \sin^2 (x/2) + \cos^2 x = 2$ (e) $\sin x = \sin (2x - \pi)$
(c) $\sin x + \cos (\pi/2 - x) = 1$ (f) $\tan x = \cot (2x + \pi)$

C.1. (a) Under what conditions (on the constant) does a basic equation involving the sine or cosine function have *no* solutions?
(b) Do part (a) for the secant and cosecant functions.
(c) Do part (a) for the tangent and cotangent functions.

C.2. Find *all* basic equations that have exactly *one* solution in $[0, 2\pi)$. [*Hint:* look at the graphs of the six trigonometric functions.]

C.3.   Let n be a fixed positive integer. Describe *all* solutions of the equation $\sin nx = \frac{1}{2}$. (*Hint:* see Exercise B.4.)

C.4.   What is wrong with this so-called solution?

$$\sin x \tan x = \sin x$$

$$\tan x = 1$$

$$x = \frac{\pi}{4} \quad \text{or} \quad \frac{5\pi}{4}$$

(*Hint:* solve the original equation by moving all terms to one side and factoring. Compare your answers with the one above.)

2. MORE TRIANGLE TRIGONOMETRY

We return now to the solution of triangles, with emphasis on oblique triangles (that is, triangles that do not contain a right angle). Throughout this section all angles are measured in degrees and all trigonometric functions are considered as functions of angles.

The following **standard notation** will be used. If a triangle has vertices A, B, C, then the side opposite vertex A is labeled a, the side opposite vertex B is labeled b, and the side opposite vertex C is labeled c, as shown in Figure 7-3.

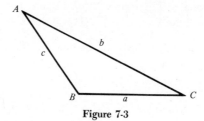

Figure 7-3

The letter A will also be used to label the *angle* of the triangle at vertex A, and similarly for B and C. Thus we shall make statements such as $A = 37°$ or $\cos B = .326$.

THE LAW OF SINES

Here is the first fact needed to solve arbitrary triangles.

LAW OF SINES

In any triangle ABC, *with sides* a, b, c *as shown in Figure 7-3,*

$$\frac{a}{\sin A} = \frac{b}{\sin B} = \frac{c}{\sin C} \; °$$

° An equality of the form $u = v = w$ is shorthand for the statement $u = v$ and $v = w$ and $w = u$.

The Law of Sines is proved at the end of this section. Here are some examples showing how it is actually used to solve triangles in these two cases:

(i)   Two angles and one side are known (AAS).
(ii) Two sides and the angle opposite one of them are known (SSA).

EXAMPLE In triangle ABC (standard notation) shown in Figure 7-4, suppose that $B = 20°$, $C = 31°$, and $b = 210$. Find the other angle and sides.

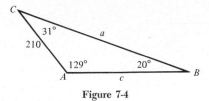

Figure 7-4

Since the sum of all three angles of any triangle is $180°$, we must have

$$A = 180° - (20° + 31°) = 180° - 51° = 129°$$

In order to find side a, we observe that we know three of the four quantities in one of the equations given by the Law of Sines:

$$\frac{a}{\sin A} = \frac{b}{\sin B}$$

$$\frac{a}{\sin 129°} = \frac{210}{\sin 20°}$$

Solving this equation for a, and using a calculator or tables, we obtain:

$$a = \frac{210(\sin 129°)}{\sin 20°} \approx \frac{210(.7771)}{.342} \approx 477.2$$

Side c is found similarly. Beginning with an equation from the Law of Sines involving c and three known quantities, we have:

$$\frac{c}{\sin C} = \frac{b}{\sin B}$$

$$\frac{c}{\sin 31°} = \frac{210}{\sin 20°}$$

$$c = \frac{210 \sin 31°}{\sin 20°} \approx \frac{210(.515)}{.342} \approx 316.2$$

The case in which two sides of a triangle and the angle opposite one of them are known is sometimes called the **ambiguous case.** For there may be two triangles that satisfy the given data. To see why, suppose sides a and b and angle A are given. Place angle A in standard position with terminal side b and think of side a as swinging from vertex C, as shown in Figure 7-5. Then there are four possibilities:

(i) *no solution:* side a is too short to reach the third side.

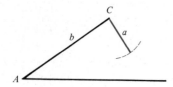

(ii) *one solution:* side a just reaches the third side and is perpendicular to it.

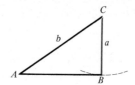

(iii) *two solutions:* an arc of radius a meets the third side at 2 points to the right of A.

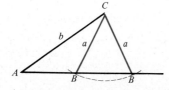

(iv) *one solution:* $a \geq b$, so that an arc of radius a meets the third side at just one point to the right of A.

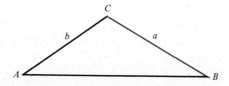

Figure 7-5

It isn't necessary to memorize these cases. And certainly don't memorize the letters here, since different ones may occur in a given problem. As the following examples illustrate, it is easy to tell just what case is involved in a particular problem and how to proceed.

EXAMPLE Given triangle ABC with $a = 6$, $b = 7$, and $A = 65°$, find angle B. We use an equation from the law of sines involving a, b, and A:

$$\frac{b}{\sin B} = \frac{a}{\sin A}$$

$$\frac{7}{\sin B} = \frac{6}{\sin 65°}$$

$$\sin B = \frac{7 \sin 65°}{6} \approx \frac{7(.9063)}{6} = 1.05735$$

There is no angle B whose sine is greater than 1. Therefore there is no triangle satisfying the given data [situation (i) above].

EXAMPLE A plane A takes off from carrier B and flies in a straight line for 12 km. At that instant, an observer on destroyer C, located 5 km from the carrier, notes that the angle determined by the carrier, the destroyer (vertex), and the plane is 37°. How far is the plane from the destroyer? The given data provide the picture in Figure 7-6.

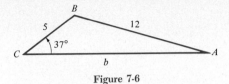

Figure 7-6

We must find side b. To do this, we first use the Law of Sines to find angle A:

$$\frac{a}{\sin A} = \frac{c}{\sin C}$$

$$\frac{5}{\sin A} = \frac{12}{\sin 37°}$$

$$\sin A = \frac{5 \sin 37°}{12} \approx \frac{5(.6018)}{12} \approx .2508$$

The methods of Section 1 show that the equation $\sin A = .2508$ has solutions in the first and second quadrants (found via calculator or tables): $A \approx 14.5°$ and $A \approx 165.5°$. But if $A = 165.5°$ and $C = 37°$, the sum of angles A, B, and C would be greater than 180°. Since this is impossible in a triangle, we conclude that $A = 14.5°$ is the only applicable solution here [situation (iv) above]. Therefore

$$B = 180° - (37° + 14.5°) = 180° - 51.5° = 128.5°$$

Using the Law of Sines again, we have

$$\frac{b}{\sin B} = \frac{c}{\sin C}$$

$$\frac{b}{\sin 128.5°} = \frac{12}{\sin 37°}$$

$$b = \frac{12 \sin 128.5°}{\sin 37°} \approx \frac{23(.7826)}{.6018} \approx 15.6$$

Thus the plane is approximately 15.6 km from the destroyer.

EXAMPLE Solve the triangle ABC with $a = 7.5$, $b = 12$, and $A = 35°$. Using the Law of Sines, we have

$$\frac{b}{\sin B} = \frac{a}{\sin A}$$

$$\frac{12}{\sin B} = \frac{7.5}{\sin 35°}$$

$$\sin B = \frac{12 \sin 35°}{7.5} \approx \frac{12(.5736)}{7.5} \approx .9178$$

Using the methods of Section 1, we find that sin $B = .9178$ has two solutions: $B \approx 66.6°$ and $B \approx 113.4°$. In each case the sum of angles A and B is less than $180°$, so there are two triangles ABC satisfying the given data, as shown in Figure 7-7 [situation (iii) above].

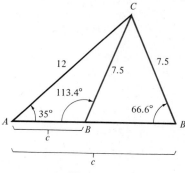

Figure 7-7

Case 1. If $B = 66.6°$, then $C = 180° - (35° + 66.6°) = 78.4°$. By the Law of Sines:

$$\frac{c}{\sin C} = \frac{a}{\sin A}$$

$$\frac{c}{\sin 78.4°} = \frac{7.5}{\sin 35°}$$

$$c = \frac{7.5 \sin 78.4°}{\sin 35°} \approx \frac{7.5(.9796)}{.5736} \approx 12.8$$

Case 2. If $B = 113.4°$, then $C = 180° - (35° + 113.4°) = 31.6°$. Consequently,

$$\frac{c}{\sin C} = \frac{a}{\sin A}$$

$$\frac{c}{\sin 31.6°} = \frac{7.5}{\sin 35°}$$

$$c = \frac{7.5 \sin 31.6°}{\sin 35°} = \frac{7.5(.5240)}{.5736} \approx 6.9$$

THE LAW OF COSINES

In order to solve triangles in cases where the Law of Sines cannot be used directly, we need another fact.

LAW OF COSINES

In any triangle ABC

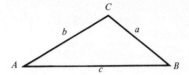

with sides a, b, c *as shown,*

$$a^2 = b^2 + c^2 - 2bc \cos A,$$
$$b^2 = a^2 + c^2 - 2ac \cos B,$$
$$c^2 = a^2 + b^2 - 2ab \cos C$$

Observe that the Pythagorean Theorem is just a special case of the Law of Cosines. For if A is a right angle, then ABC is a right triangle with hypotenuse a. Since $\cos A = \cos 90° = 0$, the first equation of the Law of Cosines becomes $a^2 = b^2 + c^2$.

It is only necessary to memorize one of the three equations above. For a careful examination shows that each of the three equations provides essentially the same thing, namely:

A description of one side of a triangle in terms
of the angle opposite it and the other two sides

If the first equation is solved for $\cos A$, it becomes

$$\cos A = \frac{a^2 - b^2 - c^2}{-2bc}$$

The other two equations can be similarly rewritten. In this form, the Law of Cosines provides

A description of each angle of a triangle
in terms of the three sides of the triangle

Consequently, the Law of Cosines can be used to solve a triangle in these two cases:

(i) Two sides and the angle between them are known (SAS).
(ii) Three sides are known (SSS).

EXAMPLE Solve the triangle ABC when $a = 16$, $b = 10$, and $C = 110°$. The right side of the third equation in the Law of Cosines involves only the known quantities a, b, and C. So we have:

$$c^2 = a^2 + b^2 - 2ab \cos C$$
$$= 16^2 + 10^2 - 2 \cdot 16 \cdot 10 \; \cos 110° = 256 + 100 - 320 \cos 110°$$
$$\approx 356 - 320 \, (-.342) \approx 465.4$$

Therefore $c \approx \sqrt{465.4} \approx 21.6$. The Law of Cosines can now be applied again to find $\cos A$. Using $c \approx 21.6$ and $c^2 \approx 465.4$ we have:

$$\cos A = \frac{a^2 - b^2 - c^2}{-2bc} \approx \frac{16^2 - 10^2 - 465.4}{-2(10)(21.6)} \approx .7162$$

A calculator or table shows that $A \approx 44.3°$. Finally, $B \approx 180° - (44.3° + 110°) = 180° - 154.3° = 25.7°$.

EXAMPLE Find the angles of triangle ABC when the sides are $a = 20$, $b = 15$, and $c = 8.3$. Using the first equation of the Law of Cosines in its rewritten version, we have:

$$\cos A = \frac{a^2 - b^2 - c^2}{-2bc} = \frac{20^2 - 15^2 - 8.3^2}{-2(15)(8.3)} = \frac{106.11}{-249} \approx -.4261$$

The equation $\cos A = -.4261$ has solutions in the second and third quadrants. Since a third-quadrant angle is greater than $180°$, we need only the second-quadrant solution here. Using a calculator or tables, we find that $A \approx 115.2°$. Using the Law of Cosines in rewritten form again yields

$$\cos B = \frac{b^2 - a^2 - c^2}{-2ac} = \frac{15^2 - 20^2 - 8.3^2}{-2(20)(8.3)} = \frac{-243.89}{-332} \approx .7346$$

$$B \approx 42.7°$$

Therefore $C \approx 180° - (115.2° + 42.7°) = 180° - 157.9° = 22.1°$.

PROOF OF THE LAW OF SINES

Given a triangle ABC, position it so that angle A is in standard position with initial side c and terminal side b. There are two possibilities, depending on whether or not angle A is greater than $90°$, as shown in Figure 7-8.

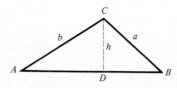

Figure 7-8

In this picture, we have drawn in the altitude h from vertex C. By definition, it is the line segment through C perpendicular to the line through A and B. Using right triangle CDB in either situation, we see that

$$\frac{h}{a} = \sin B, \quad \text{or equivalently,} \quad h = a \sin B$$

Using right triangle ADC on the left in Figure 7-8, we see that

$$\frac{h}{b} = \sin A, \qquad \text{or equivalently,} \qquad h = b \sin A$$

In the right triangle ADC on the right in Figure 7-8, the sum of angle θ and angle A is $180°$, so that $\theta = 180° - A$. Using the identity° $\sin(180° - x) = \sin x$ with $x = A$, we have:

$$\frac{h}{b} = \sin \theta = \sin(180° - A) = \sin A, \qquad \text{or equivalently,} \qquad h = b \sin A$$

Therefore, in both cases, we have $h = a \sin B$ and $h = b \sin A$, so that

$$a \sin B = b \sin A$$

Since angles A and B are angles in a triangle, both are nonzero; hence $\sin A \neq 0$ and $\sin B \neq 0$. Dividing both sides of the last equation above by $\sin A \sin B$ yields:

$$\frac{a}{\sin A} = \frac{b}{\sin B}$$

This proves the first equation in the Law of Sines. Similar arguments beginning with angles B or C in standard position prove the other equations.

PROOF OF THE LAW OF COSINES

Given triangle ABC, position it on a coordinate plane so that angle A is in standard position with initial side c and terminal side b. Depending on the size of angle A, there are two possibilities, as shown in Figure 7-9.

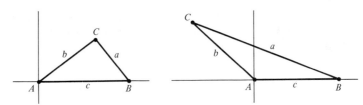

Figure 7-9

The coordinates of B are $(c, 0)$. Let (x, y) be the coordinates of C. Now C is a point on the terminal side of angle A and the distance from C to the origin A is obviously b. Therefore, according to the point-in-the-plane description of sine and cosine (p. 426), we have:

$$\frac{x}{b} = \cos A, \qquad \text{or equivalently,} \qquad x = b \cos A,$$

$$\frac{y}{b} = \sin A, \qquad \text{or equivalently,} \qquad y = b \sin A$$

° This is the "degree version" of the identity $\sin(\pi - x) = \sin x$ proved on page 341.

Using the distance formula on the coordinates of B and C, we have

$$a = \text{distance from } C \text{ to } B$$
$$= \sqrt{(x-c)^2 + (y-0)^2} = \sqrt{(b\cos A - c)^2 + (b\sin A - 0)^2}$$

Squaring both sides of this last equation and simplifying via the Pythagorean Identity yields:

$$a^2 = (b\cos A - c)^2 + (b\sin A)^2$$
$$a^2 = b^2\cos^2 A - 2bc\cos A + c^2 + b^2\sin^2 A$$
$$a^2 = b^2(\sin^2 A + \cos^2 A) + c^2 - 2bc\cos A$$
$$a^2 = b^2 + c^2 - 2bc\cos A$$

This proves the first equation in the Law of Cosines. Similar arguments beginning with angles B or C in standard position prove the other two equations.

EXERCISES

Directions.

(i)   In all exercises triangle ABC is labeled in standard notation, as explained on page 448. The word "meter" is abbreviated "m."
(ii) Find all angles to the nearest one-tenth of a degree with a calculator or to the nearest one-sixth of a degree ($=$ nearest 10 min) with tables.
(iii) Round off all lengths to one decimal place.
(iv) Because of the approximations in (ii) and (iii), especially if they are made during the calculations rather than at the end, your answers may differ slightly from those at the end of the book.

A.1.   Use the Law of Sines to solve triangle ABC, given that
(a) $A = 48°$, $B = 22°$, $a = 5$ (c) $A = 116°$, $C = 50°$, $a = 8$
(b) $B = 33°$, $C = 46°$, $b = 4$ (d) $A = 105°$, $B = 27°$, $b = 10$

A.2.   Use the Law of Cosines to solve triangle ABC, given that
(a) $A = 20°$, $b = 10$, $c = 7$ (c) $C = 118°$, $a = 6$, $b = 10$
(b) $B = 40°$, $a = 12$, $c = 20$ (d) $C = 52.5°$, $a = 6.5$, $b = 9$

A.3.   Use the Law of Cosines to solve triangle ABC, given that
(a) $a = 7$, $b = 3$, $c = 5$ (c) $a = 16$, $b = 20$, $c = 32$
(b) $a = 8$, $b = 5$, $c = 10$ (d) $a = 5.3$, $b = 7.2$, $c = 10$

B.1.   Use the Law of Sines to find all triangles ABC (if any) that satisfy the given data.
(a) $a = 25$, $b = 10$, $A = 42°$ (d) $a = 12$, $b = 5$, $B = 20°$
(b) $b = 15$, $c = 25$, $B = 47°$ (e) $b = 30$, $c = 50$, $C = 60°$
(c) $a = 20$, $c = 15$, $A = 30°$ (f) $b = 12.5$, $c = 20.1$, $B = 37.3°$

B.2.   Solve triangle ABC, given that
(a) $A = 41°$, $B = 67°$, $a = 10.5$ (d) $a = 50$, $c = 80$, $C = 45°$
(b) $a = 30$, $b = 40$, $A = 30°$ (e) $a = 6$, $b = 12$, $c = 18$
(c) $b = 4$, $c = 10$, $A = 75°$ (f) $B = 20°40'$, $C = 34°$, $b = 185$

B.3.   A surveyor marks points A and B 200 m apart on one bank of a river. She sights a point C on the opposite bank and determines that angle CAB is 80° and angle CBA is 36°. Find the distance from A to C.

B.4.   A pole tilts at an angle 9° from the vertical, away from the sun, and casts a shadow 24 ft long. The angle of elevation from the end of the pole's shadow to the top of the pole is 53°. How long is the pole?

B.5.   A vertical statue 6.3 m high stands on the top of a hill. At a point on the side of the hill 35 m from the statue's base, the angle between the hillside and a line from the top of the statue is 10°. What angle does the side of the hill make with the horizontal?

B.6.   How far is the pitcher's mound from first base on a standard baseball diamond? (*Note:* the diamond is a square with each side 90 ft long; the pitcher's mound is 60 ft from home plate along the line from home plate to second base.)

B.7.   Two trains leave a station on different tracks. The tracks make a 112° angle with the station as vertex. The first train travels at an average speed of 90 km/hr and the second at an average speed of 55 km/hr. How far apart are the trains after 2 hr and 45 min?

B.8.   The distance from Chicago to St. Louis is 490 km, from St. Louis to Atlanta 640 km, and from Atlanta to Chicago 795 km. What are the angles in the triangle with these three cities as vertices?

B.9.   The side of a hill makes an angle of 12° with the horizontal. A wire is to be run from the top of a 175-ft tower on the top of the hill to a stake located 120 ft down the hillside from the base of the tower. How long a wire is needed?

B.10.   Two straight roads meet at an angle of 40° at point D. A person walks down the first road for 18 km until he reaches town A. Now town A is 20 km from a town B on the second road. How far is town B from point D along the second road?

B.11.   A straight tunnel is to be dug through a hill. Two people stand on opposite sides of hill where the tunnel entrances are to be located. Both can see a stake located 530 m from the first person and 755 m from the second. The angle determined by the two people and the stake (vertex) is 77°. How long must the tunnel be?

B.12.   A straight path makes an angle of 6° with the horizontal. A statue at the end of the path casts a 6.5-m-long shadow straight down the path. The angle of elevation from the end of the shadow to the top of the statue is 32°. How tall is the statue?

B.13.   A plane flies in a direction of 105° from airport A. After a time, it turns and proceeds in a direction of 267°. Finally, it lands at airport B, 120 mi directly south of airport A. How far has the plane traveled? (*Note:* aerial navigation directions are explained in Exercise B.13 on p. 437.)

C.1.   Prove that the area of triangle ABC is $\frac{1}{2}bc \sin A$. (*Hint:* look at the pictures in the proof of the Law of Sines.)

3. INVERSE TRIGONOMETRIC FUNCTIONS

Three new functions are introduced here: the inverse sine, the inverse cosine, and the inverse tangent functions. As their names imply, they are closely related to the sine, cosine, and tangent functions. The definition and basic properties of these inverse trigonometric functions are special cases of a more general concept of inverse function. We shall consider this general situation in order to obtain an easy method of graphing the inverse trigonometric functions, as well as other inverse functions. Additional facts about inverse functions in the general situation are discussed in the DO IT YOURSELF! segment at the end of this section.

THE INVERSE SINE FUNCTION

The graph of the sine function from $x = -\pi/2$ to $x = \pi/2$ (shown in Figure 7-10) has this geometric property:

> Each horizontal line between $y = -1$ and $y = 1$
> intersects the graph at exactly one point

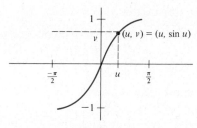

Figure 7-10

If (u, v) is a point on this graph, then $\sin u = v$. Consequently, the geometric property above can be translated into algebraic language as follows:

> For each number v between -1 and 1, there is exactly
> one number u between $-\pi/2$ and $\pi/2$ such that $\sin u = v$

We now use this fact to define a new function g.

Name of g: *inverse sine function* or *arcsine function.*
Domain of g: *all numbers in the interval* $[-1, 1]$.
Rule of g: *for each number* v *in the domain* $[-1, 1]$, *g(v) is
the unique number* u *in* $[-\pi/2, \pi/2]$ *such that*
$\sin u = v$.
Notation for the number g(x): $sin^{-1} x$ *or arcsin* x.

We shall usually use the name "inverse sine function" and the notation $g(x) = \sin^{-1} x$. But both notations mean exactly the same thing: for each number x in $[-1, 1]$,

$$\sin^{-1} x = \arcsin x = g(x) = \text{the unique number in } \left[\frac{-\pi}{2}, \frac{\pi}{2}\right] \text{ whose sine is } x$$

EXAMPLE In order to find $\sin^{-1}(\tfrac{1}{2})$ we must find the number in the interval $[-\pi/2, \pi/2]$ whose sine is $\tfrac{1}{2}$. But we know that $\sin(\pi/6) = \tfrac{1}{2}$ and $\pi/6$ is in $[-\pi/2, \pi/2]$. Therefore $\sin^{-1}(\tfrac{1}{2}) = \pi/6$.

Warning The notation $g(x) = \sin^{-1} x$ for the inverse sine function is *not* exponent notation. $\sin^{-1} x$ always denotes the value of the inverse sine function at x, and *not* the number $(\sin x)^{-1} = 1/\sin x$. This is important since $\sin^{-1} x$ and $1/\sin x$ are usually different numbers. For instance, we just saw that

$$\sin^{-1}\left(\frac{1}{2}\right) = \frac{\pi}{6} \approx .52; \quad \text{but} \quad \left(\sin \frac{1}{2}\right)^{-1} = \frac{1}{\sin\left(\dfrac{1}{2}\right)} \approx \frac{1}{.4794} \approx 2.09$$

EXAMPLE $\sin^{-1}(-\sqrt{2}/2) = -\pi/4$ since $-\pi/4$ is the unique number in $[-\pi/2, \pi/2]$ such that $\sin(-\pi/4) = -\sqrt{2}/2$. Similarly, $\sin^{-1} 1 = \pi/2$ since $\sin(\pi/2) = 1$.

EXAMPLE The inverse sine function can be evaluated by using a calculator equipped with a key labeled "$\sin^{-1}$" (or on some calculators, "inv sin" or "arc sin"). Using such a calculator in radian mode, we find that

$$\sin^{-1} .85 \approx 1.02, \quad \sin^{-1}(-.67) \approx -.73, \quad \sin^{-1}(.49) \approx .51.$$

EXAMPLE Trigonometric tables can also be used to evaluate the inverse sine function. To find $\sin^{-1}(.42)$, we look through the sine columns until we find that $\sin .4334 \approx .42$. Therefore $\sin^{-1}(.42) \approx .4334$.

EXAMPLE Find $\sin^{-1} 1.47$. Since the domain of the inverse sine function is the interval $[-1, 1]$ and 1.47 is *not* in this domain, $\sin^{-1} 1.47$ is *not defined*.

The preceding examples show that

$$\sin^{-1}\left(\frac{1}{2}\right) = \frac{\pi}{6} \qquad \text{while} \qquad \sin \frac{\pi}{6} = \frac{1}{2}$$

$$\sin^{-1}\left(\frac{-\sqrt{2}}{2}\right) = \frac{-\pi}{4} \qquad \text{while} \qquad \sin\left(\frac{-\pi}{4}\right) = \frac{-\sqrt{2}}{2}$$

$$\sin^{-1}(1) = \frac{\pi}{2} \qquad \text{while} \qquad \sin \frac{\pi}{2} = 1$$

This same pattern holds true more generally. The definition of the inverse sine function shows that for any number v in the interval $[-1, 1]$ and any number u in the interval $[-\pi/2, \pi/2]$,

$$\boxed{\sin^{-1} v = u \qquad \textit{exactly when} \qquad \sin u = v}$$

When this is the case, we have:

$$\sin^{-1}(\sin u) = \sin^{-1} v = u \qquad \text{and} \qquad \sin(\sin^{-1} v) = \sin u = v$$

In other words,

$\sin^{-1}(\sin u) = u$ *for every number* u *in* $[-\pi/2, \pi/2]$ *and*

$\sin(\sin^{-1} v) = v$ *for every number* v *in* $[-1, 1]$.

For instance, $\sin^{-1}(\sin(\pi/6)) = \sin^{-1}(\frac{1}{2}) = \pi/6$; and $\sin(\sin^{-1}(-\sqrt{2}/2)) = \sin(-\pi/4) = -\sqrt{2}/2$.

A calculator provides a visual demonstration of the facts in the box above. For example, if you enter any number between $-\pi/2$ and $\pi/2$, press the "sin" key, and then press the "sin^{-1}" key, the result will be the number with which you started.°

INVERSE FUNCTIONS

Let f and g be functions. We say that g is the **inverse** of f (or that f is the inverse of g) provided that

$$(g \circ f)(x) = x \qquad \text{for every number } x \text{ in the domain of } f,$$
$$(f \circ g)(x) = x \qquad \text{for every number } x \text{ in the domain of } g$$

This property of inverse functions may be paraphrased by saying that "g undoes what f does" and "f undoes what g does." For instance,

$$\left. \begin{array}{c} \text{begin with the} \\ \text{number } x \end{array} \right) \xrightarrow{\quad} \text{apply } f \xrightarrow{\quad} \left(\begin{array}{l} \text{then apply } g \text{ and end} \\ \text{up where you started} \end{array} \right.$$

$$x \xrightarrow{\qquad\qquad} f(x) \xrightarrow{\qquad\qquad} g(f(x)) = (g \circ f)(x) = x$$

One example of inverse functions was given on page 298, where we saw that the logarithmic function $g(x) = \log x$ is the inverse of the exponential function $f(x) = 10^x$.

As you might suspect from its name, the inverse sine function is another example of an inverse function in this sense. The inverse sine function $g(x) = \sin^{-1} x$ is the inverse of the function f whose domain is $[-\pi/2, \pi/2]$ and whose rule is $f(x) = \sin x$. To see why, we use the facts in the last box above:

$$(g \circ f)(x) = g(f(x)) = g(\sin x) = \sin^{-1}(\sin x) = x \qquad \text{for every}$$
x in $[-\pi/2, \pi/2]$, that is, for every x in the domain of f

$$(f \circ g)(x) = f(g(x)) = f(\sin^{-1} x) = \sin(\sin^{-1} x) = x \qquad \text{for every}$$
x in $[-1, 1]$, that is, for every x in the domain of g

Thus g is the inverse of f. There is a subtle but important fact to note here. The function f, whose domain is defined to be the interval $[-\pi/2, \pi/2]$, is *not* the same

---

° Except possibly for a slight error due to approximating and rounding off.

function as the sine function, whose domain consists of *all* real numbers. But f is obviously closely related to the sine function. We describe this situation by saying that f is obtained by **restricting the domain** of the sine function.

There is more than just nitpicking involved here. As shown in the DO IT YOUR-SELF! segment at the end of this section, many functions do *not* have inverses. But it is often possible to restrict the domain of such a function so as to obtain a new function that *does* have an inverse. This is just what we did above with the sine function. We shall now do it again for the cosine and tangent functions.

THE INVERSE COSINE FUNCTION

Essentially the same procedure presented above can be used to define the inverse cosine function. The graph of the cosine function from $x = 0$ to $x = \pi$ (shown in Figure 7-11) has the same geometric property as before:°

> Each horizontal line between $y = -1$ and $y = 1$
> intersects the graph at exactly one point

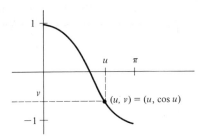

Figure 7-11

Translated into algebraic terms, this property becomes:

> For each number v between -1 and 1, there is exactly
> one number u between 0 and π such that $\cos u = v$

This fact can be used to define a new function h as follows.

Name of h: *inverse cosine function* or *arccosine function.*
Domain of h: *all numbers in the interval* $[-1, 1]$.
Rule of h: *for each number* v *in the domain* $[-1, 1]$, h(v) *is the unique number* u *in* $[0, \pi]$ *such that* $\cos u = v$.
Notation for the number h(x): $\cos^{-1} x$ *or arccos* x.

° The reason why we consider the graph from $x = 0$ to $x = \pi$, instead of from $x = -\pi/2$ to $x = \pi/2$, is explained in the DO IT YOURSELF! segment at the end of this section.

We shall usually use the name "inverse cosine function" and the notation $h(x) = \cos^{-1} x$. But both notations mean exactly the same thing: for each number x in $[-1, 1]$,

$$\cos^{-1} x = \arccos x = h(x) = \text{the unique number in } [0, \pi] \text{ whose cosine is } x$$

EXAMPLE $\cos^{-1}(\frac{1}{2}) = \pi/3$ since $\pi/3$ is the unique number in $[0, \pi]$ whose cosine is $\frac{1}{2}$. Similarly, $\cos^{-1} 0 = \pi/2$ since $\cos (\pi/2) = 0$ and $\cos^{-1}(-\sqrt{3}/2) = 5\pi/6$ since $\cos (5\pi/6) = -\sqrt{3}/2$.

EXAMPLE A calculator (in radian mode) equipped with a "$\cos^{-1}$" (or "inv cos" or "arc cos") key shows that

$$\cos^{-1}(-.7) \approx 2.35, \qquad \cos^{-1}(.15) \approx 1.42, \qquad \cos^{-1}(.87) \approx .52$$

Warning $\cos^{-1} x$ is *not* exponent notation; it doesn't mean $(\cos x)^{-1} = 1/\cos x$.

Let v be a number in the interval $[-1, 1]$ and u a number in $[0, \pi]$. The definition of the inverse cosine function shows that

$$\boxed{\cos^{-1} v = u \qquad \textit{exactly when} \qquad \cos u = v.}$$

When this is the case,

$$\cos^{-1}(\cos u) = \cos^{-1} v = u \qquad \text{and} \qquad \cos (\cos^{-1} v) = \cos u = v$$

Therefore

$$\boxed{\begin{array}{ll} \cos^{-1}(\cos u) = u & \textit{for every number } u \textit{ in } [0, \pi] \textit{ and} \\ \cos (\cos^{-1} v) = v & \textit{for every number } v \textit{ in } [-1, 1]. \end{array}}$$

These facts can sometimes be used to rewrite various expressions involving trigonometric and inverse trigonometric functions in a more convenient form.

EXAMPLE Suppose $0 \le v \le 1$. Write $\sin (\cos^{-1} v)$ as an algebraic expression in v. Now $\cos^{-1} v$ is some number, say, $\cos^{-1} v = u$. We must find $\sin (\cos^{-1} v) = \sin u$. Since $0 \le v \le 1$ we know that $u = \cos^{-1} v$ lies between 0 and $\pi/2$. Hence $\sin u$ is nonnegative. Thus by the Pythagorean identity $\sin^2 u = 1 - \cos^2 u$, we have $\sin u = \sqrt{1 - \cos^2 u}$. Furthermore,

$$\cos^{-1} v = u \qquad \text{exactly when} \qquad \cos u = v.$$

Using these facts, we see that

$$\sin (\cos^{-1} v) = \sin u = \sqrt{1 - \cos^2 u} = \sqrt{1 - v^2}$$

Therefore, for $0 \le v \le 1$, $\sin (\cos^{-1} v) = \sqrt{1 - v^2}$.

Let f be the function obtained by restricting the domain of the cosine function to the interval $[0, \pi]$. Then f has domain $[0, \pi]$ and rule $f(x) = \cos x$. The facts in the last box above can be used to show that the function $h(x) = \cos^{-1} x$ is the inverse of the function f:

$$(h \circ f)(x) = h(f(x)) = h(\cos x) = \cos^{-1}(\cos x) = x \qquad \text{for every}$$
x in $[0, \pi]$, that is, for every x in the domain of f,

$$(f \circ h)(x) = f(h(x)) = f(\cos^{-1} x) = \cos(\cos^{-1} x) = x \qquad \text{for every}$$
x in $[-1, 1]$, that is, for every x in the domain of h

THE INVERSE TANGENT FUNCTION

The graph of the tangent function from $x = -\pi/2$ to $x = \pi/2$ (Figure 7-12) shows that the tangent function has this property:

> For each real number v, there is exactly one number u strictly between $-\pi/2$ and $\pi/2$ such that $\tan u = v$

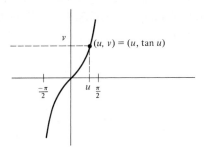

Figure 7-12

This fact is precisely what is needed to define a new function k.

Name of k: *inverse tangent function* or *arctangent function.*
Domain of k: *all real numbers.*
Rule of k: *for each real number* v, k(v) *is the unique real number*
u *in* $(-\pi/2, \pi/2)$ *such that* tan u = v.
Notation for the number k(x): tan^{-1} x *or arctan* x.

Once again, both notations mean exactly the same thing: for each real number x,

$$\tan^{-1} x = \arctan x = k(x) = \text{the unique number in } \left(\frac{-\pi}{2}, \frac{\pi}{2}\right) \text{ whose tangent is } x$$

EXAMPLES $\tan^{-1} 1 = \pi/4$ since $\pi/4$ is the unique number in $(-\pi/2, \pi/2)$ such that $\tan(\pi/4) = 1$. Similarly, $\tan^{-1}(-\sqrt{3}) = -\pi/3$ since $\tan(-\pi/3) = -\sqrt{3}$. A calculator (in radian mode) shows that

$$\tan^{-1}(-5) \approx -1.37, \qquad \tan^{-1}(1.2) \approx .88, \qquad \tan^{-1}(136) \approx 1.56$$

Warning $\tan^{-1} x$ is *not* exponent notation and does not mean $(\tan x)^{-1} = 1/\tan x$.

Let u be a number in the interval $(-\pi/2, \pi/2)$ and v any real number. According to the definition of the inverse tangent function

$$\boxed{tan^{-1} v = u \qquad exactly\ when \qquad tan\ u = v}$$

When this is the case,

$$\tan^{-1}(\tan u) = \tan^{-1} v = u \qquad \text{and} \qquad \tan(\tan^{-1} v) = \tan u = v$$

Consequently,

$$\boxed{\begin{aligned} &tan^{-1}(tan\ u) = u\ for\ every\ number\ u\ in\ (-\pi/2, \pi/2)\ and \\ &tan(tan^{-1} v) = v\ for\ every\ number\ v. \end{aligned}}$$

Just as above, these facts can be used to show that the function $k(x) = \tan^{-1} x$ is the inverse of the function f obtained by restricting the domain of the tangent function to the interval $(-\pi/2, \pi/2)$.

Many calculations involving trigonometric and inverse trigonometric functions can be made directly without a calculator or tables.

EXAMPLE   Find $\cos(\tan^{-1}(\sqrt{5}/2))$. Let $\tan^{-1}(\sqrt{5}/2) = u$, so that $-\pi/2 < u < \pi/2$ and $\tan u = \sqrt{5}/2$. We must find $\cos u$. Since $\tan u$ is positive, we must have $0 < u < \pi/2$. Construct a right triangle containing an angle of u radians whose tangent is $\sqrt{5}/2$ (Figure 7-13).

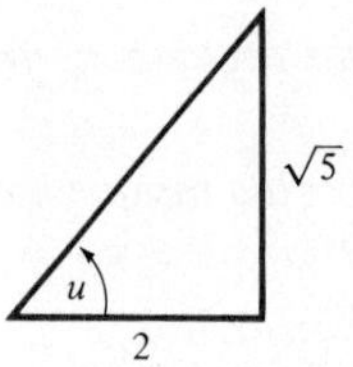

$$\tan u = \frac{\text{opposite}}{\text{adjacent}} = \frac{\sqrt{5}}{2}$$

Figure 7-13

The hypotenuse has length $\sqrt{2^2 + (\sqrt{5})^2} = \sqrt{4 + 5} = 3$. Therefore

$$\cos\left(\tan^{-1}\frac{\sqrt{5}}{2}\right) = \cos u = \frac{\text{adjacent}}{\text{hypotenuse}} = \frac{2}{3}$$

GRAPHS OF INVERSE FUNCTIONS

Since we already know the graphs of the restricted sine, cosine, and tangent functions, we shall use these graphs to obtain the graphs of their inverse functions. The method of doing this can be used to obtain the graph of *any* inverse function, providing the graph of the original function is known.

So suppose that f is a function whose graph is known (such as the restricted sine, cosine, or tangent functions whose graphs appear above). And suppose g is the inverse function of f, so that

$$(g \circ f)(x) = x \text{ for every } x \text{ in the domain of } f,$$
$$(f \circ g)(x) = x \text{ for every } x \text{ in the domain of } g$$

Under these circumstances, we claim that

> (u, v) *is on the graph of* f *exactly when* (v, u) *is on the graph of* g.

To prove this claim, let (u, v) be a point on the graph of f, so that $v = f(u)$. Applying the inverse function g to v yields:

$$g(v) = g(f(u)) = (g \circ f)(u) = u$$

Therefore $g(v) = u$ so that the point $(v, g(v))$ on the graph of g is just the point (v, u). A similar argument, beginning with a point on the graph of g, completes the proof.

The statement in the box above says that each point on the graph of the inverse function g can be obtained by taking a point on the graph of f and "reversing" its coordinates. A simple geometric way of doing just that is to reflect the plane in the line $y = x$ (that is, rotate the plane $180°$ around the line $y = x$). If you do this, the positive vertical axis ends up in the positive horizontal position and the positive horizontal axis ends up in the positive vertical position, as shown in Figure 7-14. Consequently, the point whose original coordinates were (u, v) now has first coordinate v and second coordinate u.

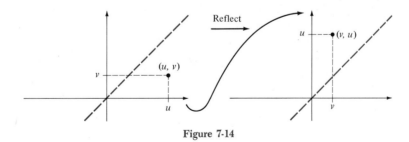

Figure 7-14

Thus reflecting the plane in the line $y = x$ moves each point (u, v) on the graph of f onto the point (v, u) on the graph of g. In summary,

> *If g is the inverse function of f, then the graph of g can be obtained by reflecting the graph of f in the line y = x.*

An informal way of thinking of this fact is this: the graph of g is the mirror image of the graph of f, with the line $y = x$ being the mirror.

Using this reflection method, we easily obtain the graphs of the inverse trigonometric functions as shown in Figure 7-15.°

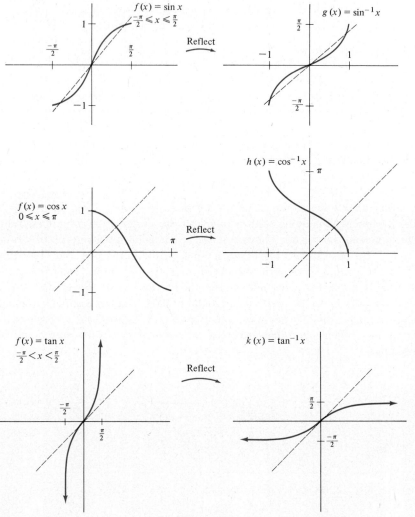

Figure 7-15

° Although it may be confusing at first, we have followed the usual custom of labeling all function variables, and hence all horizontal axes, by x.

For more examples of graphs of inverse functions, see Exercise B.4 below and Exercises B.1 and B.2 in the DO IT YOURSELF! segment at the end of this section.

EXERCISES

A.1. Find each of these numbers without using tables or calculators.

(a) $\sin^{-1}(\sqrt{2}/2)$ (d) $\sin^{-1}(\sqrt{3}/2)$ (g) $\cos^{-1}(\sqrt{3}/2)$

(b) $\cos^{-1}(\sqrt{3}/2)$ (e) $\cos^{-1} 1$ (h) $\tan^{-1}(-\sqrt{3}/3)$

(c) $\tan^{-1} 0$ (f) $\tan^{-1}(-\sqrt{3})$

A.2. Use a calculator to find the numbers below. (Read the directions for Exercise B.2.)

(a) $\sin^{-1}(.35)$ (d) $\sin^{-1}(-.795)$ (g) $\tan^{-1}(\tan(-4))$

(b) $\cos^{-1}(.76)$ (e) $\sin^{-1}(\sin 7)$ (h) $\sin^{-1}(3 \sin (-2))$

(c) $\tan^{-1}(-3.256)$ (f) $\cos^{-1}(\cos 3.5)$ (i) $\cos^{-1}(\cos 3.75)$

B.1. Find each of these numbers without using tables or calculators.

(a) $\sin^{-1}(\cos 0)$ (d) $\tan^{-1}(\sin \pi)$ (g) $\cos^{-1}(\sin (-5\pi/6))$

(b) $\cos^{-1}(\sin (\pi/6))$ (e) $\cos^{-1}(\sin (4\pi/3))$ (h) $\sin^{-1}(\cos (5\pi/4))$

(c) $\sin^{-1}(\tan (\pi/4))$ (f) $\tan^{-1}(\cos \pi)$

B.2. Since $-1 \le \sin x \le 1$ for *every* number x, the number $\sin^{-1}(\sin x)$ is *always* defined. When $-\pi/2 \le x \le \pi/2$, then $\sin^{-1}(\sin x) = x$, as shown on page 460. But for other values of x, it may *not* be true that $\sin^{-1}(\sin x) = x$. Similar remarks apply to $\cos^{-1}(\cos x)$ and $\tan^{-1}(\tan x)$. Evaluate (without using calculators or tables):

(a) $\sin^{-1}(\sin (2\pi/3))$ (c) $\tan^{-1}(\tan (5\pi/6))$ (e) $\cos^{-1}(\cos (-\pi/6))$

(b) $\cos^{-1}(\cos (5\pi/4))$ (d) $\sin^{-1}(\sin (7\pi/4))$ (f) $\tan^{-1}(\tan (-4\pi/3))$

B.3. Evaluate without using a calculator or tables. [See the Examples on page 464.]

(a) $\sin (\cos^{-1} \frac{3}{5})$ (c) $\cos (\tan^{-1}(-\frac{3}{4}))$ (e) $\tan \left(\cos^{-1} \dfrac{3\sqrt{5}}{7}\right)$

(b) $\tan (\sin^{-1} \frac{3}{5})$ (d) $\sin \left(\tan^{-1} \dfrac{\sqrt{3}}{2}\right)$ (f) $\cos (\sin^{-1} \sqrt{\frac{3}{5}})$

B.4. Sketch the graph of the inverse function of each of the functions whose graph appears in Figure 7-16.

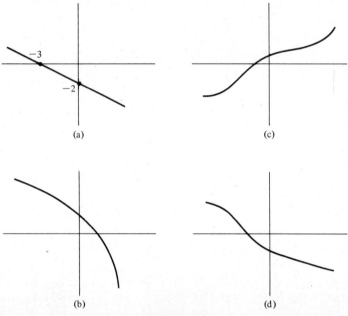

Figure 7-16

C.1. Prove that each of the following statements is an identity (that is, that the equation is true for all numbers v for which both sides are defined). (*Hint:* see the last Example on p. 462.)

(a) $\cos(\sin^{-1} v) = \sqrt{1 - v^2}$

(b) $\tan(\sin^{-1} v) = v/\sqrt{1 - v^2}$ [*Hint:* let $\sin^{-1} v = u$, so that $-\pi/2 < u < \pi/2$ and $\sin u = v$. Use the identities $1 + \tan^2 u = \sec^2 u = 1/\cos^2 u$ and $\sin^2 u + \cos^2 u = 1$ to express $\tan(\sin^{-1} v) = \tan u$ in terms of $\sin u = v$. When taking square roots, check signs carefully.]

(c) $\tan(\cos^{-1} v) = \sqrt{1 - v^2}/v$

(d) $\sin(\tan^{-1} v) = v/\sqrt{1 + v^2}$

(e) $\cos(\tan^{-1} v) = 1/\sqrt{1 + v^2}$

DO IT YOURSELF!

WHICH FUNCTIONS HAVE INVERSES?

In answering the question posed in the title, we shall restrict our attention to the type of function that actually occurs in practice, namely, functions whose graphs are smooth, connected, continuous curves. In order to understand the discussion, you should remember that an *increasing function* is one whose graph is always rising (from left to right) and a *decreasing function* is one whose graph is always falling (from left to right). For more details on these terms, see pages 135–137.

Each of the functions whose inverse function has been found thus far is either increasing or decreasing. For example, the exponential function $f(x) = 10^x$ is increasing, and its inverse function $g(x) = \log x$ is also increasing, as shown in Figure 7-17.

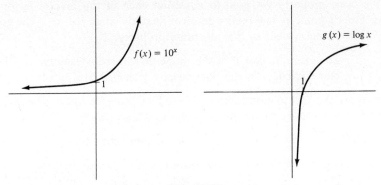

Figure 7-17

Similarly, the restricted cosine function is decreasing, as is its inverse, the inverse cosine function (see p. 466). Consequently, the following fact should seem plausible:

> *If f is either an increasing or a decreasing function, then f has an inverse function with the same property.*

In order to prove this statement, we first recall what the range of a function is. If you apply the rule of a function f to every number in its domain, the resulting set of numbers is the range of f. In graphical terms, the range of f consists of all numbers that are second coordinates of points on the graph of f. Now suppose that f is an *increasing* function and that v is a number in the range of f. Consider the horizontal line $y = v$ (the line consisting of all points with second coordinate v). For example, the situation might look like the curves in Figure 7-18.

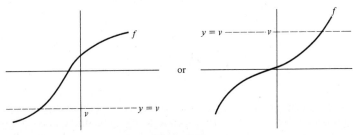

Figure 7-18

Imagine a point moving along the graph of f from left to right. Once the point crosses the horizontal line $y = v$, it keeps moving upward since the graph of f is *always rising*. The point can't move downward, and hence cannot cross the horizontal line $y = v$ again. Thus the graph of an increasing function has this geometric property:

> For each number v in the range of f, the horizontal
> line $y = v$ intersects the graph of f at exactly one point

This is the same property we used to construct each of the inverse trigonometric functions. Since a point (u, v) is on the graph of f exactly when $f(u) = v$, the geometric property above is equivalent to this algebraic property:

> For each number v in the range of f, there is exactly
> one number u in the domain of f such that $f(u) = v$

We can now use this fact to construct an inverse function g for f. Let g be the function whose domain is the same as the range of f and whose rule is:

$$g(v) = u, \quad \text{where } u \text{ is the unique number such that } f(u) = v$$

Thus for each u in the domain of f and each v in the domain of g:

$$g(v) = u \quad \text{exactly when} \quad f(u) = v$$

Therefore

$$(g \circ f)(u) = g(f(u)) = g(v) = u \quad \text{and} \quad (f \circ g)(v) = f(g(v)) = f(u) = v$$

This shows that g is the inverse function of f. To see that g is also an increasing function, remember that the graph of the inverse function g is just the reflection of the graph of f in the line $y = x$. It is easy to see geometrically that the reflection of a rising graph (such as the graph of f) is itself a rising graph. For instance, see Figure 7-19.

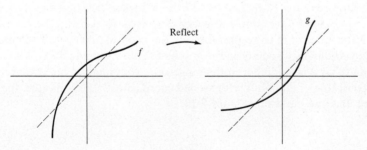

Figure 7-19

A rigorous algebraic proof of this same fact is outlined in Exercise C.1. In any case, the result is that the inverse function g is increasing. A similar argument works for decreasing functions and completes the proof of the statement in the box above.

FUNCTIONS WITHOUT INVERSES

What can be said about the inverses (if any) of functions that are neither increasing nor decreasing (that is, functions whose graphs are neither *always* rising nor *always* falling)? For example, consider Figure 7-20.

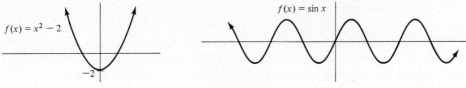

Figure 7-20

Suppose that such a function f did have an inverse g. Then the graph of g would be the reflection of the graph of f in the line $y = x$, as shown in Figure 7-21.

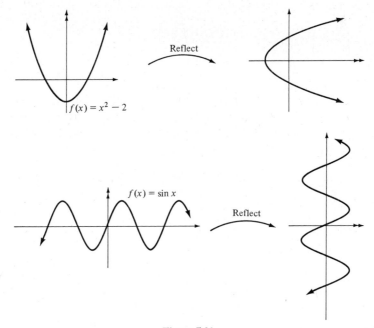

Figure 7-21

But neither of the graphs at the right in Figure 7-21 could possibly be the graph of a *function g*. To see why, remember that for any number v in the domain of a function g, there is only *one* point on the graph of g with first coordinate v, namely, the point $(v, g(v))$ (see Exercise B.7 on p. 95 and Exercise B.10 on p. 123). But this condition is certainly *not* satisfied by the right-hand graphs in Figure 7-21. For example, there are at least *two* points on each graph with first coordinate 0. A similar argument works in other cases and shows that

> *A function with a connected, continuous graph that is neither increasing nor decreasing does not have an inverse function.*

There is a technique for dealing with functions that are neither increasing nor decreasing. By restricting the domain of the given function, we can often produce a new function that is either increasing or decreasing and therefore has an inverse. The inverse trigonometric functions were constructed in this way by first restricting the domain of the trigonometric functions. Similarly, the function with rule $f(x) = x^2 - 2$ and domain the interval $[0, \infty)$ (instead of all real numbers) has a rising graph and is increasing, as shown in Figure 7-22.

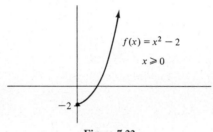

$f(x) = x^2 - 2$

$x \geqslant 0$

-2

Figure 7-22

As we have seen, this function has an inverse g whose graph is easily obtained (Figure 7-23).

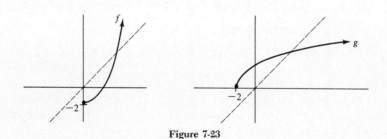

Figure 7-23

In this case, we can actually find the rule of the inverse function. We know that for any inverse function g of a function f,

$$g(v) = u \qquad \text{exactly when} \qquad f(u) = v$$

In this case, $f(u) = u^2 - 2$, so that

$$g(v) = u \qquad \text{exactly when} \qquad u^2 - 2 = v$$

Solving the equation on the right for u shows that

$$g(v) = u \qquad \text{exactly when} \qquad u = \sqrt{v + 2}°$$

In other words, $g(v) = \sqrt{v + 2}$.

EXERCISES

B.1. (a) Show that the function $f(x) = x^2$ whose domain is all real numbers has no inverse.
(b) Show that the function with domain $[0, \infty)$ and rule $h(x) = x^2$ has an inverse function g and find the rule of the inverse function g.
(c) Graph the functions h and g of part (b) on the same set of axes.

B.2. (a) Show that the inverse of the function $f(x) = x^3$ is the function $g(x) = \sqrt[3]{x}$.
(b) Graph the functions f and g on the same set of coordinate axes.

B.3. Let f be the function that assigns to each number x, the number $y = (x - 1)^3$. Show that f has an inverse function g and find its rule, by solving $y = (x - 1)^3$ for x.

B.4. (a) Express y as a function f of x by solving the equation $3x - 2y = 6$ for y.
(b) Express x as a function g of y by solving $3x - 2y = 6$ for x.
(c) Show that the functions f and g of parts (a) and (b) are inverses of each other.

B.5. Show that any linear function $f(x) = ax + b$ (with a, b fixed constants and $a \neq 0$) has an inverse function. Find the rule of the inverse function. (*Hint:* Exercise B.4 may be helpful.)

C.1. On pages 135–136, it is shown that a function f is increasing provided that

$$\text{whenever } c < d, \qquad \text{then} \qquad f(c) < f(d)$$

Let g be the inverse function of f, defined by:

$$g(v) = u \qquad \text{exactly when} \qquad f(u) = v$$

Prove that g is increasing. [*Hint:* suppose $c < d$ and $g(c) = r$ and $g(d) = s$. We must show that $r < s$. We have $f(r) = c$ and $f(s) = d$. Use the fact that f is increasing to show that $s \leq r$ is impossible.]

° We choose the positive square root here, since the graph of g shows that $g(v) = u$ is positive for each v in the domain.

The Complex Numbers

A new number system called the complex number system is introduced here. It contains the familiar real numbers, as well as solutions for equations such as $x^2 = -1$ which cannot be solved in the real number system. Although we shall consider only the mathematical properties of complex numbers, they have many practical applications in electrical engineering and other fields. Sections 2 and 3 both depend on Section 1, but Section 3 can be read before Section 2 if you wish.

1. THE COMPLEX NUMBER SYSTEM

When you began to study arithmetic in grade school, the only numbers you knew or used were the nonnegative integers. At the fifth grade level or thereabouts, fractions (that is, positive rational numbers) were introduced. In later grades other real numbers, such as π and $\sqrt{2}$ made their appearance. The number system was enlarged again in high school when negative numbers were introduced. At the beginning of this book, or even before, the entire real number system was presented.

Each time the number system is enlarged, it becomes possible to solve problems that were impossible in the original number system. For instance, a third grader, whose number system contains only integers, will tell you that the equation $7x = 19$ has no solution.° But a fifth grader, whose number system contains all positive rationals, will easily solve this equation.° Similarly, the equation $x^2 = 2$ has no solution in the system of rational numbers (see p. 2) but does have solutions in the real number system, namely, $x = \sqrt{2}$ and $x = -\sqrt{2}$.

Thus the idea of enlarging the number system in order to solve a problem that can't

---

° Of course, the equation would usually be presented to a third or fifth grader in somewhat different language, such as "Fill the box with a number that makes the sentence true: $7 \times \square = 19$."

be solved in the present number system is a perfectly natural one. You have been doing it for years, perhaps without even realizing it. And now we are about to do it again.

One of the shortcomings of the real number system is that some very simple equations have no solutions in this system. In particular, the equation $x^2 = -1$ has no solution in the real number system since the square of every real number is nonnegative. We claim that there is a larger number system that includes all the real numbers *and* a solution for the equation $x^2 = -1$. More specifically,

---

There is a number system, called the **complex number system,** *with these properties:*

(i) *The complex number system contains the real number system.*

(ii) *Addition, subtraction, multiplication, and division are defined in the complex number system and obey the same rules of arithmetic that hold in the real number system (as summarized in the boxes on pp. 550–553).*

(iii) *The complex number system contains a number (usually denoted by* i) *such that* $i^2 = -1$.

(iv) *Every complex number can be written in the standard form*

a + bi

where a *and* b *are real numbers and* $i^2 = -1$.°

(v) *Two complex numbers* a + bi *and* c + di *are equal exactly when* a = c *and* b = d.

In view of our past experience with enlarging the number system, this claim *ought* to appear plausible. Nevertheless, we are so accustomed to saying that no negative number has a square root that it may seem strange to deal with a "number" i such that $i^2 = -1$, that is, such that $i = \sqrt{-1}$. This feeling of strangeness bothered many mathematicians in the seventeenth and eighteenth centuries when the complex number system was first developed. They showed their uneasiness by the terminology they chose, terminology that is still used today. The number i and multiples of it, such as

$$5i, \quad -\tfrac{3}{5}i, \quad \sqrt{7}i, \quad \text{and more generally,} \quad bi \quad (b \text{ any real number})$$

were called **imaginary numbers.** The old familiar numbers (the integers, rationals, and irrationals) were called **real numbers.** Sums of real and imaginary numbers, such as $7 + 5i$, $\sqrt{2} - \tfrac{3}{5}i$, $-6 + \tfrac{4}{9}i$ and more generally

$$a + bi \quad (a, b \text{ any real numbers})$$

were then called **complex numbers.**

° Hereafter, whenever we write $a + bi$ or $c + di$, it is assumed without explicit mention that a, b, c, d are real numbers and $i^2 = -1$.

Despite their name, imaginary numbers are no more "unreal" than rational or irrational numbers. Complex numbers are just as valid mathematical entities as are real numbers. The actual construction of the complex number system and the proof that it has the properties claimed in the box above are outlined in Exercise C.5. For now, we shall simply accept the existence of this new system and explore the arithmetic of complex numbers.

ARITHMETIC OF COMPLEX NUMBERS

According to property (iv) in the box above, the complex numbers consist of all sums $a + bi$ with a, b real numbers and $i^2 = -1$. In particular, when a is any real number and $b = 0$, the number $a + 0i$ is a complex number. But $a + 0i$ is just the real number a since

$$a + 0i = a + 0 = a°$$

For example, $3 = 3 + 0i$ and $-\sqrt{2} = -\sqrt{2} + 0i$. This shows exactly how the complex numbers contain the real numbers. Similarly, every imaginary number bi (with b real) is also a complex number since $bi = 0 + bi$. For instance, $7i = 0 + 7i$ and $-\frac{2}{3}i = 0 + (-\frac{2}{3})i$.

When performing arithmetic with complex numbers, you should usually put your answer in the standard form $a + bi$. Ordinarily this is quite easy, since the usual laws of arithmetic are valid.

EXAMPLE $(1 + i) + (3 - 7i) = 1 + i + 3 - 7i = (1 + 3) + (i - 7i) = 4 - 6i$.

EXAMPLE $(\frac{3}{2} + 6i) - (\frac{7}{2} + 2i) = \frac{3}{2} + 6i - \frac{7}{2} - 2i = (\frac{3}{2} - \frac{7}{2}) + (6i - 2i) = -\frac{4}{2} + 4i = -2 + 4i$.

EXAMPLE $4i(2 + \frac{3}{2}i) = 4i(2) + 4i(\frac{3}{2}i) = 8i + 4(\frac{3}{2})i^2 = 8i + 6i^2 = 8i + 6(-1) = -6 + 8i$.

EXAMPLE $(2 + i)(3 - 4i) = 2 \cdot 3 + 3i - 2 \cdot 4i - 4i^2 = 6 + 3i - 8i - 4(-1) = (6 + 4) + (3i - 8i) = 10 - 5i$.

As these examples demonstrate, you can *treat all symbols just as if they were real numbers providing that you replace i^2 by -1*. In particular, all of the multiplication and factoring patterns for real numbers, such as

$$(x + y)^2 = x^2 + 2xy + y^2 \qquad \text{and} \qquad (x + y)(x - y) = x^2 - y^2$$

are also valid for complex numbers.

EXAMPLE $(3 + 2i)(3 - 2i) = 3^2 - (2i)^2 = 9 - 4i^2 = 9 - 4(-1) = 9 + 4 = 13$. In this case, the product of two complex numbers turns out to be a positive real number.

° Remember that the usual rules of arithmetic still apply so that $z + 0 = z$ and $0z = 0$ for *every* number z.

The laws of exponents, such as $r^m r^n = r^{m+n}$ and $(r^m)^n = r^{mn}$ also hold for complex numbers.

EXAMPLE   Find i^{54}. We begin with low powers of i:

$$i^2 = -1, \qquad i^3 = i^2 i = (-1)i = -i, \qquad i^4 = i^2 i^2 = (-1)(-1) = 1$$

Since $i^4 = 1$, we see that $i^8 = (i^4)^2 = 1^2 = 1$, $i^{12} = (i^4)^3 = 1^3 = 1$, and similarly, $i^{16} = 1$, $i^{20} = 1$, and so on. Now observe that $54 = 52 + 2 = 4 \cdot 13 + 2$, so that

$$i^{54} = i^{52+2} = i^{52} i^2 = i^{4 \cdot 13} i^2 = (i^4)^{13} i^2 = (1)^{13}(-1) = -1$$

CONJUGATES AND DIVISION

The **conjugate** of the complex number $a + bi$ is defined to be the number $a - bi$, and the conjugate of $a - bi$ is defined to be $a + bi$.

EXAMPLE   The conjugate of $3 + 4i$ is $3 - 4i$. Since $3i = 0 + 3i$, the conjugate of $3i$ is $-3i$. The conjugate of $-5 - 2i$ is $-5 + 2i$. Since $17 = 17 + 0i = 17 - 0i$, the conjugate of 17 is 17 itself.

Here is a useful property of conjugates; other properties are discussed below:

> *The product of a complex number and its conjugate is a nonnegative real number.*

For example, we saw above that $(3 + 2i)(3 - 2i) = 9 - 4i^2 = 13$. More generally,

$$(a + bi)(a - bi) = a^2 - (bi)^2 = a^2 - b^2 i^2 = a^2 - b^2(-1) = a^2 + b^2$$

Since a^2 and b^2 are nonnegative real numbers, so is $a^2 + b^2$. This property of conjugates is very convenient when dividing complex numbers.

EXAMPLE   Divide $3 + 4i$ by $1 + 2i$. As usual, we can write the answers as a fraction $(3 + 4i)/(1 + 2i)$. In order to express this fraction in the form $a + bi$, the first step is to *take the conjugate of the denominator*, namely, $1 - 2i$. Since $(1 - 2i)/(1 - 2i) = 1$, we have

$$\frac{3 + 4i}{1 + 2i} = \frac{3 + 4i}{1 + 2i} \cdot 1 = \frac{3 + 4i}{1 + 2i} \cdot \frac{1 - 2i}{1 - 2i} = \frac{(3 + 4i)(1 - 2i)}{(1 + 2i)(1 - 2i)}$$

$$= \frac{3 + 4i - 6i - 8i^2}{1^2 - (2i)^2} = \frac{3 + 4i - 6i - 8(-1)}{1 - 4i^2} = \frac{11 - 2i}{1 - 4(-1)}$$

$$= \frac{11 - 2i}{5} = \frac{11}{5} - \frac{2}{5}i$$

Thus we have expressed $(3 + 4i)/(1 + 2i)$ in the form $a + bi$, with $a = \frac{11}{5}$ and $b = -\frac{2}{5}$. This technique always works since the product of the denominator and its conjugate always gives a real number for the denominator of the quotient.

EXAMPLE To find $\dfrac{i}{2 + i} + \dfrac{3 + i}{1 - i}$, we first express everything in terms of a common denominator, namely, $(2 + i)(1 - i)$:

$$\frac{i}{2 + i} + \frac{3 + i}{1 - i} = \frac{i(1 - i)}{(2 + i)(1 - i)} + \frac{(2 + i)(3 + i)}{(2 + i)(1 - i)} = \frac{i(1 - i) + (2 + i)(3 + i)}{(2 + i)(1 - i)}$$

$$= \frac{i - i^2 + 6 + 3i + 2i + i^2}{2 + i - 2i - i^2} = \frac{6 + 6i}{3 - i}$$

Finally, we express this quotient in the form $a + bi$ as above:

$$\frac{i}{2 + i} + \frac{3 + i}{1 - i} = \frac{6 + 6i}{3 - i} = \frac{6 + 6i}{3 - i} \cdot \frac{3 + i}{3 + i} = \frac{18 + 18i + 6i + 6i^2}{9 - i^2}$$

$$= \frac{12 + 24i}{10} = \frac{12}{10} + \frac{24}{10}i = \frac{6}{5} + \frac{12}{5}i$$

PROPERTIES OF COMPLEX CONJUGATES

It is sometimes convenient to denote the complex number $a + bi$ by a single letter. If z is a complex number, then its conjugate is denoted by $\bar{z}$. Thus if $z = a + bi$, then $\bar{z} = a - bi$. Complex conjugation has these basic properties:

(i) *For any complex number z, both $z\bar{z}$ and $z + \bar{z}$ are real numbers.*

(ii) *For any complex numbers z and w,*

$$\overline{z + w} = \bar{z} + \bar{w}, \qquad \overline{zw} = \bar{z} \cdot \bar{w}, \qquad \overline{\left(\frac{z}{w}\right)} = \frac{\bar{z}}{\bar{w}} \quad (w \neq 0)$$

(iii) $z = \bar{z}$ *exactly when z is a real number.*

The first part of statement (i) was proved on page 477. The other proofs are equally straightforward (see Exercise C.3), so we shall only illustrate them with some examples.

EXAMPLE (i) Suppose $z = 2 + 3i$ so that $\bar{z} = 2 - 3i$. Then

$$z + \bar{z} = (2 + 3i) + (2 - 3i) = 4$$

EXAMPLE (ii) Suppose $z = 2 + 3i$ as above and $w = 5 - i$. Then

$$zw = (2 + 3i)(5 - i) = 10 + 15i - 2i - 3i^2 = 13 + 13i$$

so that $\overline{zw} = 13 - 13i$. But

$$\overline{z} \cdot \overline{w} = (2 - 3i)(5 + i) = 10 - 15i + 2i - 3i^2 = 13 - 13i$$

Hence $\overline{zw} = \overline{z} \cdot \overline{w}$.

EXAMPLE (iii) $2 = 2 + 0i$ is a real number and $\overline{2} = 2 - 0i = 2$. Thus $2 = \overline{2}$.

ABSOLUTE VALUE

The **absolute value** of the complex number $a + bi$ is denoted $|a + bi|$ and is defined to be the real number $\sqrt{a^2 + b^2}$.

EXAMPLE $|3 + 2i| = \sqrt{3^2 + 2^2} = \sqrt{13}$. Similarly, $|4 - 5i| = \sqrt{4^2 + (-5)^2} = \sqrt{41}$.

The concept of absolute value for real numbers is just a special case of absolute value for complex numbers. Remember that for every real number c, we have $\sqrt{c^2} = |c|$ and $c = c + 0i$.

EXAMPLE $|5 + 0i| = \sqrt{5^2 + 0^2} = \sqrt{5^2} = |5|$ and $|-3 + 0i| = \sqrt{(-3)^2 + 0^2} = \sqrt{(-3)^2} = |-3|$.

The properties of absolute value for complex numbers are discussed in Exercise C.4 and include these:

For any complex numbers z *and* w:

$$|z| = |\overline{z}|, \qquad |z|^2 = z\overline{z}, \qquad |zw| = |z| \cdot |w|,$$

$$\left| \frac{z}{w} \right| = \frac{|z|}{|w|} \quad (w \neq 0)$$

EXERCISES

Directions Express all numerical answers in the form $a + bi$ (a, b real numbers).

A.1. Perform the indicated addition or subtraction.

(a) $(2 + 3i) + (6 - i)$

(b) $(-5 + 7i) + (14 + 3i)$

(c) $(2 - 8i) - (4 + 2i)$

(d) $(3 + 5i) - (3 - 7i)$

(e) $\frac{5}{4} - (\frac{7}{4} + 2i)$

(f) $(\sqrt{3} + i) + (\sqrt{5} - 2i)$

(g) $\left(\frac{\sqrt{2}}{2} + i \right) - \left(\frac{\sqrt{3}}{2} - i \right)$

(h) $\left(\frac{1}{2} + \frac{\sqrt{3}i}{2} \right) + \left(\frac{3}{4} - \frac{5\sqrt{3}i}{2} \right)$

A.2. Multiply.

(a) $(2 + i)(3 + 5i)$

(b) $(2 - i)(5 + 2i)$

(c) $(-3 + 2i)(4 - i)$

(d) $(4 + 3i)(4 - 3i)$

(e) $(2 - 5i)^2$

(f) $(1 + i)(2 - i)i$

(g) $(\sqrt{3} + i)(\sqrt{3} - i)$

(h) $(\frac{1}{2} - i)(\frac{1}{4} + 2i)$

A.3. Multiply.

(a) i^{15}　(b) i^{26}　(c) i^{33}　(d) $(-i)^{53}$　(e) $(-i)^{107}$

A.4. Find $\bar{z}$, $\bar{w}$, and $\overline{zw}$.

(a) $z = 1 + i$;　$w = 2i$

(b) $z = -2 - 2i$;　$w = 3i$

(c) $z = 4 - 5i$;　$w = 2 - 3i$

(d) $z = (1 + i)^2$;　$w = 1 - 2i$

B.1. Divide.

(a) $\dfrac{1}{5 - 2i}$

(b) $\dfrac{1}{i}$

(c) $\dfrac{1}{3i}$

(d) $\dfrac{i}{2 + i}$

(e) $\dfrac{3}{4 + 5i}$

(f) $\dfrac{2 + 3i}{i}$

(g) $\dfrac{7 - 4i}{2i}$

(h) $\dfrac{2 + 3i}{1 + i}$

B.2. Divide. (*Hint:* first multiply out the denominator.)

(a) $\dfrac{1}{i(4 + 5i)}$

(b) $\dfrac{1}{(2 - i)(2 + i)}$

(c) $\dfrac{2 + 3i}{i(4 + i)}$

(d) $\dfrac{2}{(2 + 3i)(4 + i)}$

(e) $\dfrac{2 + 3i}{(3 + 2i)(3 - i)}$

(f) $\dfrac{5 - 3i}{(2 - 4i)^2}$

B.3. Compute:

(a) $\dfrac{2 + i}{1 - i} + \dfrac{1}{1 + 2i}$

(b) $\dfrac{1}{2 - i} + \dfrac{3 + i}{2 + 3i}$

(c) $\dfrac{i}{3 + i} - \dfrac{3 + i}{4 + i}$

(d) $6 + \dfrac{2i}{3 + i}$

(e) $\dfrac{1 - i}{4 + 3i} - 2i$

(f) $\dfrac{1}{(1 - i)(2 + 3i)} + \dfrac{2 + 3i}{i + 5}$

B.4. Give an example of complex numbers, z and w, such that $|z + w| \neq |z| + |w|$.

B.5. Find these absolute values.

(a) $|5 - 12i|$

(b) $|2i|$

(c) $|1 + \sqrt{2}i|$

(d) $|2 - 3i|$

(e) $|-12i|$

(f) $|i^7|$

(g) $|i(3 + i)| + |3 - i|$

(h) $|3 + 2i| - |1 + 2i|$

(i) $\left|\dfrac{1 + i}{3 - 5i}\right|$

B.6. Simplify: $i + i^2 + i^3 + \cdots + i^{15}$.

B.7. Solve for x and y. (*Example:* $2x + 3i = 4 + yi$. Remember that $a + bi = c + di$ exactly when $a = c$ and $b = d$. Hence $2x + 3i = 4 + yi$ exactly when $2x = 4$ and $3 = y$. Thus $x = 2$ and $y = 3$ are the solutions.)

(a) $3x - 4i = 6 + 2yi$

(b) $8 - 2yi = 4x + 12i$

(c) $3 + 4xi = 2y - 3i$

(d) $8 - xi = \frac{1}{2}y + 2i$

C.1. The **real part** of the complex number $z = a + bi$ is defined to be the number a and is denoted $\text{Re}(z)$. The **imaginary part** of $z = a + bi$ is defined to be the number b (*not* bi) and is denoted $\text{Im}(z)$.

(a) Show that $\text{Re}(z) = \dfrac{z + \bar{z}}{2}$ (b) Show that $\text{Im}(z) = \dfrac{z - \bar{z}}{2i}$

C.2. If $z = a + bi$ (a, b fixed real numbers), find $\dfrac{1}{z}$.

C.3. Let $z = a + bi$ and $w = c + di$ (a, b, c, d fixed real numbers).
(a) Find $\bar{z}$ and $z + \bar{z}$. Verify that $z + \bar{z}$ is a real number.
(b) Find $z + w$ and $\overline{z + w}$.
(c) Find $\bar{w}$ and $\bar{z} + \bar{w}$. Verify that $\overline{z + w} = \bar{z} + \bar{w}$.
(d) Show that $\overline{zw} = \bar{z} \cdot \bar{w}$.

C.4. Let $z = a + bi$ and $w = c + di$ (a, b, c, d fixed real numbers).
(a) Find $\bar{z}$, $|\bar{z}|$ and $|z|$. Verify that $|z| = |\bar{z}|$.
(b) Show that $|z|^2 = z\bar{z}$.
(c) Find zw and $|zw|$.
(d) Find $|w|$ and $|z| \cdot |w|$. Verify that $|zw| = |z| \cdot |w|$ [see part (c)].
(e) Assume $w \neq 0$. Find $\dfrac{z}{w}$ and $\left|\dfrac{z}{w}\right|$
(f) Find $\dfrac{|z|}{|w|}$ and verify that $\left|\dfrac{z}{w}\right| = \dfrac{|z|}{|w|}$

C.5. **Construction of the Complex Numbers.** We assume that the real number system is known. In order to construct a new number system with the desired properties, we must do the following:

(i)   Define a set C (whose elements will be called complex numbers).
(ii) The set C must contain the real numbers, or at least a copy of them.
(iii) Define addition and multiplication in the set C in such a way that the usual laws of arithmetic are valid.
(iv) Show that C has the other properties listed in the box on page 475.

We begin by defining C to be the set of all ordered pairs of real numbers. Thus $(1, 5), (-6, 0), (\frac{4}{3}, -17)$, and $(\sqrt{2}, \frac{12}{5})$ are some of the elements of the set C. More generally, a complex number (= element of C) is any pair (a, b) where a and b are real numbers. By definition, two complex numbers are *equal* exactly when they have the same first and the same second coordinate.
(a) *Addition in C* is defined by this rule:

$$(a, b) + (c, d) = (a + c, b + d)$$

For   example, $(3, 2) + (5, 4) = (3 + 5, 2 + 4) = (8, 6)$.   Similarly, $(-1, 2) + (3, -5) = ((-1) + 3, 2 + (-5)) = (2, -3)$.   Verify   that   this addition has the following properties. For any complex numbers (a, b), (c, d), (e, f) in C:

(i) $(a, b) + (c, d) = (c, d) + (a, b)$
(ii) $((a, b) + (c, d)) + (e, f) = (a, b) + ((c, d) + (e, f))$
(iii) $(a, b) + (0, 0) = (a, b)$
(iv) $(a, b) + (-a, -b) = (0, 0)$

(b) *Multiplication in C* is defined by this rule:

$$(a, b)(c, d) = (ac - bd, bc + ad)$$

For example, $(3, 2)(4, 5) = (3 \cdot 4 - 2 \cdot 5, 2 \cdot 4 + 3 \cdot 5) = (12 - 10, 8 + 15) = (2, 23)$. Similarly, $(\frac{1}{2}, 1)(-2, 1) = (\frac{1}{2}(-2) - 1 \cdot 1, 1(-2) + \frac{1}{2} \cdot 1) = (-2, -\frac{3}{2})$. Verify that this multiplication has the following properties. For any complex numbers (a, b), (c, d), (e, f) in C:

(i) $(a, b)(c, d) = (c, d)(a, b)$
(ii) $((a, b)(c, d))(e, f) = (a, b)((c, d)(e, f))$
(iii) $(a, b)(1, 0) = (a, b)$
(iv) $(a, b)(0, 0) = (0, 0)$

(c) Since C consists of ordered pairs, we can think of C as being the set of points in the coordinate plane. Verify that for any two points on the x-axis (that is, any elements of C with second coordinate zero):

(i) $(a, 0) + (c, 0) = (a + c, 0)$
(ii) $(a, 0)(c, 0) = (ac, 0)$

Now consider the x-axis as a number line in the usual way: *identify* $(t, 0)$ on the x-axis with the real number t. Statements (i) and (ii) show that when addition or multiplication in C is performed on two real numbers (that is, points on the x-axis), the result is the usual sum or product of real numbers. Thus C contains (a copy of) the real number system.

(d) *New Notation.* Since we are identifying the complex number $(a, 0)$ with the real number a, we shall hereafter denote $(a, 0)$ simply by the symbol a. Also, let i denote the complex number $(0, 1)$.

(i)   Show that $i^2 = -1$ [that is, $(0, 1)(0, 1) = (-1, 0)$].
(ii) Show that for any complex number $(0, b)$, $(0, b) = bi$ [that is, $(0, b) = (b, 0)(0, 1)$].
(iii) Show that any complex number (a, b) can be written: $(a, b) = a + bi$ [that is, $(a, b) = (a, 0) + (b, 0)(0, 1)$].

In this new notation, every complex number is of the form $a + bi$ with a, b real and $i^2 = -1$, and our construction is finished.

2. EQUATIONS AND COMPLEX NUMBERS

The system of complex numbers was constructed in order to obtain a solution for the equation $x^2 = -1$. In this section we shall see that the complex numbers contain solutions for many other equations as well.

SQUARE ROOTS

Since $i^2 = -1$, it seems reasonable to say that i is the square root of -1 and to write $i = \sqrt{-1}$. Similarly, since $(5i)^2 = 5^2 i^2 = 25(-1) = -25$, we define $\sqrt{-25}$ to be the complex number $5i$. More generally, any *positive* real number b has a real square root $\sqrt{b}$ and

$$(\sqrt{b}\, i)^2 = (\sqrt{b^2})i^2 = b(-1) = -b$$

Consequently $\sqrt{-b}$ is defined to be the complex number $\sqrt{b}\, i$. Thus

> *Every real number (in particular, every negative one) has a square root in the complex number system.*

EXAMPLE $\sqrt{-3} = \sqrt{3}\,i$ since $(\sqrt{3}\,i)^2 = (\sqrt{3})^2 i^2 = 3(-1) = -3$. Similarly, $\sqrt{-\frac{81}{4}} = \frac{9}{2}i$ since $\sqrt{\frac{81}{4}} = \frac{9}{2}$ and $(\frac{9}{2}i)^2 = (\frac{9}{2})^2 i^2 = \frac{81}{4}(-1) = -\frac{81}{4}$.

Many expressions that are not defined in the real number system now make sense in the complex number system.

EXAMPLE $\dfrac{1 + \sqrt{-3}}{2} = \dfrac{1 + \sqrt{3}\,i}{2} = \dfrac{1}{2} + \dfrac{\sqrt{3}}{2}i.$

EXAMPLE $\sqrt{-20}\,\sqrt{-5} = (\sqrt{20}\,i)(\sqrt{5}\,i) = \sqrt{20}\,\sqrt{5}\,i^2 = \sqrt{100}\,i^2 = 10(-1) = -10.$

Warning The property of radicals $\sqrt{cd} = \sqrt{c}\,\sqrt{d}$, which is valid for positive real numbers, *does not hold* when both c and d are negative. For instance, suppose $c = -20$ and $d = -5$. The preceding example shows that $\sqrt{c}\,\sqrt{d} = \sqrt{-20}\,\sqrt{-5} = -10$. But $\sqrt{cd} = \sqrt{(-20)(-5)} = \sqrt{100} = 10$. So $\sqrt{c}\,\sqrt{d} \neq \sqrt{cd}$. To avoid difficulty, always write square roots of negative numbers in terms of i *before* doing any simplifying.

POLYNOMIAL EQUATIONS WITH REAL COEFFICIENTS

Since every negative real number has a square root in the complex number system, we can now find complex solutions for equations that have no real solutions. For example, the solutions of $x^2 = -25$ are $x = \pm\sqrt{-25} = \pm 5i$ and the solutions of $x^2 = -3$ are $x = \pm\sqrt{-3} = \pm\sqrt{3}\,i$. In fact, the examples below show that

> *Every quadratic equation with real coefficients has solutions in the complex number system.*

EXAMPLE To solve the equation $2x^2 + x + 3 = 0$, we apply the quadratic formula:

$$x = \frac{-1 \pm \sqrt{1^2 - 4 \cdot 2 \cdot 3}}{2 \cdot 2} = \frac{-1 \pm \sqrt{-23}}{4}$$

Since $\sqrt{-23}$ is not a real number, this equation has no real number solutions. But $\sqrt{-23}$ *is* a complex number, namely, $\sqrt{-23} = \sqrt{23}i$. Thus the equation does have solutions in the complex number system:

$$x = \frac{-1 \pm \sqrt{-23}}{4} = \frac{-1 \pm \sqrt{23}i}{4} = -\frac{1}{4} \pm \frac{\sqrt{23}}{4}i$$

Note that the two solutions, $-\dfrac{1}{4} + \dfrac{\sqrt{23}}{4}i$ and $-\dfrac{1}{4} - \dfrac{\sqrt{23}}{4}i$, are conjugates of each other.

EXAMPLE To find *all* solutions of $x^3 = 1$, we rewrite the equation and factor:

$$x^3 = 1$$
$$x^3 - 1 = 0$$
$$(x - 1)(x^2 + x + 1) = 0$$
$$x - 1 = 0 \quad \text{or} \quad x^2 + x + 1 = 0$$

The solution of the first equation is $x = 1$. The solutions of the second can be obtained from the quadratic formula:

$$x = \frac{-1 \pm \sqrt{1^2 - 4 \cdot 1 \cdot 1}}{2 \cdot 1} = \frac{-1 \pm \sqrt{-3}}{2} = \frac{-1 \pm \sqrt{3}i}{2} = \frac{-1}{2} \pm \frac{\sqrt{3}}{2}i$$

Therefore the equation $x^3 = 1$ has one real solution $(x = 1)$ and two complex solutions $[x = -\frac{1}{2} + (\sqrt{3}/2)i$ and $x = -\frac{1}{2} - (\sqrt{3}/2)i]$. Each of these solutions is said to be a **cube root of one** or a **cube root of unity.** Observe that the two complex cube roots of unity are conjugates of each other.

The preceding examples illustrate this useful fact:

> *Let f(x) be a polynomial with real coefficients. If a complex number z is a root of f(x), then its conjugate $\bar{z}$ is also a root of f(x).*

A formal proof is outlined in Exercise C.1. Before giving examples of how this fact is used, we must first introduce polynomials whose coefficients are complex numbers, such as:

$$x + 5i, \quad x - (3 + i), \quad 3x^2 + (1 + 2i)x + (i - 1), \quad (1 - 7i)x^4 + 5ix^3 + (2 + i)x - 8$$

The discussion of polynomials with real coefficients at the beginning of Chapter 4 depends only on the fact that the arithmetic of real numbers obeys certain rules. Since the arithmetic of complex numbers obeys the same basic rules, *the entire discussion is also valid for polynomials with complex coefficients.* In particular, the Division Algorithm, the Remainder Theorem, and the Factor Theorem are true for such polynomials.

EXAMPLE Find a polynomial with *real* coefficients whose roots include the numbers 2 and $3 + i$. Since $3 + i$ is to be a root, the fact in the box above shows that its conjugate $3 - i$ must also be a root. Since a product is 0 in the complex number system only when one of its factors is 0, the polynomial

$$f(x) = (x - 2)(x - (3 + i))(x - (3 - i))$$

obviously has roots 2, $3 + i$, and $3 - i$. Now this polynomial *appears* to have nonreal coefficients. However, multiplying out and simplifying yields:

$$\begin{aligned} f(x) &= (x - 2)(x - (3 + i))(x - (3 - i)) \\ &= (x - 2)(x^2 - (3 + i)x - (3 - i)x + (3 + i)(3 - i)) \\ &= (x - 2)(x^2 - 3x - ix - 3x + ix + (3^2 - i^2)) \\ &= (x - 2)(x^2 - 6x + 10) = x^3 - 8x^2 + 22x - 20 \end{aligned}$$

Thus $f(x) = x^3 - 8x^2 + 22x - 20$ is a polynomial with real coefficients whose roots are 2, $3 + i$, and $3 - i$.

EXAMPLE Given that $z = 1 + i$ is a root of the polynomial $f(x) = x^4 - 2x^3 - x^2 + 6x - 6$, find all of its roots. According to the fact in the box above, the conjugate $\bar{z} = 1 - i$ must also be a root of $f(x)$. By the Factor Theorem (p. 177), both $x - z$ and $x - \bar{z}$ are factors of $f(x)$. Therefore the product $(x - z)(x - \bar{z})$ is a factor of $f(x)$. But

$$\begin{aligned} (x - z)(x - \bar{z}) &= (x - (1 + i))(x - (1 - i)) \\ &= x^2 - (1 + i)x - (1 - i)x + (1 + i)(1 - i) \\ &= x^2 - x - ix - x + ix + (1 - i^2) \\ &= x^2 - 2x + 2 \end{aligned}$$

Thus $x^2 - 2x + 2$ is a factor of $f(x)$. Long division shows that the other factor is $x^2 - 3$. Consequently, $f(x) = x^4 - 2x^3 - x^2 + 6x - 6 = (x^2 - 2x + 2)(x^2 - 3)$ and its roots are easily found:

$$x^4 - 2x^3 - x^2 + 6x - 6 = 0$$
$$(x^2 - 2x + 2)(x^2 - 3) = 0$$
$$x^2 - 2x + 2 = 0 \quad \text{or} \quad x^2 - 3 = 0$$

The solutions of the first equation are $1 + i$ and $1 - i$, as can be seen from our work above or from the quadratic formula. The solutions of $x^2 - 3 = 0$ are obviously $x = \pm\sqrt{3}$. Therefore $f(x)$ has roots $1 + i$, $1 - i$, $\sqrt{3}$, and $-\sqrt{3}$.

THE FUNDAMENTAL THEOREM OF ALGEBRA

The complex numbers were constructed in order to obtain a solution for the equation $x^2 = -1$. As we have just seen, they contain a good deal more—solutions for every quadratic equation with real coefficients. A natural question now arises:

> Do the complex numbers contain solutions for *every* polynomial equation with real or complex coefficients?

If the answer to this question were no, it would be necessary to enlarge the complex number system (perhaps many times) in order to obtain solutions for complicated high-degree polynomial equations. Fortunately, and perhaps surprisingly, the answer to the question turns out to be yes. Since every real number is also a complex number and since solving polynomial equations is equivalent to finding the roots of certain polynomials, the formal answer to the question is usually phrased like this:

THE FUNDAMENTAL THEOREM OF ALGEBRA

Every nonconstant polynomial with complex coefficients has a root in the complex number system.

Although this is obviously a powerful result, neither the Fundamental Theorem nor its proof provide a practical method for actually *finding* a root of a given polynomial.[*] The proof of the Fundamental Theorem will be omitted since it involves mathematical concepts beyond the scope of this book. But we shall explore some of the theorem's implications, which are sometimes useful for understanding and analyzing various problems.

Here is one consequence of the Fundamental Theorem:

Suppose $f(x)$ is a polynomial with complex coefficients with degree $n > 0$ and leading coefficient d. Then there are (not necessarily distinct) complex numbers c_1, c_2, c_3, ..., c_n such that

$$f(x) = d(x - c_1)(x - c_2)(x - c_3) \cdots (x - c_n)$$

This statement is proved below. To understand its meaning, we consider a specific example. Suppose

$$f(x) = 5x^3 - (10 + 10i)x^2 + (-5 + 20i)x + 10$$

[*] It may seem strange that it is possible to prove that a root exists without actually exhibiting one. But such "existence theorems" are quite common in mathematics. A very rough analogy is the situation that occurs when a person walking the street is killed by a sniper's bullet. The police know that there *is* a killer, but actually *finding* the killer may be difficult or even impossible.

In this case, $n = 3$, the degree of $f(x)$, and $d = 5$, the leading coefficient of $f(x)$. By multiplying out the right side below, you will see that $f(x)$ factors as:

$$f(x) = 5(x - 2)(x - i)(x - i)$$

so that $c_1 = 2$, $c_2 = i$, and $c_3 = i$.

In order to prove the statement in the box above, suppose $f(x)$ is a polynomial of positive degree n with complex coefficients. According to the Fundamental Theorem, $f(x)$ has a root in the complex number system—call it c_1. Consequently, by the Factor Theorem (p. 177), $x - c_1$ is a factor of $f(x)$. Thus

$$f(x) = (x - c_1)g(x)$$

for some polynomial $g(x)$. Since $g(x)$ is the quotient when $f(x)$ is divided by the first-degree polynomial $x - c_1$, the degree of $g(x)$ must be one less than the degree of $f(x)$. Now the Fundamental Theorem also applies to the polynomial $g(x)$: if $g(x)$ is not a constant, it must have a root in the complex number system—call this root c_2. The Factor Theorem shows that $x - c_2$ is a factor of $g(x)$, so that $g(x) = (x - c_2)h(x)$ for some polynomial $h(x)$. Just as before, $h(x)$ has degree one less than the degree of $g(x)$. Thus

$$f(x) = (x - c_1)g(x) = (x - c_1)(x - c_2)h(x)$$

If $h(x)$ is not a constant, it must have a root—call it c_3—and the same argument can be applied again. As long as the last factor is not constant, we can keep repeating the argument. At each step the degree of the last factor goes down by one. Since $f(x)$ has degree n, we see that after n steps the last factor will have degree 0. In other words, the last factor will be a nonzero constant—call it d. At this point, we have $f(x)$ factored in the desired form:

$$f(x) = (x - c_1)(x - c_2)(x - c_3) \cdots (x - c_n)d$$

Multiplying out the right side of this equation shows that

$$f(x) = dx^n + \text{lower degree terms}$$

Thus the constant d is actually the leading coefficient of $f(x)$. This completes the proof.

When a polynomial $f(x)$ of degree $n > 0$ is written in factored form as

$$f(x) = d(x - c_1)(x - c_2) \cdots (x - c)_n \quad (d \neq 0)$$

it is easy to see that the numbers $c_1, c_2, \ldots, c_n$ are roots of $f(x)$. Furthermore, every root of $f(x)$ must be one of these numbers. For if k is any root of $f(x)$, then

$$0 = f(k) = d(k - c_1)(k - c_2) \cdots (k - c_n) \quad (d \neq 0)$$

The product on the right is 0 only when one of the factors is 0, that is, when

$$\begin{array}{ccccccc}
k - c_1 = 0 & \text{or} & k - c_2 = 0 & \text{or} & \cdots & k - c_n = 0 \\
k = c_1 & \text{or} & k = c_2 & \text{or} & \cdots & k = c_n
\end{array}$$

Therefore k is indeed one of the c's. Since there may be some repetitions in the list of roots $c_1, c_1, \ldots, c_n$ we conclude that:

> *Every polynomial of degree* n > 0 *with complex coefficients has at most* n *distinct roots in the complex number system.*

Suppose a number c is a root of a polynomial $f(x)$ of degree $n > 0$. If $f(x)$ is written in factored form as above, then c appears one or more times on the list of roots c_1, $c_2, \ldots, c_n$. If c appears *exactly* k times on the list, then $f(x)$ has exactly k factors of the form $x - c$. Hence $(x - c)^k$ is a factor of $f(x)$ and no higher power of $x - c$ is a factor of $f(x)$. Therefore c is a root of *multiplicity* k (as defined on p. 192) and we see that

> *A polynomial of degree* n > 0 *with complex coefficients has exactly* n *roots in the complex number system, provided that each root is counted as many times as its multiplicity.*

EXERCISES

Directions   Express all complex answers in the form $a + bi$ unless directed otherwise.

A.1.   Express in terms of i:
   (a) $\sqrt{-36}$ (c) $\sqrt{-14}$ (e) $-\sqrt{-16}$
   (b) $\sqrt{-81}$ (d) $\sqrt{-50}$ (f) $-\sqrt{-12}$

A.2.   Express in terms of i and simplify:
   (a) $\sqrt{-16} + \sqrt{-49}$ (d) $\sqrt{-12}\sqrt{-3}$ (g) $\sqrt{-16}/\sqrt{-36}$
   (b) $\sqrt{-25} - \sqrt{-9}$ (e) $\sqrt{-8}\sqrt{-6}$ (h) $-\sqrt{-64}/\sqrt{-4}$
   (c) $\sqrt{-15} - \sqrt{-18}$ (f) $-\sqrt{-10}\sqrt{-6}$

A.3.   Simplify:
   (a) $(\sqrt{-25} + 2)(\sqrt{-49} - 3)$ (e) $\sqrt{-2}(\sqrt{2} + \sqrt{-2})$
   (b) $(5 - \sqrt{-3})(-1 + \sqrt{-9})$ (f) $(5 - \sqrt{-1})(3 + \sqrt{-2})$
   (c) $(2 + \sqrt{-5})(1 - \sqrt{-10})$ (g) $1/(1 + \sqrt{-2})$
   (d) $\sqrt{-3}(3 - \sqrt{-27})$ (h) $(1 + \sqrt{-4})/(3 - \sqrt{-9})$

B.1.   Solve these equations:
   (a) $x^2 - 2x + 5 = 0$ (c) $5x^2 + 2x + 1 = 0$ (e) $x + 6 = 2x^2$
   (b) $x^2 - 4x + 13 = 0$ (d) $x^2 + 5x = -12$ (f) $5x^2 + 3 = 2x$

B.2.   Solve these equations. (Hint: factor first.)
   (a) $x^3 - 27 = 0$ (c) $x^3 = -8$ (e) $x^4 - 1 = 0$
   (b) $x^3 + 125 = 0$ (d) $x^6 - 64 = 0$ (f) $x^4 - x^2 - 6 = 0$

B.3. Find a polynomial with *real* coefficients whose roots include the given numbers.
  (a) $2 + i$, $2 - i$ (c) 2, $2 + i$ (e) -1, i
  (b) $1 + 3i$, $1 - 3i$ (d) i, $2i$ (f) -3, $1 - i$, $1 + 2i$

B.4. Find all roots of the given polynomial. In each case, one root is given.
  (a) $x^4 - x^3 - 5x^2 - x - 6$; root i
  (b) $x^3 + x^2 + x + 1$; root i
  (c) $x^4 - 4x^3 + 6x^2 - 4x + 5$; root $2 - i$
  (d) $x^4 - 5x^3 + 10x^2 - 20x + 24$; root $2i$

B.5. Find a polynomial with *real* coefficients having the given degree and roots:
  (a) degree 2; roots $1 + 2i$ and $1 - 2i$
  (b) degree 4; roots $3i$ and $-3i$, each of multiplicity 2
  (c) degree 6; 0 is a root of multiplicity 3; and 3, $1 + i$, and $1 - i$ are each roots of multiplicity 1.

Note: Exercises B.6–B.8 deal with polynomial equations with *complex* coefficients. The techniques for solving such equations are essentially the same ones used with real coefficients.

B.6. Solve these equations. [*Hint:* first collect all terms involving the unknown x on the left side and the other terms on the right. For example, $x - 4i = 3 - 2ix$ becomes $x + 2ix = 3 + 4i$ so that $(1 + 2i)x = 3 + 4i$. Dividing both sides by $1 + 2i$ shows that $x = (3 + 4i)/(1 + 2i)$. The final step is to express this solution in the form $a + bi$.]
  (a) $3x = 2 + ix$
  (b) $2ix = 1 - ix$
  (c) $(7 + i)x - 3 = 2i + ix$
  (d) $3x + (4 + 2i) + ix = (5 + i)x$
  (e) $(3 + 2i)x + (3 + 2i) = (2 + i)x - (4 + 2i)$
  (f) $2 + 3ix - (4 + 5i) = (4i - 3)x - 5i$
  (g) $(5 - 3i)x + 4x - 1 = (1 - 4i)x + 7x - i$
  (h) $(4 + 6i)x - 7 + i = (3 + 4i)x - (5 + 2i)$

B.7. In the next section, we shall prove that every complex number has a square root. (See Exercise C.3 on p. 504.) Since the derivation of the quadratic formula depends only on the existence of square roots and the usual rules of arithmetic, the quadratic formula may be used to solve $ax^2 + bx + c = 0$, even when a, b, c are complex numbers. Solve the equations below, leaving square roots of complex numbers in the form $\sqrt{a + bi}$.
  (a) $x^2 + ix - (2 + 3i) = 0$
  (b) $x^2 + (1 + i)x - (1 + 2i) = 0$
  (c) $ix^2 + (2 - i)x - (1 + i) = 0$
  (d) $ix^2 + (1 - 2i)x + (2 + 3i) = 0$
  (e) $(1 + i)x^2 + 3ix = 1 - 3i$
  (f) $(2 - i)x^2 + (1 + i)x = -2 + 2i$

B.8. Find a polynomial with complex coefficients, having the given degree and given roots.
- **(a)** degree 2; roots i and $1 - 2i$
- **(b)** degree 2; roots $2i$ and $1 + i$
- **(c)** degree 3; roots 3, i, and $2 - i$
- **(d)** degree 4; roots $\sqrt{2}$, $-\sqrt{2}$, $1 + i$, and $1 - i$

C.1. Prove that if z is a root of a polynomial with real coefficients and degree 3, then so is $\bar{z}$, as follows. We know that $f(x) = ax^3 + bx^2 + cx + d$ for some fixed real numbers a, b, c, d and that $0 = f(z) = az^3 + bz^2 + cz + d$. Use facts (ii) and (iii) in the box on page 478 to show that:
- **(a)** $\overline{f(z)} = \bar{0} = 0$.
- **(b)** $\overline{f(z)} = \overline{az^3} + \overline{bz^2} + \overline{cz} + \bar{d}$.
- **(c)** $\overline{f(z)} = a\overline{z^3} + b\overline{z^2} + c\bar{z} + d$ [remember that a, b, c, d are real numbers].
- **(d)** $\overline{f(z)} = a\bar{z}^3 + b\bar{z}^2 + c\bar{z} + d = f(\bar{z})$.
- **(e)** $\bar{z}$ is a root of $f(x)$. [Use parts (a) and (d).]
- **(f)** Now suppose $f(x)$ is any polynomial with real coefficients and positive degree. Prove that if z is a root of $f(x)$, then so is $\bar{z}$. (*Hint:* except for notation, the proof is essentially the same as in the case of degree 3 done above.)

C.2. **(a)** Suppose z is a fixed complex number. Show that the polynomial $(x - z)(x - \bar{z})$ has *real* coefficients. [*Hint:* multiply out and use fact (i) in the box on p. 478.]
- **(b)** Let $f(x)$ be a nonconstant polynomial with real coefficients. By the Fundamental Theorem of Algebra $f(x) = d(x - c_1)(x - c_2) \cdots (x - c_n)$. If one of the roots c_i is complex (and *not* real), then some other root (say, c_j) must be its conjugate, $c_j = \bar{c_i}$ (why?). Use part (a) to show that $f(x)$ has as a factor a quadratic polynomial with *real* coefficients and roots c_i and c_j.
- **(c)** Prove that every nonconstant polynomial $f(x)$ with *real* coefficients is a product of factors, each of which is either (i) a first-degree polynomial with real coefficients or (ii) a quadratic polynomial with real coefficients and no real roots. [*Hint:* see part (b).]

3. THE GEOMETRY OF COMPLEX NUMBERS

The geometric representation of the real number system as a number line has proved very helpful. In this section, we explore a geometric representation of the complex number system as a plane. This geometric representation leads naturally to a new way of expressing complex numbers in terms of angles and distances, the so-called polar form for complex numbers. Polar form then enables us to use geometry and trigonometry to deal with various *algebraic* problems that couldn't be solved previously.

THE COMPLEX PLANE

The real number system can be represented geometrically by a straight line. Each point on such a number line corresponds to a real number, and vice versa. In an analogous fashion the complex number system can be represented geometrically by the ordinary coordinate plane:

> The complex number $a + bi$ corresponds to the point (a, b) on the plane

In other words, the point (a, b) is labeled by the complex number $a + bi$. For example, see Figure 8-1.

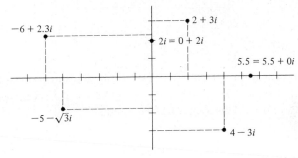

Figure 8-1

When the coordinate plane is labeled by complex numbers in this way, it is called the **complex plane.** Each real number $a = a + 0i$ corresponds to the point $(a, 0)$ on the horizontal axis. Consequently, the horizontal axis is called the **real axis.** Similarly, since each imaginary number $bi = 0 + bi$ corresponds to the point $(0, b)$, the vertical axis is called the **imaginary axis.**

Recall that the absolute value of the complex number $a + bi$ is defined to be the real number $|a + bi| = \sqrt{a^2 + b^2}$. Absolute values have a convenient geometric interpretation:

> *The distance from* a + bi *to* 0 *in the complex plane is* |a + bi|,

For the distance from $a + bi$ to 0 is just the distance from the point (a, b) to the point $(0, 0)$. It is given by the distance formula:

$$\sqrt{(a - 0)^2 + (b - 0)^2} = \sqrt{a^2 + b^2} = |a + bi|$$

Addition of complex numbers has an interesting geometric interpretation. For details, see Exercise C.2.

POLAR FORM

Let $a + bi$ be any nonzero complex number and denote $|a + bi|$ by r. In the complex plane, r is the length of the line segment joining (a, b) and $(0, 0)$ so that $r > 0$. Let θ be the angle in standard position whose terminal side is this line segment, as shown in Figure 8-2.

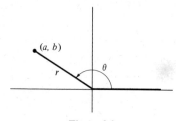

Figure 8-2

According to the point-in-the-plane description of the sine and cosine functions (p. 426),

$$\cos \theta = \frac{a}{r} \qquad \text{and} \qquad \sin \theta = \frac{b}{r}$$

so that

$$a = r \cos \theta \qquad \text{and} \qquad b = r \sin \theta$$

Consequently,

$$a + bi = r \cos \theta + (r \sin \theta)i = r(\cos \theta + i \sin \theta)^{\circ}$$

When a complex number $a + bi$ is written in this way, it is said to be in **polar form** or **trigonometric form.** The angle θ is called the **argument** and is usually expressed in radian measure. The number $r = |a + bi|$ is sometimes called the **modulus** (plural, moduli). The number 0 can also be written in polar notation by letting $r = 0$ and θ be any angle. Thus

---

Every complex number a + bi *can be written in polar form*

$$r(cos\ \theta + i\ sin\ \theta)$$

where r = |a + bi| = $\sqrt{a^2 + b^2}$,   a = r cos θ,   *and*   b = r sin θ.

When a complex number is written in polar form, the argument θ is not uniquely determined since θ, $\theta \pm 2\pi$, $\theta \pm 4\pi$, and so on, all satisfy the conditions stated in the box above.

° It is customary to place i in front of $\sin \theta$ rather than after it. Some books abbreviate $r(\cos \theta + i \sin \theta)$ as r cis θ.

EXAMPLE   Express $-\sqrt{3} + i$ in polar form $r(\cos\theta + i\sin\theta)$. Here $a = -\sqrt{3}$ and $b = 1$ so that $r = \sqrt{a^2 + b^2} = \sqrt{(-\sqrt{3})^2 + 1^2} = \sqrt{3 + 1} = 2$. Now we must find an angle θ such that

$$-\sqrt{3} + i = r(\cos\theta + i\sin\theta) = 2(\cos\theta + i\sin\theta)$$

The angle θ must satisfy:

$$\cos\theta = \frac{a}{r} = -\frac{\sqrt{3}}{2} \qquad \text{and} \qquad \sin\theta = \frac{b}{r} = \frac{1}{2}$$

Furthermore, since $-\sqrt{3} + i$ lies in the second quadrant (see Figure 8-3), θ must be a second-quadrant angle. Our knowledge of special angles and Figure 8-3 show that $\theta = 5\pi/6$ satisfies these conditions. Thus

$$-\sqrt{3} + i = 2\left(\cos\frac{5\pi}{6} + i\sin\frac{5\pi}{6}\right)$$

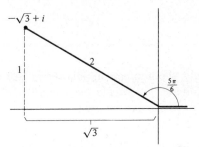

Figure 8-3

EXAMPLE   Express $-2 + 5i$ in polar form $r(\cos\theta + i\sin\theta)$. Here $a = -2$ and $b = 5$ so that $r = \sqrt{(-2)^2 + 5^2} = \sqrt{29}$. The angle θ must satisfy

$$\cos\theta = \frac{a}{r} = \frac{-2}{\sqrt{29}} \qquad \text{and} \qquad \sin\theta = \frac{b}{r} = \frac{5}{\sqrt{29}}$$

Since $-2 + 5i$ lies in the second quadrant, as shown in Figure 8-4, we can find θ by finding a second-quadrant solution to either of the equations above.

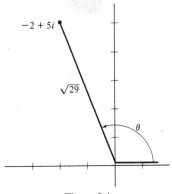

Figure 8-4

As we saw in Section 1 of Chapter 7, the equation $\cos \theta = -2/\sqrt{29} \approx -.37$ has solutions in the second and third quadrants. The second-quadrant solution is $\theta \approx 1.95$. Therefore

$$-2 + 5i \approx \sqrt{29}(\cos 1.95 + i \sin 1.95)$$

Multiplying and dividing complex numbers in polar form turns out to be quite easy:

If $z_1 = r_1(\cos \theta_1 + i \sin \theta_1)$ and $z_2 = r_2(\cos \theta_2 + i \sin \theta_2)$ are any two complex numbers, then

$$z_1 z_2 = r_1 r_2(\cos (\theta_1 + \theta_2) + i \sin (\theta_1 + \theta_2))$$

and

$$\frac{z_1}{z_2} = \frac{r_1}{r_2}(\cos (\theta_1 - \theta_2) + i \sin (\theta_1 - \theta_2)) \quad (z_2 \neq 0)$$

In other words, to multiply two numbers in polar form, just *multiply the moduli and add the arguments*. To divide, just *divide the moduli and subtract the arguments*. Before proving the statements in the box, we illustrate them with some examples.

EXAMPLE Suppose $z_1 = 2(\cos (5\pi/6) + i \sin (5\pi/6))$ and $z_2 = 3(\cos (7\pi/4) + i \sin (7\pi/4))$. Then r_1 is the number 2 and $\theta_1 = 5\pi/6$; similarly, $r_2 = 3$ and $\theta_2 = 7\pi/4$ and we have:

$$z_1 z_2 = r_1 r_2(\cos (\theta_1 + \theta_2) + i \sin (\theta_1 + \theta_2))$$

$$= 2 \cdot 3 \left(\cos \left(\frac{5\pi}{6} + \frac{7\pi}{4} \right) + i \sin \left(\frac{5\pi}{6} + \frac{7\pi}{4} \right) \right)$$

$$= 6 \left(\cos \left(\frac{10\pi}{12} + \frac{21\pi}{12} \right) + i \sin \left(\frac{10\pi}{12} + \frac{21\pi}{12} \right) \right)$$

$$= 6 \left(\cos \frac{31\pi}{12} + i \sin \frac{31\pi}{12} \right)$$

EXAMPLE Suppose $z_1 = 10(\cos (\pi/3) + i \sin (\pi/3))$ and $z_2 = 2(\cos (\pi/4) + i \sin (\pi/4))$. Since $(\pi/3) - (\pi/4) = (4\pi/12) - (3\pi/12) = \pi/12$, we have:

$$\frac{z_1}{z_2} = \frac{10 \left(\cos \frac{\pi}{3} + i \sin \frac{\pi}{3} \right)}{2 \left(\cos \frac{\pi}{4} + i \sin \frac{\pi}{4} \right)} = \frac{10}{2} \left(\cos \left(\frac{\pi}{3} - \frac{\pi}{4} \right) + i \sin \left(\frac{\pi}{3} - \frac{\pi}{4} \right) \right)$$

$$= 5 \left(\cos \frac{\pi}{12} + i \sin \frac{\pi}{12} \right)$$

Here is a proof of the polar multiplication rule. If $z_1 = r_1(\cos\theta_1 + i\sin\theta_1)$ and $z_2 = r_2(\cos\theta_2 + i\sin\theta_2)$, then

$$\begin{aligned}
z_1 z_2 &= r_1(\cos\theta_1 + i\sin\theta_1)r_2(\cos\theta_2 + i\sin\theta_2) \\
&= r_1 r_2(\cos\theta_1 + i\sin\theta_1)(\cos\theta_2 + i\sin\theta_2) \\
&= r_1 r_2(\cos\theta_1\cos\theta_2 + i\sin\theta_1\cos\theta_2 + i\cos\theta_1\sin\theta_2 + i^2\sin\theta_1\sin\theta_2) \\
&= r_1 r_2((\cos\theta_1\cos\theta_2 - \sin\theta_1\sin\theta_2) + i(\sin\theta_1\cos\theta_2 + \cos\theta_1\sin\theta_2))
\end{aligned}$$

But the addition identities for sine and cosine (p. 414) show that

$$\cos\theta_1\cos\theta_2 - \sin\theta_1\sin\theta_2 = \cos(\theta_1 + \theta_2)$$

and

$$\sin\theta_1\cos\theta_2 + \cos\theta_1\sin\theta_2 = \sin(\theta_1 + \theta_2)$$

Therefore

$$\begin{aligned}
z_1 z_2 &= r_1 r_2((\cos\theta_1\cos\theta_2 - \sin\theta_1\sin\theta_2) + i(\sin\theta_1\cos\theta_2 + \cos\theta_1\sin\theta_2)) \\
&= r_1 r_2(\cos(\theta_1 + \theta_2) + i\sin(\theta_1 + \theta_2))
\end{aligned}$$

This completes the proof of the multiplication rule. The division rule is proved similarly (Exercise C.1).

POWERS OF COMPLEX NUMBERS

Polar form provides a convenient way of calculating powers of a complex number $z = r(\cos\theta + i\sin\theta)$. For instance, the multiplication formula above shows that

$$\begin{aligned}
z^2 = z \cdot z &= r \cdot r(\cos(\theta + \theta) + i\sin(\theta + \theta)) \\
&= r^2(\cos 2\theta + i\sin 2\theta)
\end{aligned}$$

We can find z^3 similarly:

$$\begin{aligned}
z^3 = z^2 \cdot z &= r^2 \cdot r(\cos(2\theta + \theta) + i\sin(2\theta + \theta)) \\
&= r^3(\cos 3\theta + i\sin 3\theta)
\end{aligned}$$

Repeated application of the multiplication formula proves:

DeMOIVRE'S THEOREM

For any complex number $z = r(\cos\theta + i\sin\theta)$ *and any positive integer n,*

$$z^n = r^n(\cos(n\theta) + i\sin(n\theta))$$

EXAMPLE In order to compute $(-\sqrt{3} + i)^5$ we first express $-\sqrt{3} + i$ in polar form (as in the example on p. 493): $-\sqrt{3} + i = (2\cos(5\pi/6) + i\sin(5\pi/6))$. By DeMoivre's

Theorem,

$$(-\sqrt{3} + i)^5 = 2^5\left(\cos\left(5 \cdot \frac{5\pi}{6}\right) + i\sin\left(5 \cdot \frac{5\pi}{6}\right)\right) = 32\left(\cos\frac{25\pi}{6} + i\sin\frac{25\pi}{6}\right)$$

Since $25\pi/6 = (\pi/6) + (24\pi/6) = (\pi/6) + 4\pi$, we have

$$(-\sqrt{3} + i)^5 = 32\left(\cos\frac{25\pi}{6} + i\sin\frac{25\pi}{6}\right) = 32\left(\cos\frac{\pi}{6} + i\sin\frac{\pi}{6}\right)$$

$$= 32\left(\frac{\sqrt{3}}{2} + \frac{1}{2}i\right) = 16\sqrt{3} + 16i$$

EXAMPLE To find $(1 + i)^{10}$, first verify that the polar form of $1 + i$ is $1 + i = \sqrt{2}(\cos(\pi/4) + i\sin(\pi/4))$. Therefore, by DeMoivre's Theorem,

$$(1 + i)^{10} = (\sqrt{2})^{10}\left(\cos\frac{10\pi}{4} + i\sin\frac{10\pi}{4}\right)$$

$$= (2^{1/2})^{10}\left(\cos\frac{5\pi}{2} + i\sin\frac{5\pi}{2}\right) = 2^5(0 + i \cdot 1) = 32i$$

ROOTS OF UNITY

It is easy to see that the equation $x^2 = 1$ has exactly two solutions, namely, ± 1. On page 484 we saw that $x^3 = 1$ has three solutions: the real number 1 and two complex solutions. More generally, for a fixed positive integer n, the equation $x^n = 1$ has exactly n different solutions in the complex number system. These solutions (which are usually called the **nth roots of unity**) can be found via DeMoivre's Theorem, as shown in the following example.

EXAMPLE The equation $x^5 = 1$ obviously has just one *real* solution, $x = 1$. A complex solution of $x^5 = 1$ is a complex number $r(\cos\theta + i\sin\theta)$ such that $[r(\cos\theta + i\sin\theta)]^5 = 1$, or equivalently, by DeMoivre's Theorem,

$$r^5(\cos 5\theta + i\sin 5\theta) = 1 = 1(\cos 0 + i\sin 0)$$

The only way for this equation to hold is to have

$$r^5 = 1, \qquad \cos 5\theta = \cos 0 = 1, \qquad \sin 5\theta = \sin 0 = 0$$

Since r is a positive real number, we must have $r = 1$. Since the only numbers with cosine 1 and sine 0 are 0, $\pm 2\pi$, $\pm 4\pi$, and so on, we must have:

$$5\theta = 0, \pm 2\pi, \pm 4\pi, \pm 6\pi, \ldots$$

We can express this same fact in slightly different notation by saying

$$5\theta = 2k\pi \quad (k = 0, \pm 1, \pm 2, \pm 3, \ldots)$$

Solving these last equations for θ yields:

$$\theta = \frac{2k\pi}{5} \quad (k = 0, \pm 1, \pm 2, \pm 3, \ldots)$$

Substituting 0, 1, 2, 3, and 4 for k produces these possibilities for θ:

$$\theta = \frac{2 \cdot 0 \cdot \pi}{5} = 0, \qquad \theta = \frac{2 \cdot 1 \cdot \pi}{5} = \frac{2\pi}{5}, \qquad \theta = \frac{2 \cdot 2 \cdot \pi}{5} = \frac{4\pi}{5},$$

$$\theta = \frac{2 \cdot 3\pi}{5} = \frac{6\pi}{5}, \qquad \theta = \frac{2 \cdot 4\pi}{5} = \frac{8\pi}{5}$$

These five values of θ yield five different solutions of the equation $x^5 = 1$:

$$x = \cos 0 + i \sin 0 = 1 \qquad x = \cos \frac{2\pi}{5} + i \sin \frac{2\pi}{5}, \qquad x = \cos \frac{4\pi}{5} + i \sin \frac{4\pi}{5}$$

$$x = \cos \frac{6\pi}{5} + i \sin \frac{6\pi}{5}, \qquad x = \cos \frac{8\pi}{5} + i \sin \frac{8\pi}{5}$$

Any other value of k produces an angle θ whose terminal side is the same as the terminal side of one of the five angles listed above. For instance, if $k = 6$, then $\theta = 2k\pi/5 = 2 \cdot 6\pi/5 = 12\pi/5 = (2\pi/5) + 2\pi$, so that θ has the same terminal side as $2\pi/5$. Similarly, if $k = -3$, then $\theta = 2k\pi/5 = 2(-3)\pi/5 = -6\pi/5$ has the same terminal side as $4\pi/5$. Since angles with the same terminal side have the same sine and cosine, all these other values of k will lead to the same five solutions, $\cos \theta + i \sin \theta$ of $x^5 = 1$.[°] Thus we have found *all* the solutions of $x^5 = 1$, that is, all five fifth roots of unity. Since the angle θ in each solution is of the form $2k\pi/5$ with $k = 0, 1, 2, 3, 4$, we see that the five fifth roots of unity are given by the formula:

$$\cos \frac{2k\pi}{5} + i \sin \frac{2k\pi}{5}, \quad k = 0, 1, 2, 3, 4$$

If the fifth roots of unity are plotted in the complex plane, we see that they are five points equally spaced around the unit circle at intervals of $2\pi/5$, as shown in Figure 8-5.

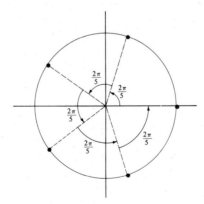

Figure 8-5

[°] We could have reached this same conclusion by noting that a fifth-degree equation, such as $x^5 = 1$, has at most five distinct solutions in the complex number system (see p. 488).

Now let n be a fixed positive integer. The equation $x^n = 1$ can be solved by exactly the same method that was used to solve $x^5 = 1$ in the preceding example: just replace 5 by n throughout the argument. There is only one difference: the five different solutions of $x^5 = 1$ were determined by letting k take all integer values from 0 to 4 (that is, from 0 to $5 - 1$). The n different solutions of $x^n = 1$ are determined by letting k take all integer values from 0 to $n - 1$. Consequently, we have this general result:

Let n *be a fixed positive integer. The* nth *roots of unity (the solutions of* $x^n = 1$*) are the* n *complex numbers*

$$\cos \frac{2k\pi}{n} + i \sin \frac{2k\pi}{n}, \qquad k = 0, 1, 2, 3, \dots, n - 1$$

When the nth *roots of unity are plotted in the complex plane, they consist of* n *points equally spaced around the unit circle at intervals of* $2\pi/n$.

EXAMPLE The cube roots of unity (the solutions of $x^3 = 1$) can be found in polar form by applying the formula in the box above with $n = 3$ and $n - 1 = 2$. Then k takes the values 0, 1, and 2 and the angles $2k\pi/n = 2k\pi/3$ are just the angles

$$\frac{2 \cdot 0 \cdot \pi}{3} = 0, \qquad \frac{2 \cdot 1 \cdot \pi}{3} = \frac{2\pi}{3}, \qquad \text{and} \qquad \frac{2 \cdot 2\pi}{3} = \frac{4\pi}{3}$$

Therefore the cube roots of unity are:

$$\cos 0 + i \sin 0 = 1, \qquad \cos \frac{2\pi}{3} + i \sin \frac{2\pi}{3}, \qquad \text{and} \qquad \cos \frac{4\pi}{3} + i \sin \frac{4\pi}{3}$$

ROOTS OF COMPLEX NUMBERS

Let $a + bi$ be any nonzero complex number. We want to solve equations such as

$$x^2 = a + bi, \qquad x^3 = a + bi, \qquad x^4 = a + bi$$

and more generally,

$$x^n = a + bi \qquad (n \text{ a fixed positive integer})$$

A solution of this last equation is often called an **nth root** of $a + bi$. If we write $a + bi$ in polar form, it's quite easy to find *one* solution of $x^n = a + bi$.

EXAMPLE To find a solution of $x^3 = 2i$, we first verify that the polar form of $2i$ is $2(\cos (\pi/2) + i \sin (\pi/2))$. We must find a cube root of this number. Now DeMoivre's Theorem tells us how to cube a number: cube the modulus and triple the argument. So we ought to be able to find cube roots by reversing this procedure: take the cube root of the modulus and one-third of the argument. Applying this idea to $2(\cos (\pi/2) +$

$i \sin(\pi/2))$ we see that the cube root of the modulus is $\sqrt[3]{2}$ and one-third of the argument is $\frac{1}{3} \cdot (\pi/2) = \pi/6$. So consider the number

$$z = \sqrt[3]{2}\left(\cos\frac{\pi}{6} + i \sin\frac{\pi}{6}\right)$$

Applying DeMoivre's Theorem, we see that

$$z^3 = (\sqrt[3]{2})^3 \left(\cos\left(3 \cdot \frac{\pi}{6}\right) + i \sin\left(3 \cdot \frac{\pi}{6}\right)\right) = 2\left(\cos\frac{\pi}{2} + i \sin\frac{\pi}{2}\right) = 2i$$

Therefore $\sqrt[3]{2}(\cos\pi/6 + i \sin\pi/6)$ is indeed a solution of $x^3 = 2i$.

The process used in this example works in the general case. To find a solution of $x^n = r(\cos\theta + i \sin\theta)$, where $r \neq 0$, take the nth root of $r°$ and divide the angle θ by n. This gives the number

$$z = \sqrt[n]{r}\left(\cos\frac{\theta}{n} + i \sin\frac{\theta}{n}\right)$$

Applying DeMoivre's Theorem to z shows that

$$z^n = (\sqrt[n]{r})^n \left(\cos\left(n \cdot \frac{\theta}{n}\right) + i \sin\left(n \cdot \frac{\theta}{n}\right)\right) = r(\cos\theta + i \sin\theta)$$

Therefore

> *One solution of* $x^n = r(\cos\theta + i \sin\theta)$ *is the number*
>
> $$\sqrt[n]{r}\left(\cos\frac{\theta}{n} + i \sin\frac{\theta}{n}\right)$$

The next step is to find the other solutions of $x^n = r(\cos\theta + i \sin\theta)$. This can be done by using the nth roots of unity. For if z is one solution of $x^n = r(\cos\theta + i \sin\theta)$, as found above, and u is an nth root of unity, then $u^n = 1$ and

$$(zu)^n = z^n u^n = r(\cos\theta + i \sin\theta) \cdot 1$$

Therefore zu is also a solution of $x^n = r(\cos\theta + i \sin\theta)$. This argument works for z and any nth root of unity. Since there are n distinct nth roots of unity, it leads to n distinct solutions of $x^n = r(\cos\theta + i \sin\theta)$. More specifically,

> *If the number* z *is one nonzero solution of* $x^n = r(\cos\theta + i \sin\theta)$ *and if* $u_1, u_2, \ldots, u_n$ *are the* n *distinct* n*th roots of unity, then the* n *different numbers*
>
> $$zu_1, \quad zu_2, \quad zu_3, \ldots, zu_n$$
>
> *are the solutions of* $x^n = r(\cos\theta + i \sin\theta)$.

° Remember that every positive real number has a unique positive nth root.

EXAMPLE In the preceding example, we saw that $z = \sqrt[3]{2}(\cos{(\pi/6)} + i \sin{(\pi/6)})$ is one solution of $x^3 = 2(\cos{(\pi/2)} + i \sin{(\pi/2)})$. On page 498 we saw that the three cube roots of unity are:

$$u_1 = \cos 0 + i \sin 0 = 1, \qquad u_2 = \cos\frac{2\pi}{3} + i \sin\frac{2\pi}{3}, \qquad u_3 = \cos\frac{4\pi}{3} + i \sin\frac{4\pi}{3}$$

Therefore the three solutions of $x^3 = 2(\cos{\pi/2} + i \sin{\pi/2})$ are

$$zu_1 = z \cdot 1 = \sqrt[3]{2}\left(\cos\frac{\pi}{6} + i \sin\frac{\pi}{6}\right)$$

$$zu_2 = \sqrt[3]{2}\left(\cos\frac{\pi}{6} + i \sin\frac{\pi}{6}\right)\left(\cos\frac{2\pi}{3} + i \sin\frac{2\pi}{3}\right)$$

$$= \sqrt[3]{2}\left(\cos\left(\frac{\pi}{6} + \frac{2\pi}{3}\right) + i \sin\left(\frac{\pi}{6} + \frac{2\pi}{3}\right)\right)$$

$$zu_3 = \sqrt[3]{2}\left(\cos\frac{\pi}{6} + i \sin\frac{\pi}{6}\right)\left(\cos\frac{4\pi}{3} + i \sin\frac{4\pi}{3}\right)$$

$$= \sqrt[3]{2}\left(\cos\left(\frac{\pi}{6} + \frac{4\pi}{3}\right) + i \sin\left(\frac{\pi}{6} + \frac{4\pi}{3}\right)\right)$$

Since the angles 0, $2\pi/3$, and $4\pi/3$ are just the angles $2k\pi/3$ with $k = 0, 1, 2$, we can summarize these results by saying that the solutions of $x^3 = 2(\cos{(\pi/2)} + i \sin{(\pi/2)})$ are the three numbers

$$\sqrt[3]{2}\left(\cos\left(\frac{\pi}{6} + \frac{2k\pi}{3}\right) + i \sin\left(\frac{\pi}{6} + \frac{2k\pi}{3}\right)\right), \qquad k = 0, 1, 2$$

A formula for the polar form of all n solutions of $x^n = r(\cos\theta + i \sin\theta)$ can be found by the same procedures used in the preceding example. We know that one solution of $x^n = r(\cos\theta + i \sin\theta)$ is

$$z = \sqrt[n]{r}\left(\cos\frac{\theta}{n} + i \sin\frac{\theta}{n}\right)$$

and that the nth roots of unity are

$$\left(\cos\frac{2k\pi}{n} + i \sin\frac{2k\pi}{n}\right), \qquad k = 0, 1, 2, \ldots, n - 1$$

Therefore all solutions can be found by taking the product of z with each root of unity:

$$\sqrt[n]{r}\left(\cos\frac{\theta}{n} + i \sin\frac{\theta}{n}\right)\left(\cos\frac{2k\pi}{n} + i \sin\frac{2k\pi}{n}\right)$$

$$= \sqrt[n]{r}\left(\cos\left(\frac{\theta}{n} + \frac{2k\pi}{n}\right) + i \sin\left(\frac{\theta}{n} + \frac{2k\pi}{n}\right)\right)$$

We have proved this result:

> Let $r(\cos \theta + i \sin \theta)$ be any nonzero complex number. The equation
> $x^n = r(\cos \theta + i \sin \theta)$ has n distinct solutions:
>
> $$\sqrt[n]{r}\left(\cos\left(\frac{\theta}{n} + \frac{2k\pi}{n}\right) + i \sin\left(\frac{\theta}{n} + \frac{2k\pi}{n}\right)\right),$$
>
> where $k = 0, 1, 2, \ldots, n - 1$

EXAMPLE To solve the equation $x^4 = 3 + 3\sqrt{3}i$ we first write $3 + 3\sqrt{3}i$ in polar form as $6(\cos(\pi/3) + i \sin(\pi/3)).$* Now apply the root formula in the box above with $r = 6$, $\theta = \pi/3$, $k = 0, 1, 2, 3$, and the solutions are:

$$\sqrt[4]{6}\left(\cos\left(\frac{\pi/3}{4} + \frac{2k\pi}{4}\right) + i \sin\left(\frac{\pi/3}{4} + \frac{2k\pi}{4}\right)\right), \qquad k = 0, 1, 2, 3$$

The angles corresponding to each value of k are as follows:

k	0	1	2	3
$\frac{1}{4}\left(\frac{\pi}{3} + 2k\pi\right)$	$\frac{1}{4}\left(\frac{\pi}{3} + 0\right) = \frac{\pi}{12}$	$\frac{1}{4}\left(\frac{\pi}{3} + 2\pi\right) = \frac{7\pi}{12}$	$\frac{1}{4}\left(\frac{\pi}{3} + 4\pi\right) = \frac{13\pi}{12}$	$\frac{1}{4}\left(\frac{\pi}{3} + 6\pi\right) = \frac{19\pi}{12}$

Therefore the four solutions of $x^4 = 3 + 3\sqrt{3}i$ are:

$$x = \sqrt[4]{6}\left(\cos\frac{\pi}{12} + i \sin\frac{\pi}{12}\right), \qquad x = \sqrt[4]{6}\left(\cos\frac{7\pi}{12} + i \sin\frac{7\pi}{12}\right),$$

$$x = \sqrt[4]{6}\left(\cos\frac{13\pi}{12} + i \sin\frac{13\pi}{12}\right), \qquad x = \sqrt[4]{6}\left(\cos\frac{19\pi}{12} + i \sin\frac{19\pi}{12}\right)$$

EXERCISES

A.1. Plot the point in the complex plane corresponding to each of these numbers:
  (a) $3 + 2i$ (d) $\sqrt{2} - 7i$ (g) $(2i)(3 - \frac{5}{2}i)$
  (b) $-7 + 6i$ (e) $(1 + i)(1 - i)$ (h) $(\frac{4}{3}i)(-6 - 3i)$
  (c) $-\frac{8}{3} - \frac{5}{3}i$ (f) $(2 + i)(1 - 2i)$

B.1. Sketch the graph of these equations in the complex plane (z denotes a complex number):
  (a) $|z| = 4$ (c) $\text{Re}(z) = -5$ (See Exercise C.1, p. 481.)
  (b) $|z| = 1$ (d) $\text{Im}(z) = 2$

° The details are omitted, but the polar form is easy to verify since $\cos(\pi/3) = \frac{1}{2}$ and $\sin(\pi/3) = \sqrt{3}/2$.

B.2. Use your knowledge of special angles to express each of these numbers in polar form:

(a) $5i$
(b) $-3i$
(c) -7
(d) $\sqrt{19}$
(e) $-\sqrt{5}i$
(f) $1 + \sqrt{3}i$
(g) $2\sqrt{3} - 2i$
(h) $-\sqrt{2} + \sqrt{2}i$

B.3. Use a calculator or tables to express these numbers in polar form:

(a) $3 + 4i$
(b) $-4 + 3i$
(c) $5 - 12i$
(d) $-\sqrt{7} - 3i$
(e) $1 + 2i$
(f) $3 - 5i$
(g) $-\frac{5}{2} + \frac{7}{2}i$
(h) $\sqrt{5} + \sqrt{11}i$

B.4. Perform the indicated multiplication or division; express your answer in both polar form $r(\cos\theta + i\sin\theta)$ *and* rectangular form $a + bi$.

(a) $3\left(\cos\dfrac{\pi}{12} + i\sin\dfrac{\pi}{12}\right) \cdot 2\left(\cos\dfrac{7\pi}{12} + i\sin\dfrac{7\pi}{12}\right)$

(b) $3\left(\cos\dfrac{\pi}{8} + i\sin\dfrac{\pi}{8}\right) \cdot 12\left(\cos\dfrac{3\pi}{8} + i\sin\dfrac{3\pi}{8}\right)$

(c) $12\left(\cos\dfrac{11\pi}{12} + i\sin\dfrac{11\pi}{12}\right) \cdot \dfrac{7}{2}\left(\cos\dfrac{\pi}{4} + i\sin\dfrac{\pi}{4}\right)$

(d) $\dfrac{8\left(\cos\dfrac{5\pi}{18} + i\sin\dfrac{5\pi}{18}\right)}{4\left(\cos\dfrac{\pi}{9} + i\sin\dfrac{\pi}{9}\right)}$

(e) $\dfrac{6\left(\cos\dfrac{7\pi}{20} + i\sin\dfrac{7\pi}{20}\right)}{4\left(\cos\dfrac{\pi}{10} + i\sin\dfrac{\pi}{10}\right)}$

(f) $\dfrac{\sqrt{54}\left(\cos\dfrac{9\pi}{4} + i\sin\dfrac{9\pi}{4}\right)}{\sqrt{6}\left(\cos\dfrac{7\pi}{12} + i\sin\dfrac{7\pi}{12}\right)}$

B.5. Convert to polar form, then multiply or divide. Express your answer in polar form.

(a) $(1 + i)(1 + \sqrt{3}i)$
(c) $\dfrac{1 + i}{1 - i}$
(e) $3i(2\sqrt{3} + 2i)$

(b) $(1 - i)(3 - 3i)$
(d) $\dfrac{2 - 2i}{-1 - i}$
(f) $\dfrac{-4i}{\sqrt{3} + i}$

B.6. Calculate each of these products; express your answer in the form $a + bi$.

(a) $\left(\cos\dfrac{\pi}{12} + i\sin\dfrac{\pi}{12}\right)^6$
(c) $3\left(\cos\dfrac{7\pi}{30} + i\sin\dfrac{7\pi}{30}\right)^5$

(b) $\left(\cos\dfrac{\pi}{5} + i\sin\dfrac{\pi}{5}\right)^{20}$
(d) $\sqrt[3]{4}\left(\cos\dfrac{7\pi}{36} + i\sin\dfrac{7\pi}{36}\right)^{12}$

B.7. Use polar form and DeMoivre's Theorem to calculate these products; express your answer in the form $a + bi$.

(a) $(1 - i)^{12}$

(c) $\left(\dfrac{\sqrt{3}}{2} + \dfrac{1}{2}i\right)^{10}$

(e) $\left(\dfrac{-1}{\sqrt{2}} + \dfrac{i}{\sqrt{2}}\right)^{14}$

(b) $(2 + 2i)^8$

(d) $\left(-\dfrac{1}{2} + \dfrac{\sqrt{3}}{2}i\right)^{20}$

(f) $(-1 + \sqrt{3}i)^8$

B.8. Find the indicated roots of unity. Express your answers in the form $a + bi$.
(a) fourth roots of unity
(b) sixth roots of unity
(c) eighth roots of unity
(d) twelfth roots of unity

B.9. Find all solutions in the complex number system of the given equation.
(a) $x^6 = -1$
(b) $x^6 + 64 = 0$
(c) $x^3 = i$
(d) $x^4 = i$
(e) $x^3 + 27i = 0$
(f) $x^6 + 729 = 0$
(g) $x^4 = -1 + \sqrt{3}i$
(h) $x^4 = -8 - 8\sqrt{3}i$

B.10. A **primitive nth root of unity** is a number z such that $z^n = 1$, but no lower positive power of z is 1.
(a) Verify that both $w = -\frac{1}{2} + (\sqrt{3}/2)i$ and w^2 are primitive cube roots of unity.
(b) Find all primitive fourth roots of unity; show that each is a power of the other.
(c) Do part (b) with "sixth" in place of "fourth."

B.11. (a) Solve the equation $x^3 + x^2 + x + 1 = 0$. (*Hint:* first find the quotient when $x^4 - 1$ is divided by $x - 1$ and then consider solutions of $x^4 - 1 = 0$.)
(b) Solve $x^5 + x^4 + x^3 + x^2 + x + 1 = 0$ [*Hint:* consider $x^6 - 1$ and $x - 1$ and see part (a).]
(c) What do you think are the solutions of $x^{n-1} + x^{n-2} + \cdots + x^3 + x^2 + x + 1 = 0$?

B.12. In the complex plane, identify each point with its complex number label. The unit circle consists of all numbers (points) z such that $|z| = 1$. Suppose v and w are two points (numbers) that move around the unit circle in such a way that $v = w^{12}$ at all times. When w has made one complete trip around the circle, how many trips has v made? (*Hint:* think polar and DeMoivre.)

B.13. Suppose u is an nth root of unity. Show that $1/u$ is also an nth root of unity. (*Hint:* use the definition, *not* polar form.)

C.1. Let $z_1 = r_1(\cos \theta_1 + i \sin \theta_1)$ and $z_2 = r_2(\cos \theta_2 + i \sin \theta_2)$. Then

$$\frac{z_1}{z_2} = \frac{r_1(\cos \theta_1 + i \sin \theta_1)}{r_2(\cos \theta_2 + i \sin \theta_2)} = \frac{r_1(\cos \theta_1 + i \sin \theta_1)}{r_2(\cos \theta_2 + i \sin \theta_2)} \cdot \frac{\cos \theta_2 - i \sin \theta_2}{\cos \theta_2 - i \sin \theta_2}$$

(a) Multiply out the denominator on the right side and use the Pythagorean Identity to show that it is just the number r_2.

(b) Multiply out the numerator on the right side above; use the subtraction identities for sine and cosine (p. 414) to show that it is $r_1(\cos(\theta_1 - \theta_2) + i \sin(\theta_1 - \theta_2))$. Therefore $z_1/z_2 = (r_1/r_2)(\cos(\theta_1 - \theta_2) + i \sin(\theta_1 - \theta_2))$.

C.2. The sum of two distinct complex numbers, $a + bi$ and $c + ci$, can be found geometrically by means of the so-called **parallelogram rule**: plot the points $a + bi$ and $c + di$ in the complex plane and form the parallelogram three of whose vertices are 0, $a + bi$, and $c + di$; for example, see Figure 8-6.

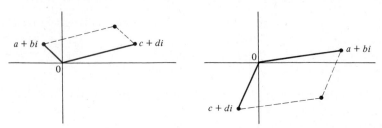

Figure 8-6

Then the fourth vertex of the parallelogram is the point whose coordinate is the sum $(a + bi) + (c + di) = (a + c) + (b + d)i$. Here is a *proof* of the parallelogram rule when $a \neq 0$ and $c \neq 0$.

(a) Find the *slope* of the line K from 0 to $a + bi$. [*Hint:* K contains the points $(0, 0)$ and (a, b).]

(b) Find the *slope* of the line N from 0 to $c + di$.

(c) Find the *equation* of the line L, through $a + bi$ and parallel to line N of part (b). [*Hint:* the point (a, b) is on L; find the slope of L by using (b) and facts about the slopes of parallel lines.]

(d) Find the *equation* of the line M, through $c + di$ and parallel to line K. [*Hint:* part (a) may be helpful.]

(e) Label the lines K, L, M, N in Figure 8-6.

(f) Show by substitution that the point $(a + c, b + d)$ satisfies both the equation of line L and the equation of line M. Therefore $(a + c, b + d)$ lies on both L and M. Since the only point on both L and M is the fourth vertex of the parallelogram (see Figure 8-6), this vertex must be $(a + c, b + d)$. Hence this vertex has coordinate $(a + c) + (b + d)i = (a + bi) + (c + di)$.

C.3. Use the root formula on page 501 and the identities

$$\cos(x + \pi) = -\cos x, \qquad \sin(x + \pi) = -\sin x$$

to show that the nonzero complex number $r(\cos\theta + i \sin\theta)$ has two square roots and that these square roots are negatives of each other.

Topics in Algebra

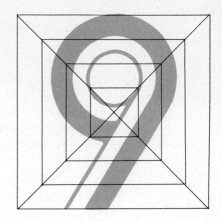

Section 1 deals with systems of linear equations. Such systems lead naturally to a discussion of matrices and determinants in Section 2. Section 3 (*The Binomial Theorem*) and Section 4 (*Mathematical Induction*) are independent of the first two sections and of each other. Either one of them can be read at any time.

1. SYSTEMS OF LINEAR EQUATIONS

In this section, we shall consider equations such as

$$2x + 3y = -7, \qquad 5x - y + 3z = 1, \qquad 4x + 2y - 8z + 23w = 0$$

that involve only the first powers of the unknowns and no higher powers (such as x^3, xy, xyz, y^2w, and so on). Such equations are called **linear equations.** Many applications of mathematics involve **systems of linear equations.** For example,

$$\begin{aligned} x + 2y &= 3, \\ 5x - 4y &= -6, \end{aligned}$$

two equations in two unknowns

$$\begin{aligned} 2x + 5y + z + w &= 0, \\ 2y - 4z + 3w &= 0, \\ 2x + 17y - 23z + 41w &= 0, \end{aligned}$$

three equations in four unknowns

$$\begin{aligned} x + 2y - 3z &= 1, \\ -3x + 2y + z &= 3, \\ 2x + 2y - 5z &= 0 \end{aligned}$$

three equations in three unknowns

A **solution of a system of equations** is a solution that satisfies *all* the equations in the system. For example, in the system of three equations in three unknowns shown at the right above,

$$x = 0, \qquad y = \tfrac{5}{4}, \qquad z = \tfrac{1}{2}$$

is a solution of all three equations (check it out) and hence is a solution for the system. On the other hand,

$$x = \tfrac{1}{2}, \qquad y = \tfrac{7}{4}, \qquad z = 1$$

is a solution of the first two equations in the system, but not of the third one (check it). Therefore this is *not* a solution of the system. It can be shown that

---

For any system of linear equations, exactly one of the following statements is true:

   (*i*)   *The system has no solution.*
  (*ii*)   *The system has exactly one solution.*
 (*iii*)   *The system has infinitely many different solutions.*

The examples below will illustrate all three possibilities.

THE SUBSTITUTION METHOD

Any system of two equations in two unknowns, as well as some larger systems, can be solved by means of substitution.

EXAMPLE   Any solution of the system

$$\begin{aligned} x + 2y &= 3 \\ 5x - 4y &= -6 \end{aligned}$$

must necessarily be a solution of the first equation. Hence x must satisfy

$$x + 2y = 3, \qquad \text{or equivalently,} \qquad x = 3 - 2y$$

Substituting this value of x in the second equation, we have:

$$\begin{aligned} 5x - 4y &= -6 \\ 5(3 - 2y) - 4y &= -6 \\ 15 - 10y - 4y &= -6 \\ -14y &= -21 \\ y &= \tfrac{-21}{-14} = \tfrac{3}{2} \end{aligned}$$

Therefore every solution of the original system must have $y = \tfrac{3}{2}$. But when $y = \tfrac{3}{2}$, we see from the first equation that:

$$\begin{aligned} x + 2y &= 3 \\ x + 2(\tfrac{3}{2}) &= 3 \\ x + 3 &= 3 \\ x &= 0 \end{aligned}$$

(We would also have found that $x = 0$ if we had substituted $y = \tfrac{3}{2}$ in the second equation.) Consequently, the original system has exactly one solution: $x = 0$, $y = \tfrac{3}{2}$.

This solution could also have been found by solving the first equation for y instead of x and substituting this value in the second equation.

Warning In order to guard against arithmetic mistakes, you should always *check your answers*. We have in fact checked the answers in all the examples. But these checks are omitted to save space.

The solutions of any system of equations in two unknowns can be interpreted geometrically by graphing each of the equations in the system. From Section 4 of Chapter 2, we know that each graph is a straight line. For instance, the graph of the equations in the preceding example is shown in Figure 9-1.

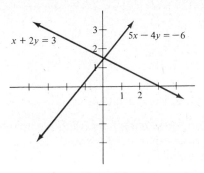

Figure 9-1

Each point on the line $x + 2y = 3$ represents a solution of this equation, and similarly for the line $5x - 4y = -6$. Since there is exactly one point on both lines, there is just one solution of the system of two equations. The geometric statement that the two lines intersect only at the point $(0, \frac{3}{2})$ is equivalent to the algebraic statement that the only solution of the system is $x = 0$, $y = \frac{3}{2}$.

THE ELIMINATION METHOD

In order to develop a systematic way of solving systems with three or more equations and unknowns, we first consider an example of a large system in which the substitution method can be used successfully.

EXAMPLE Any solution of this system of four equations in four unknowns

$$\begin{aligned}
3x - 3y + 2z + w &= 4 \\
y - 3z + 2w &= -3 \\
z - 4w &= -8 \\
2w &= 6
\end{aligned}$$

must necessarily be a solution of the last equation. Therefore w must satisfy

$$2w = 6, \quad \text{or equivalently,} \quad w = 3$$

Substituting $w = 3$ in the third equation shows that

$$z - 4w = -8$$
$$z - 4 \cdot 3 = -8$$
$$z = -8 + 12 = 4$$

Thus any solution of the original system must have $w = 3$ and $z = 4$. Substituting these values in the second equation yields:

$$y - 3z + 2w = -3$$
$$y - 3 \cdot 4 + 2 \cdot 3 = -3$$
$$y = -3 + 12 - 6 = 3$$

Finally, substituting $y = 3$, $z = 4$, and $w = 3$ in the first equation shows that

$$3x - 3y + 2z + w = 4$$
$$3x - 3 \cdot 3 + 2 \cdot 4 + 3 = 4$$
$$3x = 4 + 9 - 8 - 3 = 2$$
$$x = \tfrac{2}{3}$$

Therefore the original system has just one solution: $x = \tfrac{2}{3}$, $y = 3$, $z = 4$, $w = 3$.

The substitution method works in the preceding example because of the particular form of the system (sometimes called **echelon form**): the unknown x appears in the first equation but not in any subsequent ones; the unknown y appears in the second equation, but not in any subsequent ones; and so on. Consequently, it is possible to begin with the last equation and solve for one unknown at a time.

The fact that a system of equations in echelon form can be solved, as in the example above, suggests a possible line of attack for solving *any* system of equations: replace the given system by another system that has the same solutions *and* is in echelon form. Then solve the new system. We shall now see that this can be accomplished in several steps by changing one or two equations at a time, in such a way that the system obtained by each change has the same solutions as the original system.

One operation that can be used in such a process is to:

> *Interchange any two equations in the system.*

For the order in which the equations are listed obviously doesn't affect their solutions. Another operation is to multiply both sides of an equation in the system by a nonzero constant. It is often convenient to state this operation in different words (but with the same meaning):

> *Replace an equation in the system by a nonzero constant multiple of itself.*

As we have seen before, multiplying an equation by a nonzero constant doesn't change the solutions of the equation. Thus replacing an equation in a system by a nonzero constant multiple of itself must necessarily produce a system with the same solutions as the original system.

A third operation that can be used here is one that we haven't seen before, so we begin with an example. Suppose a system of equations includes these two equations:

$$x + 4y - 3z = 1 \qquad \text{equation A}$$
$$-3x - 6y + z = 3 \qquad \text{equation B}$$

(The system may include other equations as well, and equations A and B may not appear next to each other when the equations of the system are listed.) Any solution of the system is necessarily a solution of both equations A and B, and hence is also a solution of these two equations:

$$3x + 12y - 9z = 3 \qquad \text{3 times equation A}$$
$$-3x - 6y + z = 3 \qquad \text{equation B}$$

Now we know that if equal quantities are added to equal quantities, the sums are equal. Consequently, any solution of the two equations above must also satisfy the sum of those equations:

$$
\begin{array}{ll}
3x + 12y - 9z = 3 & \text{3 times equation A} \\
-3x - 6y + z = 3 & \text{equation B} \\
\hline
6y - 8z = 6 & \text{sum of equation B and 3 times equation A}
\end{array}
$$

Therefore any solution of the original system will also be a solution of the system obtained by

Replacing equation B by the sum of equation B and 3 times equation A

A similar argument now shows that every solution of this new system is also a solution of the original system. Thus the two systems have exactly the same solutions.

The procedures used in this example are valid in any system of equations, with any constant in place of 3; that is, you can:

> *Replace an equation in the system by the sum of itself and a constant multiple of another equation in the system.*

The same argument used in the example shows that the system obtained by this replacement will have the same solutions as the original system. You should note two important special cases of this operation. If the constant multiplier is the number 1, then the operation is:

Replace an equation in the system by the sum
of itself and another equation in the system

Since for any quantities A and B, we know that $A + (-1)B$ is just $A - B$, we see that when the constant multiplier is -1, the operation is:

Replace an equation in the system by the difference of itself and another equation in the system.

Although there are other operations that could be used, those listed above are sufficient. We can now summarize the basic plan:

THE ELIMINATION METHOD

Any system of linear equations can be transformed into a system in echelon form with the same solutions, by using a finite number of operations of these three types:

(*i*) *Interchange any two equations in the system.*
(*ii*) *Replace an equation in the system by a nonzero constant multiple of itself.*
(*iii*) *Replace an equation in the system by the sum of itself and a constant multiple of another equation in the system.*

The resulting echelon form system can then be easily solved.

The preceding discussion showed that these three types of operations do not change the solutions of a system. The following examples will show how to use these operations to transform a given system into one in echelon form.

EXAMPLE In order to transform the system

$$
\begin{aligned}
x + 4y - 3z &= 1 & &\text{equation A} \\
-3x - 6y + z &= 3 & &\text{equation B} \\
2x + 11y - 5z &= 0 & &\text{equation C}
\end{aligned}
$$

into a system in echelon form, the first step is to replace the second and third equations by equations that do not involve x. We begin by replacing equation B by the sum of itself and 3 times equation A (this was done in detail in the example above):

$$
(*)\quad
\begin{aligned}
x + 4y - 3z &= 1 & &\text{equation A} \\
6y - 8z &= 6 & &\text{sum of equation B and 3 times equation A} \\
2x + 11y - 5z &= 0 & &\text{equation C}
\end{aligned}
$$

In order to eliminate the x term in equation C, we form a new equation, the sum of equation C and -2 times equation A:

$$
\begin{aligned}
-2x - 8y + 6z &= -2 & &\text{-2 times equation A} \\
\underline{2x + 11y - 5z} &= \underline{0} & &\text{equation C} \\
3y + z &= -2 & &\text{sum of equation C and -2 times equation A}
\end{aligned}
$$

Now in the system (*) above, replace equation C by this new equation:

$$
\begin{aligned}
x + 4y - 3z &= 1 &\quad \text{equation A} \\
6y - 8z &= 6 \\
3y + z &= -2 &\quad \text{sum of equation C and } -2 \text{ times equation A}
\end{aligned}
$$

The next step is to eliminate the y term in one of the last two equations. This can be done by replacing the second equation by the sum of itself and -2 times the third equation:

$$
\begin{aligned}
x + 4y - 3z &= 1 \\
-10z &= 10 &\quad \text{sum of second equation and } -2 \text{ times third equation} \\
3y + z &= -2
\end{aligned}
$$

Finally, interchange the last two equations:

$$
\begin{aligned}
x + 4y - 3z &= 1 \\
3y + z &= -2 \\
-10z &= 10
\end{aligned}
$$

We now have a system in echelon form. Since it was obtained by using only the types of transformations discussed above, we know that it has the same solutions as the system we started with. So we can find the solutions of that original system by solving this one. Beginning with the last equation, we see that

$$
-10z = 10, \qquad \text{or equivalently,} \qquad z = -1
$$

Substituting $z = -1$ in the second equation shows that

$$
\begin{aligned}
3y + z &= -2 \\
3y + (-1) &= -2 \\
3y &= -1 \\
y &= -\tfrac{1}{3}
\end{aligned}
$$

Substituting $y = -\tfrac{1}{3}$ and $z = -1$ in the first equation yields:

$$
\begin{aligned}
x + 4y - 3z &= 1 \\
x + 4(-\tfrac{1}{3}) - 3(-1) &= 1 \\
x = 1 + \tfrac{4}{3} - 3 &= -\tfrac{2}{3}
\end{aligned}
$$

Therefore the original system has just one solution: $x = -\tfrac{2}{3}$, $y = -\tfrac{1}{3}$, $z = -1$.

EXAMPLE The first step in solving the system

$$
\begin{aligned}
x + 2y + 3z &= -2 \\
2x + 6y + z &= 2 \\
3x + 3y + 10z &= -2
\end{aligned}
$$

is to eliminate the x terms from the last two equations by performing suitable operations:

$$x + 2y + 3z = -2$$
$$2y - 5z = 6 \quad \text{sum of second equation and } -2 \text{ times first equation}$$
$$3x + 3y + 10z = -2$$

$$x + 2y + 3z = -2$$
$$2y - 5z = 6$$
$$-3y + z = 4 \quad \text{sum of third equation and } -3 \text{ times first equation}$$

In order to eliminate the y term in the last equation, we first arrange for y to have coefficient 1 in the second equation:

$$x + 2y + 3z = -2$$
$$y - \tfrac{5}{2}z = 3 \quad \text{second equation multiplied by } \tfrac{1}{2}$$
$$-3y + z = 4$$

Now it is easy to eliminate the y term in the last equation:

$$x + 2y + 3z = -2$$
$$y - \tfrac{5}{2}z = 3$$
$$-\tfrac{13}{2}z = 13 \quad \text{sum of third equation and 3 times second equation}$$

This last system is in echelon form and we can solve the third equation: $z = 13(-\tfrac{2}{13}) = -2$. Substituting $z = -2$ in the second equation shows that

$$y - \tfrac{5}{2}(-2) = 3$$
$$y = 3 - 5 = -2$$

Substituting $y = -2$ and $z = -2$ in the first equation yields

$$x + 2(-2) + 3(-2) = -2$$
$$x = -2 + 4 + 6 = 8$$

Therefore the only solution of the original system is $x = 8$, $y = -2$, $z = -2$.

EXAMPLE A system such as this one

$$2x + 5y + z + 3w = 0$$
$$2y - 4z + 3w = 0$$
$$2x + 17y - 23z + 41w = 0$$

in which all the constants on the right side are zero is called a **homogeneous system.** It obviously has at least one solution, namely, $x = 0, y = 0, z = 0, w = 0$. This solution is called the **trivial solution.** In order to see if there are any nontrivial solutions, we shall put the system into echelon form. The x term in the third equation can be eliminated by replacing the third equation by the sum of itself and -1 times the first equation:

$$2x + 5y + z + 3w = 0$$
$$2y - 4z + 3w = 0$$
$$12y - 24z + 38w = 0$$

Now we replace the third equation by the sum of itself and -6 times the second equation:

$$
\begin{aligned}
2x + 5y + z + 3w &= 0 \\
2y - 4z + 3w &= 0 \\
20w &= 0
\end{aligned}
$$

This echelon form system can now be solved. The last equation shows that $w = 0$. Substituting $w = 0$ in the second equation yields:

$$
\begin{aligned}
2y - 4z &= 0 \\
2y &= 4z \\
y &= 2z
\end{aligned}
$$

This equation obviously has an infinite number of solutions. For example, $z = 1$, $y = 2$ is a solution and $z = -\frac{5}{2}$, $y = -5$ is a solution. More generally, for each real number t, $z = t$ and $y = 2t$ is a solution of the second equation. Substituting $w = 0$, $z = t$, $y = 2t$ in the first equation shows that

$$
\begin{aligned}
2x + 5(2t) + t + 3(0) &= 0 \\
2x &= -11t \\
x &= \frac{-11t}{2}
\end{aligned}
$$

Therefore this echelon form system, and hence the original system, has an infinite number of different solutions, one for each real number t:

$$
x = \frac{-11t}{2}, \qquad y = 2t, \qquad z = t, \qquad w = 0
$$

For example, if $t = 2$, we have the solution

$$
x = \frac{-11(2)}{2} = -11, \qquad y = 2(2) = 4, \qquad z = 2, \qquad w = 0
$$

If $t = -3$, we have the solution $x = \frac{33}{2}$, $y = -6$, $z = -3$, $w = 0$; and so on.

The homogeneous system of equations in the preceding example has an infinite number of solutions. It is also quite possible for a nonhomogeneous system to have infinitely many solutions (see Exercise B.4). A system with infinitely many solutions is said to be a **dependent system.**

EXAMPLE   To solve the system

$$
\begin{aligned}
2x - 3y &= 5 \\
4x - 6y &= 1
\end{aligned}
$$

we replace the second equation by the sum of itself and -2 times the first equation:

$$
\begin{aligned}
2x - 3y &= 5 \\
0x + 0y &= -9
\end{aligned}
$$

But the left side of the equation $0x + 0y = -9$ is always 0, no matter what x and y are. Since $0 = -9$ is always false, the equation $0x + 0y = -9$ has *no solutions.* Therefore the original system cannot possibly have any solutions. This fact can be seen geometrically by graphing the two equations of the original system. The graphs are parallel lines, so there is no point which is on both graphs—that is, no solution which satisfies *both* equations in the system.

The phenomenon observed in the preceding example may occur in any size system. During the process of transforming a given system of equations into echelon form you may obtain an equation in which all the unknowns have coefficients 0, but the constant on the right side is nonzero. In this case, the system has no solutions. Such a system is called an **inconsistent system.**

SHORTCUTS

Most computer programs for solving systems of linear equations are based on the elimination method. But when you are working by hand there often are more convenient ways to find the solutions of certain systems, especially systems of two equations in two unknowns. Some of these other methods are illustrated in the examples below. You will see that they are just combinations of the substitution method and some special cases of the elimination method, in which all the steps are not written out in full. Feel free to use such shortcuts. But when in doubt, remember that the elimination method *always* works, even though it may take a bit longer.

EXAMPLE Any solution of the system

$$4x - 3y = 8$$
$$2x + 3y = -2$$

must also satisfy the equation obtained by adding these two equations:

$$4x - 3y = 8$$
$$\underline{2x + 3y = -2}$$
$$6x = 6$$

Solving this last equation shows that $x = 1$. Substituting $x = 1$ in one of the original equations—say, the first one—shows that

$$4 - 3y = 8$$
$$-3y = 4$$
$$y = -\tfrac{4}{3}$$

Therefore the solution of the original system is $x = 1$, $y = -\tfrac{4}{3}$.

EXAMPLE Any solution of the system

$$2x - 2y = 1$$
$$3x - 5y = 2$$

must also be a solution of this system:

$$6x - 6y = 3 \qquad \text{first equation multiplied by 3}$$
$$6x - 10y = 4 \qquad \text{second equation multiplied by 2}$$

(The multipliers 3 and 2 were chosen so that the coefficient of x would be the same in both new equations.) Consequently, any solution of this system must be a solution of the equation obtained by subtracting the second equation from the first:

$$\begin{array}{r} 6x - 6y = 3 \\ 6x - 10y = 4 \\ \hline 4y = -1° \end{array}$$

Solving this last equation, we see that we must have $y = -\frac{1}{4}$. Substituting this value in either of the original equations—say, the first one—shows that

$$2x - 2(-\tfrac{1}{4}) = 1$$
$$2x = 1 - \tfrac{1}{2} = \tfrac{1}{2}$$
$$x = \tfrac{1}{4}$$

EXERCISES

A.1. Which of these systems have $x = 2$, $y = -1$ as a solution?

(a) $\begin{aligned} 3x - 5y &= 11 \\ x + 8y &= 6 \end{aligned}$
    (b) $\begin{aligned} 2x + 3y &= 1 \\ x - 5y &= 7 \end{aligned}$
    (c) $\begin{aligned} \tfrac{1}{2}x + \tfrac{1}{3}y &= \tfrac{2}{3} \\ \tfrac{1}{3}x + \tfrac{1}{2}y &= \tfrac{1}{6} \end{aligned}$

A.2. Which of these systems have $x = \frac{1}{2}$, $y = 3$, $z = -1$ as a solution?

(a) $\begin{aligned} 2x - y + 4z &= -6 \\ 3y + 3z &= 6 \\ 2z &= 2 \end{aligned}$
    (b) $\begin{aligned} \tfrac{1}{2}x + 3y - z &= \tfrac{41}{4} \\ 2x - 2y + 2z &= -7 \\ \tfrac{3}{2}x + \tfrac{2}{3}y - 4z &= 9 \end{aligned}$

B.1. Solve these systems by the substitution method.

(a) $\begin{aligned} x - 2y &= 5 \\ 2x + y &= 3 \end{aligned}$
  (c) $\begin{aligned} 3x - y &= 1 \\ -x + 2y &= 4 \end{aligned}$
  (e) $\begin{aligned} -5x + 2y &= 5 \\ 2x - 4y &= -3 \end{aligned}$

(b) $\begin{aligned} 3x - 2y &= 4 \\ 2x + y &= -1 \end{aligned}$
  (d) $\begin{aligned} 2x - 3y &= 5 \\ 6x + 2y &= 2 \end{aligned}$
  (f) $\begin{aligned} \tfrac{1}{2}x + \tfrac{5}{2}y &= \tfrac{3}{4} \\ 2x - \tfrac{3}{2}y &= 1 \end{aligned}$

B.2. Solve these systems. (*Hint:* first write each equation in the form $ax + by = c$.)

(a) $\dfrac{x+y}{4} - \dfrac{x-y}{3} = 1$

    $\dfrac{x+y}{4} + \dfrac{x-y}{2} = 9$

(b) $\dfrac{y-x}{3} + \dfrac{x+y}{2} = 0$

    $\dfrac{x+y}{4} + \dfrac{x+y}{3} = 0$

(c) $\dfrac{x-y}{4} + \dfrac{x+y}{3} = 1$

    $\dfrac{x+2y}{3} + \dfrac{3x-y}{2} = -2$

(d) $\dfrac{2x-y}{3} - \dfrac{3x+y}{2} = 1$

    $\dfrac{2x+2y}{3} + \dfrac{x-3y}{2} = -2$

° If this subtraction is confusing, write it out horizontally. For instance, $-6y - (-10y) = -6y + 10y = 4y$.

B.3. Solve these systems, each of which has exactly one solution.

(a)
$$x + y \quad\;\; = 5$$
$$-x \quad\;\; + 2z = 0$$
$$2x + y - \;\; z = 7$$

(b)
$$x + 2y + 4z = 3$$
$$x \qquad\;\; + 2z = 0$$
$$2x + 4y + \;\; z = 3$$

(c)
$$2x + y - \;\; z = 4$$
$$x \qquad\;\; + 2z = 9$$
$$-3x - y + 2z = 9$$
(*Hint:* begin by interchanging the first and second equations.)

(d)
$$2x - 2y + \;\; z = -6$$
$$3x + \;\; y + 2z = \quad 2$$
$$x + \;\; y - 2z = \quad 0$$

(e)
$$-x + 3y + 2z = 0$$
$$2x - \;\; y - \;\; z = 3$$
$$x + 2y + 3z = 0$$

(f)
$$3x + 7y + 9z = 0$$
$$x + 2y + 3z = 2$$
$$x + 4y + \;\; z = 2$$

B.4. Solve these dependent systems.

(a)
$$x + \;\; y + \;\; z = 1$$
$$x - 2y + 2z = 4$$
$$2x - \;\; y + 3z = 5$$

(b)
$$2x - y + \;\; z = 1$$
$$3x + y + \;\; z = 0$$
$$7x - y + 3z = 2$$

(c)
$$11x + 10y + 9z = 5$$
$$x + \;\; 2y + 3z = 1$$
$$3x + \;\; 2y + \;\; z = 1$$

(d)
$$-x + 2y - 3z + 4w = \quad 8$$
$$2x - 4y - \;\; z + 2w = -3$$
$$5x - 4y - \;\; z + 2w = -3$$

B.5. Solve these systems.

(a)
$$x + \;\; y = 3$$
$$5x - \;\; y = 3$$
$$9x - 4y = 1$$

(b)
$$2x - \;\; y + 2z = 3$$
$$-x + 2y - \;\; z = 0$$
$$x + \;\; y - \;\; z = 1$$

(c)
$$2x - y + z = 1$$
$$x + y - z = 2$$
$$-x - y + z = 0$$

(d)
$$x + 2y + 3z = 4$$
$$2x - \;\; y + \;\; z = 3$$
$$3x + \;\; y + 4z = 7$$

(e)
$$x + \;\; y = 3$$
$$-x + 2y = 3$$
$$2x - \;\; y = 3$$

(f)
$$3x + y - 2z = \quad 4$$
$$-5x \qquad + 2z = \quad 5$$
$$-7x - y + 3z = -2$$

B.6. Carry out the elimination method far enough to determine whether the given system is dependent or inconsistent. (It isn't necessary to solve the dependent systems.)

(a)
$$x + 2y \qquad\;\; = \quad 0$$
$$y - z = \quad 2$$
$$x + \;\; y + z = -2$$

(b)
$$x + 2y + \;\; z = 0$$
$$y + 2z = 0$$
$$x + \;\; y - \;\; z = 0$$

(c)
$$x + 2y + 4z = \quad 6$$
$$y + \;\; z = \quad 1$$
$$x + 3y + 5z = 10$$

(d)
$$x + y + 2z + 3w = 1$$
$$2x + y + 3z + 4w = 1$$
$$3x + y + 4z + 5w = 2$$

(e)
$$x + 2y = 1$$
$$-x + \;\; y = 0$$
$$2x + 4y = 3$$

(f)
$$x + 2y + 3z \qquad\quad = \quad 1$$
$$3x + 2y + 4z \qquad\quad = -1$$
$$2x + 6y + 8z + \;\; w = \quad 3$$
$$2x \qquad\;\; + 2z - 2w = \quad 3$$

B.7. Solve these systems:

(a)
$$x + y + z + w = 10$$
$$x + y + 2z \quad\;\; = 11$$
$$x - 3y \quad\;\; + w = -14$$
$$y + 3z - w = 7$$

(c)
$$2x - 4y + 5z = 1$$
$$x \qquad\;\; - 3z = 2$$
$$5x - 8y + 7z = 6$$
$$3x - 4y + 2z = 3$$
$$x - 4y + 8z = -1$$

(b)
$$2x + y + z \qquad = 3$$
$$y + z + w = 5$$
$$4x \qquad + z + w = 0$$
$$3y - 2z - w = 6$$

(d)
$$4x \qquad + z + 2w + 24v = 0$$
$$2x + y \qquad\qquad + 12v = 0$$
$$3x \qquad + z + 2w + 18v = 0$$
$$4x - y \qquad + w + 24v = 0$$
$$7x - y + z + 3w + 42v = 0$$

B.8. Solve these systems:

(a)
$$\frac{1}{x} - \frac{3}{y} = 2$$
$$\frac{2}{x} + \frac{1}{y} = 3$$

(*Hint:* let $u = 1/x$ and $v = 1/y$ so that the system becomes

$$u - 3v = 2$$
$$2u + v = 3$$

Solve for u and v; then determine x and y.)

(b)
$$\frac{5}{x} + \frac{2}{y} = 0$$
$$\frac{6}{x} - \frac{4}{y} = 3$$

(e)
$$\frac{1}{x+1} - \frac{2}{y-3} + \frac{3}{z-2} = 4$$
$$\frac{5}{y-3} - \frac{10}{z-2} = -5$$
$$\frac{-3}{x+1} + \frac{4}{y-3} - \frac{1}{z-2} = -2$$

(c)
$$\frac{3}{x} - \frac{1}{y} + \frac{4}{z} = -13$$
$$\frac{1}{x} + \frac{2}{y} - \frac{1}{z} = 12$$
$$\frac{4}{x} - \frac{1}{y} + \frac{3}{z} = -7$$

(d)
$$\frac{3}{x+1} - \frac{4}{y-2} = 2$$
$$\frac{1}{x+1} + \frac{4}{y-2} = 5$$

$$\left(\text{\textit{Hint:} let } u = \frac{1}{x+1}, v = \frac{1}{y-2}.\right)$$

B.9. (a) Consider this system:

$$x + y + 4z - w = 1$$
$$y - 2z + 3w = 0$$

Verify that for each *pair* of real numbers, s and t, the system has a solution:

$$w = s, \quad z = t, \quad y = 2t - 3s, \quad x = 1 - (2t - 3s) - 4t + s = 1 - 6t + 4s$$

With this model in mind, solve these dependent systems by the elimination method.

(b)
$$\begin{aligned} x - y + 2z + 3w &= 0 \\ x \quad + z + w &= 0 \\ 3x - 2y + 5z + 7w &= 0 \end{aligned}$$

(d)
$$\begin{aligned} x + y + z - w &= 0 \\ 2x - 4y - 4z + w &= 0 \\ 4x - 2y + 2z - 3w &= 0 \\ 7x - y - z - 3w &= 0 \end{aligned}$$

(c)
$$\begin{aligned} x + 2y + z + 4w &= 1 \\ y + 3z - w &= 2 \\ x + 4y + 7z - 2w &= 5 \\ 3x + 7y + 6z + 11w &= 5 \end{aligned}$$

(e)
$$\begin{aligned} x + 2y + 3z \quad + 4v &= 0 \\ 2x + 4y + 6z + w + 9v &= 0 \\ x + 2y + 3z + w + 5v &= 0 \end{aligned}$$

B.10. Solve these systems by using the same methods as used above together with a calculator.

(a)
$$\begin{aligned} 3.25x - 2.18y &= \quad 1.96 \\ 1.92x + 6.77y &= -3.87 \end{aligned}$$

(b)
$$\begin{aligned} 463x - 801y &= 946 \\ .0375x + .912y &= -1.003 \end{aligned}$$

B.11. Find constants a, b, c, such that the three points $(-3, 2)$, $(1, 1)$, and $(2, -1)$ all lie on the graph of the quadratic function $f(x) = ax^2 + bx + c$. [*Hint:* since $(-3, 2)$ is to be on the graph, we must have $a(-3)^2 + b(-3) + c = 2$, that is, $9a - 3b + c = 2$. In a similar manner, the other two points lead to *linear* equations in a, b, c. Solve this system of three equations for a, b, c.]

B.12. In each part, find a quadratic function whose graph includes the three given points. (*Hint:* see Exercise B.11.)

(a) $(1, -2)$, $(3, 1)$, $(4, -1)$
(b) $(1, -4)$, $(-1, 6)$, $(2, -9)$
(c) $(1, 1)$, $(0, 0)$, $(-1, 2)$
(d) $(-1, 6)$, $(-2, 16)$, $(1, 4)$

B.13. The substitution method can sometimes be used to solve systems of nonlinear equations.

(a) Solve the system

$$\begin{aligned} x^2 - y &= \quad 0 \\ 2x - y &= -3 \end{aligned}$$

by solving one of the equations for y, substituting this answer in other equation, and solving it.

(b) Graph the two equations in part (a). The points where the graphs intersect are precisely the solutions of the *system* in part (a). Why?

B.14. Solve each of these systems by substitution. Then exhibit the solutions geometrically by graphing the equations in each system. (See Exercise B.13.)

(a)
$$\begin{aligned} 2x + y &= -1 \\ x^2 - y &= \quad 4 \end{aligned}$$

(b)
$$\begin{aligned} x^2 - y &= -1 \\ 2x + y &= \quad 4 \end{aligned}$$

(c)
$$\begin{aligned} x^2 + y^2 &= 25 \\ x^2 + y &= 19 \end{aligned}$$

(d)
$$\begin{aligned} x^2 + y^2 &= 20 \\ 3x - y &= -2 \end{aligned}$$

(e)
$$\begin{aligned} x^2 + y^2 + 4x - 2y &= 5 \\ x + y &= 1 \end{aligned}$$
(*Hint:* to graph the first equation, complete the square in x and y; see p. 82.)

(f)
$$\begin{aligned} 4x^2 - y^2 &= -4 \\ 16x^2 + 9y^2 &= 244 \end{aligned}$$

B.15. Find two integers whose sum is 32 and whose difference is 6. (*Hint:* if x and y denote the two integers, then x and y must satisfy $x + y = 32$ *and* $x - y = 6$.)

B.16. Find two integers whose sum is -9 and whose difference is 45. (See Exercise B.15.)

B.17. Are there two *integers* whose sum is 29 and whose difference is 8? (See Exercise B.15.)

B.18. A collection of nickels and dimes totals \$3.05. If there are 38 coins altogether, how many are nickels and how many are dimes? (*Hint:* if x is the number of nickels and y the number of dimes, then $.05x + .10y = 3.05$. Use the other given information to obtain a second linear equation, then solve the resulting system for x and y.)

B.19. A collection of nickels, dimes, and quarters totals \$6.00. If there are 52 coins altogether and twice as many dimes as nickels, how many of each kind of coin are there? (See Exercise B.18.)

B.20. At a certain store, cashews cost \$4.40/lb and peanuts \$1.20/lb. If you want to buy exactly 3 lb of nuts for \$6.00, how many pounds of each kind of nuts should you buy? (Let x be the pounds of cashews and y the pounds of peanuts you must buy. Then $x + y = 3$. Find another linear equation satisfied by x and y and solve the resulting system.)

B.21. A plane flies 3000 miles from San Francisco to Boston at a constant speed in 5 hr, flying *with* the wind all the way. The return trip, against the wind, takes 6 hr. Find the speed of the plane and the speed of the wind. [*Hint:* if x is the plane's speed and y the wind speed, then on the trip to Boston (*with* the wind), the plane travels at speed $x + y$ for 5 hr. Since it goes a distance of 3000 mi, we have $5(x + y) = 3000$. Find another equation in x and y and solve the resulting system.]

2. MATRICES AND DETERMINANTS

Matrices and determinants are two different but related concepts that arise naturally in the study of systems of linear equations. The definitions and use of matrices and determinants in solving systems of equations are discussed here. Both matrices and determinants have many other applications in mathematics. Many of these are beyond the scope of this book, but a few are presented in the exercises at the end of the section.

MATRIX METHODS FOR SOLVING SYSTEMS OF LINEAR EQUATIONS

Once you have solved several systems of linear equations by the elimination method, one fact becomes clear. The symbols used for the unknowns play no real role in the solution process, and a lot of time is wasted copying the x's, y's, z's, and so on, at each stage of the process. This fact suggests a shorthand system for representing a system of equations.

EXAMPLE This system of equations

$$
\begin{array}{rcrcrcrcr}
x & + & 2y & - & z & - & 3w & = & 2 \\
 & & 6y & + & 3z & - & 4w & = & 0 \\
6x & + & 12y & - & 3z & - & 16w & = & 3 \\
-5x & + & 2y & + & 15z & + & 10w & = & -21
\end{array}
$$

can be represented by the following rectangular array of numbers, consisting of the coefficients of the unknowns and the constants on the right side, arranged in the same order they appear in the system:

$$
\left(
\begin{array}{rrrr|r}
1 & 2 & -1 & -3 & 2 \\
0 & 6 & 3 & -4 & 0 \\
6 & 12 & -3 & -16 & 3 \\
-5 & 2 & 15 & 10 & -21
\end{array}
\right)
$$

This array is called the **augmented matrix°** of the system. Note that the second equation above has no x term, meaning that x has coefficient 0 in the equation. This is indicated in the matrix by the 0 at the beginning of the second row.

To solve this system by the elimination method, we begin by eliminating the x term in the third equation. We replace the third equation by the sum of itself and -6 times the first equation, namely, $3z + 2y = -9$. In the matrix shorthand, where a horizontal row represents an equation, this operation is carried out on the rows of the matrix.

$$
\begin{array}{rrrrr}
-6 & -12 & 6 & 18 & -12 \\
6 & 12 & -3 & -16 & 3 \\
\hline
0 & 0 & 3 & 2 & -9
\end{array}
$$

-6 times the first row
third row
sum of the third row and -6 times the first row

Replacing the third row of the original matrix by this row yields the matrix

$$
\left(
\begin{array}{rrrr|r}
1 & 2 & -1 & -3 & 2 \\
0 & 6 & 3 & -4 & 0 \\
0 & 0 & 3 & 2 & -9 \\
-5 & 2 & 15 & 10 & -21
\end{array}
\right)
$$

We continue the solution process in the same manner, using the matrix shorthand to represent the system of equations. Replace the fourth row of the last matrix above by the sum of itself and 5 times the first row (this amounts to eliminating the x term from the last equation):

$$
\left(
\begin{array}{rrrr|r}
1 & 2 & -1 & -3 & 2 \\
0 & 6 & 3 & -4 & 0 \\
0 & 0 & 3 & 2 & -9 \\
0 & 12 & 10 & -5 & -11
\end{array}
\right)
$$

° The plural of "matrix" is "matrices."

Next we must make various entries in the last row 0. (This amounts to eliminating the y and z term from the last equation.)

$$\begin{pmatrix} 1 & 2 & -1 & -3 & | & 2 \\ 0 & 6 & 3 & -4 & | & 0 \\ 0 & 0 & 3 & 2 & | & -9 \\ 0 & 0 & 4 & 3 & | & -11 \end{pmatrix}$$

replace fourth row by the sum of itself and -2 times the second row

$$\begin{pmatrix} 1 & 2 & -1 & -3 & | & 2 \\ 0 & 6 & 3 & -4 & | & 0 \\ 0 & 0 & 1 & \frac{2}{3} & | & -3 \\ 0 & 0 & 4 & 3 & | & -11 \end{pmatrix}$$

multiply third row by $\frac{1}{3}$

$$\begin{pmatrix} 1 & 2 & -1 & -3 & | & 2 \\ 0 & 6 & 3 & -4 & | & 0 \\ 0 & 0 & 1 & \frac{2}{3} & | & -3 \\ 0 & 0 & 0 & \frac{1}{3} & | & 1 \end{pmatrix}$$

replace fourth row by the sum of itself and -4 times the third row

This last matrix is just shorthand notation for this echelon form system:

$$\begin{aligned} x + 2y - z - 3w &= 2 \\ 6y + 3z - 4w &= 0 \\ z + \tfrac{2}{3}w &= -3 \\ \tfrac{1}{3}w &= 1 \end{aligned}$$

The solution is now easily found to be $x = -3$, $y = \frac{9}{2}$, $z = -5$, $w = 3$.

The matrix notation in the example above is certainly more convenient than the equation notation. When matrix notation is used, we usually change our language to suit the situation. Instead of speaking of operations on equations in the system (such as multiplying by a nonzero constant), we speak of **row operations** on the matrix. Similarly, the solution process ends when we obtain an **echelon form matrix** (such as the last matrix shown above).

EXAMPLE   The homogeneous system

$$\begin{aligned} 4x + 12y - 16z &= 0 \\ 3x + 4y + 3z &= 0 \\ x + 8y - 19z &= 0 \end{aligned}$$

has augmented matrix

$$\begin{pmatrix} 4 & 12 & -16 & | & 0 \\ 3 & 4 & 3 & | & 0 \\ 1 & 8 & -19 & | & 0 \end{pmatrix}$$

The last vertical column of this matrix consists entirely of zeros. Furthermore, any matrix obtained from this one by the usual row operations will have this same property.

For interchanging two rows, or multiplying a row by a constant, or replacing a row by the sum of itself and a constant multiple of another row, will always result in rows with last entry 0. Consequently, when dealing with homogeneous systems such as this, there is no need to write out this last column of zeros at every stage. Instead we need only deal with the **coefficient matrix:**

$$\begin{pmatrix} 4 & 12 & -16 \\ 3 & 4 & 3 \\ 1 & 8 & -19 \end{pmatrix}$$

instead of the augmented matrix of the system given above. Using the matrix of coefficients, we proceed as before to reduce it to echelon form:

$$\begin{pmatrix} 1 & 3 & -4 \\ 3 & 4 & 3 \\ 1 & 8 & -19 \end{pmatrix}$$ multiply first row by $\frac{1}{4}$

$$\begin{pmatrix} 1 & 3 & -4 \\ 0 & -5 & 15 \\ 1 & 8 & -19 \end{pmatrix}$$ replace second row by the sum of itself and -3 times the first row

$$\begin{pmatrix} 1 & 3 & -4 \\ 0 & -5 & 15 \\ 0 & 5 & -15 \end{pmatrix}$$ replace third row by the sum of itself and -1 times the first row

$$\begin{pmatrix} 1 & 3 & -4 \\ 0 & -5 & 15 \\ 0 & 0 & 0 \end{pmatrix}$$ replace third row by the sum of itself and 1 times the second row

This echelon form matrix represents the following homogeneous system:

$$x + 3y - 4z = 0$$
$$- 5y + 15z = 0$$

We don't bother to write out the trivial third equation $0x + 0y + 0z = 0$ since it is satisfied by *all* real numbers. The last equation above is equivalent to $y = 3z$ and hence has infinitely many solutions. For each real number t, there is the solution $z = t$, $y = 3t$. Substituting these values in the first equation shows that $x = -3y + 4z = -9t + 4t = -5t$. Therefore the solutions of the system are given by:

$$x = -5t, \qquad y = 3t, \qquad z = t \qquad (t \text{ any real number})$$

MATRICES

Up to now, we have used matrices only as a convenient shorthand notation for dealing with systems of equations. Matrices have many other uses, most of which have nothing to do with systems of equations. Consequently, matrices are often studied as a subject in their own right (see Exercises B.2–B.5). We shall limit our discussion here to defining some standard terms.

Let m and n be positive integers. An $m \times n$ matrix (read "m by n matrix") is a rectangular array of numbers,° with m horizontal rows and n vertical columns. For example,

$$\begin{pmatrix} 3 & 2 & -5 \\ 6 & 1 & 7 \\ -2 & 5 & 0 \end{pmatrix} \qquad \begin{pmatrix} -3 & 4 \\ 2 & 0 \\ 0 & 1 \\ 7 & 3 \\ 1 & -6 \end{pmatrix} \qquad \begin{pmatrix} 3 & 0 & 1 & 0 \\ \sqrt{2} & -\frac{1}{2} & 4 & \frac{8}{3} \\ 10 & 2 & -\frac{3}{4} & 12 \end{pmatrix} \qquad \begin{pmatrix} \sqrt{3} \\ 2 \\ 0 \\ 11 \end{pmatrix}$$

3×3 matrix	5×2 matrix	3×4 matrix	4×1 matrix
3 rows	5 rows	3 rows	4 rows
3 columns	2 columns	4 columns	1 column

In a matrix, the *rows* are horizontal and are numbered from top to bottom. The *columns* are vertical and are numbered from left to right. For example,

$$\begin{pmatrix} 11 & 3 & 14 \\ -2 & 0 & -5 \\ \frac{1}{3} & 6 & 7 \end{pmatrix} \begin{matrix} \leftarrow \text{row 1} \\ \leftarrow \text{row 2} \\ \leftarrow \text{row 3} \end{matrix}$$

column 1   column 2   column 3

A matrix with the same number of rows and columns is called a **square matrix.** Observe that each entry in a matrix can be exactly located by stating the row and column it appears in. For instance, in the 3×3 matrix above, the entry in row 1, column 2, is the number 3; the entry in row 2, column 2, is the number 0; the entry in row 3, column 1, is $\frac{1}{3}$.

The **main diagonal** of a matrix consists of the entry in row 1, column 1, the entry in row 2, column 2, the entry in row 3, column 3, and so on. For example, this matrix

$$\begin{pmatrix} 1 & 2 & -1 & -3 & 2 \\ 0 & 0 & 3 & -4 & 0 \\ 0 & 0 & -4 & \frac{2}{3} & -3 \\ \frac{4}{5} & -2 & 0 & \frac{1}{3} & 1 \end{pmatrix} \quad \text{has main diagonal} \quad \begin{matrix} 1 \\ 0 \\ -4 \\ \frac{1}{3} \end{matrix}$$

DETERMINANTS

Associated with every square matrix is a certain number, called its determinant. Determinants are used in many areas of mathematics and have applications to systems

° We shall deal only with matrices of real numbers here. But complex numbers can also be used as the entries in a matrix.

of linear equations. The **determinant of the 2 × 2 matrix** $A = \begin{pmatrix} a & b \\ c & d \end{pmatrix}$ is defined to be the number $ad - bc$. An easy way to remember this definition is to note that $ad - bc$ is just the difference of the products of diagonally opposite entries: $\begin{smallmatrix} a & b \\ c & d \end{smallmatrix}$.

Here are some examples of 2 × 2 matrices and their determinants:

Matrix	Determinant
$\begin{pmatrix} 1 & 2 \\ 6 & 4 \end{pmatrix}$	$1 \cdot 4 - 2 \cdot 6 = 4 - 12 = -8$
$\begin{pmatrix} \frac{1}{2} & 2 \\ 0 & -3 \end{pmatrix}$	$\frac{1}{2}(-3) - 2 \cdot 0 = -\frac{3}{2} - 0 = -\frac{3}{2}$
$\begin{pmatrix} 2 & -4 \\ -3 & 6 \end{pmatrix}$	$2 \cdot 6 - (-4)(-3) = 12 - 12 = 0$

The determinant of the matrix $A = \begin{pmatrix} a & b \\ c & d \end{pmatrix}$ is denoted by any one of these symbols:

$$\det A \quad \text{or} \quad |A| \quad \text{or} \quad \begin{vmatrix} a & b \\ c & d \end{vmatrix}$$

Note the straight vertical lines in this last symbol. They are *not* the same as the curved parentheses used to denote the matrix. For example,

$\begin{pmatrix} 3 & 2 \\ 4 & 5 \end{pmatrix}$ is a 2 × 2 *matrix*,

but its determinant $\begin{vmatrix} 3 & 2 \\ 4 & 5 \end{vmatrix}$ is the *number* $3 \cdot 5 - 2 \cdot 4 = 7$

Before defining determinants of 3 × 3 matrices, we must introduce a new concept. Observe that if you erase one row and one column of a 3 × 3 matrix, the remaining entries form a 2 × 2 matrix. For example, erasing the first row and second column of the 3 × 3 matrix

$\begin{pmatrix} -3 & 4 & -7 \\ 1 & 2 & 0 \\ -4 & 8 & 11 \end{pmatrix}$ produces the 2 × 2 matrix $\begin{pmatrix} 1 & 0 \\ -4 & 11 \end{pmatrix}$

If you now take the determinant of this 2 × 2 matrix, the result is a *number*. The **minor** of any entry in a 3 × 3 matrix is just such a number, namely, the determinant of the 2 × 2 matrix obtained by erasing the row and column in which the given entry appears. For instance, in the 3 × 3 matrix

$$\begin{pmatrix} a_1 & b_1 & c_1 \\ a_2 & b_2 & c_2 \\ a_3 & b_3 & c_3 \end{pmatrix}$$

the minor of c_1 is the number $a_2b_3 - b_2a_3$, obtained by erasing the row and column in which c_1 appears and taking the determinant of the result:

$$\begin{pmatrix} a_1 & b_1 & c_1 \\ a_2 & b_2 & c_2 \\ a_3 & b_3 & c_3 \end{pmatrix} \longrightarrow \begin{vmatrix} a_2 & b_2 \\ a_3 & b_3 \end{vmatrix} = a_2b_3 - b_2a_3$$

The **determinant of the 3 × 3 matrix**

$$A = \begin{pmatrix} a_1 & b_1 & c_1 \\ a_2 & b_2 & c_2 \\ a_3 & b_3 & c_3 \end{pmatrix}$$

is denoted by any one of these symbols

$$\det A \quad \text{or} \quad |A| \quad \text{or} \quad \begin{vmatrix} a_1 & b_1 & c_1 \\ a_2 & b_2 & c_2 \\ a_3 & b_3 & c_3 \end{vmatrix}$$

and is defined to be the *number*

$$\begin{vmatrix} a_1 & b_1 & c_1 \\ a_2 & b_2 & c_2 \\ a_3 & b_3 & c_3 \end{vmatrix} = a_1 \binom{\text{minor}}{\text{of } a_1} - b_1 \binom{\text{minor}}{\text{of } b_1} + c_1 \binom{\text{minor}}{\text{of } c_1}$$

$$= a_1 \begin{vmatrix} b_2 & c_2 \\ b_3 & c_3 \end{vmatrix} - b_1 \begin{vmatrix} a_2 & c_2 \\ a_3 & c_3 \end{vmatrix} + c_1 \begin{vmatrix} a_2 & b_2 \\ a_3 & b_3 \end{vmatrix}$$

$$= a_1(b_2c_3 - c_2b_3) - b_1(a_2c_3 - c_2a_3) + c_1(a_2b_3 - b_2a_3)$$

$$= a_1b_2c_3 - a_1b_3c_2 - a_2b_1c_3 + a_3b_1c_2 + a_2b_3c_1 - a_3b_2c_1$$

It isn't necessary to memorize the last line of this formula. Just remember the directions given in the top line: to find $|A|$, multiply each entry in the first row of A by its minor, insert the proper signs $(+, -, +)$, and add up the result.

EXAMPLE   The determinant of the matrix

$$A = \begin{pmatrix} 2 & 4 & 3 \\ 0 & 5 & -1 \\ 1 & -1 & 2 \end{pmatrix}$$

is just the number -1 since

$$|A| = 2 \begin{vmatrix} 5 & -1 \\ -1 & 2 \end{vmatrix} - 4 \begin{vmatrix} 0 & -1 \\ 1 & 2 \end{vmatrix} + 3 \begin{vmatrix} 0 & 5 \\ 1 & -1 \end{vmatrix}$$

$$= 2(5 \cdot 2 - (-1)(-1)) - 4(0 \cdot 2 - (-1)1) + 3(0(-1) - 5 \cdot 1)$$

$$= 2(9) - 4(1) + 3(-5) = 18 - 4 - 15 = -1$$

When the determinant of a 3 × 3 matrix is defined as above, one says that the determinant is obtained by **expanding along the first row**. Providing that the proper signs are inserted, the determinant can actually be calculated by following an analogous procedure and expanding along any row or column. See Exercise B.8 for details.

DETERMINANTS AND SYSTEMS OF LINEAR EQUATIONS

The connection between determinants and systems of linear equations can be most easily seen by looking at an arbitrary system of two equations in two unknowns. Suppose a, b, c, d, r, s are fixed real numbers. In order to solve the system

$$ax + by = r$$
$$cx + dy = s$$

we eliminate the y terms by multiplying the first equation by d and the second by $-b$:

$$
\begin{array}{ll}
\begin{aligned}
adx + bdy &= \quad rd \\
-bcx - bdy &= -bs
\end{aligned} & \quad \text{first equation multiplied by } d \\
& \quad \text{second equation multiplied by } -b \\[6pt]
\hline
\begin{aligned}
adx - bcx \quad\;\; &= rd - bs \\
(ad - bc)x &= rd - bs
\end{aligned} & \quad \text{sum}
\end{array}
$$

Now *if* the number $ad - bc$ is nonzero, we can divide both sides of the last equation by it and conclude that there is only one possible value for x:

$$x = \frac{rd - bs}{ad - bc}$$

Observe that the denominator of x is just the determinant of the **coefficient matrix** $\begin{pmatrix} a & b \\ c & d \end{pmatrix}$, while the numerator is the determinant of the matrix $\begin{pmatrix} r & b \\ s & b \end{pmatrix}$ obtained by replacing the *first* column of the coefficient matrix by the column of constants from the right side of the original equations. Thus when the determinant of the coefficient matrix is nonzero,

$$x = \frac{\begin{vmatrix} r & b \\ s & d \end{vmatrix}}{\begin{vmatrix} a & b \\ c & d \end{vmatrix}}$$

A similar argument shows that when the determinant of the coefficient matrix is nonzero, then there is only one possible value for y:

$$y = \frac{\begin{vmatrix} a & r \\ c & s \end{vmatrix}}{\begin{vmatrix} a & b \\ c & d \end{vmatrix}}$$

Note that the numerator of y is the determinant of the matrix obtained by replacing the *second* column of the coefficient matrix by the column of constants from the right side of the original equations.

EXAMPLE The determinant of the coefficient matrix of the system

$$3x - 4y = 2$$
$$7x + 7y = 3$$

is the nonzero number

$$\begin{vmatrix} 3 & -4 \\ 7 & 7 \end{vmatrix} = 3 \cdot 7 - (-4)7 = 21 + 28 = 49$$

Therefore the system has exactly one solution. It can be found by using the formulas developed above with $a = 3$, $b = -4$, $r = 2$, and $c = 7$, $d = 7$, $s = 3$:

$$x = \frac{\begin{vmatrix} 2 & -4 \\ 3 & 7 \end{vmatrix}}{\begin{vmatrix} 3 & -4 \\ 7 & 7 \end{vmatrix}} = \frac{2 \cdot 7 - (-4)3}{49} = \frac{26}{49} \qquad \text{and}$$

$$y = \frac{\begin{vmatrix} 3 & 2 \\ 7 & 3 \end{vmatrix}}{\begin{vmatrix} 3 & -4 \\ 7 & 7 \end{vmatrix}} = \frac{3 \cdot 3 - 2 \cdot 7}{49} = \frac{-5}{49}$$

We have now taken care of the case of two equations in two unknowns. Analogous but more complicated arguments work for larger systems and prove:

CRAMER'S RULE

A system of n *linear equations in* n *unknowns has exactly one solution, if the determinant of the coefficient matrix is nonzero.*

In this case, the solution of the system is given by a formula similar to the one presented above for the case of two equations.

The last example above shows that solving a system of two equations via Cramer's Rule involves about the same amount of computation as do the various other methods of solving such systems. Solving systems of three equations via Cramer's Rule often (though not always) involves more computation than does the elimination method. The solution of larger systems via Cramer's Rule almost always involves far more computation than does the elimination method. Determinants are grossly inefficient for solving such systems, even if a computer is used. However, there is one situation in which Cramer's Rule can sometimes be used effectively.

EXAMPLE The homogeneous system

$$3x + 2y + z = 0$$
$$7x + 3y + 5z = 0$$
$$-5x + y - z = 0$$

like all homogeneous systems, always has at least one solution, the trivial solution $x = 0$, $y = 0$, $z = 0$. The determinant of the coefficient matrix is

$$\begin{vmatrix} 3 & 2 & 1 \\ 7 & 3 & 5 \\ -5 & 1 & -1 \end{vmatrix} = 3 \begin{vmatrix} 3 & 5 \\ 1 & -1 \end{vmatrix} - 2 \begin{vmatrix} 7 & 5 \\ -5 & -1 \end{vmatrix} + 1 \begin{vmatrix} 7 & 3 \\ -5 & 1 \end{vmatrix}$$

$$= 3(-3 - 5) - 2(-7 + 25) + 1(7 + 15)$$
$$= 3(-8) - 2(18) + 22$$
$$= -24 - 36 + 22 = -38$$

Since this determinant is nonzero, the system has *exactly one* solution by Cramer's Rule. But we already have one solution, namely, $x = 0$, $y = 0$, $z = 0$. Therefore this trivial solution is the *only* solution of the system.

EXERCISES

A.1.   Compute these determinants

(a) $\begin{vmatrix} 3 & 5 \\ 7 & 2 \end{vmatrix}$

(d) $\begin{vmatrix} 2 & 5 \\ 1 & \frac{5}{2} \end{vmatrix}$

(g) $\begin{vmatrix} 0 & 2 & 3 \\ 1 & 7 & 9 \\ 0 & -1 & 5 \end{vmatrix}$

(b) $\begin{vmatrix} -2 & 1 \\ 6 & 4 \end{vmatrix}$

(e) $\begin{vmatrix} 1 & 0 & 2 \\ 3 & -1 & 2 \\ 1 & 2 & -3 \end{vmatrix}$

(h) $\begin{vmatrix} 3 & 2 & -3 \\ 1 & 0 & 1 \\ 0 & 4 & 0 \end{vmatrix}$

(c) $\begin{vmatrix} 3 & 5 \\ \frac{5}{2} & 7 \end{vmatrix}$

(f) $\begin{vmatrix} -1 & 2 & 3 \\ 0 & 1 & 2 \\ 4 & 0 & 5 \end{vmatrix}$

(i) $\begin{vmatrix} 3 & 2 & -3 \\ 1 & 0 & 1 \\ 0 & 4 & 1 \end{vmatrix}$

B.1.   Use matrix methods to solve these systems.

(a) $\begin{aligned} 3x - 2y + 6z &= -6 \\ -3x + 10y + 11z &= 13 \\ x - 2y - z &= -3 \end{aligned}$

(e) $\begin{aligned} x - y + 2z + 3w &= 0 \\ 3x - y + 7z + 7w &= 0 \\ 2x - 2y + 5z + 10w &= 0 \\ x - y + 3z + 7w &= 0 \end{aligned}$

(b) $\begin{aligned} x + 2y - 3z &= 9 \\ 3x - y - 4z &= 3 \\ 2x - y + 2z &= -8 \end{aligned}$

(f) $\begin{aligned} x - 3y + 4z + w + 2v &= -2 \\ 2x - 2y - z + 2w + 3v &= 7 \\ -x + y - 2z - 2w + v &= 0 \\ -2x - y + 2z + w - v &= 5 \\ 2x - 2y + z + 2w + 2v &= 3 \end{aligned}$

(c) $\begin{aligned} x - y - z &= -5 \\ 2x - 3y - 8z &= -33 \\ x - 2y - 8z &= -32 \end{aligned}$

(g) $\begin{aligned} 2x + y + 3z - 2w &= -6 \\ 4x + 3y + z - w &= -2 \\ x + y + z + w &= -5 \\ -2x - 2y + 2z + 2w &= -10 \end{aligned}$

(d) $\begin{aligned} x + 3y + 10z &= -8 \\ -x - 2y - 5z &= 3 \\ 2x + 4y + 5z &= -6 \end{aligned}$

B.2. The sum of two 2×2 matrices is the 2×2 matrix defined by this rule:

$$\begin{pmatrix} a & b \\ c & d \end{pmatrix} + \begin{pmatrix} r & s \\ t & u \end{pmatrix} = \begin{pmatrix} a+r & b+s \\ c+t & d+u \end{pmatrix}$$

For example,

$$\begin{pmatrix} 1 & 2 \\ 3 & 4 \end{pmatrix} + \begin{pmatrix} 5 & -3 \\ 2 & 6 \end{pmatrix} = \begin{pmatrix} 1+5 & 2+(-3) \\ 3+2 & 4+6 \end{pmatrix} = \begin{pmatrix} 6 & -1 \\ 5 & 10 \end{pmatrix}$$

Find these sums:

(a) $\begin{pmatrix} 3 & 2 \\ 5 & 1 \end{pmatrix} + \begin{pmatrix} 7 & -5 \\ -2 & 6 \end{pmatrix}$ (d) $\begin{pmatrix} \frac{3}{4} & 7 \\ 6 & -\frac{5}{4} \end{pmatrix} + \begin{pmatrix} \frac{1}{2} & 3 \\ -5 & \frac{3}{2} \end{pmatrix}$

(b) $\begin{pmatrix} -6 & 2 \\ 7 & -1 \end{pmatrix} + \begin{pmatrix} -8 & 4 \\ 2 & 7 \end{pmatrix}$ (e) $\begin{pmatrix} 0 & 0 \\ 0 & 0 \end{pmatrix} + \begin{pmatrix} 3 & 5 \\ 7 & 9 \end{pmatrix}$

(c) $\begin{pmatrix} \frac{3}{2} & 2 \\ 4 & \frac{7}{2} \end{pmatrix} + \begin{pmatrix} \frac{1}{2} & -\frac{3}{2} \\ \frac{5}{2} & 1 \end{pmatrix}$ (f) $\begin{pmatrix} 1 & -2 \\ -3 & 5 \end{pmatrix} + \begin{pmatrix} -1 & 2 \\ 3 & -5 \end{pmatrix}$

B.3. The product of two 2×2 matrices is the 2×2 matrix defined by this rule:

$$\begin{pmatrix} a & b \\ c & d \end{pmatrix} \begin{pmatrix} r & s \\ t & u \end{pmatrix} = \begin{pmatrix} ar+bt & as+bu \\ cr+dt & cs+du \end{pmatrix}$$

For example,

$$\begin{pmatrix} 2 & 3 \\ -1 & 5 \end{pmatrix} \begin{pmatrix} 4 & -2 \\ 3 & 0 \end{pmatrix} = \begin{pmatrix} 2 \cdot 4 + 3 \cdot 3 & 2(-2) + 3 \cdot 0 \\ (-1)4 + 5 \cdot 3 & (-1)(-2) + 5 \cdot 0 \end{pmatrix} = \begin{pmatrix} 17 & -4 \\ 11 & 2 \end{pmatrix}$$

Find these products:

(a) $\begin{pmatrix} 2 & 3 \\ 1 & 2 \end{pmatrix} \begin{pmatrix} 4 & -3 \\ 2 & 5 \end{pmatrix}$ (d) $\begin{pmatrix} -1 & 2 \\ 3 & -4 \end{pmatrix} \begin{pmatrix} 5 & 2 \\ 3 & -1 \end{pmatrix}$

(b) $\begin{pmatrix} 1 & 0 \\ 0 & 1 \end{pmatrix} \begin{pmatrix} 2 & 3 \\ 5 & 7 \end{pmatrix}$ (e) $\begin{pmatrix} 5 & 2 \\ 3 & -1 \end{pmatrix} \begin{pmatrix} -1 & 2 \\ 3 & -4 \end{pmatrix}$

  [compare your answer with (d)]

(c) $\begin{pmatrix} 0 & 1 \\ 1 & 0 \end{pmatrix} \begin{pmatrix} 3 & 5 \\ 7 & 9 \end{pmatrix}$ (f) $\begin{pmatrix} 1 & -1 \\ 2 & -2 \end{pmatrix} \begin{pmatrix} 3 & -1 \\ 3 & -1 \end{pmatrix}$

B.4. Verify that $A(B+C) = AB + AC$ and $(AB)C = A(BC)$ when

$$A = \begin{pmatrix} 1 & 2 \\ 3 & 0 \end{pmatrix}, \quad B = \begin{pmatrix} -1 & 2 \\ 3 & 4 \end{pmatrix}, \quad C = \begin{pmatrix} 2 & 3 \\ 1 & 2 \end{pmatrix}$$

B.5. Multiplication of matrices satisfies *some* of the laws that multiplication of numbers does (Exercise B.4), but *not all* of them. In each case verify that *the given statement is false* for the given matrices.

(a) $AB = BA$; $\quad A = \begin{pmatrix} 1 & 2 \\ 3 & 4 \end{pmatrix} \quad$ and $\quad B = \begin{pmatrix} -1 & 4 \\ 5 & 2 \end{pmatrix}$

(b) If $AB = AC$, then $B = C$; $\quad A = \begin{pmatrix} 1 & 2 \\ 2 & 4 \end{pmatrix}, B = \begin{pmatrix} 3 & 6 \\ -\frac{3}{2} & -3 \end{pmatrix}, C = \begin{pmatrix} 0 & 0 \\ 0 & 0 \end{pmatrix}$

(c) $A^2 = \begin{pmatrix} 0 & 0 \\ 0 & 0 \end{pmatrix}$ only if $A = \begin{pmatrix} 0 & 0 \\ 0 & 0 \end{pmatrix}$; $\quad A = \begin{pmatrix} 0 & 2 \\ 0 & 0 \end{pmatrix}$

(d) $(A + B)(A - B) = A^2 - B^2$; $\quad A = \begin{pmatrix} 3 & 1 \\ 2 & -4 \end{pmatrix} \quad$ and $\quad B = \begin{pmatrix} 2 & -1 \\ 5 & 3 \end{pmatrix}$

(e) $(A + B)(A + B) = A^2 + 2AB + B^2$; $\quad A = \begin{pmatrix} 2 & -1 \\ 3 & 5 \end{pmatrix} \quad$ and $\quad B = \begin{pmatrix} 1 & 2 \\ -3 & 4 \end{pmatrix}$

B.6. Verify each of these statements by calculating the determinant on the left side, finding the product on the right side, and comparing the two.

(a) $\begin{vmatrix} 1 & x & x^2 \\ 1 & y & y^2 \\ 1 & z & z^2 \end{vmatrix} = (x - y)(y - z)(z - x)$

(b) $\begin{vmatrix} 1 & 1 & 1 \\ u & v & w \\ u^2 & v^2 & w^2 \end{vmatrix} = (u - v)(v - w)(w - u)$

B.7. (a) Compute the determinant

$$\begin{vmatrix} 1 & x & y \\ 1 & a & b \\ 1 & c & d \end{vmatrix}$$

(b) Verify that the equation of the straight line through the distinct points (a, b) and (c, d) is

$$\begin{vmatrix} 1 & x & y \\ 1 & a & b \\ 1 & c & d \end{vmatrix} = 0$$

B.8. *Calculation of $|A|$ by expansion along a row or column.* Consider the matrix

$$A = \begin{pmatrix} a_1 & b_1 & c_1 \\ a_2 & b_2 & c_2 \\ a_3 & b_3 & c_3 \end{pmatrix}$$

(a)   Choose any row or column of the matrix (one or the other, not both). Use the entries in the chosen row or column to compute the following number

$$\pm \begin{pmatrix} \text{first entry} \\ \text{in chosen row} \\ \text{or column} \end{pmatrix} \begin{pmatrix} \text{minor of} \\ \text{this entry} \end{pmatrix} \pm \begin{pmatrix} \text{second entry} \\ \text{in chosen row} \\ \text{or column} \end{pmatrix} \begin{pmatrix} \text{minor of} \\ \text{this entry} \end{pmatrix}$$

$$\pm \begin{pmatrix} \text{third entry} \\ \text{in chosen} \\ \text{row or column} \end{pmatrix} \begin{pmatrix} \text{minor of} \\ \text{this entry} \end{pmatrix}$$

where the sign ($+$ or $-$) of each term is determined by the position of the chosen entry in the matrix, according to the following scheme:

$$\begin{pmatrix} + & - & + \\ - & + & - \\ + & - & + \end{pmatrix}$$

(b)   Verify that the number you have computed is precisely the determinant of the matrix A. (Compare your answer with $|A|$ as given on p. 525, where the computation was made using the first row.)

(c)   Make this computation five times using successively row 2, row 3, column 1, column 2, and column 3. In each case, verify that the answer is $|A|$.

B.9.   Each of the following statements is true for all $n \times n$ matrices and determinants. Give a specific numerical example of each statement, using 3×3 matrices and determinants.

(a)   If the matrix B is obtained from the matrix A by interchanging two rows, then $|B| = -|A|$.

(b)   If the matrix B is obtained from the matrix A by multiplying a row of A by a constant c, then $|B| = c|A|$.

(c)   If the matrix B is obtained from the matrix A by adding a constant multiple of one row to another row, then $|B| = |A|$.

(d)   If a matrix A contains a row of zeros, then $|A| = 0$.

(e)   Statements (a)–(d) are true with "row" replaced by "column."

B.10.   Use Cramer's Rule to solve these systems of equations:

(a) $-3x + 5y = 2$
      $2x + 7y = 1$

(b) $7x - 12y = 4$
      $3x - 5y = 2$

(c) $\frac{3}{2}x + 2y = \frac{5}{2}$
      $5x - 7y = 1$

(d) $x - \frac{5}{3}y = 2$
      $6x + \frac{4}{3}y = 1$

(e) $7x - 12y = 4$
      $3x - 5y = 2$

(f) $\sqrt{5}x - 2\sqrt{3}y = 2$

(e) $\sqrt{3}x + 2\sqrt{5}y = 3$

B.11.   Use Cramer's Rule to determine which of these homogeneous systems have exactly one solution (namely, the trivial one $x = 0$, $y = 0$, $z = 0$).

(a) $2x + y - 2z = 0$
      $3x + 2y - z = 0$
      $4x + y - 3z = 0$

(b) $x + 2y - 3z = 0$
      $3x - 5y - 9z = 0$
      $2x + 4y - 6z = 0$

(c) $2x + 4y + z = 0$
      $6x - y + 3z = 0$
      $4x + 6y + 2z = 0$

(d) $3x + 2y - z = 0$
      $2x + y + z = 0$
      $5x - 2y - z = 0$

3. THE BINOMIAL THEOREM

The Binomial Theorem provides a formula for calculating the product $(x + y)^n$ for any positive integer n. Before we can state the Binomial Theorem, certain preliminary definitions are needed.

FACTORIALS AND BINOMIAL COEFFICIENTS

Let n be a positive integer. The symbol $n!$ (read n **factorial**) denotes the product of all the integers from 1 to n. For example,

$$2! = 1 \cdot 2 = 2, \quad 3! = 1 \cdot 2 \cdot 3 = 6, \quad 4! = 1 \cdot 2 \cdot 3 \cdot 4 = 24,$$
$$5! = 1 \cdot 2 \cdot 3 \cdot 4 \cdot 5 = 120, \quad 10! = 1 \cdot 2 \cdot 3 \cdot 4 \cdot 5 \cdot 6 \cdot 7 \cdot 8 \cdot 9 \cdot 10 = 3{,}628{,}800$$

and in general,

$$\boxed{n! = 1 \cdot 2 \cdot 3 \cdot 4 \cdots (n - 2)(n - 1)n}$$

As you can see, $n!$ may be very large even when n is relatively small. In fact, 15! is larger than a trillion and 20! is larger than $2 \cdot 10^{18}$. It will be convenient for certain calculations later to give a meaning to the symbol 0! (read zero factorial). We *define* 0! to be the number 1.

We now use factorials to define some numbers that will turn out to be quite important.

> If r and n are integers with $0 \le r \le n$, then
>
> the symbol $\binom{n}{r}$ denotes the number $\dfrac{n!}{r!(n - r)!}$.
>
> Each of the numbers $\binom{n}{r}$ is called a **binomial coefficient**.

For example,

$$\binom{5}{3} = \frac{5!}{3!(5 - 3)!} = \frac{5}{3!2!} = \frac{1 \cdot 2 \cdot 3 \cdot 4 \cdot 5}{(1 \cdot 2 \cdot 3)(1 \cdot 2)} = \frac{4 \cdot 5}{2} = 10,$$

$$\binom{4}{2} = \frac{4!}{2!(4 - 2)!} = \frac{4!}{2!2!} = \frac{1 \cdot 2 \cdot 3 \cdot 4}{(1 \cdot 2)(1 \cdot 2)} = \frac{3 \cdot 4}{2} = 6,$$

$$\binom{3}{0} = \frac{3!}{0!(3 - 0)!} = \frac{3!}{0!3!} = \frac{3!}{3!} = 1 \quad \text{and} \quad \binom{3}{3} = \frac{3!}{3!(3 - 3)!} = \frac{3!}{3!0!} = \frac{3!}{3!} = 1$$

More generally, for any nonnegative integer n we have

$$\binom{n}{0} = \frac{n!}{0!(n-0)!} = \frac{n!}{0!n!} = \frac{n!}{n!} = 1 \quad \text{and} \quad \binom{n}{n} = \frac{n!}{n!(n-n)!} = \frac{n!}{n!0!} = \frac{n!}{n!} = 1$$

Therefore

$$\binom{n}{0} = 1 \quad \text{and} \quad \binom{n}{n} = 1 \quad \text{for each integer } n \geq 0$$

The preceding examples illustrate this important fact about binomial coefficients:

Every binomial coefficient $\binom{n}{r}$ is an integer.

This fact is proved in Exercise C.4, page 549. We shall assume it for now. If we list the binomial coefficients for each value of n in this manner,

$n = 0$ $\qquad\qquad \binom{0}{0}$

$n = 1$ $\qquad\qquad \binom{1}{0} \quad \binom{1}{1}$

$n = 2$ $\qquad\qquad \binom{2}{0} \quad \binom{2}{1} \quad \binom{2}{2}$

$n = 3$ $\qquad\qquad \binom{3}{0} \quad \binom{3}{1} \quad \binom{3}{2} \quad \binom{3}{3}$

$n = 4$ $\qquad\qquad \binom{4}{0} \quad \binom{4}{1} \quad \binom{4}{2} \quad \binom{4}{3} \quad \binom{4}{4}$

$\vdots$

and then calculate each of them, we obtain the following array of numbers:

```
row 0                1
row 1              1   1
row 2            1   2   1
row 3          1   3   3   1
row 4        1   4   6   4   1
```
$\vdots$

This array is called **Pascal's triangle.** Some of its properties are discussed in Exercise C.1(d).

THE BINOMIAL THEOREM

We want to find a formula for calculating products such as

$$(x + y)^4, \qquad (x + y)^7, \qquad (x + y)^{24}$$

and more generally,

$$(x + y)^n, \qquad n \text{ any nonnegative integer}$$

We begin by calculating these products for small values of n to see if we can find some kind of pattern. Verify that the following computations are accurate:

$$
\begin{array}{lll}
n = 0 & (x + y)^0 = & 1 \\
n = 1 & (x + y)^1 = & x + y \\
n = 2 & (x + y)^2 = & x^2 + 2xy + y^2 \\
n = 3 & (x + y)^3 = & x^3 + 3x^2y + 3xy^2 + y^3 \\
n = 4 & (x + y)^4 = & x^4 + 4x^3y + 6x^2y^2 + 4xy^3 + y^4
\end{array}
$$

For these small values of n, some parts of the pattern are already clear. For each positive n, the first term is x^n and the last term is y^n. Beginning with the second term,

The successive exponents of y are $1, 2, 3, \ldots, n$

In each term before the last one, the exponent of x is 1 less than the preceding term.

Suppose this pattern holds true for larger values of n as well. Then for a fixed n, the expansion of $(x + y)^n$ would have first term x^n. In the second term, the exponent of x would be 1 less than n, namely, $n - 1$, and the exponent of y would be 1. So the second term would be of the form $(\text{constant})x^{n-1}y$. In the next term, the exponent of x would be 1 less again, namely, $n - 2$, and the exponent of y would be 2. Continuing in this fashion, we would have

$$(x + y)^n = x^n + (^\circ)x^{n-1}y + (^\circ)x^{n-2}y^2 + (^\circ)x^{n-3}y^3 + \cdots + (^\circ)xy^{n-1} + y^n$$

where the symbols $(^\circ)$ indicate the various constant coefficients.

In order to determine the constant coefficients in the expansion of $(x + y)^n$, we return to the computations made above for $n = 0, 1, 2, 3, 4$. The terms x, y, x^2, y^2, and so on, each have coefficient 1. If we omit the x's and y's and just list the coefficients that appear in the computations above, we obtain this array of numbers:

$$
\begin{array}{ccccccccc}
n = 0 & & & & 1 & & & & \\
n = 1 & & & 1 & & 1 & & & \\
n = 2 & & 1 & & 2 & & 1 & & \\
n = 3 & & 1 & 3 & & 3 & & 1 & \\
n = 4 & 1 & & 4 & & 6 & & 4 & & 1
\end{array}
$$

But this is just the top of Pascal's triangle. In the case $n = 4$, it means that the coefficients of the expansion of $(x + y)^4$ are just the binomial coefficients $\binom{4}{0}, \binom{4}{1}, \binom{4}{2},$

$\binom{4}{3}$, $\binom{4}{4}$; and similarly for the other small values of n. If this pattern holds true for larger n as well, then the coefficients of the expansion of $(x + y)^n$ are just the binomial coefficients

$$\binom{n}{0}, \ \binom{n}{1}, \ \binom{n}{2}, \ \binom{n}{3}, \dots, \binom{n}{n-1}, \ \binom{n}{n}$$

Since $\binom{n}{0} = 1$ and $\binom{n}{n} = 1$ for every n, the first and last coefficients on this list are 1. This is consistent with the fact that the first and last terms are x^n and y^n.

The preceding discussion suggests that the following result is true:

THE BINOMIAL THEOREM

For each positive integer n,

$$(x + y)^n = x^n + \binom{n}{1}x^{n-1}y + \binom{n}{2}x^{n-2}y^2 +$$

$$\binom{n}{3}x^{n-3}y^3 + \cdots + \binom{n}{n-1}xy^{n-1} + y^n$$

The Binomial Theorem will be proved in the next section by means of mathematical induction. We shall assume its truth for now and illustrate some of its uses.

EXAMPLE In order to compute $(x + y)^8$ we apply the Binomial Theorem in the case $n = 8$:

$$(x + y)^8 = x^8 + \binom{8}{1}x^7y + \binom{8}{2}x^6y^2 + \binom{8}{3}x^5y^3$$

$$+ \binom{8}{4}x^4y^4 + \binom{8}{5}x^3y^5 + \binom{8}{6}x^2y^6 + \binom{8}{7}xy^7 + y^8$$

Now verify that

$$\binom{8}{1} = \frac{8!}{1!7!} = 8, \quad \binom{8}{2} = \frac{8!}{2!6!} = 28, \quad \binom{8}{3} = \frac{8!}{3!5!} = 56, \quad \binom{8}{4} = \frac{8!}{4!4!} = 70$$

Using these facts, we see that

$$\binom{8}{5} = \frac{8!}{5!3!} = \binom{8}{3} = 56, \quad \binom{8}{6} = \frac{8!}{6!2!} = \binom{8}{2} = 28, \quad \binom{8}{7} = \frac{8!}{7!1!} = \binom{8}{1} = 8$$

Substituting these values in the expansion above, we have

$$(x + y)^8$$
$$= x^8 + 8x^7y + 28x^6y^2 + 56x^5y^3 + 70x^4y^4 + 56x^3y^5 + 28x^2y^6 + 8xy^7 + y^8$$

EXAMPLE   To find $(1 - z)^6$, we note that $1 - z = 1 + (-z)$ and apply the Binomial Theorem with $x = 1$, $y = -z$, and $n = 6$:

$(1 - z)^6$

$= 1^6 + \binom{6}{1}1^5(-z) + \binom{6}{2}1^4(-z)^2 + \binom{6}{3}1^3(-z)^3 + \binom{6}{4}1^2(-z)^4 + \binom{6}{5}1(-z)^5 + (-z)^6$

$= 1 - \binom{6}{1}z + \binom{6}{2}z^2 - \binom{6}{3}z^3 + \binom{6}{4}z^4 - \binom{6}{5}z^5 + z^6$

$= 1 - 6z + 15z^2 - 20z^3 + 15z^4 - 6z^5 + z^6$

EXAMPLE   Show that $(1.001)^{1000} > 2$ without using a calculator. We write 1.001 as $1 + .001$ and apply the Binomial Theorem with $x = 1$, $y = .001$, and $n = 1000$:

$$(1.001)^{1000} = (1 + .001)^{1000} = 1^{1000} + \binom{1000}{1}1^{999}(.001) + \text{other positive terms}$$

$$= 1 + \binom{1000}{1}(.001) + \text{other positive terms}$$

But $\binom{1000}{1} = \dfrac{1000!}{1!999!} = \dfrac{1000 \cdot 999!}{999!} = 1000$. Therefore $\binom{1000}{1}(.001) = 1,000(.001) = 1$ and

$$(1.001)^{1000} = 1 + 1 + \text{other positive terms} = 2 + \text{other positive terms}$$

Hence $(1.001)^{1000} > 2$.

Sometimes we only need to know one term in the expansion of $(x + y)^n$. If you examine the expansion given by the Binomial Theorem, you will see that in the second term y has exponent 1, in the third term y has exponent 2, and so on. Thus

---

In the binomial expansion of $(x + y)^n$,

   The exponent of y *is always one less than the number of the term.*

Furthermore, in each of the middle terms of the expansion,

   The coefficient of the term containing y^r *is* $\binom{n}{r}$.

   The sum of the x-exponent and the y-exponent is n.

For instance, in the *ninth* term of the expansion of $(x + y)^{13}$, y has exponent 8, the coefficient is $\binom{13}{8}$ and x must have exponent 5 (since $8 + 5 = 13$). Thus the ninth term is $\binom{13}{8}x^5y^8$.

EXAMPLE What is the ninth term of the expansion of $\left(2x^2 + \dfrac{\sqrt[4]{y}}{\sqrt{6}}\right)^{13}$? We shall use

the Binomial Theorem with $n = 13$ and with $2x^2$ in place of x and $\sqrt[4]{y}/\sqrt{6}$ in place of y. The remarks above show that the ninth term is

$$\binom{13}{8}(2x^2)^5\left(\frac{\sqrt[4]{y}}{\sqrt{6}}\right)^8$$

Since $\sqrt[4]{y} = y^{1/4}$ and $\sqrt{6} = \sqrt{3}\sqrt{2} = 3^{1/2}2^{1/2}$, we can simplify as follows:

$$\binom{13}{8}(2x^2)^5\left(\frac{\sqrt[4]{y}}{\sqrt{6}}\right)^8 = \binom{13}{8}2^5(x^2)^5\frac{(y^{1/4})^8}{(3^{1/2})^8(2^{1/2})^8} = \binom{13}{8}2^5x^{10}\frac{y^2}{3^4\cdot2^4}$$

$$= \binom{13}{8}\frac{2}{3^4}x^{10}y^2 = \frac{13\cdot12\cdot11\cdot10\cdot9}{5\cdot4\cdot3\cdot2}\cdot\frac{2}{3^4}x^{10}y^2 = \frac{286}{9}x^{10}y^2$$

EXERCISES

A.1. Evaluate:
 (a) $6!$ (b) $\dfrac{11!}{8!}$ (c) $\dfrac{12!}{9!3!}$ (d) $\dfrac{9!-8!}{7!}$

A.2. Evaluate
 (a) $\binom{5}{3}+\binom{5}{2}-\binom{6}{3}$ (e) $\binom{100}{96}$

 (b) $\binom{12}{11}-\binom{11}{10}+\binom{7}{0}$ (f) $\binom{75}{72}$

 (c) $\binom{6}{0}+\binom{6}{1}+\binom{6}{2}+\binom{6}{3}+\binom{6}{4}+\binom{6}{5}+\binom{6}{6}$

 (d) $\binom{6}{0}-\binom{6}{1}+\binom{6}{2}-\binom{6}{3}+\binom{6}{4}-\binom{6}{5}+\binom{6}{6}$

A.3. Expand:
 (a) $(x+y)^5$ (c) $(a-b)^5$ (e) $(2x+y^2)^5$
 (b) $(a+b)^7$ (d) $(c-d)^8$ (f) $(3u-v^3)^6$

B.1. (a) Verify that $\binom{9}{1} = 9$ and $\binom{9}{8} = 9$.

 (b) Prove that for each positive integer n, $\binom{n}{1} = n$ and $\binom{n}{n-1} = n$. [*Note:* part (a) is just the case when $n = 9$ and $n - 1 = 8$.]

B.2. Expand and simplify (where possible).

(a) $(\sqrt{x} + 1)^6$ (c) $(1 - c)^{10}$ (e) $(x^{-3} + x)^4$

(b) $(2 - \sqrt{y})^5$ (d) $\left(\sqrt{c} + \dfrac{1}{\sqrt{c}}\right)^7$ (f) $(3x^{-2} - x^2)^6$

B.3. Use the Binomial Theorem to compute:

(a) $(1 + \sqrt{3})^4 + (1 - \sqrt{3})^4$ (c) $(1 + i)^6$, where $i^2 = -1$

(b) $(\sqrt{3} + 1)^6 - (\sqrt{3} - 1)^6$ (d) $(\sqrt{2} - i)^4$, where $i^2 = -1$

B.4. Find the indicated term of the expansion of the given expression.

(a) third, $(x + y)^5$ (d) third, $(a + 2)^8$

(b) fourth, $(a + b)^6$ (e) fourth, $\left(u^{-2} + \dfrac{u}{2}\right)^7$

(c) fifth, $(c - d)^7$ (f) fifth, $(\sqrt{x} - \sqrt{2})^7$

B.5. Find the coefficient of

(a) $x^5 y^8$ in the expansion of $(2x - y^2)^9$

(b) $x^{12} y^6$ in the expansion of $(x^3 - 3y)^{10}$

(c) $\dfrac{1}{x^3}$ in the expansion of $\left(2x + \dfrac{1}{x^2}\right)^6$

B.6. Find the constant term in the expansion of $\left(y - \dfrac{1}{2y}\right)^{10}$.

B.7. (a) Verify that $\dbinom{7}{2} = \dbinom{7}{5}$.

(b) Let r and n be integers with $0 \le r \le n$. Prove that $\dbinom{n}{r} = \dbinom{n}{n-r}$. [Note: part (a) is just the case when $n = 7$ and $r = 2$.]

B.8. For any positive integer n,

(a) Prove that $2^n = \dbinom{n}{0} + \dbinom{n}{1} + \dbinom{n}{2} + \cdots + \dbinom{n}{n}$. [Hint: $2^n = (1 + 1)^n$.]

(b) Prove that $\dbinom{n}{0} - \dbinom{n}{1} + \dbinom{n}{2} - \dbinom{n}{3} + \dbinom{n}{4} - \cdots + (-1)^k \dbinom{n}{k} + \cdots + (-1)^n \dbinom{n}{n} = 0$.

B.9. (a) Use the Binomial Theorem with $x = \sin\theta$ and $y = \cos\theta$ to find $(\cos\theta + i\sin\theta)^4$ where $i^2 = -1$.

(b) Use DeMoivre's Theorem to find $(\cos\theta + i\sin\theta)^4$.

(c) Use the fact that the two expressions obtained in parts (a) and (b) must be equal to express $\cos 4\theta$ and $\sin 4\theta$ in terms of $\sin\theta$ and $\cos\theta$.

B.10. (a) Let f be the function given by $f(x) = x^5$. Let h be a nonzero number and compute $f(x + h) - f(x)$ [but leave all binomial coefficients in the form $\binom{5}{r}$ here and below].

(b) Use part (a) to show that h is a factor of $f(x + h) - f(x)$ and find $\dfrac{f(x + h) - f(x)}{h}$.

(c) If h is *very* close to 0, find a simple approximation of the quantity $\dfrac{f(x + h) - f(x)}{h}$. [See part (b).]

(d) Do parts (a)–(c) with $f(x) = x^8$ in place of $f(x) = x^5$.

(e) Do parts (a)–(c) with $f(x) = x^{12}$ in place of $f(x) = x^5$.

(f) Let n be a fixed positive integer. Do parts (a)–(c) with $f(x) = x^n$ in place of $f(x) = x^5$.

C.1. Let r and n be integers such that $0 \le r \le n$.

(a) Verify that $(n - r)! = (n - r)(n - (r + 1))!$

(b) Verify that $(n - r)! = ((n + 1) - (r + 1))!$

(c) Prove that $\binom{n}{r + 1} + \binom{n}{r} = \binom{n + 1}{r + 1}$ for any $r \le n - 1$. [*Hint:* write out the terms on the left side and use parts (a) and (b) to express each of them as a fraction with denominator $(r + 1)!(n - r)!$. Then add these two fractions, simplify the numerator and compare the result with $\binom{n + 1}{r + 1}$.]

(d) Interpret the result of part (c) as a statement about how the entries of one row of Pascal's triangle are related to the entries in the next row.

(e) Use part (d) to write out rows 2 to 10 of Pascal's triangle *without* computing any binomial coefficients.

C.2. (a) Find these numbers and write them one *below* the next: 11^0, 11^1, 11^2, 11^3, 11^4.

(b) Compare the list in part (a) with rows 0 to 4 of Pascal's triangle. What's the explanation?

(c) What can be said about 11^5 and row 5 of Pascal's triangle?

(d) Calculate all integer powers of 101 from 101^0 to 101^8, list the results one under the other, and compare the list with rows 0 to 8 of Pascal's triangle. What's the explanation? What happens with 101^9?

4. MATHEMATICAL INDUCTION

In earlier parts of this book, we have used several results like the Binomial Theorem that were only verified for small values of n. In these cases, there was usually a clear pattern and it seemed plausible that the results would be valid for all values of n. But at some stage of mathematical development such plausible statements must be backed up by *proof*. We have now reached that stage. In this section, we shall study mathematical induction, which can be used to prove the Binomial Theorem, DeMoivre's Theorem, and many other statements, such as

The sum of the first n positive integers is the number $\dfrac{n(n + 1)}{2}$.

$2^n > n$ for every positive integer n.

For each positive integer n, 4 is a factor of $7^n - 3^n$.

All of the statements above have a common property. For example, a statement such as

the sum of the first n positive integers is the number $\dfrac{n(n + 1)}{2}$,

or, in symbols,

$$1 + 2 + 3 + \cdots + n = \frac{n(n + 1)}{2}$$

is really an infinite sequence of statements, one for each possible value of n:

$$n = 1: \qquad\qquad 1 = \frac{1(2)}{2}$$

$$n = 2: \qquad\qquad 1 + 2 = \frac{2(3)}{2}$$

$$n = 3: \qquad\qquad 1 + 2 + 3 = \frac{3(4)}{2}$$

$$n = 4: \qquad 1 + 2 + 3 + 4 = \frac{4(5)}{2}$$

and so on. Obviously, there isn't time enough to verify every one of the statements on this list, one at a time. But we can find a workable method of proof by examining how each statement on the list is *related* to the *next* statement on the list.

For instance, for $n = 50$, the statement is:

$$1 + 2 + 3 + \cdots + 50 = \frac{50(51)}{2}$$

At the moment, we don't know whether or not this statement is true. But just *suppose* that it were true. What could then be said about the next statement, the one for $n = 51$:

$$1 + 2 + 3 + \cdots + 50 + 51 = \frac{51(52)}{2}?$$

Well, *if* it is true that

$$1 + 2 + 3 + \cdots + 50 = \frac{50(51)}{2}$$

then adding 51 to both sides and simplifying the right side would yield these equalities:

$$1 + 2 + 3 + \cdots + 50 + 51 = \frac{50(51)}{2} + 51$$

$$1 + 2 + 3 + \cdots + 50 + 51 = \frac{50(51)}{2} + \frac{2(51)}{2} = \frac{50(51) + 2(51)}{2}$$

$$1 + 2 + 3 + \cdots + 50 + 51 = \frac{(50 + 2)51}{2}$$

$$1 + 2 + 3 + \cdots + 50 + 51 = \frac{51(52)}{2}$$

Since this last equality is just the original statement for $n = 51$, we conclude that

If the statement is true for $n = 50$, *then* it is also true for $n = 51$

We have *not* proved that the statement actually *is* true for $n = 50$, but only that *if* it is, then it is also true for $n = 51$.

We claim that this same conditional relationship holds for any two consecutive values of n. In other words, we claim that for any positive integer k,

(°)   *If* the statement is true for $n = k$, *then* it is also true for $n = k + 1$.

The proof of this claim is the same argument used above (with k and $k + 1$ in place of 50 and 51): *if* it is true that

$$1 + 2 + 3 + \cdots + k = \frac{k(k + 1)}{2} \qquad \text{(original statement for } n = k\text{),}$$

then adding $k + 1$ to both sides and simplifying the right side produces these equalities:

$$1 + 2 + 3 + \cdots + k + (k + 1) = \frac{k(k + 1)}{2} + (k + 1)$$

$$1 + 2 + 3 + \cdots + k + (k + 1) = \frac{k(k + 1)}{2} + \frac{2(k + 1)}{2} = \frac{k(k + 1) + 2(k + 1)}{2}$$

$$1 + 2 + 3 + \cdots + k + (k + 1) = \frac{(k + 2)(k + 1)}{2}$$

$$1 + 2 + 3 + \cdots + k + (k + 1) = \frac{(k + 1)((k + 1) + 1)}{2} \qquad \text{(original statement for } n = k + 1\text{)}$$

We have proved that claim (°) is valid for each positive integer k. We have *not* proved that the original statement is true for any value of n, but only that *if* it is true for

$n = k$, then it is also true for $n = k + 1$. Applying this fact when $k = 1, 2, 3, \ldots$, we see that $(^{\circ\circ})$

$$\begin{cases} \textit{if} & \text{the statement is true for } n = 1, & \textit{then} \text{ it is also true for } n = 1 + 1 = 2; \\ \textit{if} & \text{the statement is true for } n = 2, & \textit{then} \text{ it is also true for } n = 2 + 1 = 3; \\ \textit{if} & \text{the statement is true for } n = 3, & \textit{then} \text{ it is also true for } n = 3 + 1 = 4; \\ & \vdots \\ \textit{if} & \text{the statement is true for } n = 50, & \textit{then} \text{ it is also true for } n = 50 + 1 = 51; \\ \textit{if} & \text{the statement is true for } n = 51, & \textit{then} \text{ it is also true for } n = 51 + 1 = 52; \\ & \vdots \end{cases}$$

and so on.

We are finally in a position to *prove* the original statement: $1 + 2 + 3 + \cdots + n = n(n + 1)/2$. Obviously, it *is true* for $n = 1$ since $1 = 1(2)/2$. Now apply in turn each of the propositions on list $(^{\circ\circ})$ above. Since the statement *is* true for $n = 1$, it must also be true for $n = 2$, and hence for $n = 3$, and hence for $n = 4$, and so on, for every value of n. Therefore the original statement is true for *every* positive integer n.

The preceding proof is an illustration of the following principle:

PRINCIPLE OF MATHEMATICAL INDUCTION

Suppose there is given a statement involving the positive integer n *and that:*

(i) *The statement is true for* n = 1.
(ii) *If the statement is true for* n = k *(where* k *is a positive integer), then the statement is also true for* n = k + 1.

Then the statement is true for every *positive integer* n.

Property (i) in this principle is simply a statement of fact. To verify that it holds, you must prove the given statement is true for $n = 1$. This is usually easy, as in the preceding example. Property (ii) is a *conditional* property. It does not assert that the given statement *is* true for $n = k$, but only that *if* it is true for $n = k$, then it is also true for $n = k + 1$. So to verify that property (ii) holds, you need only prove this conditional proposition:

If the statement is true for $n = k$, *then* it is also true for $n = k + 1$

In order to prove this, or any conditional proposition, you must proceed as in the example above: assume the "if" part and use this assumption to prove the "then" part. As we saw above, the same argument will usually work for any possible k. Once this conditional proposition has been proved, you can use it *together with* property (i) to conclude that the given statement is necessarily true for every n, just as in the preceding example.

Thus proof by mathematical induction reduces to two steps:

Step 1:   Prove that the given statement is true for $n = 1$.
Step 2:   Let k be a positive integer. Assume that the given statement is true for $n = k$. Use this assumption to prove that the statement is true for $n = k + 1$.

Step 2 may be performed before step 1 if you wish. Step 2 is sometimes referred to as the **inductive step.** The assumption that the given statement is true for $n = k$ in this inductive step is called the **induction hypothesis.** Here are some more examples of inductive proofs.

EXAMPLE   Here's how mathematical induction is used to prove that

$$2^n > n \quad \text{for every positive integer } n$$

In this case, the statement involving n is: $2^n > n$.

Step 1   When $n = 1$, we have the statement $2^1 > 1$. This is obviously true.

Step 2   Let k be any positive integer. We assume that the statement is true for $n = k$, that is, we assume that $2^k > k$. We shall use this assumption to prove that the statement is true for $n = k + 1$, that is, that $2^{k+1} > k + 1$. We begin with the induction hypothesis:[*] $2^k > k$. Multiplying both sides of this inequality by 2 yields:

$$2 \cdot 2^k > 2k$$
(°)
$$2^{k+1} > 2k$$

Since k is a positive integer, we know that $k \geq 1$. Adding k to each side of the inequality $k \geq 1$, we have

$$k + k \geq k + 1$$
$$2k \geq k + 1$$

Combining this result with inequality (°) above, we see that

$$2^{k+1} > 2k \geq k + 1$$

The first and last terms of this inequality show that $2^{k+1} > k + 1$. Therefore the statement is true for $n = k + 1$. This argument works for any positive integer k. Thus we have completed the inductive step. By the Principle of Mathematical Induction, we conclude that $2^n > n$ for every positive integer n.

EXAMPLE   Simple arithmetic shows that:

$$7^2 - 3^2 = 49 - 9 = 40 = 4 \cdot 10 \quad \text{and} \quad 7^3 - 3^3 = 343 - 27 = 316 = 4 \cdot 79$$

---

[*] This is the point at which you usually must do some work. Remember that what follows is the "finished proof." It does not include all the thought, scratch work, false starts, and so on, that were done before this proof was actually found.

In each case, 4 is a factor. These examples suggest that

For each positive integer n, 4 is a factor of $7^n - 3^n$

This conjecture can be proved by induction as follows.

Step 1 When $n = 1$, the statement is: 4 is a factor of $7^1 - 3^1$. Since $7^1 - 3^1 = 4 = 4 \cdot 1$, the statement is true for $n = 1$.

Step 2 Let k be a positive integer and assume that the statement is true for $n = k$, that is, that 4 is a factor of $7^k - 3^k$. Let us denote the other factor by D, so that the induction hypothesis is: $7^k - 3^k = 4D$. We must use this assumption to prove that the statement is true for $n = k + 1$, that is, that 4 is a factor of $7^{k+1} - 3^{k+1}$. Here is the proof:

$$
\begin{aligned}
7^{k+1} - 3^{k+1} &= 7^{k+1} - 7 \cdot 3^k + 7 \cdot 3^k - 3^{k+1} && (\text{since } -7 \cdot 3^k + 7 \cdot 3^k = 0) \\
&= 7(7^k - 3^k) + (7 - 3)3^k && (\text{factor}) \\
&= 7(4D) + (7 - 3)3^k && (\text{induction hypothesis}) \\
&= 7(4D) + 4 \cdot 3^k && (7 - 3 = 4) \\
&= 4(7D + 3^k) && (\text{factor out 4})
\end{aligned}
$$

From this last line, we see that

$$7^{k+1} - 3^{k+1} = 4(7D + 3^k), \quad \text{that is,} \quad 4 \text{ is a factor of } 7^{k+1} - 3^{k+1}$$

Thus the statement is true for $n = k + 1$ and the inductive step is complete. Therefore by the Principle of Mathematical Induction the conjecture is actually true for every positive integer n.

Another example of mathematical induction, the proof of the Binomial Theorem, is given at the end of this section.

Sometimes a statement involving the integer n may be false for $n = 1$ and (possibly) other small values of n, but true for all values of n beyond a particular number. For instance, the statement $2^n > n^2$ is false for $n = 1, 2, 3, 4$. But it is true for $n = 5$ and all larger values of n. A variation on the Principle of Mathematical Induction can be used to prove this fact and similar statements. See Exercise B.11 for details.

A COMMON MISTAKE WITH INDUCTION

It is often tempting to omit step 2 of an inductive proof when the given statement can easily be verified for small values of n, especially if a clear pattern seems to be developing. In fact, we have omitted step 2 in earlier sections of this book when presenting plausible results such as DeMoivre's Theorem and the Binomial Theorem. Although this approach simplified the discussion of those results, it may have left some students with the mistaken impression that step 2 is not really *necessary* in such cases. As the next example shows, however, the truth of a statement for many small values of n does *not* guarantee that it is true for every positive integer n. In other words, *omitting step 2 may lead to error.*

EXAMPLE A positive integer is said to be **prime** if its only positive integer factors are itself and 1. For instance, 11 is prime since its only positive integer factors are 11 and 1. But 15 is not prime since it has factors other than 15 and 1, namely, 3 and 5. For each positive integer n, consider the number

$$f(n) = n^2 - n + 11$$

You can readily verify that

$$f(1) = 11, \quad f(2) = 13, \quad f(3) = 17, \quad f(4) = 23, \quad f(5) = 31$$

and that *each of these numbers is prime.* Furthermore, as n increases there is a clear pattern: the first two numbers differ by 2, the next two differ by 4, the next two differ by 6, the next two by 8, and so on. Based on this evidence, we might make the conjecture:

For each positive integer n, the number $f(n) = n^2 - n + 11$ is prime

We have seen that this conjecture is true for $n = 1, 2, 3, 4, 5$. Unfortunately, however, it is *false* for some values of n. For instance, when $n = 11$,

$$f(11) = 11^2 - 11 + 11 = 11^2 = 121$$

But 121 is obviously *not* prime since it has a factor other than 121 and 1, namely, 11. You can verify that the statement is also false for $n = 12$ but true for $n = 13$.

In the preceding example, the proposition

If the statement is true for $n = k$, then it is true for $n = k + 1$

is false when $k = 10$ and $k + 1 = 11$. If you were not aware of this and tried to complete step 2 of an inductive proof, you would not have been able to find a valid proof for it. Of course, the fact that you can't find a proof of a proposition doesn't always mean that no proof exists. But when you are unable to complete step 2, you are warned that there is a possibility that the given statement may be false for some values of n. This warning should prevent you from drawing any wrong conclusions.

PROOF OF THE BINOMIAL THEOREM (Optional)

We shall use induction to prove that for every positive integer n,

$$(x + y)^n = x^n + \binom{n}{1}x^{n-1}y + \binom{n}{2}x^{n-2}y^2 + \binom{n}{3}x^{n-3}y^3 + \cdots + \binom{n}{n-1}xy^{n-1} + y^n$$

This theorem was discussed and its notation explained in Section 3.

Step 1. When $n = 1$, there are only two terms on the right side of the equation above and the statement reads $(x + y)^1 = x^1 + y^1$. This is certainly true.

Step 2. Let k be any positive integer and assume that the theorem is true for $n = k$, that is, that

$$(x + y)^k$$
$$= x^k + \binom{k}{1}x^{k-1}y + \binom{k}{2}x^{k-2}y^2 + \cdots + \binom{k}{r}x^{k-r}y^r + \cdots + \binom{k}{k-1}xy^{k-1} + y^k$$

On the right side above, we have included a typical middle term $\binom{k}{r}x^{k-r}y^r$: the sum of the exponents is k and the bottom part of the binomial coefficient is the same as the y-exponent. We shall use this assumption to prove that the theorem is true for $n = k + 1$, that is, that

$$(x + y)^{k+1}$$
$$= x^{k+1} + \binom{k+1}{1}x^k y + \binom{k+1}{2}x^{k-1}y^2 + \cdots + \binom{k+1}{r+1}x^{k-r}y^{r+1} + \cdots$$

$$+ \binom{k+1}{k}xy^k + y^{k+1}$$

We have simplified some of the terms on the right side; for instance, $(k + 1) - 1 = k$ and $(k + 1) - (r + 1) = k - r$. But this is the correct statement for $n = k + 1$: the coefficients of the middle terms are $\binom{k+1}{1}, \binom{k+1}{2}, \binom{k+1}{3}$, and so on; the sum of the exponents of each middle term is $k + 1$ and the bottom part of each binomial coefficient is the same as the y-exponent.

In order to prove the theorem for $n = k + 1$, we shall need this fact about binomial coefficients: for any integers r and k with $0 \leq r < k$,

$$(^\circ) \qquad\qquad \binom{k}{r+1} + \binom{k}{r} = \binom{k+1}{r+1}$$

A proof of this fact is outlined in Exercise C.1 on page 539.

To prove the theorem for $n = k + 1$, we first note that

$$(x + y)^{k+1} = (x + y)(x + y)^k$$

Applying the induction hypothesis to $(x + y)^k$, we see that

$$(x + y)^{k+1}$$
$$= (x + y)\left[x^k + \binom{k}{1}x^{k-1}y + \binom{k}{2}x^{k-2}y^2 + \cdots + \binom{k}{r}x^{k-r}y^r + \binom{k}{r+1}x^{k-(r+1)}y^{r+1} \right.$$
$$\left. + \cdots + \binom{k}{k-1}xy^{k-1} + y^k \right]$$
$$= x\left[x^k + \binom{k}{1}x^{k-1}y + \cdots + y^k \right] + y\left[x^k + \binom{k}{1}x^{k-1}y + \cdots + y^k \right]$$

Next we multiply out the right-hand side. Remember that multiplying by x increases the x-exponent by 1 and multiplying by y increases the y-exponent by 1.

$$(x + y)^{k+1}$$
$$= \left[x^{k+1} + \binom{k}{1}x^k y + \binom{k}{2}x^{k-1}y^2 + \cdots + \binom{k}{r}x^{k-r+1}y^r + \binom{k}{r+1}x^{k-r}y^{r+1} \right.$$
$$\left. + \cdots + \binom{k}{k-1}x^2 y^{k-1} + xy^k \right] + \left[x^k y + \binom{k}{1}x^{k-1}y^2 + \binom{k}{2}x^{k-2}y^3 \right.$$

$$+ \cdots + \binom{k}{r}x^{k-r}y^{r+1} + \binom{k}{r+1}x^{k-(r+1)}y^{r+2} + \cdots$$

$$+ \binom{k}{k-1}xy^k + y^{k+1}\Big]$$

$$= x^{k+1} + \Big[\binom{k}{1} + 1\Big]x^ky + \Big[\binom{k}{2} + \binom{k}{1}\Big]x^{k-1}y^2 + \cdots$$

$$+ \Big[\binom{k}{r+1} + \binom{k}{r}\Big]x^{k-r}y^{r+1} + \cdots + \Big[1 + \binom{k}{k+1}\Big]xy^k + y^{k+1}$$

Now apply statement (*) above to each of the coefficients of the middle terms. For instance, with $r = 1$, statement (*) shows that $\binom{k}{2} + \binom{k}{1} = \binom{k+1}{2}$. Similarly, with $r = 0$, $\binom{k}{1} + 1 = \binom{k}{1} + \binom{k}{0} = \binom{k+1}{1}$; and so on. Then the expression above for $(x + y)^{k+1}$ becomes:

$$(x + y)^{k+1} = x^{k+1} + \binom{k+1}{1}x^ky + \binom{k+1}{2}x^{k-1}y^2$$

$$+ \cdots + \binom{k+1}{r+1}x^{k-r}y^{r+1} + \cdots + \binom{k+1}{k}xy^k + y^{k+1}$$

Since this last statement says the theorem is true for $n = k + 1$, the inductive step is complete. By the Principle of Mathematical Induction the theorem is true for every positive integer n.

EXERCISES

Directions for Exercises B.1–B.7 Use mathematical induction to prove that each of the given statements is true for every positive integer n.

B.1. (a) $1 + 2 + 2^2 + 2^3 + 2^4 + \cdots + 2^{n-1} = 2^n - 1$

(b) $1 + 3 + 3^2 + 3^3 + 3^4 + \cdots + 3^{n-1} = \dfrac{3^n - 1}{2}$

(c) $1 + 3 + 5 + 7 + \cdots + (2n - 1) = n^2$

(d) $2 + 4 + 6 + 8 + \cdots + 2n = n^2 + n$

(e) $1^2 + 2^2 + 3^2 + \cdots + n^2 = \dfrac{n(n + 1)(2n + 1)}{6}$

B.2. (a) $\dfrac{1}{2} + \dfrac{1}{4} + \dfrac{1}{8} + \cdots + \dfrac{1}{2^n} = 1 - \dfrac{1}{2^n}$

(b) $\dfrac{1}{1 \cdot 2} + \dfrac{1}{2 \cdot 3} + \dfrac{1}{3 \cdot 4} + \cdots + \dfrac{1}{n(n + 1)} = \dfrac{n}{n + 1}$

B.3. $\left(1 + \dfrac{1}{1}\right)\left(1 + \dfrac{1}{2}\right)\left(1 + \dfrac{1}{3}\right) \cdots \left(1 + \dfrac{1}{n}\right) = n + 1$

B.4. (a) $n + 2 > n$ (d) $3^n \geq 1 + 2n$
(b) $2n + 2 > n$ (e) $3n > n + 1$
(c) $3^n \geq 3n$ (f) $(\frac{3}{2})^n > n$

B.5. Let c and d be fixed real numbers. Prove that $c + (c + d) + (c + 2d) +$

$(c + 3d) + \cdots + (c + (n - 1)d) = \dfrac{n(2c + (n - 1)d)}{2}.$

B.6. Let r be a fixed real number with $r \neq 1$. Prove that $1 + r + r^2 + r^3 +$

$\cdots + r^{n-1} = \dfrac{r^n - 1}{r - 1}.$ (Remember that $1 = r^0$; so when $n = 1$ the left side
reduces to $r^0 = 1$.)

B.7. (a) 3 is a factor of $2^{2n+1} + 1$ (c) 64 is a factor of $3^{2n+2} - 8n - 9$
(b) 5 is a factor of $2^{4n-2} + 1$ (d) 64 is a factor of $9^n - 8n - 1$

B.8. (a) Write *each* of $x^2 - y^2$, $x^3 - y^3$, and $x^4 - y^4$ as a product of $x - y$ and
another factor.
(b) Make a conjecture as to how $x^n - y^n$ can be written as a product of $x - y$
and another factor. Use induction to prove your conjecture.

B.9. Let $x_1 = \sqrt{2}$; $x_2 = \sqrt{2 + \sqrt{2}}$; $x_3 = \sqrt{2 + \sqrt{2 + \sqrt{2}}}$; and so on. Prove that
$x_n < 2$ for every positive integer n.

B.10. Is the given statement true or false? If it is true, prove it. If it is false, give a
counterexample.
(a) Every odd positive integer is prime.
(b) The number $n^2 + n + 17$ is prime for every positive integer n.
(c) $(n + 1)^2 > n^2 + 1$ for every positive integer n.
(d) 3 is a factor of the number $n^3 - n + 3$ for every positive integer n.
(e) 4 is a factor of the number $n^4 - n + 4$ for every positive integer n.

B.11. Let q be a *fixed* integer. Suppose a statement involving the integer n has these
two properties:

(i) The statement is true for $n = q$.
(ii) *If* the statement is true for $n = k$ (where k is an integer with $k \geq q$), then
the statement is also true for $n = k + 1$.

Then we can claim that the statement is true for every integer n greater than or
equal to q.

(a) Give an informal explanation that shows why the claim above should be
valid. Note that when $q = 1$, this claim is precisely the Principle of Mathematical Induction.
(b) The claim made above will be called the Extended Principle of Mathematical Induction. State the two steps necessary to use this principle to
prove that a given statement is true for all $n \geq q$. (See the discussion on
pages 542–543.)

(c)   Use the Extended Principle of Mathematical Induction to prove that $2n - 4 > n$ for every $n \geq 5$. (Use 5 for q here.)

(d)   Let r be a fixed real number with $r > 1$. Prove that $(1 + r)^n > 1 + nr$ for every integer $n \geq 2$. (Use 2 for q here.)

In parts (e)–(h) prove that the given statement is true for the indicated integer values of n.

(e) $n^2 > n$ for all $n \geq 2$

(f) $2^n > n^2$ for all $n \geq 5$

(g) $3^n > 2^n + 10n$ for all $n \geq 4$

(h) $2n < n!$ for all $n \geq 4$

C.1.   Let n be a positive integer. Suppose that there are three pegs and on one of them n rings are stacked, with each ring being smaller in diameter than the one below it (see Figure 9-2).

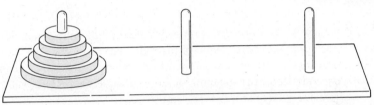

Figure 9-2

We want to transfer the stack of rings to another peg according to these rules: (i) only one ring may be moved at a time; (ii) a ring can be moved to any peg, provided it is never placed on top of a smaller ring; (iii) the final order of the rings on the new peg must be the same as the original order on the first peg.

(a)   What is the smallest possible number of moves when $n = 2$? $n = 3$? $n = 4$?

(b)   Make a conjecture as to the smallest possible number of moves required for any n. Prove your conjecture by induction.

C.2.   The basic formula for compound interest $T(x) = P(1 + r)^x$ was discussed on page 266. Prove by induction that the formula is valid whenever x is a positive integer. (*Note: P* and *r* are assumed to be constant.)

C.3.   Use induction to prove DeMoivre's Theorem: for any complex number $z = r(\cos \theta + i \sin \theta)$ and any positive integer n,

$$z^n = r^n(\cos (n\theta) + i \sin (n\theta))$$

C.4.   Let r and n be integers with $0 \leq r \leq n$. Prove that $\binom{n}{r}$ is an integer. [*Hint:* use induction to prove this statement for every positive integer n: the number $\binom{n}{r}$ is an integer for all r with $0 \leq r \leq n$. For $n = 1$, this statement says that $\binom{1}{0}$ and $\binom{1}{1}$ are integers. Exercise C.1(c) on p. 539 is needed for the inductive step.]

Algebra Review

Here is a review of the fundamental algebraic facts and techniques that are used frequently in this book and in calculus. You must be able to handle these various algebraic manipulations quickly and efficiently.

ARITHMETIC

Here is a brief review of the important properties of addition, subtraction, multiplication, and division of real numbers. Some of the following properties (or "laws") have names. The names are sometimes convenient for reference purposes, but the important thing is that you understand the meaning of each of these properties.

NEGATIVES

For each real number c, the negative of c is denoted $-c$ and

$$c + (-c) = 0$$

Negatives have the following properties:

$$(i) \qquad -(-c) = c$$
$$(i) \qquad (-1) \cdot c = -c$$
$$(iii) \quad b + (-c) = b - c$$
$$(iv) \quad -(b + c) = -b - c$$
$$(v) \quad b - (-c) = b + c$$

EXAMPLES (i) $\quad -(-5) = 5$
$\qquad\qquad\qquad -(-\frac{3}{2}) = \frac{3}{2}$

EXAMPLES (ii) $\quad (-1) \cdot 5 = -5$
$\qquad\qquad\qquad (-1)(-6) = 6 = -(-6)$

EXAMPLES (iii) $\quad 5 + (-7) = 5 - 7 = -2$
$\qquad\qquad\qquad -2 + (-3) = -2 - 3 = -5$

EXAMPLES (iv) $\quad -(5 + 9) = -5 - 9 = -14$
$\qquad\qquad\qquad -(-7 + 3) = -(-7) - 3 = 7 - 3 = 4$

550

EXAMPLES (v) $3 - (-2) = 3 + 2 = 5$

$\frac{3}{4} - (-\frac{19}{4}) = \frac{3}{4} + \frac{19}{4} = \frac{22}{4} = \frac{11}{2}$

SIGNS

If b *and* c *are any real numbers, then*

(*i*) $(-b)(-c) = bc$

(*ii*) $(-b)c = b(-c) = -(bc)$

(*iii*) $\dfrac{-b}{-c} = \dfrac{b}{c}$ $(c \neq 0)$

(*iv*) $\dfrac{-b}{c} = \dfrac{b}{-c} = -\dfrac{b}{c}$ $(c \neq 0)$

EXAMPLES (i) $(-4)(-6) = 24 = 4 \cdot 6$

$(-\frac{3}{8})(-\frac{9}{4}) = \frac{27}{32} = \frac{3}{8} \cdot \frac{9}{4}$

EXAMPLES (ii) $(-3)5 = 3(-5) = -(3 \cdot 5) = -15$

$(-\frac{1}{2})(\frac{3}{5}) = \frac{1}{2}(-\frac{3}{5}) = -(\frac{1}{2} \cdot \frac{3}{5}) = -\frac{3}{10}$

EXAMPLES (iii) $\dfrac{-11}{-6} = \dfrac{11}{6}$

$\dfrac{-\pi}{-2} = \dfrac{\pi}{2}$

EXAMPLES (iv) $\dfrac{-2}{7} = \dfrac{2}{-7} = -\dfrac{2}{7}$

$\dfrac{-\pi}{4} = \dfrac{\pi}{-4} = -\dfrac{\pi}{4}$

COMMUTATIVE LAWS

If b *and* c *are real numbers, then*

$b + c = c + b$ *and* $bc = cb$

EXAMPLES $3 + 5 = 8 = 5 + 3$

$\pi + (-6) = -6 + \pi$

EXAMPLES $3 \cdot 5 = 15 = 5 \cdot 3$

$(-7)6 = -42 = 6(-7)$

There is no commutative law for subtraction, because order *does* make a difference in subtraction. For instance,

$$5 - 2 = 3, \qquad \text{but} \qquad 2 - 5 = -3$$

ASSOCIATIVE LAWS

If b, c, *and* d *are real numbers, then*

$$b + (c + d) = (b + c) + d \qquad and \qquad b(cd) = (bc)d$$

EXAMPLES

$3 + (5 + 7) = (3 + 5) + 7 \qquad$ since $3 + (5 + 7) = 3 + 12 = 15$

$\qquad\qquad\qquad\qquad\qquad\qquad$ and $\qquad (3 + 5) + 7 = 8 + 7 = 15$

$\quad 2(3 \cdot 5) = (2 \cdot 3)5 \qquad$ since $2(3 \cdot 5) = 2 \cdot 15 = 30 \qquad$ and $\qquad (2 \cdot 3)5 = 6 \cdot 5 = 30$

DISTRIBUTIVE LAWS

If b, c, *and* d *are real numbers, then*

$$b(c + d) = bc + bd \qquad and \qquad (c + d)b = cb + db$$
$$b(c - d) = bc - bd \qquad and \qquad (c - d)b = cb - db$$

It really isn't necessary to state four versions of the distributive law. The first one includes all the rest as special cases since $b(c + d) = (c + d)b$ by commutativity and $c + (-d) = c - d$ by the properties of negatives.

EXAMPLES

$3(7 + 4) = 3 \cdot 7 + 3 \cdot 4 \qquad$ since $3(7 + 4) = 3 \cdot 11 = 33$

$\qquad\qquad\qquad\qquad\qquad\qquad$ and $\qquad 3 \cdot 7 + 3 \cdot 4 = 21 + 12 = 33$

$(3 - 7)6 = 3 \cdot 6 + (-7) \cdot 6 \qquad$ since $(3 - 7)6 = (-4)6 = -24$

$\qquad\qquad\qquad\qquad\qquad\qquad$ and $\qquad 3 \cdot 6 + (-7)6 = 18 - 42 = -24$

IDENTITIES

For every real number b

$$b + 0 = b \qquad and \qquad b \cdot 1 = b$$

EXAMPLES $\quad (-7) + 0 = -7 \qquad$ and $\qquad (-7) \cdot 1 = -7.$

ZERO PRODUCTS

(i) *For every real number b,*
$$b \cdot 0 = 0 \cdot b$$

(ii) *If a product is zero, then at least one of the factors is zero; in other words,*
$$\text{if } cd = 0, \qquad \text{then } c = 0 \qquad \text{or} \qquad d = 0 \quad (\text{or both})$$

An immediate consequence of statement (ii) is that

A product of nonzero factors is nonzero; that is,

$$\text{if } c \neq 0 \qquad \text{and} \qquad d \neq 0, \qquad \text{then } cd \neq 0$$

Finally, you should remember that the quotient of two real numbers c/d is defined only when $d \neq 0$. Expressions such as $1/0$, $\pi/0$, $0/0$, $-3/0$, and so on, have no meaning; they are *not* real numbers. In other words,

Division by 0 is not defined.

There are good reasons for excluding division by zero. Some of them are considered in Exercise C.2.

EXPONENTS

Exponents provide a shorthand notation for certain products. If b is a real number, then

The product bb is written b^2; the product bbb is written b^3

Similarly, for any positive integer n,

The product $b \cdot b \cdot b \cdots b$ (n *factors*) *is written* b^n.

In this notation b^1 is just b, so we usually omit the exponent 1. It will be convenient later if we now define b^0, when $b \neq 0$:

> If $b \neq 0$, then b^0 is defined to be the number 1.

Note that 0^0 is not defined.

EXAMPLE $3^4 = 3 \cdot 3 \cdot 3 \cdot 3 = 9 \cdot 3 \cdot 3 = 27 \cdot 3 = 81.$

EXAMPLE $a^3 b^4 = aaabbbb.$

EXAMPLE $3^0 = 1; \ 7^0 = 1; \ \pi^0 = 1; \ a^0 = 1 \ (a \neq 0).$

Since exponents are just shorthand, it is fairly easy to figure out the rules they obey, as we now see.

> In order to find the product of c^m and c^n, add the exponents:
> $$c^m \cdot c^n = c^{m+n}$$

EXAMPLE $3^2 \cdot 3^3 = (3 \cdot 3)(3 \cdot 3 \cdot 3) = 3 \cdot 3 \cdot 3 \cdot 3 \cdot 3 = 3^5 = 3^{2+3}.$

EXAMPLE $a^7 a^{16} = a^{7+16} = a^{23}.$

EXAMPLE $7^2 \cdot 3^2 \cdot 7^6 \cdot 3^7 = 7^2 \cdot 7^6 \cdot 3^2 \cdot 3^7 = (7^{2+6})(3^{2+7}) = 7^8 \cdot 3^9.$

> In order to compute $(c^m)^n$, multiply the exponents:
> $$(c^m)^n = c^{mn}$$

EXAMPLE $(5^3)^2 = (5^3)(5^3) = (5 \cdot 5 \cdot 5)(5 \cdot 5 \cdot 5) = 5^6 = 5^{3 \cdot 2}.$

EXAMPLE $((-7)^4)^5 = (-7)^{20}.$

> In order to find the quotient, when c^m is divided by c^n, subtract the exponents:
> $$\frac{c^m}{c^n} = c^{m-n} \quad (c \neq 0)$$

EXAMPLES $3^5/3^2 = 3 \cdot 3 \cdot 3 \cdot 3 \cdot 3/3 \cdot 3 = 3 \cdot 3 \cdot 3 = 3^3 = 3^{5-2}$ and $7^6/7^6 = 7^{6-6} = 7^0 = 1$.

EXAMPLE $d^{12}/d^7 = d^{12-7} = d^5$ $(d \neq 0)$.

ALGEBRAIC EXPRESSIONS

The key idea in algebra is to use letters (or other symbols) to stand for numbers. For example, consider the statement

The sum of one number and 3 times the square of a second number is 12

If we denote the two numbers by b and c, then this statement can be written:

$$b + 3c^2 = 12$$

Similarly,

$\dfrac{x^2 + y}{2y + z} + 7\pi$ indicates that the sum of the square of the number x and
the number y is to be divided by the sum of 2 times y and the number z;
then the result is to be added to the product of 7 and π

Expressions such as

$$b + 3c^2 \qquad \text{and} \qquad \frac{x^2 + y}{2y + z} + 7\pi$$

are called **algebraic expressions.** For now you should interpret all algebraic expressions as statements about numbers, in which some of the numbers are denoted by letters.

It is essential that you be able to do arithmetic with algebraic expressions. Since such expressions are just shorthand for numbers, *the same rules of arithmetic discussed above for numbers are also valid for algebraic expressions.* Here are some examples. Note that it is usually necessary to rearrange and "collect similar terms" in order to put the final answer in as simple a form as possible.

ADDITION AND SUBTRACTION OF ALGEBRAIC EXPRESSIONS

EXAMPLE $(a^2b - 3c) + (4ab + 7c - 6ab) + 7a^2b$
$$= a^2b - 3c + 4ab + 7c - 6ab + 7a^2b$$
$$= (a^2b + 7a^2b) + (-3c + 7c) + (4ab - 6ab)$$
$$= a^2b(1 + 7) + (-3 + 7)c + (4 - 6)ab$$
$$= 8a^2b + 4c - 2ab$$

You should be careful about minus signs and subtraction of terms in parentheses. For instance,

EXAMPLE $-(b + 3) = -b - 3$, and *not* $-b + 3$.

EXAMPLE $-(c + 7 - y) = -c - 7 - (-y) = -c - 7 + y.$

EXAMPLE $(x^4 - 6x^2 + x - 11) - (3x^2 - 5x + 2)$
$$= x^4 - 6x^2 + x - 11 - 3x^2 + 5x - 2$$
$$= x^4 + (-6x^2 - 3x^2) + (x + 5x) + (-11 - 2)$$
$$= x^4 + (-6 - 3)x^2 + (1 + 5)x + (-13)$$
$$= x^4 - 9x^2 + 6x - 13.$$

In actual practice, addition and subtraction of algebraic expressions is seldom done in such detail. You usually do many of the intermediate steps in your head.

MULTIPLICATION OF ALGEBRAIC EXPRESSIONS

When multiplying algebraic expressions, the commutative laws are frequently used without explicit mention. For example,

$$(2x)(3x) = 2 \cdot 3 \cdot x \cdot x = 6x^2 \qquad \text{or} \qquad 7b(4a) = 28ab$$

One way to multiply two expressions such as $(a + b)$ and $(c + d + e)$ is to use the distributive laws repeatedly:

$$(a + b)(c + d + e) = a(c + d + e) + b(c + d + e)$$
$$= ac + ad + ae + bc + bd + be$$

What this amounts to in practice is to

> Multiply every term in the first expression by *every* term in the second expression and add the results

EXAMPLE $(7 + 2)(3 + 4) = 7(3 + 4) + 2(3 + 4) = 7 \cdot 3 + 7 \cdot 4 + 2 \cdot 3 + 2 \cdot 4$
$$= 21 + 28 + 6 + 8 = 63$$

EXAMPLE $(2x - 5)(3x + 4) = 2x(3x + 4) - 5(3x + 4)$
$$= 2x \cdot 3x + 2x \cdot 4 - 5 \cdot 3x - 5 \cdot 4$$
$$= 6x^2 + 8x - 15x - 20 = 6x^2 - 7x - 20$$

EXAMPLE $(y - 2)(3y^2 - 7y + 4) = y(3y^2 - 7y + 4) - 2(3y^2 - 7y + 4)$
$$= 3y^3 - 7y^2 + 4y - 6y^2 + 14y - 8$$
$$= 3y^3 - 13y^2 + 18y - 8$$

FACTORING

In a sense, factoring is the reverse of multiplication. In multiplication we take two expressions and find their product. In factoring we start with the *product* and must find the expressions (factors) which multiply together to produce the given product.

One nice thing about factoring is that you can always check your answers. Just multiply the factors you've found to see if the result is the expression you began with.

Nevertheless, it takes a good deal of practice to become reasonably efficient at factoring and even then trial and error plays a role. So don't expect to be good at factoring just by glancing at examples. You will need to practice by doing many exercises as well.

Several multiplication patterns appear quite frequently. You should learn to recognize them in order to improve your factoring skills. Each of these patterns is listed below in a box, followed by several examples.

DIFFERENCE OF SQUARES

$$(c + d)(c - d) = c^2 - d^2$$

EXAMPLE $y^2 - 25 = y^2 - 5^2 = (y + 5)(y - 5)$.

EXAMPLE $t^2 - 6 = t^2 - (\sqrt{6})^2 = (t + \sqrt{6})(t - \sqrt{6})$.

EXAMPLE $3u^2 - 10v^2 = (\sqrt{3}u)^2 - (\sqrt{10}v)^2 = (\sqrt{3}u + \sqrt{10}v)(\sqrt{3}u - \sqrt{10}v)$.

EXAMPLE $x^{10} - y^{10} = (x^5)^2 - (y^5)^2 = (x^5 + y^5)(x^5 - y^5)$.

PERFECT SQUARES

$$(c + d)^2 = (c + d)(c + d) = c^2 + 2cd + d^2$$
$$(c - d)^2 = (c - d)(c - d) = c^2 - 2cd + d^2$$

Notice that each of the "end terms" of $c^2 \pm 2cd + d^2$ is a square.

EXAMPLE $y^2 + 6y + 9 = y^2 + 6y + 3^2 = y^2 + 2 \cdot y \cdot 3 + 3^2 = (y + 3)^2$.

EXAMPLE $4x^2 - 36x + 81 = (2x)^2 - 36x + 9^2 = (2x)^2 - 2(2x)9 + 9^2$
$$= (2x - 9)^2.$$

$$x^2 + (c + d)x + cd = (x + c)(x + d)$$

You should notice two *key facts* about $x^2 + (c + d)x + cd$:

The constant term is the product cd.
The coefficient of x is the sum c + d.

EXAMPLE In order to factor $x^2 + 9x + 18$ as $(x + c)(x + d)$ we must find numbers c and d whose *product* is the constant term 18 and whose *sum* is the coefficient of x, namely, 9. By taking pairs of integers whose product is 18 we find some possibilities for c and d:

c	1	-1	2	-2	3	-3
d	18	-18	9	-9	6	-6
$c+d$	19	-19	11	-11	9	-9

Clearly, we should use $c = 3$ and $d = 6$. We verify that we have factored $x^2 + 9x + 18$ correctly by multiplying:

$$(x + 3)(x + 6) = x^2 + 3x + 6x + 18 = x^2 + 9x + 18$$

EXAMPLE To factor $x^2 - 2x - 24$ we must find numbers c and d whose *product* is the constant term -24 and whose *sum* is the coefficient of x, namely, -2. By taking pairs of integers whose product is -24 we obtain a list of possibilities:

c	1	-1	2	-2	3	-3	4	-4
d	-24	24	-12	12	-8	8	-6	6
$c+d$	-23	23	-10	10	-5	5	-2	2

Clearly, $c = 4$ and $d = -6$ should work. And it does, as you can see by multiplying:

$$(x + 4)(x - 6) = x^2 + 4x - 6x - 24 = x - 2x - 24$$

EXAMPLE To factor $y^2 - 8y + 7$, we want numbers c and d whose product is the constant term 7 and whose sum is -8, the coefficient of y. The obvious choices are -7 and -1:

$$(y - 7)(y - 1) = y^2 - 7y - y + 7 = y^2 - 8y + 7$$

EXAMPLE To factor $z^2 + 6zv - 40v^2$, we look for two numbers whose product is $-40v^2$ and whose sum is $6v$. Trial and error suggests $10v$ and $-4v$:

$$(z + 10v)(z - 4v) = z^2 + 10zv - 4zv - 40v^2 = z^2 + 6zv - 40v^2$$

Here is a similar but slightly more complicated pattern:

$$(ax + c)(bx + d) = abx^2 + (ad + bc)x + cd$$

The key facts about $abx^2 + (ad + bc)x + cd$ are:

The constant term is the product cd.
The coefficient of x^2 is the product ab.
The coefficient of x is the number ad + bc.

EXAMPLE In order to factor $6x^2 + 11x + 4$ as $(ax + c)(bx + d)$, we must find numbers a and b whose product is 6, the coefficient of x^2, and numbers c and d whose product is the constant term 4. Here are some possibilities:

$ab = 6$

a	1	−1	2	−2	3	−3	6	−6
b	6	−6	3	−3	2	−2	1	−1

$cd = 4$

c	1	−1	2	−2	4	−4
d	4	−4	2	−2	1	−1

Notice that for *each* choice of a and b there are *six* possibilities for c and d. Trial and error and a good deal of checking via multiplication finally show that

$$(2x + 1)(3x + 4) = 6x^2 + 3x + 8x + 4 = 6x^2 + 11x + 4$$

EXAMPLE To factor $24y^2 - 14y - 5$, we consider pairs a and b whose product is 24, the coefficient of y^2, and pairs c and d whose product is the constant term -5. For example,

$$a = 3, \quad b = 8, \quad c = 5, \quad d = -1, \quad \text{which yields } (3y + 5)(8y - 1)$$

However, $(3y + 5)(8y - 1) = 24y^2 + 40y - 3y - 5 = 24y^2 + 37y - 5$, which isn't quite right. So we must try other possibilities, such as $a = 6, b = 4, c = -5, d = 1$. This gives

$$(6y - 5)(4y + 1) = 24y^2 - 20y + 6y - 5 = 24y^2 - 14y - 5$$

which is just what we want.

DIFFERENCE OF CUBES

$$(c - d)(c^2 + cd + d^2) = c^3 - d^3$$

EXAMPLE $8 - x^3 = 2^3 - x^3 = (2 - x)(2^2 + 2x + x^2) = (2 - x)(4 + 2x + x^2)$

EXAMPLE $z^3 - 5 = z^3 - (\sqrt[3]{5})^3 = (z - \sqrt[3]{5})(z^2 + \sqrt[3]{5}z + (\sqrt[3]{5})^2)$.

SUM OF CUBES

$$(c + d)(c^2 - cd + d^2) = c^3 + d^3$$

EXAMPLE $x^3 + 125 = x^3 + 5^3 = (x + 5)(x^2 - 5x + 5^2) = (x + 5)(x^2 - 5x + 25)$.

EXAMPLE $x^3 + 8y^3 = x^3 + (2y)^3 = (x + 2y)(x^2 - 2xy + (2y)^2)$
$$= (x + 2y)(x^2 - 2xy + 4y^2).$$

PERFECT CUBES

$$(c - d)^3 = c^3 - 3c^2d + 3cd^2 - d^3$$
$$(c + d)^3 = c^3 + 3c^2d + 3cd^2 + d^3$$

EXAMPLE $x^3 - 12x^2 + 48x - 64$
$$= x^3 - 12x^2 + 48x - 4^3 = x^3 - 3x^2 \cdot 4 + 3x \cdot 4^2 - 4^3 = (x - 4)^3.$$

Factoring expressions involving exponents larger than 3 is usually quite difficult. But sometimes it can be done.

EXAMPLE $x^6 - y^6 = (x^3)^2 - (y^3)^2 = (x^3 + y^3)(x^3 - y^3)$
$$= (x + y)(x^2 - xy + y^2)(x - y)(x^2 + xy + y^2).$$

EXAMPLE $x^8 - 1 = (x^4)^2 - 1 = (x^4 + 1)(x^4 - 1)$
$$= (x^4 + 1)(x^2 + 1)(x^2 - 1) = (x^4 + 1)(x^2 + 1)(x + 1)(x - 1).$$

Certain expressions can be factored by using an appropriate substitution.

EXAMPLE To factor $x^4 - 2x^2 - 3$, let $u = x^2$. Then $x^4 - 2x^2 - 3$
$$= (x^2)^2 - 2x^2 - 3 = u^2 - 2u - 3 = (u - 3)(u + 1) = (x^2 - 3)(x^2 + 1)$$
$$= (x + \sqrt{3})(x - \sqrt{3})(x^2 + 1).$$

EXAMPLE To factor $x^6 + 2x^3 + 1$, let $u = x^3$. Then $x^6 + 2x^3 + 1$
$$= (x^3)^2 + 2x^3 + 1 = u^2 + 2u + 1 = (u + 1)^2 = (x^3 + 1)^2$$
$$= ((x + 1)(x^2 - x + 1))^2 = (x + 1)^2(x^2 - x + 1)^2.$$

Occasionally an expression can be regrouped and the distributive law used to factor out a common factor.

EXAMPLE $3x^3 + 3x^2 + x + 1 = 3x^2(x + 1) + (x + 1) = (3x^2 + 1)(x + 1)$.

EXAMPLE $x^3 + 5x^2 - 3x - 15 = x^2(x + 5) - 3(x + 5) = (x^2 - 3)(x + 5)$
$$= (x + \sqrt{3})(x - \sqrt{3})(x + 5).$$

COMPLETING THE SQUARE

Observe that if we add 9 to the expression $x^2 + 6x$ it becomes a perfect square:

$$x^2 + 6x + 9 = (x + 3)^2$$

Similarly, the expression $x^2 - 7x$ can be made into a perfect square by adding $(\frac{7}{2})^2 = \frac{49}{4}$:

$$x^2 - 7x + (\tfrac{7}{2})^2 = (x - \tfrac{7}{2})^2$$

The process of adding a constant to an algebraic expression of the form $x^2 + bx$ in order to make it a perfect square is called **completing the square.** It is a useful technique in several situations, so you should know how to do it.

In order to see just what number should be added to $x^2 + bx$ in order to get a perfect square, consider what a perfect square looks like:

$$(x + c)^2 = (x + c)(x + c) = x^2 + 2cx + c^2$$

Thus in *any* perfect square, the constant term is of the form c^2 for some real number c. Consequently, in order to make $x^2 + bx$ a perfect square, we must add c^2, with c chosen in such a way that

$$x^2 + bx + c^2 = (x + c)^2 = x^2 + 2cx + c^2$$

The only time this is true is when $b = 2c$; that is, $c = b/2$. In other words, if we add $c^2 = (b/2)^2$ to $x^2 + bx$, we get a perfect square:

$$x^2 + bx + \left(\frac{b}{2}\right)^2 = \left(x + \frac{b}{2}\right)^2 \qquad \text{(check this multiplication!)}$$

In summary,

> *In order to complete the square in the expression* $x^2 + bx$, *add the square of half the coefficient of* x, *namely,* $(b/2)^2$.

EXAMPLE   To complete the square in $x^2 + 9x$, just take half the coefficient of x, namely, $\frac{1}{2}(9) = \frac{9}{2}$, and square it: $(\frac{9}{2})^2 = \frac{81}{4}$. This is the number to add:

$$x^2 + 9x + \tfrac{81}{4} = x^2 + 9x + (\tfrac{9}{2})^2 = (x + \tfrac{9}{2})^2$$

EXAMPLE   To complete the square in the expression $y^2 - 8y$, just add $(\frac{1}{2}(-8))^2 = (-4)^2 = 16$:

$$y^2 - 8y + 16 = y^2 - 8y + (-4)^2 = (y - 4)^2$$

Warning   The technique described above is only valid in expressions where the coefficient of x^2 is 1. It won't work in such expressions as $3x^2 - 15x$. However, in such cases we can factor:

$$3x^2 - 15x = 3(x^2 - 5x)$$

and then complete the square in $x^2 - 5x$ by adding $(-\frac{5}{2})^2 = \frac{25}{4}$:

$$3(x^2 - 5x + \tfrac{25}{4}) = 3(x^2 - 5x + (-\tfrac{5}{2})^2) = 3(x - \tfrac{5}{2})^2$$

FRACTIONS

When we divide one real number by another, the quotient is usually written as a fraction. For example, if 7 is divided by 5, the quotient is $\frac{7}{5}$. Similarly, if the number π is divided by the number $3 + \sqrt{2}$, the quotient is $\pi/(3 + \sqrt{2})$.

As you know, the same number can be expressed as a fraction in many different ways. For instance,

$$\tfrac{1}{2} = \tfrac{2}{4} = \tfrac{7}{14} = \tfrac{16}{32}, \quad \text{and so on} \qquad \tfrac{7}{3} = \tfrac{14}{6} = \tfrac{35}{15}, \quad \text{and so on}$$

There are two basic rules for equality of fractions, the first of which is:

$$\frac{a}{b} = \frac{c}{d} \qquad exactly\ when \qquad ad = bc \qquad (b \neq 0,\ d \neq 0)$$

Here and below, whenever we write fractions such as a/b or c/d we *always* assume $b \neq 0$ and $d \neq 0$. Otherwise, we would just have meaningless expressions $a/0$ or $b/0$.

EXAMPLES $\tfrac{1}{2} = \tfrac{3}{6}$ since $1 \cdot 6 = 2 \cdot 3$; similarly $\tfrac{7}{3} = \tfrac{14}{6}$ since $7 \cdot 6 = 42 = 3 \cdot 14$.

EXAMPLE $\dfrac{rst}{rs^3} = \dfrac{st^2}{s^3 t}$ since $(rst)(s^3 t) = rs^4 t^2 = (rs^3)(st^2)$.

EXAMPLE $\dfrac{x^2 + 2x}{x^2 + x - 2} = \dfrac{x}{x - 1}$ since $(x^2 + 2x)(x - 1) = x^3 + x^2 - 2x$

$$= (x^2 + x - 2)x.$$

CANCELLATION

For every nonzero number k,

$$\frac{ka}{kb} = \frac{a}{b} \qquad (b \neq 0).$$

On an informal level we usually say that we **cancel** the k in the top and bottom of the fraction. Notice that cancellation is just a special case of the preceding rule since $(ka)b = (kb)a$, so that $ka/kb = a/b$.

EXAMPLES $\dfrac{2}{4} = \dfrac{2 \cdot 1}{2 \cdot 2} = \dfrac{1}{2}$; $\dfrac{27}{15} = \dfrac{3 \cdot 9}{3 \cdot 5} = \dfrac{9}{5}$; $\dfrac{7 \cdot 9 \cdot 59}{7 \cdot 9 \cdot 71} = \dfrac{9 \cdot 59}{9 \cdot 71} = \dfrac{59}{71}$.

EXAMPLE $\dfrac{x^4 - 1}{x^2 + 1} = \dfrac{(x^2 + 1)(x^2 - 1)}{(x^2 + 1)} = \dfrac{(x^2 - 1)}{1} = x^2 - 1.$

A fraction is said to be in **lowest terms** if its numerator (top) and denominator (bottom) have no common factors. Cancellation is used to express a particular fraction in lowest terms.

EXAMPLE $\frac{8}{14}$ is *not* in lowest terms since 2 is a factor of both 8 and 14. In order to express $\frac{8}{14}$ in lowest terms, we cancel the common factor 2: $\frac{8}{14} = \frac{2 \cdot 4}{2 \cdot 7} = \frac{4}{7}$. The fraction $\frac{4}{7}$ *is* in lowest terms since 4 and 7 have no common factors (except, of course, the trivial factor 1).

EXAMPLE In order to express $\frac{x^2 + x - 6}{x^2 - 3x + 2}$ in lowest terms, we first factor both top and bottom to see if there is a common factor which can be canceled:

$$\frac{x^2 + x - 6}{x^2 - 3x + 2} = \frac{(x - 2)(x + 3)}{(x - 2)(x - 1)} = \frac{x + 3}{x - 1}$$

The fraction $(x + 3)/(x - 1)$ is in lowest terms..

ADDITION AND SUBTRACTION OF FRACTIONS

Adding (or subtracting) two fractions which have the same denominator (bottom) is easy. It amounts to using the distributive law as follows:

$$\frac{a}{b} + \frac{c}{b} = a\left(\frac{1}{b}\right) + c\left(\frac{1}{b}\right) = (a + c)\left(\frac{1}{b}\right) = \frac{a + c}{b} \qquad (b \neq 0)$$

Subtraction is done similarly. In summary,

$$\boxed{\frac{a}{b} + \frac{c}{b} = \frac{a + c}{b} \qquad and \qquad \frac{a}{b} - \frac{c}{b} = \frac{a - c}{b} \qquad (b \neq 0)}$$

EXAMPLES $\frac{7}{3} + \frac{13}{3} = \frac{7 + 13}{3} = \frac{20}{3}$ and $\frac{4}{5} - \frac{7}{5} = \frac{4 - 7}{5} = \frac{-3}{5}$.

EXAMPLE $\frac{7x^2 + 2}{x^2 + 3} - \frac{4x^2 + 2x - 5}{x^2 + 3}$

$$= \frac{(7x^2 + 2) - (4x^2 + 2x - 5)}{x^2 + 3}$$

$$= \frac{7x^2 + 2 - 4x^2 - 2x + 5}{x^2 + 3} = \frac{3x^2 - 2x + 7}{x^2 + 3}.$$

In order to add or subtract two fractions with different denominators (bottoms), we must first find a **common denominator;** that is, we must express both fractions with the same number on the bottom.

EXAMPLE   In order to add $\frac{1}{2}$ and $\frac{2}{3}$ we first write $\frac{1}{2} = \frac{3}{6}$ and $\frac{2}{3} = \frac{4}{6}$. Then

$$\frac{1}{2} + \frac{2}{3} = \frac{3}{6} + \frac{4}{6} = \frac{3+4}{6} = \frac{7}{6}$$

One way to find a common denominator for a/b and c/d (with $b \neq 0$, $d \neq 0$) is to use the product of the two denominators bd. Note that

$$\frac{a}{b} = \frac{ad}{bd} \qquad \text{and} \qquad \frac{c}{d} = \frac{bc}{bd}$$

Consequently,

$$\frac{a}{b} + \frac{c}{d} = \frac{ad}{bd} + \frac{bc}{bd} = \frac{ad + bc}{bd} \qquad (b \neq 0, \quad d \neq 0)$$

$$\frac{a}{b} - \frac{c}{d} = \frac{ad}{bd} - \frac{bc}{bd} = \frac{ad - bc}{bd} \qquad (b \neq 0, \quad d \neq 0)$$

EXAMPLE   In order to find $\frac{6}{5} + \frac{4}{7}$, we use $7 \cdot 5 = 35$ as a denominator:

$$\frac{6}{5} + \frac{4}{7} = \frac{6 \cdot 7}{5 \cdot 7} + \frac{5 \cdot 4}{5 \cdot 7} = \frac{42}{35} + \frac{20}{35} = \frac{42 + 20}{35} = \frac{62}{35}$$

EXAMPLE
$$\frac{2x + 1}{3x} - \frac{x^2 - 2}{x - 1} = \frac{(2x + 1)(x - 1)}{3x(x - 1)} - \frac{3x(x^2 - 2)}{3x(x - 1)}$$

$$= \frac{(2x + 1)(x - 1) - 3x(x^2 - 2)}{3x(x - 1)} = \frac{2x^2 - x - 1 - 3x^3 + 6x}{3x^2 - 3x}$$

$$= \frac{-3x^3 + 2x^2 + 5x - 1}{3x^2 - 3x}.$$

EXAMPLE   In order to find $\dfrac{1}{z} + \dfrac{3z}{z + 1} - \dfrac{z^2}{(z + 1)^2}$ we could use the product $z(z + 1)(z + 1)^2$ as denominator. But it will be more efficient to use just $z(z + 1)^2$, since all three fractions can be expressed with this denominator:

$$\frac{1}{z} = \frac{(z + 1)^2}{z(z + 1)^2}, \qquad \frac{3z}{z + 1} = \frac{3z \cdot z(z + 1)}{(z + 1)z(z + 1)} = \frac{3z^2(z + 1)}{z(z + 1)^2},$$

$$\frac{z^2}{(z + 1)^2} = \frac{z^2 \cdot z}{(z + 1)^2 z} = \frac{z^3}{z(z + 1)^2}$$

Therefore

$$\frac{1}{z} + \frac{3z}{z+1} - \frac{z^2}{(z+1)^2} = \frac{(z+1)^2}{z(z+1)^2} + \frac{3z^2(z+1)}{z(z+1)^2} - \frac{z^3}{z(z+1)^2}$$

$$= \frac{(z+1)^2 + 3z^2(z+1) - z^3}{z(z+1)^2} = \frac{z^2 + 2z + 1 + 3z^3 + 3z^2 - z^3}{z(z+1)^2}$$

$$= \frac{2z^3 + 4z^2 + 2z + 1}{z(z+1)^2} = \frac{2z^3 + 4z^2 + 2z + 1}{z^3 + 2z^2 + z}$$

MULTIPLICATION OF FRACTIONS

Multiplication of fractions is easier than addition since you don't have to find common denominators. Just multiply top by top and bottom by bottom:

$$\frac{a}{b} \cdot \frac{c}{d} = \frac{ac}{bd} \qquad (b \neq 0, \quad d \neq 0)$$

EXAMPLES $\quad \frac{3}{4} \cdot \frac{7}{5} = \frac{3 \cdot 7}{4 \cdot 5} = \frac{21}{20}; \quad \frac{2}{3} \cdot \frac{5}{6} = \frac{2 \cdot 5}{3 \cdot 6} = \frac{10}{18} = \frac{5}{9}.$

EXAMPLE $\quad \frac{x^2-1}{x^2+2} \cdot \frac{3x-4}{x+1} = \frac{(x^2-1)(3x-4)}{(x^2+2)(x+1)}.$ In order to reduce this answer to lowest terms, we do some factoring and canceling:

$$\frac{(x^2-1)(3x-4)}{(x^2+2)(x+1)} = \frac{(x-1)(x+1)(3x-4)}{(x^2+2)(x+1)} = \frac{(x-1)(3x-4)}{x^2+2} = \frac{3x^2-7x+4}{x^2+2}$$

DIVISION OF FRACTIONS

The basic rule for simplifying a complicated fraction, such as $\frac{a/b}{c/d}$ (with $b \neq 0, c \neq 0,$ $d \neq 0$), is *invert the denominator (bottom) and multiply by the numerator (top)*. The two simplest cases are:

$$\frac{a}{\frac{c}{d}} = \frac{a}{1} \cdot \frac{d}{c} = \frac{ad}{c} \qquad and \qquad \frac{\frac{a}{b}}{c} = \frac{a}{b} \cdot \frac{1}{c} = \frac{a}{bc} \qquad (b \neq 0, \quad c \neq 0, \quad d \neq 0)$$

EXAMPLES $\dfrac{7}{3/4} = \dfrac{7}{1} \cdot \dfrac{4}{3} = \dfrac{28}{3}$ and $\dfrac{5/6}{4} = \dfrac{5}{6} \cdot \dfrac{1}{4} = \dfrac{5}{24}.$

EXAMPLE $\dfrac{x+2}{\dfrac{x^2+1}{x-3}} = \dfrac{x+2}{1} \cdot \dfrac{x-3}{x^2+1} = \dfrac{(x+2)(x-3)}{x^2+1} = \dfrac{x^2-x-6}{x^2+1}.$

EXAMPLE $\dfrac{\dfrac{y^2}{y+2}}{y^3+y} = \dfrac{y^2}{y+2} \cdot \dfrac{1}{y^3+y} = \dfrac{y^2}{(y+2)(y^3+y)} = \dfrac{y^2}{(y+2)y(y^2+1)}$

$$= \dfrac{y}{(y+2)(y^2+1)} = \dfrac{y}{y^3+2y^2+y+2}.$$

The same rule (invert and multiply) applies to more complicated fractions:

$$\boxed{\dfrac{\dfrac{a}{b}}{\dfrac{c}{d}} = \dfrac{a}{b} \cdot \dfrac{d}{c} = \dfrac{ad}{bc} \qquad (b \neq 0, \quad c \neq 0, \quad d \neq 0)}$$

EXAMPLES $\dfrac{3/4}{2/3} = \dfrac{3}{4} \cdot \dfrac{3}{2} = \dfrac{9}{8}$ and $\dfrac{7/2}{15/8} = \dfrac{7}{2} \cdot \dfrac{8}{15} = \dfrac{56}{30} = \dfrac{28}{15}.$

EXAMPLE $\dfrac{16y^2z/8yz^2}{yz/6y^3z^3} = \dfrac{16y^2z}{8yz^2} \cdot \dfrac{6y^3z^3}{yz} = \dfrac{16 \cdot 6 \cdot y^5z^4}{8y^2z^3} = 2 \cdot 6 \cdot y^{5-2}z^{4-3} = 12y^3z.$

It is often necessary to simplify complicated algebraic expressions which involve all four operations on fractions (addition, subtraction, multiplication, and division).

EXAMPLE In order to simplify

$$\dfrac{\dfrac{3}{x^2-4} + \dfrac{1}{x+2}}{5 - \dfrac{6}{x-2}}$$

we first use the fact that $x^2 - 4 = (x+2)(x-2)$ to find a common denominator on the top and then continue:

$$\dfrac{\dfrac{3}{x^2-4} + \dfrac{1}{x+2}}{5 - \dfrac{6}{x-2}} = \dfrac{\dfrac{3}{x^2-4} + \dfrac{x-2}{x^2-4}}{\dfrac{5(x-2)}{x-2} - \dfrac{6}{x-2}} = \dfrac{\dfrac{3+x-2}{x^2-4}}{\dfrac{5(x-2)-6}{x-2}} = \dfrac{\dfrac{1+x}{x^2-4}}{\dfrac{5x-16}{x-2}}$$

$$= \frac{1 + x}{x^2 - 4} \cdot \frac{x - 2}{5x - 16}$$

$$= \frac{(1 + x)(x - 2)}{(x + 2)(x - 2)(5x - 16)} = \frac{1 + x}{(x + 2)(5x - 16)} = \frac{x + 1}{5x^2 - 6x - 32}$$

RADICALS

The basic properties of radicals are often useful for simplifying algebraic expressions. These properties are:

$$
\begin{array}{ll}
(\sqrt{a})^2 = a \quad (a \geq 0), & (\sqrt[3]{a})^3 = a \\
\sqrt{ab} = \sqrt{a}\sqrt{b} \quad (a \geq 0, \ b \geq 0), & \sqrt[3]{ab} = \sqrt[3]{a}\sqrt[3]{b} \\
\sqrt{\dfrac{a}{b}} = \dfrac{\sqrt{a}}{\sqrt{b}} \quad (a \geq 0, \ b > 0), & \sqrt[3]{\dfrac{a}{b}} = \dfrac{\sqrt[3]{a}}{\sqrt[3]{b}} \quad (b \neq 0)
\end{array}
$$

Also remember that

$$\sqrt{a + b} \neq \sqrt{a} + \sqrt{b} \quad and \quad \sqrt[3]{a + b} \neq \sqrt[3]{a} + \sqrt[3]{b}$$

EXAMPLE $\sqrt[3]{16x^3y^6} = \sqrt[3]{16}\sqrt[3]{x^3}\sqrt[3]{y^6} = \sqrt[3]{8 \cdot 2}\sqrt[3]{x^3}\sqrt[3]{(y^2)^3} = \sqrt[3]{8}\sqrt[3]{2}\sqrt[3]{x^3}\sqrt[3]{(y^2)^3}$
$$= 2\sqrt[3]{2}xy^2.$$

EXAMPLE $\sqrt{\dfrac{48}{t^{12}u^4}} = \dfrac{\sqrt{48}}{\sqrt{t^{12}u^4}} = \dfrac{\sqrt{16 \cdot 3}}{\sqrt{t^{12}u^4}} = \dfrac{\sqrt{16}\sqrt{3}}{\sqrt{(t^6)^2}\sqrt{(u^2)^2}} = \dfrac{4\sqrt{3}}{t^6u^2}.$

EXAMPLE $(3 + 4\sqrt{2})(7 - \sqrt{2}) = 3(7 - \sqrt{2}) + 4\sqrt{2}(7 - \sqrt{2})$
$$= 21 - 3\sqrt{2} + 28\sqrt{2} - 4\sqrt{2}\sqrt{2}$$
$$= 21 + 25\sqrt{2} - 4 \cdot 2 = 13 + 25\sqrt{2}.$$

EXAMPLE $\dfrac{\left(\dfrac{3x}{\sqrt{x} + \sqrt{y}}\right)^2}{\dfrac{1}{x + y}} = \left(\dfrac{3x}{\sqrt{x} + \sqrt{y}}\right)^2 \cdot \dfrac{(x + y)}{1}$

$$= \frac{9x^2}{x + 2\sqrt{x}\sqrt{y} + y} \cdot \frac{(x + y)}{1} = \frac{9x^3 + 9x^2y}{x + y + 2\sqrt{x}\sqrt{y}}.$$

RATIONALIZING THE DENOMINATOR

It is sometimes convenient to write fractions in a form which does not have any radicals in the denominator. Consider, for example, the fraction $1/\sqrt{2}$. Using the fact that $\sqrt{2}/\sqrt{2} = 1$, we have:

$$\frac{1}{\sqrt{2}} = \frac{1}{\sqrt{2}} \cdot \frac{\sqrt{2}}{\sqrt{2}} = \frac{\sqrt{2}}{2}$$

One advantage of the form $\sqrt{2}/2$ over $1/\sqrt{2}$ becomes apparent when you need to use a decimal approximation for $\sqrt{2}$, say $\sqrt{2} \approx 1.414$. In order to find the decimal approximation of $1/\sqrt{2}$ we must compute $1/1.414$, which requires some nontrivial long division. But if we use $\sqrt{2}/2$ it is easy to see that $\sqrt{2}/2 \approx 1.414/2 = .707$.

Here are some more examples of this process, which is usually called **rationalizing the denominator.** The basic idea is to multiply both numerator and denominator by the "radical quantity" appearing in the denominator:

EXAMPLE $\dfrac{1}{\sqrt{3}} = \dfrac{1}{\sqrt{3}} \cdot \dfrac{\sqrt{3}}{\sqrt{3}} = \dfrac{\sqrt{3}}{3}.$

EXAMPLE $\sqrt{\dfrac{5}{2x+1}} = \dfrac{\sqrt{5}}{\sqrt{2x+1}} = \dfrac{\sqrt{5}}{\sqrt{2x+1}} \cdot \dfrac{\sqrt{2x+1}}{\sqrt{2x+1}} = \dfrac{\sqrt{10x+5}}{2x+1}.$

Sometimes a slightly different procedure is needed to rationalize the denominator.

EXAMPLE   In order to rationalize the denominator of $\dfrac{7}{\sqrt{5}+\sqrt{3}}$, we must multiply both top and bottom by something which will eliminate the radicals in the denominator. Observe that $(\sqrt{5}+\sqrt{3})(\sqrt{5}-\sqrt{3}) = (\sqrt{5})^2 - (\sqrt{3})^2 = 5 - 3 = 2$. Thus

$$\frac{7}{\sqrt{5}+\sqrt{3}} = \left(\frac{7}{\sqrt{5}+\sqrt{3}}\right)\left(\frac{\sqrt{5}-\sqrt{3}}{\sqrt{5}-\sqrt{3}}\right) = \frac{7(\sqrt{5}-\sqrt{3})}{(\sqrt{5}+\sqrt{3})(\sqrt{5}-\sqrt{3})} = \frac{7(\sqrt{5}-\sqrt{3})}{2}$$

EQUATIONS

You should be able to solve simple equations, as illustrated in the examples below. The examples are presented in a careful step-by-step fashion. The reason for each step is listed on the right side of the page. Usually these reasons are just particular cases of the following:

BASIC PRINCIPLES FOR SOLVING EQUATIONS

(i)   *Adding or subtracting the same quantity from both sides of an equation produces an equation with the same solutions as the original equation.*

(ii) *Multiplying or dividing both sides of an equation by the same nonzero quantity produces an equation with the same solutions as the original equation.*

The general idea is to use the basic principles repeatedly to transform a given equation into a simpler equation with the same solutions.

EXAMPLE   Here is the solution to the equation $3x - 6 = 7x + 4$.

$$3x - 6 = 7x + 4$$
$$3x = 7x + 10 \qquad \text{(add 6 to both sides)}$$
$$-4x = 10 \qquad \text{(subtract } 7x \text{ from both sides)}$$
$$x = \frac{10}{-4} = -\frac{5}{2} \qquad \text{(divide both sides by } -4)$$

By using the basic principles, the equation $3x - 6 = 7x + 4$ has been transformed into the equation $x = -\frac{5}{2}$, whose solution is obvious.

EXAMPLE   Here is a step-by-step solution of the equation $\dfrac{t + 7}{2} = \dfrac{5t - 4}{7}$.°

$$\frac{t + 7}{2} = \frac{5t - 4}{7}$$
$$14\left(\frac{t + 7}{2}\right) = 14\left(\frac{5t - 4}{7}\right) \qquad \text{(multiply both sides by 14)}$$
$$7(t + 7) = 2(5t - 4) \qquad \text{(simplify)}$$
$$7t + 49 = 10t - 8 \qquad \text{(multiply out both sides)}$$
$$7t = 10t - 57 \qquad \text{(subtract 49 from both sides)}$$
$$-3t = -57 \qquad \text{(subtract } 10t \text{ from both sides)}$$
$$t = \frac{-57}{-3} = \frac{57}{3} = 19 \qquad \text{(divide both sides by } -3)$$

° Although the letter x is most commonly used to denote the unknown in an equation, other letters may just as well be used (such as t here).

EXAMPLE Temperatures may be measured in two different units, Fahrenheit (F) or Celsius (C). The relationship between the two is given by the formula $F = \frac{9}{5}C + 32$. In order to *express C in terms of F*, we must solve the equation $F = \frac{9}{5}C + 32$ for C:

$$F = \frac{9C}{5} + 32$$

$$F - 32 = \frac{9C}{5} \qquad \text{(subtract 32 from both sides)}$$

$$5(F - 32) = 9C \qquad \text{(multiply both sides by 5)}$$

$$\frac{5(F - 32)}{9} = C, \text{ or equivalently,} \qquad \text{(divide both sides by 9)}$$

$$C = \frac{5}{9}(F - 32)$$

You may have noticed that basic principle (ii) for solving equations specifies that you must multiply or divide only by *nonzero* quantities. In some cases this requires a bit of care.

EXAMPLE The equation $\dfrac{3x + 1}{x - 7} = 5$ only makes sense when $x - 7 \neq 0$, that is, when $x \neq 7$. Consequently, we must keep an *extra condition* in mind when we solve this equation, (namely, $x \neq 7$):

$$\frac{3x + 1}{x - 7} = 5$$

$$3x + 1 = 5(x - 7) \qquad \text{(multiply both sides by } x - 7; \text{ the}$$
$$\qquad\qquad\qquad\qquad \text{quantity } x - 7 \text{ is } nonzero \text{ since } x \neq 7)$$

$$3x + 1 = 5x - 35 \qquad \text{(multiply out right side)}$$

$$-2x + 1 = -35 \qquad \text{(subtract } 5x \text{ from both sides)}$$

$$-2x = -36 \qquad \text{(subtract 1 from both sides)}$$

$$x = 18 \qquad \text{(divide both sides by } -2)$$

Since $x = 18$ certainly satisfies the extra condition that $x \neq 7$, $x = 18$ must be the solution of the original equation.

EXAMPLE The equation $\dfrac{4}{x + 8} = \dfrac{3}{5x - 10}$ only makes sense when $x + 8 \neq 0$ *and* $5x - 10 \neq 0$, that is, when $x \neq -8$ and $5x \neq 10$ (equivalently, $x \neq 2$).

$$\frac{4}{(x + 8)} = \frac{3}{(5x - 10)}$$

$$\frac{4}{x + 8}(x + 8)(5x - 10) = \frac{3}{5x - 10} \cdot (x + 8)(5x - 10) \qquad \begin{array}{l}\text{(multiply both sides by} \\ (x + 8)(5x - 10); \text{ this quan-} \\ \text{tity is } nonzero \text{ since} \\ x \neq -8 \text{ and } x \neq 2)\end{array}$$

$$4(5x - 10) = 3(x + 8) \qquad \text{(simplify)}$$
$$20x - 40 = 3x + 24 \qquad \text{(multiply out both sides)}$$
$$17x - 40 = 24 \qquad \text{(subtract } 3x \text{ from both sides)}$$
$$17x = 64 \qquad \text{(add 40 to both sides)}$$
$$x = \frac{64}{17} \qquad \text{(divide both sides by 17)}$$

Since $x = \frac{64}{17}$ satisfies the extra conditions that $x \neq -8$ and $x \neq 2$, $x = \frac{64}{17}$ must be the solution to the original equation.

EXAMPLE When solving the equation $\dfrac{x - 3}{x - 7} = \dfrac{3x - 17}{x - 7}$ we note that $x - 7$ must be nonzero in order for the equation to be meaningful (that is, $x \neq 7$).

$$\frac{x - 3}{x - 7} = \frac{3x - 17}{x - 7}$$
$$x - 3 = 3x - 17 \qquad \text{(multiply both sides by the nonzero quantity } x - 7)$$
$$-2x - 3 = -17 \qquad \text{(subtract } 3x \text{ from both sides)}$$
$$-2x = -14 \qquad \text{(add 3 to both sides)}$$
$$x = 7 \qquad \text{(divide both sides by } -2)$$

But now there's a problem: $x = 7$ obviously does *not* satisfy the condition $x \neq 7$, which is necessary for the original equation to make sense. Therefore *the original equation has no solution.*

EXAMPLE Solve the following equation *for b:*

$$3b(a + 4t) = 6a - t(a - 3b)$$
$$3ab + 12tb = 6a - ta + 3tb \qquad \text{(multiply out both sides)}$$
$$3ab + 12tb - 3tb = 6a - ta \qquad \text{(subtract } 3tb \text{ from both sides)}$$
$$3ab + 9tb = 6a - ta \qquad \text{(combine like terms)}$$
$$(3a + 9t)b = 6a - ta \qquad \text{(factor out } b)$$
$$b = \frac{6a - ta}{3a + 9t} \qquad \text{(divide both sides by } 3a + 9t)$$

Note that we can only divide by $3a + 9t$ if the quantity $3a + 9t$ is nonzero. Therefore the solution of the original equation is

$$b = \frac{6a - ta}{3a + 9t} \qquad \text{provided } 3a + 9t \neq 0$$

All of the equations discussed above are called **first-degree** (or **linear**) **equations.** This means that the unknown appears only with exponent 1, in contrast, for instance, to the equation $x^3 - 3x^2 + 6x - 5 = x^2 + 2$ in which higher powers of the unknown x occur.

Higher degree equations are discussed in detail in Chapter 4. For now we look at only one very simple type of **second-degree** (or **quadratic**) **equation;** for instance: $x^2 + 5 = 41$. Subtracting 5 from each side produces the equation $x^2 = 36$. A solution of this equation is any number whose square is 36. There are exactly *two* such numbers: $x = \sqrt{36} = 6$ and $x = -\sqrt{36} = -6$.

EXAMPLE   In order to solve $(x + 5)^2 = 18$, we note that $x + 5$ must be a number whose square is 18. The only numbers whose squares are 18 are $\sqrt{18}$ and $-\sqrt{18}$. Therefore

$$x + 5 = \sqrt{18} \quad \text{or} \quad x + 5 = -\sqrt{18}$$

so that

$$x = \sqrt{18} - 5 \quad \text{or} \quad x = -\sqrt{18} - 5$$

Note that $\sqrt{18} = \sqrt{9 \cdot 2} = \sqrt{9} \cdot \sqrt{2} = 3\sqrt{2}$. Thus the solutions of $(x + 5)^2 = 18$ are $x = 3\sqrt{2} - 5$ and $x = -3\sqrt{2} - 5$.

EXERCISES

A.1.   Express each of the following numbers as a single integer.
   (a) $14 - 10 + 3$
   (b) $2 - (3 - 5) + 3 - (-7)$
   (c) $(1 - 8 + 7) - (3 - 5 + 8(2 - 5))$
   (d) $(-5)(6)$
   (e) $(7)(-2)(-5)$
   (f) $(-1)(-2)(-3)(-4)(-5)$
   (g) $(-6)(5 - 1 + 4) - 3.5 + 1.5$
   (h) $(7 - (-6))(4 + (3 - (-7)))$

A.2.   Express each of the following numbers as a single integer.

   (a) $\dfrac{17 + 3}{4} - 5$
   (b) $(-6)\left(\dfrac{9 + 3}{9 - 3} - 2\right)$
   (c) $\dfrac{\dfrac{5 + 7}{5 - 7} + 10}{\dfrac{18 - 3}{5(-3)} - 3}$

A.3.   Find the negative of each of the following numbers.
   (a) -5.7
   (b) 5.7
   (c) $-5 + 5.7$
   (d) $\dfrac{5.7}{5}$
   (e) $\frac{1}{2} - \frac{3}{5}$
   (f) $2\pi(7 - \pi)$
   (g) $(-2\pi)(-7 - \pi)$
   (h) 0

A.4.   Express each of the following numbers as a single integer.

   (a) $(-6)^2$
   (b) -6^2
   (c) $5 + 4(3^2 + 2^3)$
   (d) $\dfrac{(-3)^2 + (-2)^4}{-2^2 - 1}$
   (e) $\dfrac{(-4)^2 + 2}{(-4)^2 - 7} + 1$
   (f) $(-3)2^2 + 4^2 - 1$

A.5. Evaluate:

(a) $\left(\dfrac{-5}{4}\right)^3$ (c) $-\left(\dfrac{7}{4}+\dfrac{3}{4}\right)^2$ (e) $\left(\dfrac{5}{7}\right)^2+\left(\dfrac{2}{7}\right)^2$

(b) $\left(-\dfrac{1}{2}\right)^5$ (d) $\left(\dfrac{1}{3}\right)^3+\left(\dfrac{2}{3}\right)^3$ (f) $\left(-\dfrac{7}{4}+\dfrac{3}{4}-\dfrac{6}{4}\right)^2$

A.6. Simplify the following expressions. Each letter should appear at most once in your answer. For example, $6a^2b^5/3a^2b = 2b^4$.

(a) $x \cdot x^2 \cdot x^5$ (d) $(3x^2y)^2$ (g) $(2w)^3(3w)(4w)^2$

(b) $(.03)y^2 \cdot y^7$ (e) $(1.3)z^2(2y)$ (h) $\dfrac{4w^6}{w^4}$

(c) $(2x^2)^33x$ (f) $(a^2)(7a)(-3a^3)$ (i) $\dfrac{(x^5)^2}{x^3}$

A.7. Simplify (as in Exercise A.6).

(a) $\dfrac{x^4(x^2)^3}{x^3}$ (c) $\dfrac{(u^4)^3(v^2)^2}{(v^2)^3}$ (e) $\left(\dfrac{x^7}{y^6}\right)^2 \cdot \left(\dfrac{y^2}{x}\right)^4$

(b) $\left(\dfrac{z^2}{t^3}\right)^4 \cdot \left(\dfrac{z^3}{t}\right)^5$ (d) $\left(\dfrac{e^6}{c^4}\right)^2 \cdot \left(\dfrac{c^3}{e}\right)^3$ (f) $\left(\dfrac{x}{y}\right)^2\left(\dfrac{y}{z}\right)^3\left(\dfrac{z}{t}\right)^4$

A.8. Simplify (as in Exercise A.6).

(a) $\left(\dfrac{ab^2c^3d^4}{abc^2d}\right)^2$ (c) $\dfrac{(3x)^2(y^2)^3x^2}{\dfrac{x}{y^3} \cdot \left(\dfrac{y}{x^2}\right)^3}$

(b) $(3x^2y - 2xy^2)xy^3 + (7xy^3 + 2y^4)x^2y$ (d) $\left(x^2z + \dfrac{1}{xz}\right)(2x)^3z^2$

A.9. Perform the indicated addition or subtraction. For example, $6x^2y + (3x)(-5xy) = 6x^2y - 15x^2y = -9x^2y$ and $(x^6 + x^3) - (3x^6 - 2x^3 + x^2 + 5) = -2x^6 + 3x^3 - x^2 - 5$.

(a) $x + 2x$ (d) $(x^2 + 2x + 1) - (x^3 - 3x^2 + 4)$

(b) $5w + 7w - 3w$ (e) $\left(u^4 - (-3)u^3 + \dfrac{u}{2} + 1\right) + \left(u^4 - 2u^3 + 5 - \dfrac{u}{2}\right)$

(c) $6a^2b + (-8b)a^2$ (f) $\left(u^4 - (-3)u^3 + \dfrac{u}{2} + 1\right) - \left(u^4 - 2u^3 + 5 - \dfrac{u}{2}\right)$

A.10. Perform the indicated addition or subtraction (as in Exercise A.9).

(a) $(6a^2b + 3ac - 5abc) + (-6ab^2 - 3ab + 6abc)$
(b) $(x^5y - 2x + 3xy^3) - (-2x - x^5y + 2xy^3)$
(c) $(9x - x^3 + 1) - (2x^3 + (-6)x + (-7))$
(d) $(4z - 6z^2w - (-2)z^3w^2) + (8 - 6z^2w - zw^3 + 4z^3w^2)$
(e) $(x^2 - 3xy) - (x + xy) - (x^2 + xy)$
(f) $(x - y - z) - (x + y + z) - (y + z - x)$

A.11. Multiply the following expressions. For example, $3x(4x^2 - 6) = 12x^3 - 18x$.
  (a) $2x(x^2 + 2)$
  (b) $(-5y)(-3y^2 + 1)$
  (c) $x^2y(xy - 6xy^2)$
  (d) $3ax(4ax - 2a^2y + 2ay)$
  (e) $6z^3(2z + 5)$
  (f) $-3z^2(12z^6 - 7z^5)$
  (g) $3ab(4a - 6b + 2a^2b)$
  (h) $(-3ax)(4ax - 5x)$

A.12. Multiply the following expressions. For example, $(3x + 2)(3x - 2) = (3x)^2 - 2^2 = 9x^2 - 4$.
  (a) $(2x + 5)(2x - 5)$
  (b) $(y + 7)(y - 7)$
  (c) $(3x - y)(3x + y)$
  (d) $(x + 4y)(x - 4y)$
  (e) $(2x^2 - 9y)(2x^2 + 9y)$
  (f) $(2a + 3b)(2a - 3b)$
  (g) $(4x^3 - 5y^2)(4x^3 + 5y^2)$
  (h) $(3x^2 + 2y^4)(3x^2 - 2y^4)$

A.13. Multiply the following expressions. For example, $(3x + 2)(x - 6) = 3x^2 - 16x - 12$.
  (a) $(x + 1)(x - 2)$
  (b) $(y - 6)(2y + 2)$
  (c) $(y + 3)(y - 3)$
  (d) $(x + 2)(2x - 5)$
  (e) $(w - 4)^2$
  (f) $(3x + 7)(-x + 5)$
  (g) $(-2x + 4)(-x - 3)$
  (h) $(ab + 1)(a - 2)$
  (i) $(x + 1)(y + 2)$

A.14. Multiply the following expressions. For example, $(3x + 2)(x - 6x + 4) = (3x + 2)x + (3x + 2)(-6x) + (3x + 2)4 = 3x^2 + 2x - 18x^2 - 12x + 12x + 8 = -15x^2 + 2x + 8$.
  (a) $(c - 2)(2c^2 - 3c + 1)$
  (b) $(2y + 3)(y^2 + 3y - 1)$
  (c) $(x + 2y)(2x^2 - xy + y^2)$
  (d) $(ab + 1)(a - z)$
  (e) $(r + s + 1)(a + r)$
  (f) $(x - 2)(3x^2 + 2x + 1)$
  (g) $(5w + 6)(-3w^2 + 4w - 3)$
  (h) $(5x - 2y)(x^2 - 2xy + 3y^2)$

A.15. Multiply the following expressions. For example, $(x + 1)(x + 5)(x - 7) = (x + 1)(x^2 - 2x - 35) = x(x^2 - 2x - 35) + 1(x^2 - 2x - 35) = x^3 - 2x^2 - 35x + x^2 - 2x - 35 = x^3 - x^2 - 37x - 35$. One could also obtain the same answer by proceeding $(x + 1)(x + 5)(x - 7) = (x^2 + 6x + 5)(x - 7)$ and so on.
  (a) $2x(3x + 1)(4x - 2)$
  (b) $3y(-y + 2)(3y + 1)$
  (c) $(x - 1)(x - 2)(x - 3)$
  (d) $(y - 2)(3y + 2)(y + 2)$
  (e) $(2x + 3y)(x - 2y)(3x - y)$
  (f) $(x + 4y)(2y - x)(3x - y)$

A.16. Factor the following expressions. For example, $x^2 - 9 = (x + 3)(x - 3)$ and $y^2 - 10y + 25 = (y - 5)^2$.
  (a) $x^2 - 4$
  (b) $x^2 + 6x + 9$
  (c) $9y^2 - 25$
  (d) $y^2 - 4y + 4$
  (e) $81x^2 + 36x + 4$
  (f) $4x^2 - 12x + 9$
  (g) $5 - x^2$
  (h) $1 - 36u^2$
  (i) $49 + 28z + 4z^2$
  (j) $25u^2 - 20uv + 4v^2$
  (k) $x^4 - y^4$
  (l) $18y^2 - 2x^2$

A.17. Factor the following expressions.
  (a) $x^2 + x - 6$
  (b) $y^2 + 11y + 30$
  (c) $z^2 + 4z + 3$
  (d) $x^2 - 8x + 15$
  (e) $y^2 + 5y - 36$
  (f) $z^2 - 9z + 14$
  (g) $x^2 - 6x + 9$
  (h) $4y^2 - 81$
  (i) $x^2 + 7x + 10$
  (j) $w^2 - 6w - 16$
  (k) $x^2 + 11x + 18$
  (l) $x^2 + 3xy - 28y^2$

A.18. Factor the following expressions.
(a) $3x^2 + 4x + 1$ (e) $9x^2 - 72x$ (i) $8u^2 + 6u - 9$
(b) $4y^2 + 4y + 1$ (f) $4x^2 - 4x - 3$ (j) $2y^2 - 4y + 2$
(c) $2z^2 + 11z + 12$ (g) $10x^2 - 8x - 2$ (k) $4x^2 + 20xy + 25y^2$
(d) $10x^2 - 17x + 3$ (h) $7z^2 + 23z + 6$ (l) $63u^2 - 46uv + 8v^2$

A.19. Factor the following expressions.
(a) $x^3 - 125$ (e) $8 + x^3$ (i) $x^3 + 1$
(b) $y^3 + 64$ (f) $z^3 - 9z^2 + 27z - 27$ (j) $x^3 - 1$
(c) $x^3 + 6x^2 + 12x + 8$ (g) $-x^3 + 15x^2 - 75x + 125$ (k) $8x^3 - y^3$
(d) $y^3 - 3y^2 + 3y - 1$ (h) $27 - t^3$ (l) $(x - 1)^3 + 1$

A.20. Factor the following expressions.
(a) $x^6 - 64$ (f) $x^6 + 16x^3 + 64$
(b) $x^5 - 8x^2$ (g) $z^6 - 1$
(c) $y^4 + 7y^2 + 10$ (h) $y^6 + 26y^3 - 27$
(d) $z^4 - 5z^2 - 6$ (i) $x^4 + 2x^2y - 3y^2$
(e) $81 - y^4$ (j) $x^8 - 17x^4 + 16$

A.21. Factor the following expressions by regrouping and using the distributive law.
(a) $x^2 - yz + xz - xy$ (d) $u^2v - 2w^2 - 2uvw + uw$
(b) $x^6 - 2x^4 - 8x^2 + 16$ (e) $x^3 + 4x^2 - 8x - 32$
(c) $a^3 - 2b^2 + 2a^2b - ab$ (f) $z^8 - 5z^7 + 2z - 10$

A.22. Complete the square:
(a) $x^2 + 4x$ (d) $x^2 - x$ (g) $2x^2 + 4x$ (i) $2z^2 + 14z$
(b) $y^2 - 6y$ (e) $x^2 + 12x$ (h) $3y^2 - 30y$ (j) $3x^2 + 5x$
(c) $z^2 + 3z$ (f) $y^2 - 11y$

A.23. Express these fractions in lowest terms.

(a) $\dfrac{63}{49}$ (d) $\dfrac{x^2 - 4}{x + 2}$ (g) $\dfrac{a^2 - b^2}{a^3 - b^3}$

(b) $\dfrac{121}{33}$ (e) $\dfrac{x^2 - x - 2}{x^2 + 2x + 1}$ (h) $\dfrac{x^4 - 3x^2}{x^3}$

(c) $\dfrac{13 \cdot 27 \cdot 22 \cdot 10}{6 \cdot 4 \cdot 11 \cdot 12}$ (f) $\dfrac{z + 1}{z^3 + 1}$ (i) $\dfrac{(x + c)(x^2 - cx + c^2)}{x^4 + c^3x}$

A.24. Perform the indicated addition or subtraction.

(a) $\dfrac{3}{7} + \dfrac{2}{5}$ (d) $\dfrac{1}{a} - \dfrac{2a}{b}$ (g) $\dfrac{b}{c} - \dfrac{c}{b}$

(b) $\dfrac{7}{8} - \dfrac{5}{6}$ (e) $\dfrac{c}{d} + \dfrac{3c}{e}$ (h) $\dfrac{a}{b} + \dfrac{2a}{b^2} + \dfrac{3a}{b^3}$

(c) $\left(\dfrac{19}{7} + \dfrac{1}{2}\right) - \dfrac{1}{3}$ (f) $\dfrac{r}{s} + \dfrac{s}{t} + \dfrac{t}{r}$

A.25. Perform the indicated addition or subtraction.

(a) $\dfrac{1}{x+1} - \dfrac{1}{x}$

(b) $\dfrac{1}{2x+1} + \dfrac{1}{2x-1}$

(c) $\dfrac{1}{x+4} + \dfrac{2}{(x+4)^2} - \dfrac{3}{x^2+8x+16}$

(d) $\dfrac{1}{x} + \dfrac{1}{xy} + \dfrac{1}{xy^2}$

(e) $\dfrac{1}{x} - \dfrac{1}{3x-4}$

(f) $\dfrac{3}{x-1} + \dfrac{4}{x+1}$

(g) $\dfrac{1}{x+y} + \dfrac{x+y}{x^3+y^3}$

(h) $\dfrac{6}{5(x-1)(x-2)^2} + \dfrac{x}{3(x-1)^2(x-2)}$

(i) $\dfrac{1}{4x(x+1)(x+2)^3} - \dfrac{6x+2}{4(x+1)^3}$

(j) $\dfrac{x+y}{(x^2-xy)(x-y)^2} - \dfrac{2}{(x^2-y^2)^2}$

A.26. Multiply and express in lowest terms:

(a) $\dfrac{3}{4} \cdot \dfrac{12}{5} \cdot \dfrac{10}{9}$

(b) $\dfrac{10}{45} \cdot \dfrac{6}{14} \cdot \dfrac{1}{2}$

(c) $\dfrac{3a^2c}{4ac} \cdot \dfrac{8ac^3}{9a^2c^4}$

(d) $\dfrac{6x^2y}{2x} \cdot \dfrac{y}{21xy}$

(e) $\dfrac{7x}{11y} \cdot \dfrac{66y^2}{14x^3}$

(f) $\dfrac{ab}{c^2} \cdot \dfrac{cd}{a^2b} \cdot \dfrac{ad}{bc^2}$

A.27. Multiply and express in lowest terms:

(a) $\dfrac{3x+9}{2x} \cdot \dfrac{8x^2}{(x^2-9)}$

(b) $\dfrac{4x+16}{3x+15} \cdot \dfrac{2x+10}{x+4}$

(c) $\dfrac{5y-25}{3} \cdot \dfrac{y^2}{y^2-25}$

(d) $\dfrac{6x-12}{6x} \cdot \dfrac{8x^2}{x-2}$

(e) $\dfrac{u}{u-1} \cdot \dfrac{u^2-1}{u^2}$

(f) $\dfrac{t^2-t-6}{t^2-6t+9} \cdot \dfrac{t^2+4t-5}{t^2-25}$

(g) $\dfrac{2u^2+uv-v^2}{4u^2-4uv+v^2} \cdot \dfrac{8u^2+6uv-9v^2}{4u^2-9v^2}$

(h) $\dfrac{2x^2-3xy-2y^2}{6x^2-5xy-4y^2} \cdot \dfrac{6x^2+6xy}{x^2-xy-2yz}$

A.28. Compute the quotient and express in lowest terms.

(a) $\dfrac{\dfrac{5}{12}}{\dfrac{4}{14}}$

(b) $\dfrac{\dfrac{100}{52}}{\dfrac{27}{26}}$

(c) $\dfrac{\dfrac{uv}{v^2w}}{\dfrac{vw}{u^2v}}$

(d) $\dfrac{\dfrac{3x^2y}{(xy)^2}}{\dfrac{3xyz}{x^2y}}$

(e) $\dfrac{\dfrac{x+3}{x+4}}{\dfrac{2x}{x+4}}$

(f) $\dfrac{\dfrac{(x+2)^2}{(x-2)^2}}{\dfrac{x^2+2x}{x^2-4}}$

(g) $\dfrac{\dfrac{x+y}{x+2y}}{\left(\dfrac{x+y}{xy}\right)^2}$

(h) $\dfrac{\dfrac{u^3+v^3}{u^2-v^2}}{\dfrac{u^2-uv+v^2}{u+v}}$

(i) $\dfrac{\dfrac{(c+d)^2}{c^2-d^2}}{cd}$

A.29. Assume all letters represent positive numbers and simplify:

(a) $\sqrt{144}$

(b) $\sqrt[3]{125}$

(c) $\sqrt{\dfrac{64}{49}}$

(d) $\sqrt[3]{\dfrac{27}{64}}$

(e) $\sqrt{\dfrac{y^2}{4}}$

(f) $\sqrt{9a^6}$

(g) $\sqrt[3]{8a^3b^{12}}$

(h) $\sqrt[3]{64x^6y^{12}}$

(i) $\sqrt{u}\sqrt{u^8}$

(j) $\sqrt{a}(1+\sqrt{a})$

(k) $\sqrt{2}\sqrt{xy}\sqrt{8xy}$

(l) $\dfrac{\sqrt{6x}}{\sqrt{3x^5}}$

(m) $\sqrt{uv}\left(\sqrt{\dfrac{u}{v}}+\dfrac{\sqrt{v}}{\sqrt{u}}\right)$

(n) $\dfrac{\sqrt[3]{27c^4d^3}}{\sqrt[3]{8c^5d^6}}$

(o) $\left(\dfrac{\sqrt{3c}}{\sqrt[3]{6c^2}}\right)^4$

A.30. Rationalize each denominator:

(a) $\dfrac{2}{\sqrt{5}}$

(b) $\sqrt{\dfrac{16}{5}}$

(c) $\sqrt{\dfrac{7}{10}}$

(d) $\sqrt{\dfrac{9x^4}{23}}$

(e) $\dfrac{1}{\sqrt{x}}$

(f) $\dfrac{\sqrt{6}}{\sqrt{6}+\sqrt{2}}$

(g) $\dfrac{\sqrt{r}+\sqrt{s}}{\sqrt{r}-\sqrt{s}}$

(h) $\dfrac{1}{\sqrt{a}-2\sqrt{b}}$

(i) $\dfrac{u^2-v^2}{\sqrt{u+v}-\sqrt{u-v}}$

A.31. Solve the following equations.

(a) $3x+2=26$

(b) $\dfrac{y}{5}-3=14$

(c) $3z+2=9z+7$

(d) $-7(t+2)=3(4t+1)$

(e) $2\left(\dfrac{x}{3}+1\right)+5\left(\dfrac{4x}{3}-2\right)=2$

(f) $2x-\dfrac{x+5}{4}=\dfrac{3x-1}{2}+1$

A.32. Solve the following equations.

(a) $\dfrac{3x + 2}{x - 6} = 4$

(f) $\dfrac{-5}{t} = \dfrac{7}{t}$

(b) $\dfrac{3x + 2}{x - 6} = \dfrac{2}{x - 6} + 3$

(g) $\dfrac{2x - 7}{x + 4} = \dfrac{5}{x + 4} - 2$

(c) $\dfrac{2x - 1}{2x + 1} = \dfrac{1}{4}$

(h) $\dfrac{z + 4}{z + 5} = \dfrac{-1}{z + 5}$

(d) $\dfrac{4}{3y + 1} = \dfrac{1}{y}$

(i) $1 + \dfrac{1}{x} = \dfrac{2 + x}{x} + 1$

(e) $\dfrac{x - 3}{x - 7} = \dfrac{x - 4}{x - 7}$

A.33. Solve each equation for the indicated letter.

(a) $A = \dfrac{h}{2}(b + b')$ for b 　(d) $a(y - 7) + 2a\left(\dfrac{y}{2} + a\right) = 16$ 　 for y

(b) $V = \pi b^2 c$ 　 for c 　(e) $ax + (4a - b)(x + 3b) = 2x - a$ for x
(c) $A = 2\pi r(r + h)$ for h 　(f) $at^2 - (3b + t)t - 4a + 7 = 0$ 　 for a

A.34. Solve the following equations.

(a) $x^2 = 64$ 　　(d) $(x - 7)^2 = 16$ 　　(g) $y^2 - 6y + 9 = 0$
(b) $x^2 + 7 = 32$ 　(e) $(y + 2)^2 = 5$ 　　(h) $9x^2 + 12x + 4 = 14$
(c) $z^2 = 56$ 　　(f) $x^2 + 2x + 1 = 9$

B.1. In the expression $7y^4 - 5y^3 + 17y^2 + 2$, the **coefficient** of y^4 is the number 7, the coefficient of y^3 is -5, the coefficient of y^2 is 17, and the coefficient of y is 0. In each of the following find the coefficient of y^2 (or x^2 or z^2 as the case may be). Don't do any more multiplying than you have to. *Example:* $(y + 3)(y^2 + y) = y^3 + 4y^2 + 3y$, so the coefficient of y^2 is 4.

(a) $(y^2 + 3y + 1)(2y - 3)$ 　　(e) $(y + 2)^3$
(b) $(y^3 + 2y - 6)(y^2 + 1)$ 　　(f) $(x^2 + x + 1)(x - 1)$
(c) $(z^2 - 1)(z + 1)$ 　　　　(g) $(x^2 + x + 1)(x^2 - x + 1)$
(d) $(3 + y)(3 - y)$ 　　　　(h) $(2x^2 + 1)(4x^2 - 1)$

B.2. Multiply the following. Arrange the terms of your answer according to descending powers of x, with each power of x appearing at most once. For example, $(ax + b)(4x - c) = 4ax^2 + (4b - ac)x - bc$.

(a) $(ax + b)(3x + 2)$ 　　(d) $rx(3rx + 1)(4x - r)$
(b) $(4x - c)(dx + c)$ 　　(e) $(x - a)(x - b)(x - c)$
(c) $(ax + b)(bx + a)$ 　　(f) $(x + b)(cx + d)$

B.3. Use the difference of squares [namely, $(c + d)(c - d) = c^2 - d^2$] to perform the following multiplications. For example, $25 \cdot 15 = (20 + 5)(20 - 5) = 20^2 - 5^2 = 400 - 25 = 375$.

(a) $21 \cdot 19$ 　　(c) $41 \cdot 39$ 　　(e) $53 \cdot 47$ 　　(g) $52 \cdot 68$
(b) $31 \cdot 29$ 　　(d) $73 \cdot 67$ 　　(f) $63 \cdot 77$ 　　(h) $31 \cdot 49$

B.4. Assume all exponent letters represent nonnegative integers and perform the indicated multiplication. Keep in mind the properties of exponents. For example, $(2x^k)(3x + x^{r+1}) = (2x^k)(3x) + (2x^k)(x^{r+1}) = 6x^{k+1} + 2x^{k+r+1}$.

(a) $3^r \cdot 3^1 \cdot 3^t$
(b) $(2x^n)(8x^r)$
(c) $(3y^k)(2y^{k+2})$
(d) $(x^m + 2)(x^n - 3)$
(e) $(y^r + 1))(y^s - 4)$
(f) $(2z^{2n} - 5)(z^{3n} + 3z^n + 1)$

B.5. Factor the following expressions by regrouping and using the distributive law.

(a) $2x^2 + 5xy - 3y^2 + 6x - 3y$
(b) $x^2 - 9xy + 14y^2 + 3xy^2 - 6y^3$
(c) $x^3 + x - 3y - 27y^3$
(d) $8u^3 + 10u + v^3 + 5v$

B.6. Factor the following expressions. For example, $x^2 - x/4 - \frac{3}{8} = (x + \frac{1}{2})(x - \frac{3}{4})$.

(a) $x^2 - \dfrac{1}{64}$
(b) $x^3 - \dfrac{1}{8}$
(c) $y^2 - \dfrac{2y}{3} - \dfrac{5}{36}$
(d) $x^2 + x - \dfrac{3}{4}$
(e) $z^2 + 3z + \dfrac{35}{16}$
(f) $9t^2 - 3t - 2$

B.7. Use a calculator to solve the following equations. *Check your answers.*

(a) $2.37x + 3.1288 = 6.93x - 2.48$
(b) $18.923y - 15.4228 = 10.003y + 18.161$
(c) $6.31(x - 3.53) = 5.42(x + 1.07) - 21.1584$

B.8. Each of the following statements is *false*. Find the mistake and correct it. For example, *Statement:* $(x/6)^2 = x^2/6$; *Mistake:* not squaring the denominator 6; *Correct statement:* $(x/6)^2 = x^2/36$.

(a) $3(y + 2) = 3y + 2$

(b) $x - (3y + 4) = x - 3y + 4$
(c) $(x + y)^2 = x + y^2$

(d) $(2x)^3 = 2x^3$

(e) $(y^5)^6 = y^{11}$

(f) $\sqrt{7x} = 7\sqrt{x}$

(g) $\left(\dfrac{1}{3}x\right)\left(\dfrac{1}{3}y\right) = \dfrac{1}{3}xy$

(h) $\dfrac{1}{a} + \dfrac{1}{b} = \dfrac{1}{a + b}$

(i) $\sqrt{r + s} = \sqrt{r} + \sqrt{s}$
(j) $(x + y)^2 = x^2 + y^2$

(k) $\dfrac{x^2}{x^2 + x^6} = 1 + x^3$

(l) $\left(\dfrac{1}{\sqrt{a} + \sqrt{b}}\right)^2 = \dfrac{1}{a + b}$

(m) $(a - b)^2 = a^2 - b^2$

(n) $y + y + y = y^3$

B.9. Each of the following statements is *false*. Find the mistake and correct it (as in Exercise B.8).

(a) $x^2 - 5x - 6 = (x - 3)(x - 2)$
(b) $\sqrt[3]{27 + 125} = 3 + 5 = 8$
(c) $(-2)^5 = 2^5$
(d) $\dfrac{\frac{1}{x}}{\frac{1}{y}} = \dfrac{1}{xy}$

(e) $3 + \dfrac{4}{x} = \dfrac{7}{x}$

(i) $\sqrt{2x + 1} + \sqrt{3x + 2} = \sqrt{5x + 3}$

(f) $(a + b)^3 = a^3 + 3ab + b^3$

(j) $(a + b)^3 = a^3 + b^3$

(g) $a^3 + b^3 = (a + b)(a^2 + b^2)$

(k) $\dfrac{u}{v} + \dfrac{v}{u} = 1$

(h) $(\sqrt{x} + \sqrt{y}) \dfrac{1}{\sqrt{x} + \sqrt{y}} = x + y$

C.1. Explain algebraically why each of the following parlor tricks always works.

(a) Write down a nonzero number. Add 1 to it and square the result. Subtract 1 from the original number and square the result. Subtract the second square from the first square. Divide by the number you started with. The answer is 4.

(b) Write down a positive number. Add 4 to it. Multiply the result by the original number. Add 4 to this result and then take the square root. Subtract the number you started with. The answer is 2.

(c) Invent a similar parlor trick in which the answer is always the number you started with.

C.2. Suppose that division by zero *was* defined. Then $\frac{1}{0}$ would be a real number. Hence $\frac{1}{0}$ would either be a nonzero number or $\frac{1}{0}$ would be 0. This exercise shows that the first of these possibilities leads to a logical contradiction and that the second is highly unreasonable, if we want division by zero to "behave" like division by other numbers. Consequently, it is better to leave division by zero undefined.

(a) Assuming that division by zero obeys the usual rules of arithmetic of fractions, show that $\dfrac{1}{1/0} = 0$. (Remember: $\frac{0}{1}$ is defined and $\frac{0}{1} = 0$.)

(b) Suppose $\frac{1}{0}$ is a nonzero number, say, $\frac{1}{0} = c$ with $c \neq 0$. Use part (a) to show that $1/c = 0$. Consequently,

$$c\left(\dfrac{1}{c}\right) = c \cdot 0$$

$$1 = 0$$

This is a logical contradiction. Therefore $1/0$ cannot be a nonzero number.

(c) Suppose that $1/0 = 0$. If division by 0 is going to behave like division by other numbers, then all four of the following statements *should* be true:

(i)   Whenever a is very close to 5, then $1/a$ is very close to $\frac{1}{5} = .2$.
(ii) Whenever a is very close to 2, then $1/a$ is very close to $\frac{1}{2} = .5$.
(iii) Whenever a is very close to 1, then $1/a$ is very close to $\frac{1}{1} = 1.0$.
(iv) Whenever a is very close to 0, then $1/a$ is very close to $\frac{1}{0} = 0$.

Use a calculator (if necessary) to verify that the first three statements are true but that *the last statement is false*. For example, $a = 4.999$ and $a = 5.0001$ are very close to 5 and $\frac{1}{4.999} = .2000040$ and $\frac{1}{5.00001} = .1999996$ are very close to $\frac{1}{5} = .2$. Since statement (iv) is false, we conclude that $\frac{1}{0} = 0$ is an unreasonable definition if we want division by 0 to behave like division by other numbers.

Geometry Review

The few facts from high school geometry that are needed to read this book are reviewed here. We shall frequently deal with **triangles** such as those shown below. Each triangle has three sides (straight-line segments) and three angles, formed at the points where the various sides meet. When angles are measured in degrees, the sum of the measures of all three angles of a triangle is *always* 180°. For instance, see Figure A-1.

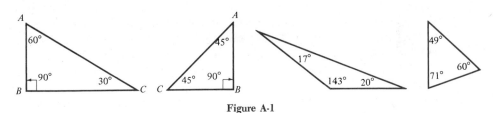

Figure A-1

A **right angle** is an angle that measures 90°. A right triangle is a triangle, one of whose angles is a right angle, such as the first two triangles shown in Figure A-1. The side of a right triangle that lies opposite the right angle is called the **hypotenuse**. In each of the right triangles in Figure A-1, side AC is the hypotenuse. A famous result of antiquity is the following theorem:

PYTHAGOREAN THEOREM

If the sides of a right triangle have lengths a *and* b *and the hypotenuse has length* c, *then*

$$c^2 = a^2 + b^2$$

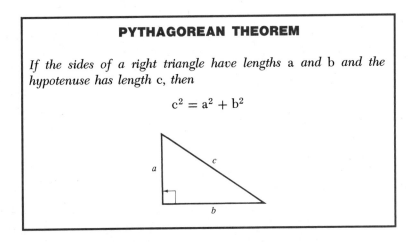

EXAMPLE   Consider the right triangle with sides of lengths 5 and 12, as shown in Figure A-2.

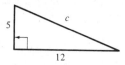

Figure A-2

According to the Pythagorean Theorem the length c of the hypotenuse satisfies the equation: $c^2 = 5^2 + 12^2 = 25 + 144 = 169$.   Since $169 = 13^2$, we see that c must be 13.

SPECIAL TRIANGLES

There are two theorems of plane geometry that are often helpful when dealing with right triangles.

THEOREM I

If two angles of a triangle are equal, then the two sides opposite these angles have the same length.

EXAMPLE   Suppose the hypotenuse of the right triangle shown in Figure A-3 has length 1 and that angles B and C each measure $45°$.

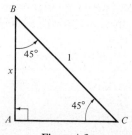

Figure A-3

Then by Theorem I, sides AB and AC have the same length. If x is the length of side AB, then by the Pythagorean Theorem:

$$x^2 + x^2 = 1^2$$
$$2x^2 = 1$$
$$x^2 = \tfrac{1}{2}$$

$$x = \sqrt{\frac{1}{2}} = \frac{1}{\sqrt{2}} = \frac{\sqrt{2}}{2}$$

(We ignore the other solution of this equation, namely, $x = -\sqrt{\frac{1}{2}}$, since x represents a length here and thus must be nonnegative.) Therefore the sides of a $90°$-$45°$-$45°$ triangle with hypotenuse 1 are each of length $\sqrt{2}/2$.

THEOREM II

In a right triangle that has an angle of $30°$, the length of the side opposite the $30°$ angle is one-half the length of the hypotenuse.

EXAMPLE   Suppose that in the right triangle shown in Figure A-4 angle B is $30°$ and the length of hypotenuse BC is 2.

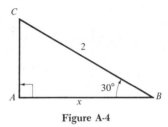

Figure A-4

By Theorem II the side opposite the $30°$ angle, namely, side AC, has length 1. If x denotes the length of side AB, then by the Pythagorean Theorem:

$$1^2 + x^2 = 2^2$$
$$x^2 = 3$$
$$x = \sqrt{3}$$

EXAMPLE   The right triangle shown in Figure A-5 has a $30°$ angle at C and side AC has length $\sqrt{3}/2$.

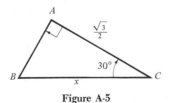

Figure A-5

Let x denote the length of the hypotenuse BC. By Theorem II, side AB has length $\frac{1}{2}x$. By the Pythagorean Theorem:

$$\left(\frac{1}{2}x\right)^2 + \left(\frac{\sqrt{3}}{2}\right)^2 = x^2$$

$$\frac{x^2}{4} + \frac{3}{4} = x^2$$

$$\frac{3}{4} = \frac{3}{4}x^2$$

$$x^2 = 1$$

$$x = 1$$

Therefore the triangle has hypotenuse of length 1 and sides of lengths $1/2$ and $\sqrt{3}/2$.

SIMILAR TRIANGLES

Two triangles, as in Figure A-6,

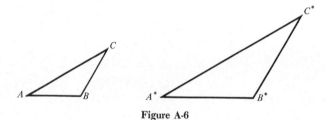

Figure A-6

are said to be **similar** if their corresponding angles are equal (that is, $\angle A = \angle A^*$; $\angle B = \angle B^*$; and $\angle C = \angle C^*$). Thus similar triangles have the same *shape* but not necessarily the same *size*. Here is the key fact about similar triangles:

THEOREM III

Suppose triangle ABC *with sides* a, b, c *is similar to triangle* A*B*C* *with sides* a*, b*, c* *(that is,* $\angle A = \angle A^*$; $\angle B = \angle B^*$; $\angle C = \angle C^*$*).*

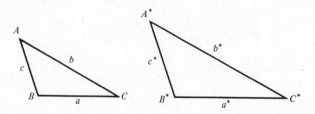

Then

$$\frac{a}{a^{\circ}} = \frac{b}{b^{\circ}} = \frac{c}{c^{\circ}}$$

These equalities are equivalent to:

$$\frac{a}{b} = \frac{a^{\circ}}{b^{\circ}}, \qquad \frac{b}{c} = \frac{b^{\circ}}{c^{\circ}}, \qquad \frac{a}{c} = \frac{a^{\circ}}{c^{\circ}}$$

The equivalence of the equalities in the conclusion of the theorem is easily verified. For example, since

$$\frac{a}{a^{\circ}} = \frac{b}{b^{\circ}}$$

we have

$$ab^{\circ} = a^{\circ}b$$

Dividing both sides of this equation by bb° yields:

$$\frac{ab^{\circ}}{bb^{\circ}} = \frac{a^{\circ}b}{bb^{\circ}}$$

$$\frac{a}{b} = \frac{a^{\circ}}{b^{\circ}}$$

The other equivalences are proved similarly.

EXAMPLE Suppose the triangles in Figure A-7

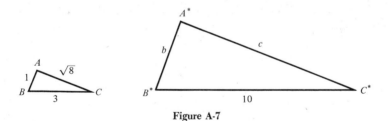

Figure A-7

are similar and that the sides have the lengths indicated. Then by Theorem III,

$$\frac{\text{length } AC}{\text{length } A^{\circ}C^{\circ}} = \frac{\text{length } BC}{\text{length } B^{\circ}C^{\circ}}$$

In other words,

$$\frac{\sqrt{8}}{c} = \frac{3}{10}$$

so that

$$3c = 10\sqrt{8}$$

$$c = \left(\frac{10}{3}\right)\sqrt{8}$$

Similarly, by Theorem III,

$$\frac{\text{length } AB}{\text{length } A^\circ B^\circ} = \frac{\text{length } BC}{\text{length } B^\circ C^\circ}$$

so that

$$\frac{1}{b} = \frac{3}{10}$$

$$3b = 10$$

$$b = \frac{10}{3}$$

Therefore the sides of triangle $A^\circ B^\circ C^\circ$ are of lengths 10, $\frac{10}{3}$, and $\frac{10}{3}\sqrt{8}$.

Table of
Powers, Roots,
and Reciprocals

N	N²	N³	$\sqrt{N}$	$\sqrt[3]{N}$	1/N	N	N²	N³	$\sqrt{N}$	$\sqrt[3]{N}$	1/N
1	1	1	1.000	1.000	1.0000	51	2601	132651	7.141	3.708	.0196
2	4	8	1.414	1.260	.5000	52	2704	140608	7.211	3.733	.0192
3	9	27	1.732	1.442	.3333	53	2809	148877	7.280	3.756	.0189
4	16	64	2.000	1.587	.2500	54	2916	157464	7.348	3.780	.0185
5	25	125	2.236	1.710	.2000	55	3025	166375	7.416	3.803	.0182
6	36	216	2.449	1.817	.1667	56	3136	175616	7.483	3.826	.0179
7	49	343	2.646	1.913	.1429	57	3249	185193	7.550	3.848	.0175
8	64	512	2.828	2.000	.1250	58	3364	195112	7.616	3.871	.0172
9	81	729	3.000	2.080	.1111	59	3481	205379	7.681	3.893	.0169
10	100	1000	3.162	2.154	.1000	60	3600	216000	7.746	3.915	.0167
11	121	1331	3.317	2.224	.0909	61	3721	226981	7.810	3.936	.0164
12	144	1728	3.464	2.289	.0833	62	3844	238328	7.874	3.958	.0161
13	169	2197	3.606	2.351	.0769	63	3969	250047	7.937	3.979	.0159
14	196	2744	3.742	2.410	.0714	64	4096	262144	8.000	4.000	.0156
15	225	3375	3.873	2.466	.0667	65	4225	274625	8.062	4.021	.0154
16	256	4096	4.000	2.520	.0625	66	4356	287496	8.124	4.041	.0152
17	289	4913	4.123	2.571	.0588	67	4489	300763	8.185	4.062	.0149
18	324	5832	4.243	2.621	.0556	68	4624	314432	8.246	4.082	.0147
19	361	6859	4.359	2.668	.0526	69	4761	328509	8.307	4.102	.0145
20	400	8000	4.472	2.714	.0500	70	4900	343000	8.367	4.121	.0143
21	441	9261	4.583	2.759	.0476	71	5041	357911	8.426	4.141	.0141
22	484	10648	4.690	2.802	.0455	72	5184	373248	8.485	4.160	.0139
23	529	12167	4.796	2.844	.0435	73	5329	389017	8.544	4.179	.0137
24	576	13824	4.899	2.884	.0417	74	5476	405224	8.602	4.198	.0135
25	625	15625	5.000	2.924	.0400	75	5625	421875	8.660	4.217	.0133
26	676	17576	5.099	2.962	.0385	76	5776	438976	8.718	4.236	.0132
27	729	19683	5.196	3.000	.0370	77	5929	456533	8.775	4.254	.0130
28	784	21952	5.292	3.037	.0357	78	6084	474552	8.832	4.273	.0128
29	841	24389	5.385	3.072	.0345	79	6241	493039	8.888	4.291	.0127
30	900	27000	5.477	3.107	.0333	80	6400	512000	8.944	4.309	.0125
31	961	29791	5.568	3.141	.0323	81	6561	531441	9.000	4.327	.0123
32	1024	32768	5.657	3.175	.0313	82	6724	551368	9.055	4.344	.0122
33	1089	35937	5.745	3.208	.0303	83	6889	571787	9.110	4.362	.0120
34	1156	39304	5.831	3.240	.0294	84	7056	592704	9.165	4.380	.0119
35	1225	42875	5.916	3.271	.0286	85	7225	614125	9.220	4.397	.0118
36	1296	46656	6.000	3.302	.0278	86	7396	636056	9.274	4.414	.0116
37	1369	50653	6.083	3.332	.0270	87	7569	658503	9.327	4.431	.0115
38	1444	54872	6.164	3.362	.0263	88	7744	681472	9.381	4.448	.0114
39	1521	59319	6.245	3.391	.0256	89	7921	704969	9.434	4.465	.0112
40	1600	64000	6.325	3.420	.0250	90	8100	729000	9.487	4.481	.0111
41	1681	68921	6.403	3.448	.0244	91	8281	753571	9.539	4.498	.0110
42	1764	74088	6.481	3.476	.0238	92	8464	778688	9.592	4.514	.0109
43	1849	79507	6.557	3.503	.0233	93	8649	804357	9.644	4.531	.0108
44	1936	85184	6.633	3.530	.0227	94	8836	830584	9.695	4.547	.0106
45	2025	91125	6.708	3.557	.0222	95	9025	857375	9.747	4.563	.0105
46	2116	97336	6.782	3.583	.0217	96	9216	884736	9.798	4.579	.0104
47	2209	103823	6.856	3.609	.0213	97	9409	912673	9.849	4.595	.0103
48	2304	110592	6.928	3.634	.0208	98	9604	941192	9.899	4.610	.0102
49	2401	117649	7.000	3.659	.0204	99	9801	970299	9.950	4.626	.0101
50	2500	125000	7.071	3.684	.0200	100	10000	1000000	10.000	4.642	.0100

Table of
Common
Logarithms

x	0	1	2	3	4	5	6	7	8	9
1.0	.0000	.0043	.0086	.0128	.0170	.0212	.0253	.0294	.0334	.0374
1.1	.0414	.0453	.0492	.0531	.0569	.0607	.0645	.0682	.0719	.0755
1.2	.0792	.0828	.0864	.0899	.0934	.0969	.1004	.1038	.1072	.1106
1.3	.1139	.1173	.1206	.1239	.1271	.1303	.1335	.1367	.1399	.1430
1.4	.1461	.1492	.1523	.1553	.1584	.1614	.1644	.1673	.1703	.1732
1.5	.1761	.1790	.1818	.1847	.1875	.1903	.1931	.1959	.1987	.2014
1.6	.2041	.2068	.2095	.2122	.2148	.2175	.2201	.2227	.2253	.2279
1.7	.2304	.2330	.2355	.2380	.2405	.2430	.2455	.2480	.2504	.2529
1.8	.2553	.2577	.2601	.2625	.2648	.2672	.2695	.2718	.2742	.2765
1.9	.2788	.2810	.2833	.2856	.2878	.2900	.2923	.2945	.2967	.2989
2.0	.3010	.3032	.3054	.3075	.3096	.3118	.3139	.3160	.3181	.3201
2.1	.3222	.3243	.3263	.3284	.3304	.3324	.3345	.3365	.3385	.3404
2.2	.3424	.3444	.3464	.3483	.3502	.3522	.3541	.3560	.3579	.3598
2.3	.3617	.3636	.3655	.3674	.3692	.3711	.3729	.3747	.3766	.3784
2.4	.3802	.3820	.3838	.3856	.3874	.3892	.3909	.3927	.3945	.3962
2.5	.3979	.3997	.4014	.4031	.4048	.4065	.4082	.4099	.4116	.4133
2.6	.4150	.4166	.4183	.4200	.4216	.4232	.4249	.4265	.4281	.4298
2.7	.4314	.4330	.4346	.4362	.4378	.4393	.4409	.4425	.4440	.4456
2.8	.4472	.4487	.4502	.4518	.4533	.4548	.4564	.4579	.4594	.4609
2.9	.4624	.4639	.4654	.4669	.4683	.4698	.4713	.4728	.4742	.4757
3.0	.4771	.4786	.4800	.4814	.4829	.4843	.4857	.4871	.4886	.4900
3.1	.4914	.4928	.4942	.4955	.4969	.4983	.4997	.5011	.5024	.5038
3.2	.5051	.5065	.5079	.5092	.5105	.5119	.5132	.5145	.5159	.5172
3.3	.5185	.5198	.5211	.5224	.5237	.5250	.5263	.5276	.5289	.5302
3.4	.5315	.5328	.5340	.5353	.5366	.5378	.5391	.5403	.5416	.5428
3.5	.5441	.5453	.5465	.5478	.5490	.5502	.5514	.5527	.5539	.5551
3.6	.5563	.5575	.5587	.5599	.5611	.5623	.5635	.5647	.5658	.5670
3.7	.5682	.5694	.5705	.5717	.5729	.5740	.5752	.5763	.5775	.5786
3.8	.5798	.5809	.5821	.5832	.5843	.5855	.5866	.5877	.5888	.5899
3.9	.5911	.5922	.5933	.5944	.5955	.5966	.5977	.5988	.5999	.6010
4.0	.6021	.6031	.6042	.6053	.6064	.6075	.6085	.6096	.6107	.6117
4.1	.6128	.6138	.6149	.6159	.6170	.6180	.6191	.6201	.6212	.6222
4.2	.6232	.6243	.6253	.6263	.6274	.6284	.6294	.6304	.6314	.6325
4.3	.6335	.6345	.6355	.6365	.6375	.6385	.6395	.6405	.6415	.6425
4.4	.6435	.6444	.6454	.6464	.6474	.6484	.6493	.6503	.6513	.6522
4.5	.6532	.6542	.6551	.6561	.6571	.6580	.6590	.6599	.6609	.6618
4.6	.6628	.6637	.6646	.6656	.6665	.6675	.6684	.6693	.6702	.6712
4.7	.6721	.6730	.6739	.6749	.6758	.6767	.6776	.6785	.6794	.6803
4.8	.6812	.6821	.6830	.6839	.6848	.6857	.6866	.6875	.6884	.6893
4.9	.6902	.6911	.6920	.6928	.6937	.6946	.6955	.6964	.6972	.6981
5.0	.6990	.6998	.7007	.7016	.7024	.7033	.7042	.7050	.7059	.7067
5.1	.7076	.7084	.7093	.7101	.7110	.7118	.7126	.7135	.7143	.7152
5.2	.7160	.7168	.7177	.7185	.7193	.7202	.7210	.7218	.7226	.7235
5.3	.7243	.7251	.7259	.7267	.7275	.7284	.7292	.7300	.7308	.7316
5.4	.7324	.7332	.7340	.7348	.7356	.7364	.7372	.7380	.7388	.7396
x	0	1	2	3	4	5	6	7	8	9

x	0	1	2	3	4	5	6	7	8	9
5.5	.7404	.7412	.7419	.7427	.7435	.7443	.7451	.7459	.7466	.7474
5.6	.7482	.7490	.7497	.7505	.7513	.7520	.7528	.7536	.7543	.7551
5.7	.7559	.7566	.7574	.7582	.7589	.7597	.7604	.7612	.7619	.7627
5.8	.7634	.7642	.7649	.7657	.7664	.7672	.7679	.7686	.7694	.7701
5.9	.7709	.7716	.7723	.7731	.7738	.7745	.7752	.7760	.7767	.7774
6.0	.7782	.7789	.7796	.7803	.7810	.7818	.7825	.7832	.7839	.7846
6.1	.7853	.7860	.7868	.7875	.7882	.7889	.7896	.7903	.7910	.7917
6.2	.7924	.7931	.7938	.7945	.7952	.7959	.7966	.7973	.7980	.7987
6.3	.7993	.8000	.8007	.8014	.8021	.8028	.8035	.8041	.8048	.8055
6.4	.8062	.8069	.8075	.8082	.8089	.8096	.8102	.8109	.8116	.8122
6.5	.8129	.8136	.8142	.8149	.8156	.8162	.8169	.8176	.8182	.8189
6.6	.8195	.8202	.8209	.8215	.8222	.8228	.8235	.8241	.8248	.8254
6.7	.8261	.8267	.8274	.8280	.8287	.8293	.8299	.8306	.8312	.8319
6.8	.8325	.8331	.8338	.8344	.8351	.8357	.8363	.8370	.8376	.8382
6.9	.8388	.8395	.8401	.8407	.8414	.8420	.8426	.8432	.8439	.8445
7.0	.8451	.8457	.8463	.8470	.8476	.8482	.8488	.8494	.8500	.8506
7.1	.8513	.8519	.8525	.8531	.8537	.8543	.8549	.8555	.8561	.8567
7.2	.8573	.8579	.8585	.8591	.8597	.8603	.8609	.8615	.8621	.8627
7.3	.8633	.8639	.8645	.8651	.8657	.8663	.8669	.8675	.8681	.8686
7.4	.8692	.8698	.8704	.8710	.8716	.8722	.8727	.8733	.8739	.8745
7.5	.8751	.8756	.8762	.8768	.8774	.8779	.8785	.8791	.8797	.8802
7.6	.8808	.8814	.8820	.8825	.8831	.8837	.8842	.8848	.8854	.8859
7.7	.8865	.8871	.8876	.8882	.8887	.8893	.8899	.8904	.8910	.8915
7.8	.8921	.8927	.8932	.8938	.8943	.8949	.8954	.8960	.8965	.8971
7.9	.8976	.8982	.8987	.8993	.8998	.9004	.9009	.9015	.9020	.9025
8.0	.9031	.9036	.9042	.9047	.9053	.9058	.9063	.9069	.9074	.9079
8.1	.9085	.9090	.9096	.9101	.9106	.9112	.9117	.9122	.9128	.9133
8.2	.9138	.9143	.9149	.9154	.9159	.9165	.9170	.9175	.9180	.9186
8.3	.9191	.9196	.9201	.9206	.9212	.9217	.9222	.9227	.9232	.9238
8.4	.9243	.9248	.9253	.9258	.9263	.9269	.9274	.9279	.9284	.9289
8.5	.9294	.9299	.9304	.9309	.9315	.9320	.9325	.9330	.9335	.9340
8.6	.9345	.9350	.9355	.9360	.9365	.9370	.9375	.9380	.9385	.9390
8.7	.9395	.9400	.9405	.9410	.9415	.9420	.9425	.9430	.9435	.9440
8.8	.9445	.9450	.9455	.9460	.9465	.9469	.9474	.9479	.9484	.9489
8.9	.9494	.9499	.9504	.9509	.9513	.9518	.9523	.9528	.9533	.9538
9.0	.9542	.9547	.9552	.9557	.9562	.9566	.9571	.9576	.9581	.9586
9.1	.9590	.9595	.9600	.9605	.9609	.9614	.9619	.9624	.9628	.9633
9.2	.9638	.9643	.9647	.9652	.9657	.9661	.9666	.9671	.9675	.9680
9.3	.9685	.9689	.9694	.9699	.9703	.9708	.9713	.9717	.9722	.9727
9.4	.9731	.9736	.9741	.9745	.9750	.9754	.9759	.9763	.9768	.9773
9.5	.9777	.9782	.9786	.9791	.9795	.9800	.9805	.9809	.9814	.9818
9.6	.9823	.9827	.9832	.9836	.9841	.9845	.9850	.9854	.9859	.9863
9.7	.9868	.9872	.9877	.9881	.9886	.9890	.9894	.9899	.9903	.9908
9.8	.9912	.9917	.9921	.9926	.9930	.9934	.9939	.9943	.9948	.9952
9.9	.9956	.9961	.9965	.9969	.9974	.9978	.9983	.9987	.9991	.9996
x	0	1	2	3	4	5	6	7	8	9

Table of Trigonometric Functions

DEG	MIN	RADIANS	SIN	COS	TAN	COT	SEC	CSC			
0	0	.0000	.0000	1.0000	.0000		1.000		1.5709	90	0
0	10	.0029	.0029	1.0000	.0029	343.773	1.000	343.774	1.5680	89	50
0	20	.0058	.0058	1.0000	.0058	171.885	1.000	171.888	1.5651	89	40
0	30	.0087	.0087	1.0000	.0087	114.589	1.000	114.593	1.5622	89	30
0	40	.0116	.0116	.9999	.0116	85.940	1.000	85.946	1.5593	89	20
0	50	.0145	.0145	.9999	.0145	68.750	1.000	68.757	1.5564	89	10
1	0	.0175	.0175	.9998	.0175	57.290	1.000	57.299	1.5534	89	0
1	10	.0204	.0204	.9998	.0204	49.104	1.000	49.114	1.5505	88	50
1	20	.0233	.0233	.9997	.0233	42.964	1.000	42.976	1.5476	88	40
1	30	.0262	.0262	.9997	.0262	38.188	1.000	38.202	1.5447	88	30
1	40	.0291	.0291	.9996	.0291	34.368	1.000	34.382	1.5418	88	20
1	50	.0320	.0320	.9995	.0320	31.242	1.001	31.258	1.5389	88	10
2	0	.0349	.0349	.9994	.0349	28.636	1.001	28.654	1.5360	88	0
2	10	.0378	.0378	.9993	.0378	26.432	1.001	26.451	1.5331	87	50
2	20	.0407	.0407	.9992	.0407	24.542	1.001	24.562	1.5302	87	40
2	30	.0436	.0436	.9990	.0437	22.904	1.001	22.926	1.5273	87	30
2	40	.0465	.0465	.9989	.0466	21.470	1.001	21.494	1.5244	87	20
2	50	.0495	.0494	.9988	.0495	20.206	1.001	20.230	1.5214	87	10
3	0	.0524	.0523	.9986	.0524	19.081	1.001	19.107	1.5185	87	0
3	10	.0553	.0552	.9985	.0553	18.075	1.002	18.103	1.5156	86	50
3	20	.0582	.0581	.9983	.0582	17.169	1.002	17.198	1.5127	86	40
3	30	.0611	.0610	.9981	.0612	16.350	1.002	16.380	1.5098	86	30
3	40	.0640	.0640	.9980	.0641	15.605	1.002	15.637	1.5069	86	20
3	50	.0669	.0669	.9978	.0670	14.924	1.002	14.958	1.5040	86	10
4	0	.0698	.0698	.9976	.0699	14.301	1.002	14.336	1.5011	86	0
4	10	.0727	.0727	.9974	.0729	13.727	1.003	13.763	1.4982	85	50
4	20	.0756	.0756	.9971	.0758	13.197	1.003	13.235	1.4953	85	40
4	30	.0785	.0785	.9969	.0787	12.706	1.003	12.746	1.4924	85	30
4	40	.0814	.0814	.9967	.0816	12.251	1.003	12.291	1.4894	85	20
4	50	.0844	.0843	.9964	.0846	11.826	1.004	11.868	1.4865	85	10
5	0	.0873	.0872	.9962	.0875	11.430	1.004	11.474	1.4836	85	0
5	10	.0902	.0901	.9959	.0904	11.059	1.004	11.105	1.4807	84	50
5	20	.0931	.0929	.9957	.0934	10.712	1.004	10.759	1.4778	84	40
5	30	.0960	.0958	.9954	.0963	10.385	1.005	10.433	1.4749	84	30
5	40	.0989	.0987	.9951	.0992	10.078	1.005	10.128	1.4720	84	20
5	50	.1018	.1016	.9948	.1022	9.788	1.005	9.839	1.4691	84	10
6	0	.1047	.1045	.9945	.1051	9.514	1.006	9.567	1.4662	84	0
6	10	.1076	.1074	.9942	.1080	9.255	1.006	9.309	1.4633	83	50
6	20	.1105	.1103	.9939	.1110	9.010	1.006	9.065	1.4604	83	40
6	30	.1134	.1132	.9936	.1139	8.777	1.006	8.834	1.4574	83	30
6	40	.1164	.1161	.9932	.1169	8.556	1.007	8.614	1.4545	83	20
6	50	.1193	.1190	.9929	.1198	8.345	1.007	8.405	1.4516	83	10
7	0	.1222	.1219	.9925	.1228	8.144	1.008	8.206	1.4487	83	0
7	10	.1251	.1248	.9922	.1257	7.953	1.008	8.016	1.4458	82	50
7	20	.1280	.1276	.9918	.1287	7.770	1.008	7.834	1.4429	82	40
7	30	.1309	.1305	.9914	.1317	7.596	1.009	7.661	1.4400	82	30
7	40	.1338	.1334	.9911	1346	7.429	1.009	7.496	1.4371	82	20
7	50	.1367	.1363	.9907	.1376	7.269	1.009	7.337	1.4342	82	10
8	0	.1396	.1392	.9903	.1405	7.115	1.010	7.185	1.4313	82	0
8	10	.1425	.1421	.9899	.1435	6.968	1.010	7.040	1.4284	81	50
8	20	.1454	.1449	.9894	.1465	6.827	1.011	6.900	1.4255	81	40
8	30	.1484	.1478	.9890	.1495	6.691	1.011	6.765	1.4225	81	30
8	40	.1513	.1507	.9886	.1524	6.561	1.012	6.636	1.4196	81	20
8	50	.1542	.1536	.9881	.1554	6.435	1.012	6.512	1.4167	81	10
9	0	.1571	.1564	.9877	.1584	6.314	1.012	6.392	1.4138	81	0
			COS	SIN	COT	TAN	CSC	SEC	RADIANS	DEG	MIN

DEG	MIN	RADIANS	SIN	COS	TAN	COT	SEC	CSC			
9	0	.1571	.1564	.9877	.1584	6.314	1.012	6.392	1.4138	81	0
9	10	.1600	.1593	.9872	.1614	6.197	1.013	6.277	1.4109	80	50
9	20	.1629	.1622	.9868	.1644	6.084	1.013	6.166	1.4080	80	40
9	30	.1658	.1650	.9863	.1673	5.976	1.014	6.059	1.4051	80	30
9	40	.1687	.1679	.9858	.1703	5.871	1.014	5.955	1.4022	80	20
9	50	.1716	.1708	.9853	.1733	5.769	1.015	5.855	1.3993	80	10
10	0	.1745	.1736	.9848	.1763	5.671	1.015	5.759	1.3964	80	0
10	10	.1774	.1765	.9843	.1793	5.576	1.016	5.665	1.3935	79	50
10	20	.1804	.1794	.9838	.1823	5.485	1.016	5.575	1.3905	79	40
10	30	.1833	.1822	.9833	.1853	5.396	1.017	5.487	1.3876	79	30
10	40	.1862	.1851	.9827	.1883	5.309	1.018	5.403	1.3847	79	20
10	50	.1891	.1880	.9822	.1914	5.226	1.018	5.320	1.3818	79	10
11	0	.1920	.1908	.9816	.1944	5.145	1.019	5.241	1.3789	79	0
11	10	.1949	.1937	.9811	.1974	5.066	1.019	5.164	1.3760	78	50
11	20	.1978	.1965	.9805	.2004	4.989	1.020	5.089	1.3731	78	40
11	30	.2007	.1994	.9799	.2035	4.915	1.020	5.016	1.3702	78	30
11	40	.2036	.2022	.9793	.2065	4.843	1.021	4.945	1.3673	78	20
11	50	.2065	.2051	.9787	.2095	4.773	1.022	4.876	1.3644	78	10
12	0	.2094	.2079	.9781	.2126	4.705	1.022	4.810	1.3615	78	0
12	10	.2123	.2108	.9775	.2156	4.638	1.023	4.745	1.3585	77	50
12	20	.2153	.2136	.9769	.2186	4.574	1.024	4.682	1.3556	77	40
12	30	.2182	.2164	.9763	.2217	4.511	1.024	4.620	1.3527	77	30
12	40	.2211	.2193	.9757	.2247	4.449	1.025	4.560	1.3498	77	20
12	50	.2240	.2221	.9750	.2278	4.390	1.026	4.502	1.3469	77	10
13	0	.2269	.2250	.9744	.2309	4.331	1.026	4.445	1.3440	77	0
13	10	.2298	.2278	.9737	.2339	4.275	1.027	4.390	1.3411	76	50
13	20	.2327	.2306	.9730	.2370	4.219	1.028	4.336	1.3382	76	40
13	30	.2356	.2334	.9724	.2401	4.165	1.028	4.284	1.3353	76	30
13	40	.2385	.2363	.9717	.2432	4.113	1.029	4.232	1.3324	76	20
13	50	.2414	.2391	.9710	.2462	4.061	1.030	4.182	1.3295	76	10
14	0	.2443	.2419	.9703	.2493	4.011	1.031	4.134	1.3265	76	0
14	10	.2473	.2447	.9696	.2524	3.962	1.031	4.086	1.3236	75	50
14	20	.2502	.2476	.9689	.2555	3.914	1.032	4.039	1.3207	75	40
14	30	.2531	.2504	.9681	.2586	3.867	1.033	3.994	1.3178	75	30
14	40	.2560	.2532	.9674	.2617	3.821	1.034	3.950	1.3149	75	20
14	50	.2589	.2560	.9667	.2648	3.776	1.034	3.906	1.3120	75	10
15	0	.2618	.2588	.9659	.2679	3.732	1.035	3.864	1.3091	75	0
15	10	.2647	.2616	.9652	.2711	3.689	1.036	3.822	1.3062	74	50
15	20	.2676	.2644	.9644	.2742	3.647	1.037	3.782	1.3033	74	40
15	30	.2705	.2672	.9636	.2773	3.606	1.038	3.742	1.3004	74	30
15	40	.2734	.2700	.9628	.2805	3.566	1.039	3.703	1.2975	74	20
15	50	.2763	.2728	.9621	.2836	3.526	1.039	3.665	1.2946	74	10
16	0	.2793	.2756	.9613	.2867	3.487	1.040	3.628	1.2916	74	0
16	10	.2822	.2784	.9605	.2899	3.450	1.041	3.592	1.2887	73	50
16	20	.2851	.2812	.9596	.2931	3.412	1.042	3.556	1.2858	73	40
16	30	.2880	.2840	.9588	.2962	3.376	1.043	3.521	1.2829	73	30
16	40	.2909	.2868	.9580	.2994	3.340	1.044	3.487	1.2800	73	20
16	50	.2938	.2896	.9572	.3026	3.305	1.045	3.453	1.2771	73	10
17	0	.2967	.2924	.9563	.3057	3.271	1.046	3.420	1.2742	73	0
17	10	.2996	.2952	.9555	.3089	3.237	1.047	3.388	1.2713	72	50
17	20	.3025	.2979	.9546	.3121	3.204	1.048	3.356	1.2684	72	40
17	30	.3054	.3007	.9537	.3153	3.172	1.049	3.326	1.2655	72	30
17	40	.3083	.3035	.9528	.3185	3.140	1.049	3.295	1.2626	72	20
17	50	.3113	.3062	.9520	.3217	3.108	1.050	3.265	1.2596	72	10
18	0	.3142	.3090	.9511	.3249	3.078	1.051	3.236	1.2567	72	0
			COS	SIN	COT	TAN	CSC	SEC	RADIANS	DEG	MIN

TABLE OF TRIGONOMETRIC FUNCTIONS **595**

DEG	MIN	RADIANS	SIN	COS	TAN	COT	SEC	CSC			
18	0	.3142	.3090	.9511	.3249	3.078	1.051	3.236	1.2567	72	0
18	10	.3171	.3118	.9502	.3281	3.047	1.052	3.207	1.2538	71	50
18	20	.3200	.3145	.9492	.3314	3.018	1.053	3.179	1.2509	71	40
18	30	.3229	.3173	.9483	.3346	2.989	1.054	3.152	1.2480	71	30
18	40	.3258	.3201	.9474	.3378	2.960	1.056	3.124	1.2451	71	20
18	50	.3287	.3228	.9465	.3411	2.932	1.057	3.098	1.2422	71	10
19	0	.3316	.3256	.9455	.3443	2.904	1.058	3.072	1.2393	71	0
19	10	.3345	.3283	.9446	.3476	2.877	1.059	3.046	1.2364	70	50
19	20	.3374	.3311	.9436	.3508	2.850	1.060	3.021	1.2335	70	40
19	30	.3403	.3338	.9426	.3541	2.824	1.061	2.996	1.2306	70	30
19	40	.3432	.3365	.9417	.3574	2.798	1.062	2.971	1.2276	70	20
19	50	.3462	.3393	.9407	.3607	2.773	1.063	2.947	1.2247	70	10
20	0	.3491	.3420	.9397	.3640	2.747	1.064	2.924	1.2218	70	0
20	10	.3520	.3448	.9387	.3673	2.723	1.065	2.901	1.2189	69	50
20	20	.3549	.3475	.9377	.3706	2.699	1.066	2.878	1.2160	69	40
20	30	.3578	.3502	.9367	.3739	2.675	1.068	2.855	1.2131	69	30
20	40	.3607	.3529	.9356	.3772	2.651	1.069	2.833	1.2102	69	20
20	50	.3636	.3557	.9346	.3805	2.628	1.070	2.812	1.2073	69	10
21	0	.3665	.3584	.9336	.3839	2.605	1.071	2.790	1.2044	69	0
21	10	.3694	.3611	.9325	.3872	2.583	1.072	2.769	1.2015	68	50
21	20	.3723	.3638	.9315	.3906	2.560	1.074	2.749	1.1986	68	40
21	30	.3752	.3665	.9304	.3939	2.539	1.075	2.729	1.1957	68	30
21	40	.3782	.3692	.9293	.3973	2.517	1.076	2.709	1.1927	68	20
21	50	.3811	.3719	.9283	.4006	2.496	1.077	2.689	1.1898	68	10
22	0	.3840	.3746	.9272	.4040	2.475	1.079	2.669	1.1869	68	0
22	10	.3869	.3773	.9261	.4074	2.455	1.080	2.650	1.1840	67	50
22	20	.3898	.3800	.9250	.4108	2.434	1.081	2.632	1.1811	67	40
22	30	.3927	.3827	.9239	.4142	2.414	1.082	2.613	1.1782	67	30
22	40	.3956	.3854	.9228	.4176	2.394	1.084	2.595	1.1753	67	20
22	50	.3985	.3881	.9216	.4210	2.375	1.085	2.577	1.1724	67	10
23	0	.4014	.3907	.9205	.4245	2.356	1.086	2.559	1.1695	67	0
23	10	.4043	.3934	.9194	.4279	2.337	1.088	2.542	1.1666	66	50
23	20	.4072	.3961	.9182	.4314	2.318	1.089	2.525	1.1637	66	40
23	30	.4102	.3987	.9171	.4348	2.300	1.090	2.508	1.1607	66	30
23	40	.4131	.4014	.9159	.4383	2.282	1.092	2.491	1.1578	66	20
23	50	.4160	.4041	.9147	.4417	2.264	1.093	2.475	1.1549	66	10
24	0	.4189	.4067	.9135	.4452	2.246	1.095	2.459	1.1520	66	0
24	10	.4218	.4094	.9124	.4487	2.229	1.096	2.443	1.1491	65	50
24	20	.4247	.4120	.9112	.4522	2.211	1.097	2.427	1.1462	65	40
24	30	.4276	.4147	.9100	.4557	2.194	1.099	2.411	1.1433	65	30
24	40	.4305	.4173	.9088	.4592	2.177	1.100	2.396	1.1404	65	20
24	50	.4334	.4200	.9075	.4628	2.161	1.102	2.381	1.1375	65	10
25	0	.4363	.4226	.9063	.4663	2.145	1.103	2.366	1.1346	65	0
25	10	.4392	.4253	.9051	.4699	2.128	1.105	2.352	1.1317	64	50
25	20	.4421	.4279	.9038	.4734	2.112	1.106	2.337	1.1287	64	40
25	30	.4451	.4305	.9026	.4770	2.097	1.108	2.323	1.1258	64	30
25	40	.4480	.4331	.9013	.4806	2.081	1.109	2.309	1.1229	64	20
25	50	.4509	.4358	.9001	.4841	2.066	1.111	2.295	1.1200	64	10
26	0	.4538	.4384	.8988	.4877	2.050	1.113	2.281	1.1171	64	0
26	10	.4567	.4410	.8975	.4913	2.035	1.114	2.268	1.1142	63	50
26	20	.4596	.4436	.8962	.4950	2.020	1.116	2.254	1.1113	63	40
26	30	.4625	.4462	.8949	.4986	2.006	1.117	2.241	1.1084	63	30
26	40	.4654	.4488	.8936	.5022	1.991	1.119	2.228	1.1055	63	20
26	50	.4683	.4514	.8923	.5059	1.977	1.121	2.215	1.1026	63	10
27	0	.4712	.4540	.8910	.5095	1.963	1.122	2.203	1.0997	63	0
			COS	SIN	COT	TAN	CSC	SEC	RADIANS	DEG	MIN

DEG	MIN	RADIANS	SIN	COS	TAN	COT	SEC	CSC			
27	0	.4712	.4540	.8910	.5095	1.963	1.122	2.203	1.0997	63	0
27	10	.4741	.4566	.8897	.5132	1.949	1.124	2.190	1.0967	62	50
27	20	.4771	.4592	.8884	.5169	1.935	1.126	2.178	1.0938	62	40
27	30	.4800	.4617	.8870	.5206	1.921	1.127	2.166	1.0909	62	30
27	40	.4829	.4643	.8857	.5243	1.907	1.129	2.154	1.0880	62	20
27	50	.4858	.4669	.8843	.5280	1.894	1.131	2.142	1.0851	62	10
28	0	.4887	.4695	.8829	.5317	1.881	1.133	2.130	1.0822	62	0
28	10	.4916	.4720	.8816	.5354	1.868	1.134	2.118	1.0793	61	50
28	20	.4945	.4746	.8802	.5392	1.855	1.136	2.107	1.0764	61	40
28	30	.4974	.4772	.8789	.5430	1.842	1.138	2.096	1.0735	61	30
28	40	.5003	.4797	.8774	.5467	1.829	1.140	2.085	1.0706	61	20
28	50	.5032	.4823	.8760	.5505	1.816	1.142	2.074	1.0677	61	10
29	0	.5061	.4848	.8746	.5543	1.804	1.143	2.063	1.0648	61	0
29	10	.5091	.4874	.8732	.5581	1.792	1.145	2.052	1.0618	60	50
29	20	.5120	.4899	.8718	.5619	1.780	1.147	2.041	1.0589	60	40
29	30	.5149	.4924	.8704	.5658	1.767	1.149	2.031	1.0560	60	30
29	40	.5178	.4950	.8689	.5696	1.756	1.151	2.020	1.0531	60	20
29	50	.5207	.4975	.8675	.5735	1.744	1.153	2.010	1.0502	60	10
30	0	.5236	.5000	.8660	.5774	1.732	1.155	2.000	1.0473	60	0
30	10	.5265	.5025	.8646	.5812	1.720	1.157	1.990	1.0444	59	50
30	20	.5294	.5050	.8631	.5851	1.709	1.159	1.980	1.0415	59	40
30	30	.5323	.5075	.8616	.5890	1.698	1.161	1.970	1.0386	59	30
30	40	.5352	.5100	.8601	.5930	1.686	1.163	1.961	1.0357	59	20
30	50	.5381	.5125	.8587	.5969	1.675	1.165	1.951	1.0328	59	10
31	0	.5411	.5150	.8572	.6009	1.664	1.167	1.942	1.0298	59	0
31	10	.5440	.5175	.8557	.6048	1.653	1.169	1.932	1.0269	58	50
31	20	.5469	.5200	.8542	.6088	1.643	1.171	1.923	1.0240	58	40
31	30	.5498	.5225	.8526	.6128	1.632	1.173	1.914	1.0211	58	30
31	40	.5527	.5250	.8511	.6168	1.621	1.175	1.905	1.0182	58	20
31	50	.5556	.5274	.8496	.6208	1.611	1.177	1.896	1.0153	58	10
32	0	.5585	.5299	.8480	.6249	1.600	1.179	1.887	1.0124	58	0
32	10	.5614	.5324	.8465	.6289	1.590	1.181	1.878	1.0095	57	50
32	20	.5643	.5348	.8450	.6330	1.580	1.184	1.870	1.0066	57	40
32	30	.5672	.5373	.8434	.6371	1.570	1.186	1.861	1.0037	57	30
32	40	.5701	.5398	.8418	.6412	1.560	1.188	1.853	1.0008	57	20
32	50	.5730	.5422	.8403	.6453	1.550	1.190	1.844	.9978	57	10
33	0	.5760	.5446	.8387	.6494	1.540	1.192	1.836	.9949	57	0
33	10	.5789	.5471	.8371	.6536	1.530	1.195	1.828	.9920	56	50
33	20	.5818	.5495	.8355	.6577	1.520	1.197	1.820	.9891	56	40
33	30	.5847	.5519	.8339	.6619	1.511	1.199	1.812	.9862	56	30
33	40	.5876	.5544	.8323	.6661	1.501	1.202	1.804	.9833	56	20
33	50	.5905	.5568	.8307	.6703	1.492	1.204	1.796	.9804	56	10
34	0	.5934	.5592	.8290	.6745	1.483	1.206	1.788	.9775	56	0
34	10	.5963	.5616	.8274	.6787	1.473	1.209	1.781	.9746	55	50
34	20	.5992	.5640	.8258	.6830	1.464	1.211	1.773	.9717	55	40
34	30	.6021	.5664	.8241	.6873	1.455	1.213	1.766	.9688	55	30
34	40	.6050	.5688	.8225	.6916	1.446	1.216	1.758	.9658	55	20
34	50	.6080	.5712	.8208	.6959	1.437	1.218	1.751	.9629	55	10
35	0	.6109	.5736	.8192	.7002	1.428	1.221	1.743	.9600	55	0
35	10	.6138	.5760	.8175	.7046	1.419	1.223	1.736	.9571	54	50
35	20	.6167	.5783	.8158	.7089	1.411	1.226	1.729	.9542	54	40
35	30	.6196	.5807	.8141	.7133	1.402	1.228	1.722	.9513	54	30
35	40	.6225	.5831	.8124	.7177	1.393	1.231	1.715	.9484	54	20
35	50	.6254	.5854	.8107	.7221	1.385	1.233	1.708	.9455	54	10
36	0	.6283	.5878	.8090	.7265	1.376	1.236	1.701	.9426	54	0
			COS	SIN	COT	TAN	CSC	SEC	RADIANS	DEG	MIN

DEG	MIN	RADIANS	SIN	COS	TAN	COT	SEC	CSC			
36	0	.6283	.5878	.8090	.7265	1.376	1.236	1.701	.9426	54	0
36	10	.6312	.5901	.8073	.7310	1.368	1.239	1.695	.9397	53	50
36	20	.6341	.5925	.8056	.7355	1.360	1.241	1.688	.9368	53	40
36	30	.6370	.5948	.8039	.7400	1.351	1.244	1.681	.9339	53	30
36	40	.6400	.5972	.8021	.7445	1.343	1.247	1.675	.9309	53	20
36	50	.6429	.5995	.8004	.7490	1.335	1.249	1.668	.9280	53	10
37	0	.6458	.6018	.7986	.7536	1.327	1.252	1.662	.9251	53	0
37	10	.6487	.6041	.7969	.7581	1.319	1.255	1.655	.9222	52	50
37	20	.6516	.6065	.7951	.7627	1.311	1.258	1.649	.9193	52	40
37	30	.6545	.6088	.7934	.7673	1.303	1.260	1.643	.9164	52	30
37	40	.6574	.6111	.7916	.7720	1.295	1.263	1.636	.9135	52	20
37	50	.6603	.6134	.7898	.7766	1.288	1.266	1.630	.9106	52	10
38	0	.6632	.6157	.7880	.7813	1.280	1.269	1.624	.9077	52	0
38	10	.6661	.6180	.7862	.7860	1.272	1.272	1.618	.9048	51	50
38	20	.6690	.6202	.7844	.7907	1.265	1.275	1.612	.9019	51	40
38	30	.6720	.6225	.7826	.7954	1.257	1.278	1.606	.8989	51	30
38	40	.6749	.6248	.7808	.8002	1.250	1.281	1.601	.8960	51	20
38	50	.6778	.6271	.7790	.8050	1.242	1.284	1.595	.8931	51	10
39	0	.6807	.6293	.7771	.8098	1.235	1.287	1.589	.8902	51	0
39	10	.6836	.6316	.7753	.8146	1.228	1.290	1.583	.8873	50	50
39	20	.6865	.6338	.7735	.8195	1.220	1.293	1.578	.8844	50	40
39	30	.6894	.6361	.7716	.8243	1.213	1.296	1.572	.8815	50	30
39	40	.6923	.6383	.7698	.8292	1.206	1.299	1.567	.8786	50	20
39	50	.6952	.6406	.7679	.8342	1.199	1.302	1.561	.8757	50	10
40	0	.6981	.6428	.7660	.8391	1.192	1.305	1.556	.8728	50	0
40	10	.7010	.6450	.7642	.8441	1.185	1.309	1.550	.8699	49	50
40	20	.7039	.6472	.7623	.8491	1.178	1.312	1.545	.8669	49	40
40	30	.7069	.6494	.7604	.8541	1.171	1.315	1.540	.8640	49	30
40	40	.7098	.6517	.7585	.8591	1.164	1.318	1.535	.8611	49	20
40	50	.7127	.6539	.7566	.8642	1.157	1.322	1.529	.8582	49	10
41	0	.7156	.6561	.7547	.8693	1.150	1.325	1.524	.8553	49	0
41	10	.7185	.6583	.7528	.8744	1.144	1.328	1.519	.8524	48	50
41	20	.7214	.6604	.7509	.8796	1.137	1.332	1.514	.8495	48	40
41	30	.7243	.6626	.7490	.8847	1.130	1.335	1.509	.8466	48	30
41	40	.7272	.6648	.7470	.8899	1.124	1.339	1.504	.8437	48	20
41	50	.7301	.6670	.7451	.8952	1.117	1.342	1.499	.8408	48	10
42	0	.7330	.6691	.7431	.9004	1.111	1.346	1.494	.8379	48	0
42	10	.7359	.6713	.7412	.9057	1.104	1.349	1.490	.8349	47	50
42	20	.7389	.6734	.7392	.9110	1.098	1.353	1.485	.8320	47	40
42	30	.7418	.6756	.7373	.9163	1.091	1.356	1.480	.8291	47	30
42	40	.7447	.6777	.7353	.9217	1.085	1.360	1.476	.8262	47	20
42	50	.7476	.6799	.7333	.9271	1.079	1.364	1.471	.8233	47	10
43	0	.7505	.6820	.7314	.9325	1.072	1.367	1.466	.8204	47	0
43	10	.7534	.6841	.7294	.9380	1.066	1.371	1.462	.8175	46	50
43	20	.7563	.6862	.7274	.9435	1.060	1.375	1.457	.8146	46	40
43	30	.7592	.6884	.7254	.9490	1.054	1.379	1.453	.8117	46	40
43	40	.7621	.6905	.7234	.9545	1.048	1.382	1.448	.8188	46	20
43	50	.7650	.6926	.7214	.9601	1.042	1.386	1.444	.8059	46	10
44	0	.7679	.6947	.7193	.9657	1.036	1.390	1.440	.8030	46	0
44	10	.7709	.6967	.7173	.9713	1.030	1.394	1.435	.8000	45	50
44	20	.7738	.6988	.7153	.9770	1.024	1.398	1.431	.7971	45	40
44	30	.7767	.7009	.7133	.9827	1.018	1.402	1.427	.7942	45	30
44	40	.7796	.7030	.7112	.9884	1.012	1.406	1.423	.7913	45	20
44	50	.7825	.7050	.7092	.9942	1.006	1.410	1.418	.7884	45	10
45	0	.7854	.7071	.7071	1.0000	1.000	1.414	1.414	.7855	45	0
			COS	SIN	COT	TAN	CSC	SEC	RADIANS	DEG	MIN

Answers to Selected Exercises

CHAPTER 1: PRELIMINARIES

SECTION 1 PAGES 7–9

A.2. (a) $7 > 5$ (c) $x \geq 0$ (e) $-3 < z < -2$ (g) $c < 4 \leq d$ **A.3.**
(a) $-1000 < .01$ (c) $77.77 < 77.777$ (e) $\sqrt{19} > 4$ (g) $\pi > 3.1415$
(i) $-\frac{1}{10} > -6$ **A.5.** π is the negative of $-\pi$; the negative of $(4 + (-6))$ is $-(4 + (-6)) = -4 + 6 = 2.$ **B.4.** (a) $1/a \geq 1/b$ (c) $a^2 \leq b^2$ (e) $a^3 \leq b^3$
B.5. (b) The point a lies on or to the right of b.   (g) $a < b$ **B.6.** (a) 48
(c) .42   (g) 34

PAGES 10–11

A.1. (a) $.777\cdots$ (c) $1.6428571428571\cdots$ (e) $.0526315789473684210521052\cdots$
C.1. (a) $\frac{37}{99}$ (c) $\frac{758,679}{9900}$

SECTION 2 PAGE 13

B.1. (a) Q, A (c) Z, Q, A (e) A (g) A

SECTION 3 PAGES 18–20

A.1. (a) -7 (c) 169 (e) -1 (g) $\pi - \sqrt{2}$ **A.2.** (a) $<$ (c) $>$
(e) $<$ (g) $=$ **A.4.** (a) 14.5   (c) 100   (e) $\pi + 3$ **B.3.** (d) b is closer to
0 than c is to 3.   (f) $|x - c| = 5$ **B.4.** (a) c can't be within 2 units of 1 *and* within 3 units of 12 at the same time.   **B.5.** (a) all real numbers x (c) $x \leq 0$
B.6. (a) $x = 1$ or -1 (c) $x = 1$ or 3 (e) $x = \pi - 4$ or $\pi + 4$ **B.7.** (a) $x = 0$ or
-4 (c) $z = 2$ or -4 **B.8.** (a) $x = -\frac{7}{2}$ or $\frac{1}{2}$ (c) $x = \frac{5}{2}$ or -1 (e) $x = \frac{5}{6}$
or $-\frac{1}{2}$ **B.9.** $|x - 1| = 2$ or 4; $|x| = 1$ or 5   **B.10.** $|x - 2| = 1$ or 7
B.11. (a) $-7 < x < 7$ (c) $-4 < x < -2$ (e) $x \geq 5$ or $x \leq -5$

SECTION 4 PAGES 25–26

A.1. (a) $(-\infty, -\frac{8}{5}]$ (c) $(1, \infty)$ (e) $(-\infty, \frac{4}{7})$ **B.1.** (a) $[-\frac{7}{17}, \infty)$
(c) $(-\infty, \frac{9}{5})$ **B.2.** (a) $(-\infty, 3)$ (c) $[1, \infty)$ **B.3.** (a) $[4, \infty)$ (c) $(-20, \infty)$

B.4. (a) $(2, 4)$ (c) $[-2, -\frac{3}{4}]$ (e) $[-1, \frac{1}{8})$ **B.5.** (a) $[-4, 8]$ (c) no solutions
B.6. (a) $[-\frac{4}{3}, 0]$ (c) $(\frac{7}{6}, \frac{11}{6})$ (e) $(-\frac{5}{2}, -\frac{1}{2})$ **B.7.** (a) $x < -2$ or $x > -1$
(c) $x \le -\frac{11}{20}$ or $x \ge -\frac{1}{4}$ (e) $x < \frac{3}{7}$ or $x > \frac{5}{7}$ **B.8.** (a) $x < -\frac{53}{40}$ or $x > -\frac{43}{40}$
(c) $x < -\frac{7}{4}$ or $x > \frac{13}{4}$ (e) $[-\frac{4}{3}, \frac{8}{3}]$ **B.9.** (a) approximately $[.602, \infty)$ (c) approximately $(-\infty, -1.053)$ **B.10.** (a) All real numbers are solutions since $|2x + 1| \ge 0$ no matter what x is.

CHAPTER 2: INTRODUCTION TO ANALYTIC GEOMETRY

SECTION 1 PAGES 34–36

A.1. A $(-3, 3)$; C $(-2.3, 0)$; E $(0, 2)$; G $(2, 0)$; I $(3, -1)$
A.3. (a) $(-1.5, 4)$ (c) $(2.5, 6)$ (e) $(\frac{13}{4}, -\frac{1}{6})$ **A.4.** (a) 13 (c) 5
(e) $\sqrt{2}|a - b|$ **B.4.** (a) The other vertices of possible squares are: $(6, 1)$, $(6, 5)$ and $(-2, 1)$, $(-2, 5)$ and $(4, 3)$, $(0, 3)$. **B.6.** (a) vertical straight line through $(3, 0)$
(c) straight line passing through $(0, 0)$ and bisecting the first and third quadrants (e) all points in the first and third quadrants (coordinate axes *not* included) (g) same as (e)

B.7.

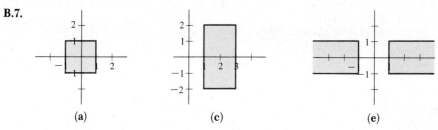

(a)   (c)   (e)

B.9. $10 + 5\sqrt{2}$ **B.10** (a) $(0, 0)$, $(6, 0)$ (c) $(3, -5 + \sqrt{11})$, $(3, -5 - \sqrt{11})$
C.1. (a) $(0, -5)$ goes to $(0, 0)$; $(2, 2)$ goes to $(2, 7)$ (c) $(a, b + 5)$ (e) $(-4a, b - 5)$

SECTION 2 PAGES 39–45

C.1. (a) 7 (c) 6 (e) 8 **C.2.** (a) 16.6 million (c) 1972 and 1985
(e) 1982, decreased; 1995, increased. **C.3.** (a) Computronics (c) From March 10 to March 20 the price of Synergistics *rose* approximately $23; the price of Computronics never rose more than $20 in any 30-day period. **C.4.** (a) 35 (c) $\frac{4500}{25} = \$180$ for the first 25. (e) The first 20 cost the most. **C.5.** (a) Car A travels approximately 32 yd from 0 to 5 sec; car B travels approximately 50 yd from 10 to 12 sec. (c) At 10 sec car B leads by 100 yd, but car A is traveling at a faster rate of speed. **C.6.** (a) approximately 13 mi (c) Car B went approximately 12 mi in the first 15 min and approximately 21 mi in the first 30 min; so car B went approximately $21 - 12 = 9$ mi between 15 and 30 min after start. (e) Car A traveled 40 mi in 60 min, so its average speed was $\frac{40}{60} = \frac{2}{3}$ mi/min. (g) The steeper the graph is over an interval, the greater the average speed over that interval (why?). **C.7.** (a) (iii) no change (c) The water level *rose* the fastest from noon to 3:00 P.M. (a 10-ft rise in 3 hr). It rose more slowly from 7:30 P.M. to midnight (a 10-ft rise in 4.5 hr). (e) The average *rate* of change in water level from 3:00 to 8:00 P.M. was $\frac{26}{5} = 5.2$ ft/hr; from approximately 5:00 to 6:00 P.M. the level fell approx. 5.2 ft, so the rate during that 1 hr was also 5.2 ft/hr.

SECTION 3   PAGES 55–57

A.1. (a) $\frac{5}{2}$ (c) 4   (e) $-\frac{8}{7}$ (g) $(\sqrt{5} + \sqrt{2})/7$ (i) $(1 - \pi)/(1 + \pi)$
A.4. (a) parallel   (c) perpendicular   (e) parallel   **B.3.** approximately
60.208 ft   **B.5.** (b) no   **B.6.** (a) yes   **B.8.** (a) 22   (c) -5 (e) 24

SECTION 4   PAGES 65–68

A.1. (a) yes, since $3 \cdot 1 - (-2) - 5 = 0$ (c) yes, since $3 \cdot 2 + 6 = 12$ (e) no, since
$(3 - 2)^2 + (4 + 5)^2 \neq 4$ **A.2.** (a) $y - 5 = 1(x - 3)$, or equivalently, $y = x + 2$
(c) $y = -x + 8$ (e) $y = x/2 + \frac{5}{2}$ **A.3.** (a) $y = -x - 5$
(c) $y = -7x/3 + \frac{34}{9}$ (e) $y = x/8.7$, or equivalently, $y = 10x/87$ (g) $y - \sqrt{2} =$
$\dfrac{\sqrt{6} - \sqrt{2}}{-\sqrt{2} - \sqrt{8}}(x - \sqrt{8})$ **A.4.** (a) $y = x + 2$ (c) $y = -4x + 2$ (e) $y = x/2 + 3$
A.5. $m =$ slope, $b = y$-intercept: (a) $m = 2$, $b = 5$ (c) $m = -4$, $b = -\frac{7}{2}$
(e) $m = -\frac{3}{7}$, $b = -\frac{11}{7}$ **B.1.** (a) $y = 3x + 7$ (c) $y = 3x/2$ (e) $y = x - 5$
(g) $y = -x + 2$ (i) $y = 2$ (k) $y = x/3$ **B.2.** (a) parallel   (c) perpendicular   (e) perpendicular   **B.3.** (a) $k = -\frac{11}{3}$

B.8.

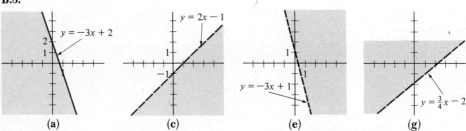

(a)   (c)   (e)   (g)

B.9.

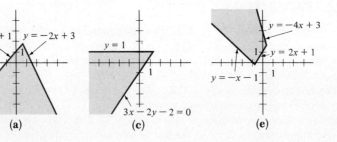

(a)   (c)   (e)

C.2. (a) The distance from P to Q is 9400 ft.   (b) B is 88 ft from the road.

SECTION 5   PAGES 78–81

A.1. (a) $(x + 3)^2 + (y - 4)^2 = 4$ (c) $x^2 + y^2 = 2$ (e) $(x - 4)^2 + (y - 7)^2 = \frac{1}{4}$
(g) $(x - 3)^2 + (y - \frac{8}{3})^2 = 3$

A.2.

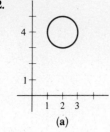

(a)

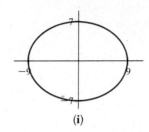

(c)

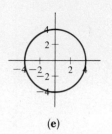

(e)

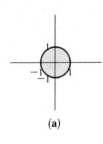

(g)

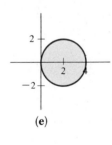

(i)

A.3. **(a)** $x^2/49 + y^2/4 = 1$ **(c)** $x^2/81 + y^2/100 = 1$ **A.4.** **(a)** 8π **(c)** $2\pi\sqrt{3}$
(e) $7\pi/\sqrt{3}$ **A.5.** **(a)** $x^2/4 - y^2/9 = 1$ **B.1.** **(a)** $(x-2)^2 + (y-2)^2 = 8$
(c) $(x-1)^2 + (y-2)^2 = 8$ **(e)** $(x-4)^2 + (y-3)^2 = 81$ **(g)** $(x+5)^2 +$
$(y-4)^2 = 16$ **B.2.** **(a)** inside   **(c)** outside   **(e)** outside

B.3.

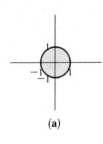

(a)

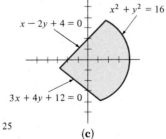

(c)

(e)

B.4.

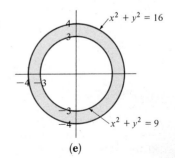

(a)   **(c)**   **(e)**

B.4.

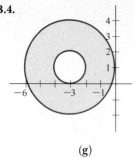

(g)

B.6.

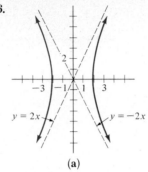

$y = 2x$ $y = -2x$

(a)

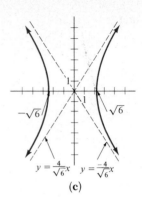

$y = \frac{4}{\sqrt{6}}x$ $y = \frac{-4}{\sqrt{6}}x$

(c)

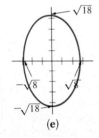

(e)

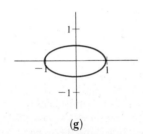

(g)

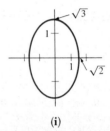

(i)

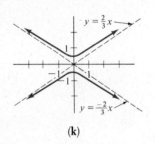

$y = \frac{2}{3}x$ $y = \frac{-2}{3}x$

(k)

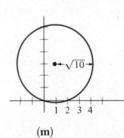

(m)

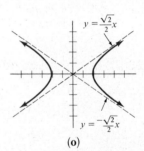

$y = \frac{\sqrt{2}}{2}x$ $y = \frac{-\sqrt{2}}{2}x$

(o)

B.7. (a) $\Rightarrow$ means "implies". $a > b > 0 \Rightarrow \sqrt{a^2 - b^2} > 0$ and $a > 0 \Rightarrow e = \dfrac{\sqrt{a^2 - b^2}}{a}$ > 0. Furthermore, $a > 0$ and $b > 0 \Rightarrow b/a > 0 \Rightarrow b^2/a^2 > 0 \Rightarrow 1 + (b^2/a^2) > 1 \Rightarrow 1$ $> 1 - (b^2/a^2) \Rightarrow 1 > (a^2 - b^2)/a^2 = e^2 \Rightarrow 1 > e$. **C.2.** $\sqrt{300} \approx 17.32$ ft

PAGES 85–86

A.1. (a) center $(-4, 3)$, radius $2\sqrt{10}$ (c) center $(-3, 2)$, radius $2\sqrt{7}$ (e) center $(-12.5, -5)$, radius $\sqrt{169.25}$ (g) center $(-\frac{1}{4}, \frac{1}{4})$, radius $\sqrt{\frac{13}{8}}$ **B.1.** (a) $x^2 + y^2 - 3y/2 - 1 = 0$ (c) $x^2 + y^2 - 2x - 4y - 5 = 0$

B.2.

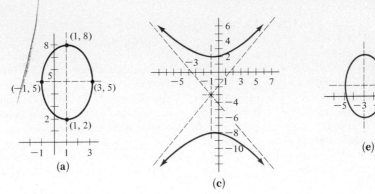

(a)   (c)   (e)

CHAPTER 3: FUNCTIONS

SECTION 1 PAGES 93–95

A.1. **(b)** $[-\tfrac{4}{3}] = -2$, $[-16\tfrac{1}{2}] = -17$, $[6.75] = 6$, $[\tfrac{2}{3}] = 0$ **B.2.** **(a)** tax on \$500 is 0; tax on \$6783 is \$119.15; tax on \$12,500 is \$405   **B.4.** **(a)** The domain is all *nonnegative* numbers. Let c and d be *different* numbers in the domain. The rule assigns $16c^2$ to c and $16d^2$ to d (see p. 88). If we had $16c^2 = 16d^2$, then we would have $c^2 = d^2$, so that $c = \pm\sqrt{d^2} = \pm d$. Since c and d are both nonnegative, they can't have opposite signs. So we would have $c = d$. This is impossible since c and d are different numbers. So it can't happen that $16c^2 = 16d^2$ when $c \neq d$. Therefore the function is one-to-one.   **B.6.** **(a)** domain $[-3, 3]$, range $[-4, 4]$ **(c)** domain $[-2.5, 2]$, range $[-3, 2.5]$

SECTION 2 PAGES 103–107

A.1. **(a)** $\sqrt{3} + 1$ **(c)** $\sqrt{\tfrac{11}{2}} - \tfrac{3}{2}$ **(e)** $\sqrt{\sqrt{2} + 3} - \sqrt{2} + 1$ **(g)** 4
A.2. **(a)** $\tfrac{34}{3}$ **(c)** $\tfrac{59}{12}$ **(e)** $(a + k)^2 + 1/(a + k) + 2$ **(g)** $(2 - x)^2 +$
$(2 - x) + 2$ **A.3.** **(a)** 8   **(c)** -1 **(e)** $s^2 + 2s$ **(g)** $t^2 - 1$
(i) $1/t^2 - 1$ **A.4.** **(a)** 4   **(c)** $270 + \sqrt{14}$ **(e)** 11   **(g)** $|2 + x| +$
$\sqrt{-x - 2} + x^2$ **(i)** $28 + \sqrt{3} - |2 - x| - \sqrt{x - 2} - x^2$

A.5.

	$f(2)$	$f(\tfrac{16}{3})$	$f(2) - f(\tfrac{16}{3})$	$f(r)$	$f(r) - f(x)$	$\dfrac{f(r) - f(x)}{r - x}$
(a)	2	$\tfrac{16}{3}$	$-\tfrac{10}{3}$	r	$r - x$	1
(c)	12	12	0	12	0	0
(e)	13	23	-10	$3r + 7$	$3(r - x)$	3
(g)	-2	$-\tfrac{208}{9}$	$\tfrac{190}{9}$	$r - r^2$	$r - r^2 - x + x^2$	$1 - r - x$
(i)	$\tfrac{1}{2}$	$\tfrac{3}{16}$	$\tfrac{5}{16}$	$1/r$	$1/r - 1/x$	$1/rx$

A.6. (a) $f(-3) \approx .7$; $f(-3/2) \approx 2.1$; $f(0) \approx -2.8$; $f(1) \approx 0$; $f(5/2) \approx 1.8$; $f(4) \approx 1.6$
A.7. (a) $f(-3) + f(-3/2) \approx .7 + 2.1 = 2.8$; $\qquad\qquad$ $f(0) - f(2) \approx -3 - 2.8 = -5.8$
(c) $f(5/2) - f(3) = 3 - (-1) = 4$; $\qquad\qquad$ $f(4) + 3f(-2) \approx .5 + 3(-1.8) = -4.9$
A.8. (b) $T(3.6) = .18$; $T(0.6) = .03$ $\quad$ **B.1.** (b) (i) is true; (iii) is false. $\quad$ **B.2.** (a) (i),
(ii), (iv) are true; (iii) is false for all $x \neq 0$. $\quad$ (c) all are true $\quad$ **B.3.** (a) all real numbers
(c) all real numbers $\quad$ (e) all real numbers $\quad$ (g) $[0, \infty)$ $\quad$ (i) all real numbers
(k) all nonzero real numbers $\quad$ **B.6.** (a) all real numbers $\quad$ (c) all real numbers
(e) all real numbers $\quad$ (g) $(-\infty, 1]$ $\quad$ (i) all real numbers *except* -2 and 3
(k) $[6, 12]$ $\quad$ **B.7.** (a) $x + \Delta x$ $\quad$ (c) 12 $\quad$ (e) $1/(x + \Delta x)$ $\quad$ (g) $x + \Delta x + 5$
(i) $x^2 + 2x\,\Delta x + (\Delta x)^2 + 3x + 3\Delta x - 7$ $\quad$ **B.8.** (a) 1 $\quad$ (c) 0 $\quad$ (e) $-1/(x(x + \Delta x))$
(g) 1 $\quad$ (i) $2x + \Delta x + 3$

B.9. (a)

	$\Delta x = 2$	$\Delta x = 1$	$\Delta x = .5$	$\Delta x = .1$	$\Delta x = .01$
(a)	1	1	1	1	1
(c)	0	0	0	0	0
(e)	$\dfrac{-1}{x^2 + 2x}$	$\dfrac{-1}{x^2 + x}$	$\dfrac{-1}{x^2 + .5x}$	$\dfrac{-1}{x^2 + .1x}$	$\dfrac{-1}{x^2 + .01x}$
(g)	1	1	1	1	1
(i)	$2x + 5$	$2x + 4$	$2x + 3.5$	$2x + 3.1$	$2x + 3.01$

(b): (a) 1 $\quad$ (c) 0 $\quad$ (e) $-1/x^2$ $\quad$ (g) 1 $\quad$ (i) $2x + 3$ $\quad$ **C.1.** (a) $d(0) = 0$;
$d(2) = 64$ $\quad$ (c) $d(2) = 64; d(4) = 256$ $\quad$ (e) 128 ft $\quad$ (g) $d(t_2) - d(t_1)$ $\quad$ **C.2.** (a) (i)
42, (v) many possibilities, including 4:00 P.M., (vii) lower

PAGES 111–112

A.1. $D(t) = \begin{cases} 55t & \text{if } t \le 2 \\ 100 + 45(t - 2) & \text{if } t > 2 \end{cases}$ $\quad$ **A.4.** $P(t) = .7t - 1800$, where P = profit and
t = number of pounds sold. $\quad$ **B.1.** (a) $c(x) = 5.75x + (45,000/x)$ $\quad$ **B.2.** (a) $N(p) = 17{,}500 - 100p$, where p is the price per hamburger in *cents* (not dollars) and
N is the number of hamburgers sold. $\quad$ (c) $D(p) = (17{,}500 - 100p)p - 40(17{,}500 - 100p) - 110{,}000$, where p is the price in cents and D is the daily profit in cents. $\quad$ **B.4.** $V(t) = t(16 - 2t)(10 - 2t)$; maximum volume when $t = 2$.

SECTION 3 PAGES 121–123

A.1. $\qquad\qquad$ **A.2.** see $\quad$ **A.3.** $\qquad\qquad\qquad$ **A.5.**
$\qquad\qquad\qquad$ page 201

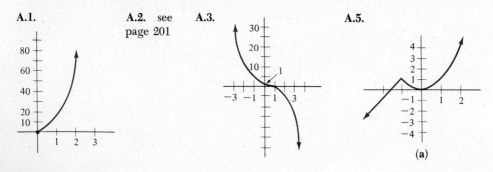

(a)

A.5.

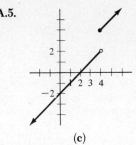

(c)   (e)

B.1.

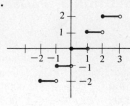

B.2.

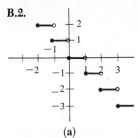

(a)   (c)

B.4. see page 219

B.5.

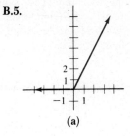

(a)

B.7.

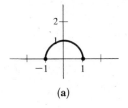

(a)   (c)

B.8.

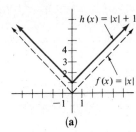

(a)

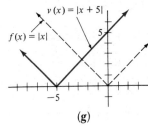

(c)   (e)   (g)

B.11.

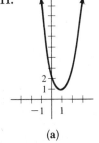

(a)

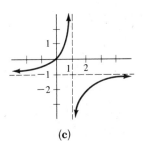

(c)

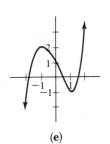

(e)

SECTION 4   PAGES 127–130

B.1. (a) 4   (c) 3.5   (e) 4.5   (g) 4   (i) 1, 5   (k) $[-3, 3]$
B.3. (a) $[-8, 9]$ (c) $-6.9,\ -3,\ 0,\ 3,\ 7$ (e) 1, 3, 5, and others   (g) $-3,\ 7$
(i) 1   **B.4.** (a) domain $f = [-6, 7]$ (c) $-1.4,\ -.2$ (e) $x = 3$ (g) $[-2, -1]$
and $[3, 7]$ **B.5.** (a) approximately $5000 - 17{,}000 = -12{,}000$ dollars (negative profit
means a *loss* of \$12,000)   (c) 12,000   (f) Profit on 5000 widgets is approximately
$5000 - 22{,}000 = -17{,}000$ dollars (a loss); the smallest number manufactured without a loss is
approximately 14,500.   **B.6.** (a) $\text{tem}(10) = 47°$ (c) 10:54 A.M. and 8:24 P.M.
$(= 20.4\,\text{hr})$ (e) $\text{tem}(10) = 47°$, so (10, 47) is on the graph; $\text{tem}(16) = 63.5°$, so (16, 63.5) is
on the graph. The point (10, 47) lies 16.5 units lower than (16, 63.5).   (g) $\text{tem}(6 \cdot 2) =$
$\text{tem}(12) = 57$, but $\text{tem}(2) = 38$, so that $6 \cdot \text{tem}(2) = 6(38) = 228$ (i) 10:54 A.M. $(= 10.9\,\text{hr})$
and 8:24 P.M. $(= 20.4\,\text{hr})$

SECTION 5   PAGES 137–141

A.1. (a), (c), (d), (f), (h) are symmetric with respect to the y-axis.   **A.2.** (a) (i) yes (ii) no
(iii) yes (iv) no   (c) (i)–(iv) no   (e) (i) yes, (iv) no   **A.3.** (a) increasing on $[-2.5, 0]$
and $[1.5, 3]$, decreasing on $[-6, -2.5]$ and $[0, 1.5]$ **B.2.** (a) Some possible answers are: (i)
4; (iii) 2; (v) all five.   (b) (i) 3, 5 and other pairs (iii) 4, 2 and other pairs (v) 1, 5 and
other pairs   **B.5.** (a) $f(x + c) = f(x + 2c) = f(x + 3c) = f(x + 4c) = f(x + 5c) = k$
(c) $f(x + nc) = k$ **B.6.** (a) 8   (c) 6   **B.8.** (a) If $0 < c < d \le 10$, then $c^2 < d^2$.
Hence $c^2 + 3 < d^2 + 3$. But this says $f(c) < f(d)$. Thus if $0 < c < d \le 10$, then $f(c) < f(d)$.
Therefore f is increasing on $(0, 10]$.   (c) If $0 < c < d \le 10$, then $c^2 < d^2$. But $c^2 < d^2$
and $c < d$ imply that $c^2 + c < d^2 + d$. Hence $c^2 + c + 5 < d^2 + d + 5$. But this means
$h(c) < h(d)$. Therefore h is increasing on $(0, 10]$.   **B.9.** (a) $(c, -d)$ (c) symmetric
with respect to the x-axis

PAGES 143–144

A.1. (a) and (c)   **A.2.** (a), (b), (d), and (g)   **A.3.** (a) even (c) even (e) neither

SECTION 6   PAGES 154–156

A.1.

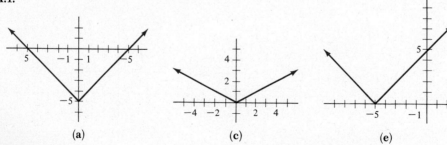

(a)   (c)   (e)

A.1.

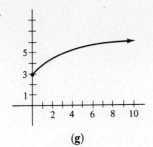

(g)

A.3.

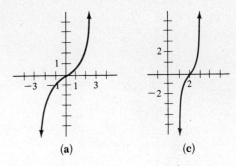

(a)   (c)

A.3.

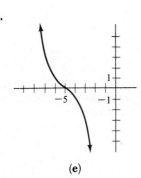

(e)

A.5.

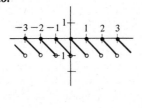

B.1.

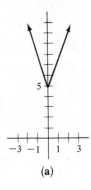

(a)

B.1.

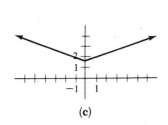

(c)   (e)   (g)

B.2.

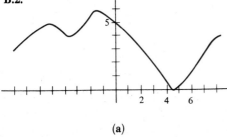

(a)

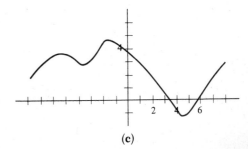

(c)

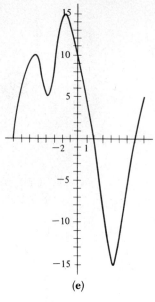

(e)

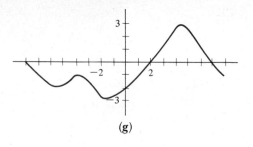

(g)

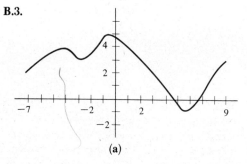

(i)

B.3.

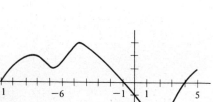

(k)

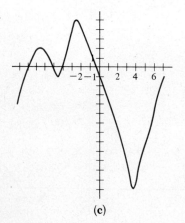

(a)

B.3.

(c)

C.1.

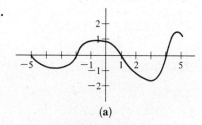

(a)

C.1.

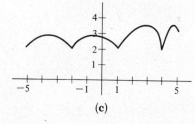

(c)

C.2.

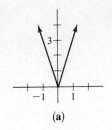

(a)

(c)

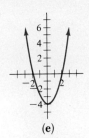

(e)

PAGES 161–162

A.1. (a) $(5, 2)$, upward (c) $(1, 2)$, downward (e) $(-\frac{3}{2}, \frac{3}{2})$, downward **A.2.** (a) $(\frac{2}{3}, \frac{19}{3})$, downward (c) $(\frac{1}{2}, \frac{1}{4})$, downward (e) $(-\frac{1}{2}, \frac{3}{4})$, upward (g) $(\frac{1}{2}, -\frac{1}{2})$, downward **A.3.** 7500

B.1.

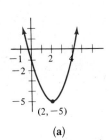

(a)

(b)

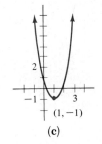

(c)

B.2. 0 **B.3.** 16 **C.1.** $t = \frac{5}{2}$, $h = 196$ **C.3.** $t = \frac{125}{8}$, $h = (125)^2/4 = 3906.25$
C.5. $A = $ area $= hb/2$. Since $h + b = 30$, we have $A = hb/2 = (30 - b)b/2 = 15b - b^2/2$.
Maximum area corresponds to the highest point (b, A) on the graph of the function
$A = 15b - b^2/2$. This occurs when $b = 15$; hence $h = 30 - 15 = 15$. **C.7.** If the base
has side x and the height is y, then surface area $S = 2x^2 + 4xy$. Since $8x + 4y = 8$, we have
$y = 2 - 2x$ and hence $S = 2x^2 + 4x(2 - 2x) = -6x^2 + 8x$. Maximum area occurs when
$x = \frac{2}{3}$. Hence $y = 2 - 2(\frac{2}{3}) = \frac{2}{3}$.

SECTION 7 PAGES 167–171

A.1. (a) 0 (c) 30 (e) 42 **A.2.** (a) $49, 1, -8$ (c) $-25, 1, 4$ (e) $5, 1,$
0 **A.3.** (a) $x^2 + 6x + 8$ (c) $x + 6$ **A.4.** (a) $(g \circ f)(x) = |2x^2 + 2x - 2| + 2$;
$(f \circ g)(x) = 2(x - 1)^2 + 10|x - 1| + 11$ (c) $(g \circ f)(x) = (-3x + 2)^3$; $(f \circ g)(x) = -3x^3$
$+ 2$ (e) $(g \circ f)(x) = (\sqrt[3]{x})^2 - 1$; $(f \circ g)(x) = \sqrt[3]{x^2 - 1}$ **A.7.** (a) $(f + g)(x) =$
$2x^2 + 2x + |x - 1| + 1$; $(f - g)(x) = 2x^2 + 2x - |x - 1| - 3$; $(g - f)(x) = |x - 1| -$
$2x^2 - 2x + 3$ (c) $(f + g)(x) = x^3 - 3x + 2$; $(f - g)(x) = -x^3 - 3x + 2$; $(g - f)(x) =$
$x^3 + 3x - 2$ (e) $(f + g)(x) = \sqrt[3]{x} + x^2 - 1$; $(f - g)(x) = \sqrt[3]{x} - x^2 + 1$; $(g - f)(x)$
$= x^2 - \sqrt[3]{x} - 1$ **A.8.** (a) $(f + g)(x + h) = 2(x + h)^2 + 2(x + h) + |x + h - 1| + 1$;
$(fg)(x + h) = |x + h - 1|(2(x + h)^2 + 2(x + h) - 1) + 4(x + h)^2 + 4(x + h) - 2$
(c) $(f + g)(x + h) = (x + h)^3 - 3(x + h) + 2$; $(fg)(x + h) = -3(x + h)^4 + 2(x + h)^3$
(e) $(f + g)(x + h) = \sqrt[3]{x + h} + (x + h)^2 - 1$; $(fg)(x + h) = \sqrt[3]{x + h}(x + h)^2 -$

$\sqrt[3]{x+h}$ **A.9.** (a) $(fg)(x) = |x-1|(2x^2 + 2x - 1) + 4x^2 + 4x - 2;$ $\left(\dfrac{f}{g}\right)(x) =$

$\dfrac{2x^2 + 2x - 1}{|x-1| + 2}$; $\left(\dfrac{g}{f}\right)(x) = \dfrac{|x-1| + 2}{2x^2 + 2x - 1}$ (c) $(fg)(x) = -3x^4 + 2x^3;$ $\left(\dfrac{f}{g}\right)(x) =$

$(-3x + 2)/x^3;$ $\left(\dfrac{g}{f}\right)(x) = x^3/(-3x + 2)$ (e) $(fg)(x) = (\sqrt[3]{x})x^2 - \sqrt[3]{x};$ $\left(\dfrac{f}{g}\right)(x) =$

$\sqrt[3]{x}/(x^2 - 1);$ $\left(\dfrac{g}{f}\right)(x) = (x^2 - 1)/\sqrt[3]{x}$ **B.1.** (a) $f(-4) = -3,$ $g(-4) = -1;$

$f(0) = 1,\, g(0) = 1.3;\, f(2) = 1,\, g(2) = 1.3;\, f(4) = -2,\, g(4) = 0$

B.2. (a)

x	1	2	3	4	5
$(g \circ f)(x)$	4	2	5	4	4

(c)

x	1	2	3	4	5
$(f \circ f)(x)$	1	3	3	5	1

B.3. The given function is $B \circ A$, where A and B are the functions listed here. In some cases, other correct answers are possible. (a) $A(x) = x^2 + 2,\, B(x) = \sqrt[3]{x}$ (c) $A(x) = 7x^3 - 10x + 17,\, B(x) = x^7$ (e) $A(x) = x^2 - \sqrt{x} + 2, B(x) = |x|$ (g) $A(t) = t + 2, B(t) = t\sqrt{t^2 - 5}$ (i) $A(x) = 3x^2 + 5x - 7,$ $B(x) = 1/x$ **B.5.** (a) $(f \circ g)(x) = (\sqrt{x})^3$, domain $[0, \infty)$; $(g \circ f)(x) = \sqrt{x^3}$, domain $[0, \infty)$ (c) $(f \circ g)(x) = \sqrt{5x + 10}$, domain $[-2, \infty)$; $(g \circ f)(x) = 5\sqrt{x} + 10$, domain $[-10, \infty)$ (e) $(f \circ g)(x) = 1/(x^2 + 1)$, domain all real numbers; $(g \circ f)(x) = (1/x^2) + 1$, domain all nonzero real numbers   **B.7.** (a) $f(x^2) = 2x^6 + 5x^2 - 1$ **C.1.** (a) $A(81) = 50,000;$ $T(11.5) = 85°;$ 65,000 at 4:00 P.M.; $T(8.5) = 72;$ $A(T(12)) = A(88) \approx 62,500$ (c) approximately 9:30–11:30 A.M.

CHAPTER 4: POLYNOMIAL AND RATIONAL FUNCTIONS

SECTION 1 PAGES 178–179

A.1. (a) polynomial; 1, 1, 3   (c) polynomial; $-7, -7, 0$ (e) not a polynomial since 10 has a nonconstant exponent   (g) not a polynomial since square root of $x - 7$ is involved   **A.2.** $q(x)$ is the quotient and $r(x)$ the remainder: (a) $q(x) = 3x^3 - 3x^2 + 5x - 11,$ $r(x) = 12$ (c) $q(x) = x^2 + 2x - 6,$ $r(x) = -7x + 7$ (e) $q(x) = 5x^2 + 5x + 5,$ $r(x) = 0$ **A.3.** (a) no   (c) yes   **A.4.** (a) 2   (c) 23   (e) 54 **A.5.** (a) no   (c) no   (e) yes   **B.1.** (a) $(x + 2)(x - 1) = x^2 + x - 2;$ $(3x^2 - 2)(2x^3 - 3x^2 + x - 1) = 6x^5 - 9x^4 - x^3 + 3x^2 - 2x + 2$ **B.4.** (a) 1   (c) 1

PAGES 184–185

A.1. (a)

```
2 |  3  -8    0     9      5
         6   -4    -8      2
      3  -2   -4     1  | 7
```

quotient $3x^3 - 2x^2 - 4x + 1$; remainder 7

(c)

```
-3 | 2    5    0    -2   -8
        -6    3    -9    33
      2  -1    3   -11  | 25
```

quotient $2x^3 - x^2 + 3x - 11$; remainder 25

(e)

```
7 | 5   0   -3    -4        6
       35  245  1,694  11,830
    5  35  242  1,690 | 11,836
```

quotient $5x^3 + 35x^2 + 242x + 1690$; remainder 11,836

(g)

```
2 | 1   -6    4     2    -7
        2   -8   -8   -12
    1  -4   -4   -6  | -19
```

quotient $x^3 - 4x^2 - 4x - 6$; remainder -19

A.2. (a) $f(10) = 170{,}802$ (c) $f(20) = 5{,}935{,}832$ (e) $f(\frac{1}{2}) = \frac{55}{16}$ **B.1.** (a) quotient $3x^3 + \frac{3}{4}x^2 - \frac{29}{16}x - \frac{29}{64}$; remainder $\frac{483}{256}$ (c) quotient $2x^3 - 6x^2 + 2x + 2$; remainder 1 (e) quotient $x^3 + 5x^2 - 6x + 1$; remainder 1   **B.2.** (a) $g(x) = (x + 4)(3x^2 - 3x + 1)$ (c) $g(x) = (x - \frac{1}{2})(2x^4 - 6x^3 + 12x^2 - 10)$ **B.3.** (a) quotient $x^2 - 2.15x + 4$; remainder 2.25

SECTION 2 PAGES 192–195

A.1. (a) $7, -3$ (c) $2, 0$ **A.2.** (a) $\frac{3}{10}$ (c) $-\frac{7}{4}$ (e) $\frac{2}{3}$ **A.3.** (a) $0, 4$ (c) $-100, -7, 2, 7$ **A.4.** (a) $-3, 4$ (c) $-2, -7$ (e) $\frac{1}{2}, -3$ (g) $-5, -\frac{1}{5}$ **A.5.** (a) $-3, 5$ (c) $-4, 8$ (e) $-1, \frac{1}{3}$ **A.6.** (a) $-5, -3$ (c) $-1 - \sqrt{2}/2$, $-1 + \sqrt{2}/2$ (e) $(-4 + \sqrt{6})/5$, $(-4 - \sqrt{6})/5$ (g) $(3 + \sqrt{89})/8$, $(3 - \sqrt{89})/8$ **A.7.** (a) 12; two real roots   (c) 180; two real roots   (e) 0; one real root **B.1.** (a) 1   (c) $0, 5$ (e) $\sqrt{3}, -\sqrt{3}$ (g) $0, -\frac{1}{3}, \frac{1}{3}$ **B.2.** (a) $2, (3 + \sqrt{5})/2$, $(3 - \sqrt{5})/2$ (c) $-3, -\frac{1}{4}, \frac{1}{2}$ (e) $0, -7, 1$ **B.3.** (a) $k = \pm 10$ (c) $k = 16$ (e) $k = 25$ **B.4.** (a) $k = 4$ **B.5.** many correct answers, including: (a) $(x - 1)(x - 7)(x + 4)$ (c) $(x - 1)^2(x - 2)^3(x - 3)$ **B.7.** (a) $x = (-y \pm \sqrt{49y^2 + 108})/6$ **B.9.** $-13, -12$ or $12, 13$ **B.10.** 4 in.   **C.1.** (a) small; large   (c) (i) $x = \frac{1}{8}$ in. (for example, a baby shrew) (ii) $x = \frac{5}{2}$ in. (for example, a field mouse) (iii) $x = 60$ in. (for example, a human)   (e) It could eat its own weight once every 600 days. For comparison, an average size human who ate his or her own weight once every 600 days would eat approximately $1\frac{1}{2}$ lb of food per week. If this food were 1 lb of bread and $\frac{1}{2}$ lb of cheese, it would provide about 2800 calories per week, that is, 400 calories per day. But the minimum subsistence level for such a human is approximately 1500 calories per day.

PAGES 199–200

A.1. (a) $1, -1, -3$ (c) $1, -1, -5$ (e) $-4, 1, \frac{1}{2}$ **A.2.** (a) $1, 2, -\frac{1}{2}$ (c) $1, \frac{1}{2}$, $\frac{1}{3}$ (e) $-1, 2$ (g) $\frac{2}{3}$ **A.3.** (a) $-3, 2$ (c) -2 (e) no rational roots **A.4.** (a) between 2 and 3   (c) between -3 and -2 (e) between -5 and -4; between -1 and 0; between 0 and 1; between 2 and 3

SECTION 3 PAGES 211–213

A.3.

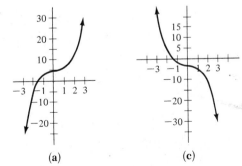

(a)

(c)

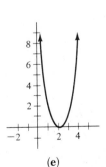

(e)

B.1.

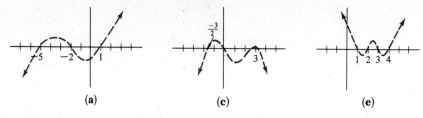

(a)   (c)   (e)

B.3. (a) increasing
        (c) increasing

B.5.

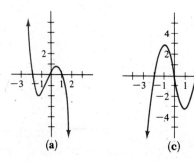
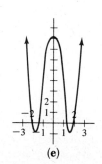

(a)   (c)   (e)

B.7.

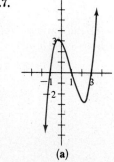
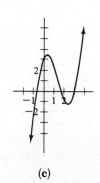
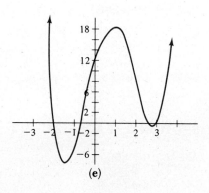

(a)   (c)   (e)

PAGE 215

A.1. (a)

x	10	50	100	250
$f(x) = x^3$	1000	125,000	1,000,000	15,625,000
$g(x) = 10x^2$	1000	25,000	100,000	625,000

(c)

x	3,000	5,000	10,000
$k(x) = 50x^3$	1,350,000,000,000	6,250,000,000,000	50,000,000,000,000
$t(x) = x^4/100$	810,000,000,000	6,250,000,000,000	100,000,000,000,000

B.1. yes, at $(100, 100{,}000)$

SECTION 4   PAGES 232–234

A.1. (a) all real numbers except -2 (c) all real numbers except -1 (e) all real numbers except $0, 1, -1$

A.3.

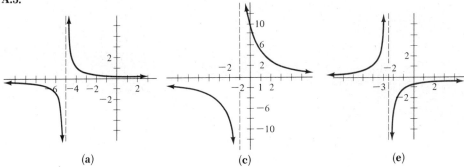

(a)   (c)   (e)

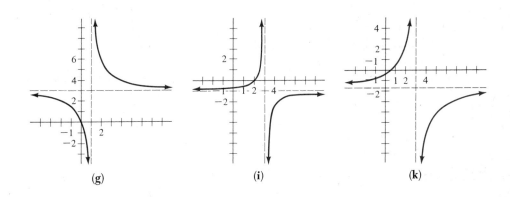

(g)   (i)   (k)

A.4. (a) 3, 2   (c) 5, 2   (e) $(-3 + \sqrt{17})/4$, $(-3 - \sqrt{17})/4$, 0   **B.1.** One such function is $f(x) = (4x - 4)/(x - 4)$. More generally, for any number a, the graph of $f(x) = (ax - a)/((a/4)x - a)$ passes through the given points.

B.3.

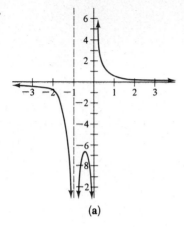

(a)

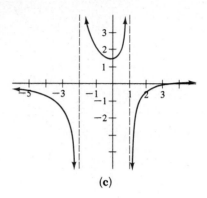

(c)

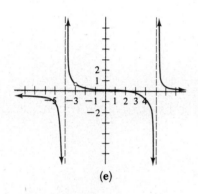

(e)

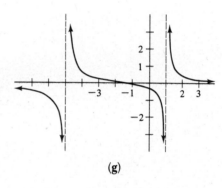

(g)

B.4.

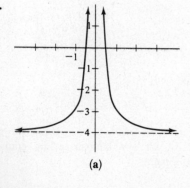

(a)

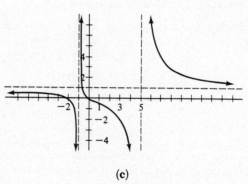

(c)

B.5.

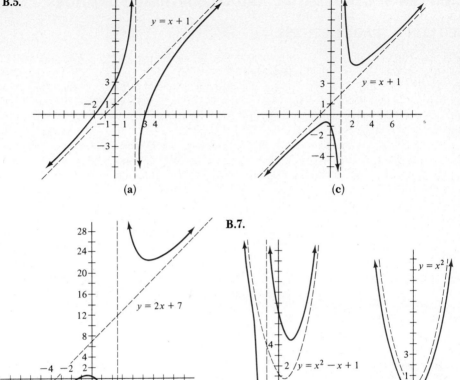

(a)

(c)

B.7.

(e)

(a)

(c)

C.1. (a) $4 \cdot 10^2/(6.4)^2 = 9.765625$ (c) no roots

SECTION 5   PAGES 241–243

A.1. (a) $x < -2$ or $x > 1$ (c) $-1 \leq x \leq 0$ or $x \geq 1$ (e) $x < -\sqrt{7}$ or $x > \sqrt{7}$
(g) $x \leq -1$ or $0 \leq x \leq 5$ **B.1.** (a) $x > 1$ (c) $-\frac{3}{2} < x < -1$ (e) $x \leq -\frac{9}{2}$ or
$x > -3$ **B.2.** (a) $-3 < x < 1$ or $x > 2$ (c) $x \leq -3$ or $-2 \leq x < 0$ or $x > 1$
(e) $x < -5$ or $x = -3$ or $1 < x < 2$ **B.3.** (a) $-3 < x < 1$ or $x \geq 5$
(c) $0 < x < 1$ **B.4.** (a) $-3 < x < -2$ (c) $x = 0$ or $1 < x < 2$ or $x > 3$
B.5. (a) $x < (-7 - \sqrt{37})/2$ or $x > (-7 + \sqrt{37})/2$ (c) $x = -\frac{5}{3}$
(e) $(-1 + \sqrt{33})/(-4) \leq x \leq (1 + \sqrt{33})/4$ **B.6.** (a) $x \leq -\frac{7}{2}$ or $x \geq -\frac{5}{4}$
(c) $x < -5$ or $-5 < x < -\frac{4}{3}$ or $x > 6$ (e) $-\frac{1}{7} < x < 3$ **B.7.** (a) $-\sqrt{3} < x < -1$
or $1 < x < \sqrt{3}$ (c) $-3 < x < 3$ (e) $x < -\sqrt{6}$ or $x > \sqrt{6}$
B.8. (a) $-3 \leq x \leq 2$ (c) $-1 \leq x \leq 0$ or $1 \leq x \leq 2$ **C.3.** (a) all real numbers
(c) all real numbers

CHAPTER 5: EXPONENTIAL AND LOGARITHMIC FUNCTIONS

SECTION 1   PAGES 251–253

A.1. (a) 10^5 (c) $\frac{1}{32}$ (e) $100,000$ (g) $\frac{1}{3}$ (i) $\frac{17}{4}$ **A.2.** (a) -112
(c) $6 + 4\sqrt{2}$ (e) $\frac{3}{16}$ **A.3.** (a) 2^6 (c) $2^{4/3}$ (e) 2^9 (g) $2^{5/6}$
(i) $2^{-15/2}$ **A.4.** (a) c^3 (c) $(14abcd)^{1/5}$ (e) $\sqrt[6]{32}$ **A.5.** (a) $3^s/3^r$
(c) $1/a^3$ (e) $(6s)^8/r^t$ (g) $8c$ (i) $2b^2/a^2$ **A.6.** (a) $2ab$ (c) $2c^2d^3$
(e) $(a^2 + b^2)^{1/3}$ (g) $a^{3/16}$ (i) $x^{23/12}$ **B.1.** (a) $x^{7/6} - x^{11/6}$ (c) $x - y$
(e) $x + y - (x + y)^{3/2}$ **B.2.** (a) $(x^{1/3} + 3)(x^{1/3} - 2)$ (c) $(x^{1/2} + 3)(x^{1/2} + 1)$
(e) $(x^{2/5} + 9)(x^{1/5} + 3)(x^{1/5} - 3)$ **B.3.** There are many correct answers, including
these: (a) $2^2 + 3^2 \neq (2 + 3)^2$ (c) $2^1 \cdot 3^2 \neq (2 \cdot 3)^{1+2}$ (e) $3^{-2} \neq -3^2$ **B.4.**
(a) $\frac{1}{2}$ (c) 1 (e) $1/49ab$ (g) a^x (i) $b^{x/4}$ **B.5.** (a) $1/x^{1/5}y^{2/5}$ (c) 1
(e) $c^{5/3} + (1/c^{5/3}) - 2$ **B.6.** (a) $(\sqrt{7} - \sqrt{2})^2 = 7 - 2\sqrt{7}\sqrt{2} + 2 = 9 - 2\sqrt{14}$
(c) $(\sqrt{5} - \sqrt{2})^3 = (\sqrt{5} - \sqrt{2})^2(\sqrt{5} - \sqrt{2}) = (7 - 2\sqrt{10})(\sqrt{5} - \sqrt{2}) = 7\sqrt{5} - 2\sqrt{50} - 7\sqrt{2} + 2\sqrt{20} = 7\sqrt{5} - 2(5\sqrt{2}) - 7\sqrt{2} + 2(2\sqrt{5}) = -17\sqrt{2} + 11\sqrt{5}$ **B.9.** (a)
$(7.9327)10^4$ (c) $(2.0)10^{-3}$ (e) $(5.963)10^{12}$ (g) $(2.0)10^{-11}$

PAGES 256–257

A.1. (a) 3, -3 (c) $-\frac{23}{11}$ (e) 9 **B.1.** (a) $3 + \sqrt{8}$, $3 - \sqrt{8}$
(c) 2, -1 (e) -1, -4 **B.2.** (a) 6 (c) 3, 7 (e) no solutions   (g) $\frac{1}{3}$,
-1 **B.3.** (a) $\pm a/\sqrt{A^2 - 1}$ (c) $\pm x/\sqrt{1 - K^2}$ **B.4.** (a) 8, $-\frac{27}{8}$ (c) 16
(e) 1, 81 **B.5.** (a) $\frac{1}{3}$, $-\frac{1}{2}$ (c) 1

SECTION 2   PAGES 263–265

A.1. Many possible answers, including $x = 1.07$ and $x = 60$ **A.2.** (a) 512 (c) 2
(e) $\sqrt[4]{3}$

A.3.

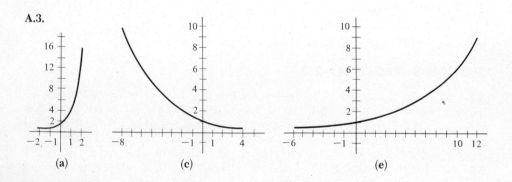

(a)   (c)   (e)

A.4. (a) $\dfrac{10^x(10^{\Delta x} - 1)}{\Delta x}$ (c) $3^{-x}\left(\dfrac{3^{-\Delta x} - 1}{\Delta x}\right)$ (e) $2^x\left(\dfrac{(2^{\Delta x} - 1)}{\Delta x}\right) + 2^{-x}\left(\dfrac{2^{-\Delta x} - 1}{\Delta x}\right)$
A.5. (a) e^{3x+1} (c) $5^{2x} - 5^{-2x}$

B.1.

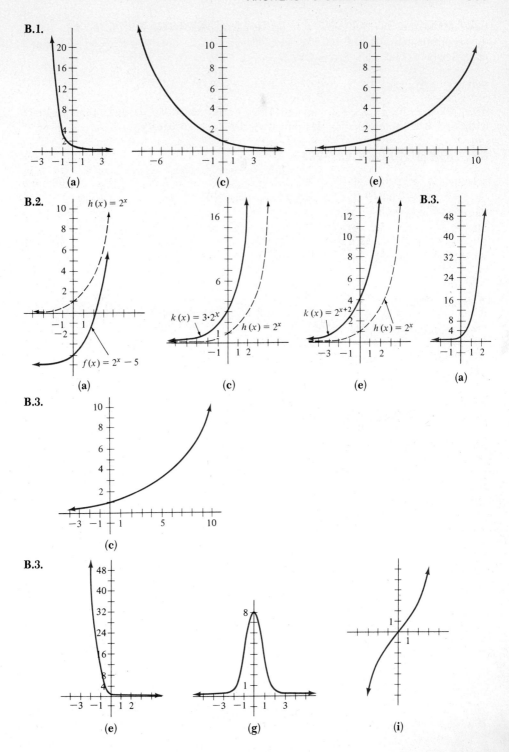

(a)

(c)

(e)

B.2.

$h(x) = 2^x$

$f(x) = 2^x - 5$

(a)

$k(x) = 3 \cdot 2^x$

$h(x) = 2^x$

(c)

$k(x) = 2^{x+2}$

$h(x) = 2^x$

(e)

B.3.

(a)

B.3.

(c)

B.3.

(e)

(g)

(i)

B.6. (a) neither   (c) even   (e) even   **B.8.** 13 hr   **B.10.** (a) 100 kg
(b) approximately 139.3 kg, 163.2 kg, and 186.5 kg   **C.2.** (c) slower   (e) $t(x)$ grows
more slowly   **C.4.** (a) $P(x) = 2^x/100$

PAGES 270–271

A.1. $\sqrt{2}/2 \approx .707$ g   **A.3.** (a) \$814.45   (c) \$824.35   **B.2.** approximately
6.992%   **B.4.** approximately .444 billion yr, that is, 444,000,000 yr

SECTION 3   PAGES 282–285

A.1. (a) 4   (c) 6   (e) $-\frac{2}{3}$ (g) 7   **A.2.** (a) 4   (c) $-\frac{3}{2}$
(e) -6 (g) -8 **A.3.** (a) $\log_{10}(.01) = -2$ (c) $\log_{10}\sqrt[3]{10} = \frac{1}{3}$
(e) $\log_{10}27.3 \approx 1.4367$ (g) $\log_{10}r = 7k$ (i) $\log_{10}y = x^2 + 2$ **A.4.** (a)
$\log_5 625 = 4$ (c) $\log_2(\frac{1}{8}) = -3$ (e) $\log_b 3379 = 14$ (g) $\log_e .0183 \approx -4$
(i) $\log_a c = -b$ **A.5.** (a) $10^4 = 10,000$ (c) $10^{2.86} \approx 750$ (e) $10^{-.097} \approx .8$
(g) $10^b = a$ (i) $10^{z+w} = x^2 + 2y$ **A.6.** (a) $2^{1/2} = \sqrt{2}$ (c) $8^{-2/3} = \frac{1}{4}$
(e) $e^{1.099} \approx 3$ **A.7.** (a) $\sqrt{43}$ (c) 4.7   (e) $\sqrt{x^2 + y^2}$ (g) 14   (i) 931
A.8. (a) -4 (c) 6   (e) -1 (g) $\frac{5}{4}$ (i) -4 **A.9.** (a) 4   (c) $\frac{3}{4}$
(e) $\frac{3}{2}$ **A.10.** (a) .43   (c) .26   (e) .10   (g) .53   (i) .7
A.11. (a) $\log_e(x^2 y^3)$ (c) $\log_e(x - 3)$ (e) $\log_e(x^{-7})$ **B.1.** (a) $-.039$
(c) $-.595$ (e) 1.32   (g) .878   (i) 1.88   **B.2.** (a) true   (c) true
(e) false   (g) false   **B.4.** (a) 5   (c) 2   (e) 1   (g) $\sqrt{1.01}$
B.5. (a) $\pm\sqrt{10,001}$ (c) 3   (e) $\sqrt{11}/3$ **B.6.** (a) 1 and 100   (c) $\frac{1}{10}$ and
1000   (e) 5^5 and $\frac{1}{5}$ **B.8.** (a) 4.0969   (c) 7.301   (e) 5.699

PAGES 292–293

A.2. Each characteristic is underlined: (a) $\underline{0}.0043$ (c) $\underline{1}.6590$ (e) $\underline{3}.8965$
(g) $0.9614 - 4$ (i) $\underline{7}.7582$ **A.3.** (a) .3702   (c) 2.6701   (e) $0.9276 - 4$
(g) $0.8826 - 1$ (i) 6.7262   **A.4.** (a) 2.22   (c) 89.7   (e) .0662
A.5. (a) 7.544   (c) 197,500   (e) .00005485   **A.6.** (a) $x \approx 438$ (c) $x \approx 29.75$
(e) $z \approx .02329$ **A.7.** (a) 3.149   (c) 1.285   (e) 24,230,000   (g) 9.68
B.1. (a) 2728   (c) .007762   (e) 15.11

SECTION 4   PAGES 299–301

A.1. (a) $b = 10$ (c) $b = 2$ **A.2.** (a) $(-1, \infty)$ (c) $(-\infty, 0)$ (e) $(0, \infty)$
(g) all x such that $x < -1$ or $x > 1$ (i) $(2, \infty)$

A.3.

B.2. (a) all x
except $x = 0$
(c) all positive x
(e) all x greater
than 1

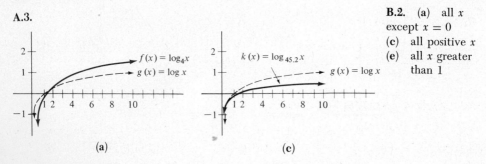

(a)   (c)

B.3.

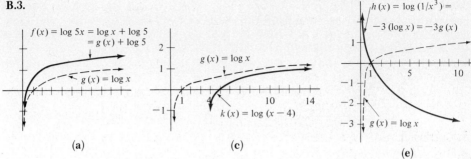

(a)   (c)   (e)

B.4.

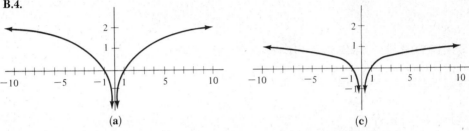

(a)   (c)

B.4.   **B.5.**

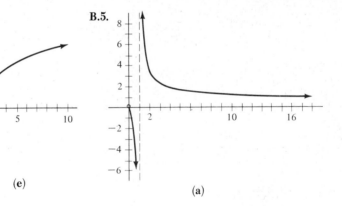

(e)   (a)

B.7. $A = -9$, $B = 10$ **C.1.** (a) x must be doubled   (c) The bill increases \$1.28 from $L = 8$ to $L = 9$ and increases \$2,621.44 from $L = 19$ to $L = 20$.

PAGES 303–304

A.1. (a) 2   (c) approximately 2.54   **B.1.** (a) 71.2, 66.8, 64.0, 62.0   (b) 16 months   **B.2.** (a) 77   (c) 13 weeks, 34 weeks, 90 weeks   **B.3.** approximately 199.53 and 7.94

SECTION 5 PAGES 309–311

A.1. (a) $x = 4$ (c) $x = 1$ (e) $x = -\frac{1}{9}$ (g) $x = \frac{1}{2}$ or $x = -3$ (i) $x = -2$ or $x = -\frac{1}{2}$ **A.3.** (a) $x = \log 5/\log 3 \approx 1.465$ (c) $x = \log 2 \approx .301$ (e) $x =$

$(\log 20 - \log 4)/\log 7 = \log 5/\log 7 \approx .827$ **A.4.** (a) $x = \log 3/(\log 3 - \log 2) =$
$\log 3/\log 1.5 \approx 2.7095$ (c) $x = (\log 3 - 5 \log 5)/(\log 5 + 2 \log 3) \approx -1.825$ (e) $x =$
$(\log 2 - \log 3)/(3 \log 2 + \log 3) \approx -.1276$ **A.5.** (a) $x = 0$ or $x = 1$ (c) $x = 0$
(e) $x = \log 2/\log 4 = \log 2/(2 \log 2) = \frac{1}{2}$ or $x = \log 3/\log 4 \approx .792$ (g) $x = 0$
or $x = (2 \log 5)/\log 2 \approx 4.644$ **B.1.** (a) $x = \log_5(3 + 2\sqrt{2})$ or $x = \log_5(3 - 2\sqrt{2})$
(c) $x = \log_5(t + \sqrt{t^2 - 1})$ or $x = \log_5(t - \sqrt{t^2 - 1})$ **B.2.** (a) $x = \ln(2 + \sqrt{3})$
or $x = \ln(2 - \sqrt{3})$ (c) $x = \frac{1}{2}\ln((t + 1)/(t - 1))$ **B.3.** (a) $\log 2/\log 1.07 \approx$
10.245 yr   (c) $\dfrac{\log 2}{365[\log(1 + (.07/365))]} \approx 9.9$ yr   **B.5.** (a) $\log 3/\log 1.05 \approx 22.5$ yr
B.7. $5000/(1.0125)^{36} \approx \$3197.05$ **B.9.** $(5730)(\log .64)/(-\log 2) \approx 3689.3$ yr old
B.11. $(3.6)(\log .15)/(-\log 2) \approx 9.853$ days

CHAPTER 6: TRIGONOMETRIC FUNCTIONS

SECTION 1   PAGES 320–323

A.1. (a) $40°$ **A.3.** (a) $90°$ (c) $45°$ (e) $30°$ (g) $4°$ (i) $75°$
(k) $40°$ **A.4.** (a) $5\pi/3$ (c) $3\pi/4$ (e) $3\pi/5$ (g) $11\pi/7$ (i) $7 - 2\pi$
(k) $27.43 - 8\pi$ **A.5.** Many correct answers, including: (a) $9\pi/4,\ 17\pi/4,\ -17\pi/4,$
$-15\pi/4$ (c) $11\pi/6,\ 23\pi/6,\ -13\pi/6,\ -25\pi/6$ **B.1.** (a) $-4\pi/3$ (c) $-7\pi/6$
(e) $-41\pi/6$ **B.2.** (a) $s = rt$ (c) $A = rs/2$ **B.3.** (a) 4.25 (b) 4.8
B.4. (a) $50/9$ radians   (c) 2000 radians   **B.5.** (a) 5 (c) 8.75 **C.1.** (a) 2π
(c) 7π (e) $9\pi/4$ **C.2.** (a) 4π radians in 1 min; 14π radians in 3.5 min; $9\pi/2$ radians in
$\frac{9}{8}$ min   (c) $2k\pi$ radians in 1 min; $7k\pi$ radians in 3.5 min; $9k\pi/4$ radians in $\frac{9}{8}$ min

PAGE 324

A.1. (a) $17\pi/180$ (c) $2\pi/5$ **A.2.** (a) $324°$ (c) $1170/\pi$ degrees $\approx 372.42°$
A.3. (a) $27.5125°$ (c) $13.705°$ **B.1.** (a) approximately $114.6°$ (c) approximately $228.2°$ (e) approximately $16.6°$

SECTION 2   PAGES 331–334

A.1. (a) -1 (c) 0 (e) 0 (g) 0 (i) 0 **A.2.** (b) $\sin t = 1/\sqrt{5},\ \cos t =$
$-2/\sqrt{5}$ **A.3.** (a) $\sin t = -3/\sqrt{10},\qquad \cos t = 1/\sqrt{10}$ (c) $\sin t = -.8,$
$\cos t = .6$ **A.4.** (a) $(fg)(t) = 3\sin^2 t + 6\sin t\cos t$ (c) $(fg)(t) = 3\sin t + 3\cos t +$
$\sin^3 t + \sin^2 t\cos t$ **A.5.** (a) $(\cos t + 2)(\cos t - 2)$ (c) $(\sin t + \cos t)(\sin t - \cos t)$
(e) $(\sin t + 3)^2$ (g) $(3\sin t + 1)(2\sin t - 1)$ (i) $(\cos^2 t + 5)(\cos t + 1)(\cos t - 1)$
A.6. (a) $(f \circ g)(t) = \cos(2t + 4)$; $(g \circ f)(t) = 2\cos t + 4$ (c) $(f \circ g)(t) = \sin(t^2 + 2)$;
$(g \circ f)(t) = \sin^2(t + 3) - 1$ **B.3.** (a) all numbers t except $t = \pi/2$ **B.5.** (a) $t =$
$\pi/2 + 2n\pi$, where n can be any integer (positive, zero, or negative)   (c) $t = \pi/2 + n\pi$,
where n can be any integer   (e) $t = \pi/2 + n\pi$, where n can be any integer
C.3. (a) During a full minute, each horse moves through an angle of 2π radians. The
angle between horses A and B is $\pi/4$ radians (which is $\frac{1}{8}$ of 2π radians). Thus the position
occupied by B at time t will be occupied by A exactly $\frac{1}{8}$ min later, that is, at time $t + \frac{1}{8}$.
Therefore the distance from B to the x-axis at time t, namely, $B(t)$, will be the *same* as the

distance from A to the x-axis at the time $t + \frac{1}{8}$, namely, $A(t + \frac{1}{8})$. Hence $B(t) = A(t + \frac{1}{8})$.
(c) $E(t) = D(t + \frac{1}{8})$; $F(t) = D(t + \frac{1}{3})$ (e) $E(t) = 5A(t + \frac{1}{8})$; $F(t) = 5A(t + \frac{1}{3})$
(g) $B(t) = \sin(2\pi t + \pi/4)$; $C(t) = \sin(2\pi t + 2\pi/3)$

SECTION 3   PAGES 342–344

A.1. (a) yes   (c) no   (e) yes   **A.2.** (a) $\sin t$ is negative for t in $(-\pi, 0)$ or in $(\pi, 2\pi)$ **B.1.** (a) $\cos t = -\frac{3}{5}$ (c) $\cos t = \sqrt{3}/2$ (e) $\cos t = 0$ **B.2.** (a) $\cos(\pi/4) = \sqrt{2}/2$; $\cos(\pi/3) = \frac{1}{2}$ (c) $\sin(-4\pi/3) = \sin(-((\pi/3) + \pi)) = -\sin((\pi/3) + \pi)$
$= -(-\sin \pi/3) = \sqrt{3}/2$ (e) $-\sqrt{2}/2$ **B.3.** (a) 0 (c) 0 **B.4.** (a) $k = \pi$ (c) $k = \pi$ (e) $k = \pi$ **B.5.** (a) $1 - 2\cos^2 t$ (c) $(\sin t)^{3/2} |\cos t|$ (e) $(\cos t)/4$
(g) $(3\cos^2 t + 8\cos t + 5)/4$ **B.6.** (a) $(2\sin^2 t)/(11\cos t)$ (c) $\sin t - 1$
(e) $\sin t - 1$ **B.7.** (a) $\sin(t - \pi) = \sin(-(\pi - t)) = -\sin(\pi - t)$ since $\sin(-v) = -\sin v$ for any v. But $-\sin(\pi - t) = -\sin t$ since $\sin(\pi - t) = \sin t$.   (c) $\cos(\pi - t) = \cos((-t) + \pi) = -\cos(-t) = -\cos t$ (e) $\cos(2\pi - t) = \cos((-t) + 2\pi) = \cos(-t) = \cos t$ **B.8.** (a) $\sin^4 t - \cos^4 t = (\sin^2 t + \cos^2 t)(\sin^2 t - \cos^2 t) = 1(\sin^2 t - \cos^2 t) = \sin^2 t - (1 - \sin^2 t) = 2\sin^2 t - 1$ (c) *Hint:* when $B \neq 0$, $D \neq 0$, the statement $A/B = C/D$ is true exactly when $AD = BC$. So in order to verify an identity of the form $A/B = C/D$, you need only verify the equivalent statement $AD = BC$.   (e) See (c).

SECTION 4   PAGES 355–359

A.2. (a) $\sin t = \frac{1}{2}$, $\cos t = -\sqrt{3}/2$ (c) $\sin t = -\sqrt{2}/2$, $\cos t = -\sqrt{2}/2$
(e) $\sin t = 0, \cos t = -1$ (g) $\sin t = -\frac{1}{2}, \cos t = -\sqrt{3}/2$ **A.3.** (a) $\sin t = 7/\sqrt{53}$,
$\cos t = 2/\sqrt{53}$ (c) $\sin t = -10/\sqrt{103}$, $\cos t = \sqrt{3}/\sqrt{103}$ (e) $\sin t = 4/\sqrt{17}$,
$\cos t = 1/\sqrt{17}$ **A.4.** (a) $\sin \theta = \sqrt{2}/\sqrt{11}$, $\cos \theta = 3/\sqrt{11}$ (c) $\sin \theta = \sqrt{3}/\sqrt{7}$,
$\cos \theta = 2/\sqrt{7}$ (e) $\sin \theta = h/m$, $\cos \theta = d/m$ **B.1.** (a) $-\sqrt{3}/2$ (c) $-\sqrt{2}/2$
(e) $\sqrt{2}(1 - \sqrt{3})/4$ **B.2.** (a) $\sin t = -3/\sqrt{10}, \cos t = 1/\sqrt{10}$ (c) $\sin t = -5/\sqrt{34}, \cos t = 3/\sqrt{34}$ **B.3.** (a) .92 radians   (c) $b = 8.0$ **B.5.** (a) .72 radians
(c) $a = 17.05$ **B.6.** (a) $t = .85$ radians   (c) $t = .85$ radians   (e) $t = .55$ radians   **B.7.** (a) $h = 25\sqrt{2}/2$ (c) $h = 300$ (e) $h = 50\sqrt{3}$ **C.1.** $\pi/6$

PAGE 364

A.1. (a) .2136   (c) .9605   (e) .7353   **A.2.** (a) $-.9367$ (c) $-.9872$
(e) .9903   **A.3.** (a) .9848   (c) .3584   (e) $-.6316$ **A.4.** (a) .9696
(c) $-.4041$ **B.1.** (a) .9914   (c) $-.3961$ (e) $-.6539$ **B.2.** (a) .7742
(c) .3685   (e) .3335   **B.3.** (a) .4780   (c) .9915

SECTION 5   PAGES 379–384

A.1. (a) 2   (c) 1   (e) 3
A.2.

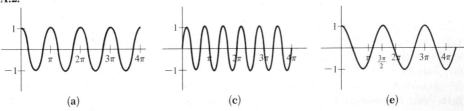

(a)   (c)   (e)

A.3.

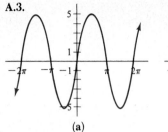

(a)

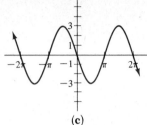

(c)

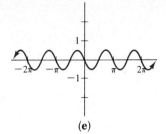

(e)

B.1.

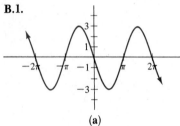

(a)

(c)

B.2.   (a) 1
(c) $f(t) = 1$ when
$t = \frac{1}{4}, \frac{5}{4}, \frac{9}{4}, \frac{13}{4},$
$\frac{17}{4}, \frac{21}{4}, \frac{25}{4}$ and
$f(t) = -1$ when $t = \frac{3}{4},$
$\frac{7}{4}, \frac{11}{4}, \frac{15}{4}, \frac{19}{4}, \frac{23}{4}$

B.3.

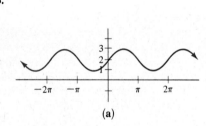

(a)

(c)

B.4.

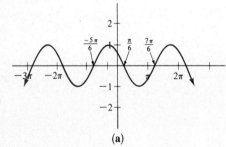

(a)

(c)

B.5. (a) $0, \pi, 2\pi$ (c) 0   **B.6.** (a) amplitude 3, period π, phase shift $\pi/2$ (c) 7,
$2\pi/7, -1/49$ (e) 1, 1, 0   (g) $6, \frac{2}{3}, -1/3\pi$ **B.7.** Many correct answers, including:
(a) $3 \sin (8t - 8\pi/5)$ (c) $\frac{2}{3} \cos 2\pi t$ (e) $7 \sin ((6\pi/5)(t + \pi/2))$ **B.8.** (a) $2 \sin 4t$
(c) $\frac{3}{2} \cos (t/2)$ (e) $\frac{1}{2} \cos 8t$ (g) $(-12) \cos 10t$ (i) $(-10) \cos 2\pi t$
B.10. (a) $h(t) = 6 \sin (\pi t/2)$ (c) $h(t) = 6 \cos (\pi t/2)$

C.1.

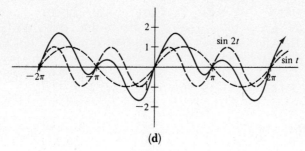

(d)

C.2.

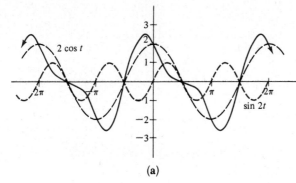

(a)

C.2.

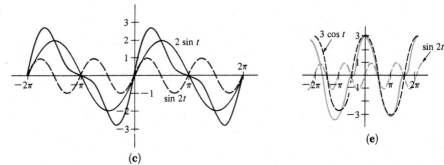

(c)   (e)

C.3. (a) $a(t) = 4 \cos 4\pi t$ (c) S is located at the point where the terminal side of an angle of $-5\pi/6$ radians in standard position meets the circle containing A. T is located at the point where the terminal side of an angle of $\pi/3$ radians meets the circle containing E.

SECTION 6 PAGES 391–394

A.1. (a) $-\sqrt{3}/3$ (c) $-\sqrt{3}/3$ (e) -1 **A.2.** (a) $\frac{4}{3}$ (c) $\frac{1}{2}$
A.3. (a) $\sin t = 3/\sqrt{13}$, $\cos t = 2/\sqrt{13}$, $\tan t = \frac{3}{2}$ (c) $9/\sqrt{117}$, $6/\sqrt{117}$, $\frac{3}{2}$ (e) $\frac{1}{6}$,
$\sqrt{35}/6$, $1/\sqrt{35}$ **A.4.** (a) 1 (c) 1 **B.1.** (a) $\tan(\pi - t) = \dfrac{\sin(\pi - t)}{\cos(\pi - t)} = \dfrac{\sin t}{-\cos t}$
$= -\tan t$ (c) $1 + \tan^2 t = 1 + \dfrac{\sin^2 t}{\cos^2 t} = \dfrac{\cos^2 t + \sin^2 t}{\cos^2 t} = \dfrac{1}{\cos^2 t}$ **B.3.** (a) $4\sqrt{3}/3$

(c) $10\sqrt{3}$ **B.4.** (a) $h \approx 6.58$ (c) 2.04 (e) 20.6 **B.5.** (a) $.65$ (c) $.85$
B.6. (a) $y = (\sqrt{3}/3)x$ (c) $y = -\sqrt{3}x$ (e) $y = \sqrt{3}x$ **B.7.** (a) 1.4807 radians (c) $.9513$ radians **B.8.** (a) $\tan t \cos t = (\sin t/\cos t)(\cos t) = \sin t$ (c) Begin

by multiplying $\dfrac{\cos t}{1 - \sin t}$ by $1 = \dfrac{1 + \sin t}{1 + \sin t}$; then: $\dfrac{\cos t}{1 - \sin t} = \dfrac{\cos t}{1 - \sin t} \cdot \dfrac{1 + \sin t}{1 + \sin t} =$

$\dfrac{\cos t + \cos t \sin t}{1 - \sin^2 t} = \dfrac{\cos t + \cos t \sin t}{\cos^2 t} = \dfrac{1}{\cos t} + \dfrac{\sin t}{\cos t} = \dfrac{1}{\cos t} + \tan t$

SECTION 7 PAGES 404–408

A.1. (a) fourth (c) second (e) fourth

A.2.

	$\tan t$	$\cot t$	$\sin t$	$\cos t$	$\sec t$	$\csc t$
(a)	$4/3$	$3/4$	$4/5$	$3/5$	$5/3$	$5/4$
(c)	$-12/5$	$-5/12$	$12/13$	$-5/13$	$-13/5$	$13/12$
(e)	-5	$-1/5$	$5/\sqrt{26}$	$-1/\sqrt{26}$	$-\sqrt{26}$	$\sqrt{26}/5$
(g)	$\sqrt{3}/\sqrt{2}$	$\sqrt{2}/\sqrt{3}$	$\sqrt{3/5}$	$\sqrt{2/5}$	$\sqrt{5/2}$	$\sqrt{5/3}$
(i)	$-1/2$	-2	$1/\sqrt{5}$	$-2/\sqrt{5}$	$-\sqrt{5}/2$	$\sqrt{5}$

A.3.

	$\tan t$	$\cot t$	$\sin t$	$\cos t$	$\sec t$	$\csc t$
(a)	$1/2$	2	$1/\sqrt{5}$	$2/\sqrt{5}$	$\sqrt{5}/2$	$\sqrt{5}$
(c)	$5/3$	$3/5$	$5/\sqrt{34}$	$3/\sqrt{34}$	$\sqrt{34}/3$	$\sqrt{34}/5$

A.5. (a) $\sin t - \sec t$ (c) $\csc^2 t + 2\cot t$ (e) $\cot^3 t - \tan^3 t$ **B.1.** (a) $\sin t = -\sqrt{3}/2$, $\cos t = -\frac{1}{2}$, $\tan t = \sqrt{3}$, $\cot t = \sqrt{3}/3$, $\sec t = -2$, $\csc t = -2\sqrt{3}/3$
(c) $\sin t = \frac{1}{2}$, $\cos t = -\sqrt{3}/2$, $\tan t = -\sqrt{3}/3$, $\cot t = -\sqrt{3}$, $\sec t = -2\sqrt{3}/3$,
$\csc t = 2$ **B.3.** (a) $(\csc t)(\sec t - \csc t)$ (c) $-\tan^2 t - \sec^2 t$

(e) $(\cos t - \sec t)(\cos^2 t + \sec^2 t + 1)$ **B.4.** (a) $\cot t$ (c) $\dfrac{2\tan t + 1}{3\sin t + 1}$

(e) $4 - \tan t$ **B.5.** (a) 5.49 (c) 1.48 (e) 2.53 (g) 4.32 **B.6.** (a) 1.1
(c) $.5$ (e) 1.0 **B.8.** (a) $\cot(-t) = \cos(-t)/\sin(-t) = \cos t/(-\sin t) = -\cot t$
(c) $\csc(-t) = 1/\sin(-t) = 1/(-\sin t) = -\csc t$ **B.9.** (a) $\sec(\pi - t) = 1/\cos(\pi - t)$
$= 1/(-\cos t) = -\sec t$ (c) $\cot(\pi - t) = \cos(\pi - t)/\sin(\pi - t) = -\cos t/\sin t =$
$-\cot t$ (e) $\csc(\pi - t) = 1/\sin(\pi - t) = 1/\sin t = \csc t$ **B.10.** (a) $\cos t \csc t =$

$(\cos t)(1/\sin t) = \cos t/\sin t = \cot t$ (c) $\dfrac{\sec t}{\csc t} = \dfrac{1/(\cos t)}{1/(\sin t)} = \dfrac{\sin t}{\cos t} = \tan t$

(e) $\tan t \sin t + \cos t = (\sin t/\cos t)(\sin t) + \cos t = (\sin^2 t/\cos t) + \cos t = (\sin^2 t + \cos^2 t)/\cos t = 1/\cos t = \sec t$ **B.11.** (a) $\sec^4 t - \sec^2 t = (\sec^2 t)(\sec^2 t - 1) =$

$\sec^2 t \tan^2 t = \dfrac{1}{\cos^2 t} \cdot \dfrac{\sin^2 t}{\cos^2 t} = \dfrac{\sin^2 t}{\cos^4 t}$ (c) $\cot t + \tan t = \dfrac{\cos t}{\sin t} + \dfrac{\sin t}{\cos t} = \dfrac{\cos^2 t + \sin^2 t}{\sin t \cos t} =$

$\dfrac{1}{\sin t \cos t} = \dfrac{1}{\sin t} \cdot \dfrac{1}{\cos t} = \sec t \csc t$

B.12.

	$\tan t$	$\cot t$	$\sin t$	$\cos t$	$\sec t$	$\csc t$
(a)	$-\sqrt{3}$	$1/\sqrt{3}$	$\sqrt{3}/2$	$-1/2$	-2	$2/\sqrt{3}$
(c)	$-2/\sqrt{5}$	$-\sqrt{5}/2$	$-2/3$	$\sqrt{5}/3$	$3/\sqrt{5}$	$-3/2$
(e)	$-1/\sqrt{63}$	$-\sqrt{63}$	$1/8$	$-\sqrt{63}/8$	$-8/\sqrt{63}$	8

C.1. Using triangle OAC, $\sin t = $ opposite/hypotenuse $= CA/CO = CA/1 = CA$. Using triangle OBD, $\tan t = $ opposite/adjacent $= BD/BO = BD/1 = BD$. Triangles OAC and OFE are similar (why?). Hence $\cot t = AO/CA = FE/FO = FE/1 = FE$.

SECTION 8   PAGES 422–425

A.1.

	$\tan t$	$\cot t$	$\sin t$	$\cos t$	$\sec t$	$\csc t$
(a)	$2-\sqrt{3}$	$\dfrac{1}{2-\sqrt{3}}$	$\dfrac{\sqrt{6}-\sqrt{2}}{4}$	$\dfrac{\sqrt{6}+\sqrt{2}}{4}$	$\dfrac{4}{\sqrt{6}+\sqrt{2}}$	$\dfrac{4}{\sqrt{6}-\sqrt{2}}$
(c)	$-2-\sqrt{3}$	$\dfrac{-1}{2+\sqrt{3}}$	$\dfrac{\sqrt{6}+\sqrt{2}}{4}$	$\dfrac{\sqrt{2}-\sqrt{6}}{4}$	$\dfrac{4}{\sqrt{2}-\sqrt{6}}$	$\dfrac{4}{\sqrt{6}+\sqrt{2}}$

A.2.   (a) $\cos x$ (c) $-2\sin x \sin y$

A.3.

	$\tan t$	$\cot t$	$\sin t$	$\cos t$	$\sec t$	$\csc t$
(a)	$-1+\sqrt{2}$	$\dfrac{1}{-1+\sqrt{2}}$	$\dfrac{\sqrt{2-\sqrt{2}}}{2}$	$\dfrac{\sqrt{2+\sqrt{2}}}{2}$	$\dfrac{2}{\sqrt{2+\sqrt{2}}}$	$\dfrac{2}{\sqrt{2-\sqrt{2}}}$
(c)	$-\sqrt{3+2\sqrt{2}}$	$\dfrac{-1}{\sqrt{3+2\sqrt{2}}}$	$\dfrac{\sqrt{2+\sqrt{2}}}{2}$	$-\dfrac{\sqrt{2-\sqrt{2}}}{2}$	$\dfrac{-2}{\sqrt{2-\sqrt{2}}}$	$\dfrac{2}{\sqrt{2+\sqrt{2}}}$

A.4. (a) $\cos x$ (c) $-\sin x$ (e) $\sec x$ **B.1.** (a) $\sec^2 x - \csc^2 x = (1 + \tan^2 x) - (1 + \cot^2 x) = \tan^2 x - \cot^2 x$ (c) $\sec^4 x - \tan^4 x = (\sec^2 x - \tan^2 x)(\sec^2 x + \tan^2 x) = 1(\sec^2 x + \tan^2 x) = (1 + \tan^2 x) + \tan^2 x = 1 + 2\tan^2 x$

(e) $\dfrac{1 + \cos x}{\sin x} + \dfrac{\sin x}{1 + \cos x} = \dfrac{(1 + \cos x)^2 + \sin^2 x}{(\sin x)(1 + \cos x)} = \dfrac{(1 + 2\cos x + \cos^2 x) + \sin^2 x}{(\sin x)(1 + \cos x)} = \dfrac{2 + 2\cos x}{(\sin x)(1 + \cos x)} = \dfrac{2(1 + \cos x)}{(\sin x)(1 + \cos x)} = \dfrac{2}{\sin x} = 2\csc x$

(g) $\dfrac{\sec x + \csc x}{1 + \tan x} = \dfrac{\dfrac{1}{\cos x} + \dfrac{1}{\sin x}}{1 + \dfrac{\sin x}{\cos x}} = \dfrac{\dfrac{\sin x + \cos x}{\cos x \sin x}}{\dfrac{\cos x + \sin x}{\cos x}} = \dfrac{\sin x + \cos x}{\cos x \sin x} \cdot \dfrac{\cos x}{\cos x + \sin x} = \dfrac{\cos x}{\cos x \sin x} = \dfrac{1}{\sin x} = \csc x$ **B.3.** (a) The

given identity is equivalent to $(1 - \sin x)(\sec x + \tan x) = \cos x$, which is proved as

follows: $(1 - \sin x)(\sec x + \tan x) = \sec x - \sin x \sec x + \tan x - \sin x \tan x =$
$\dfrac{1}{\cos x} - \sin x \cdot \dfrac{1}{\cos x} + \dfrac{\sin x}{\cos x} - \sin x \cdot \dfrac{\sin x}{\cos x} = \dfrac{1 - \sin^2 x}{\cos x} = \dfrac{\cos^2 x}{\cos x} = \cos x$

(c) $\dfrac{1}{\sin x \cos x} - \cot x = \dfrac{1}{\sin x \cos x} - \dfrac{\cos x}{\sin x} = \dfrac{1 - \cos^2 x}{\sin x \cos x} = \dfrac{\sin^2 x}{\sin x \cos x} = \dfrac{\sin x}{\cos x};$ on the

other hand, $\dfrac{\sin x \cos x}{1 - \sin^2 x} = \dfrac{\sin x \cos x}{\cos^2 x} = \dfrac{\sin x}{\cos x};$ since all steps are reversible, we have:

$\dfrac{1}{\sin x \cos x} - \cot x = \dfrac{\sin x}{\cos x} = \dfrac{\sin x \cos x}{1 - \sin^2 x}$ (e) $(\csc x - \cot x)^4(\csc x + \cot x)^4 =$

$[(\csc x - \cot x)(\csc x + \cot x)]^4 = [\csc^2 x - \cot^2 x]^4 = 1^4 = 1$ (g) $\dfrac{1 + \sec x}{\sin x + \tan x} =$

$\dfrac{1 + \dfrac{1}{\cos x}}{\sin x + \dfrac{\sin x}{\cos x}} = \dfrac{\dfrac{\cos x + 1}{\cos x}}{\dfrac{\sin x \cos x + \sin x}{\cos x}} = \dfrac{\cos x + 1}{\cos x} \cdot \dfrac{\cos x}{(\sin x)(\cos x + 1)} = \dfrac{1}{\sin x} = \csc x$

B.4. $\cos 3x = 4\cos^3 x - 3\cos x$ **B.5.** $\dfrac{\tan x + \tan y}{\tan x - \tan y} = \dfrac{\dfrac{\sin x}{\cos x} + \dfrac{\sin y}{\cos y}}{\dfrac{\sin x}{\cos x} - \dfrac{\sin y}{\cos y}} =$

$\dfrac{\dfrac{\sin x \cos y + \cos x \sin y}{\cos x \cos y}}{\dfrac{\sin x \cos y - \cos x \sin y}{\cos x \cos y}} = \dfrac{\sin x \cos y + \cos x \sin y}{\sin x \cos y - \cos x \sin y} = \dfrac{\sin (x + y)}{\sin (x - y)}$ **B.7.** (a) $.3 -$

$.8\sqrt{.75} \approx -.3928$ (c) $.3 + .8\sqrt{.75} \approx .9928$ (e) $(.4 + .6\sqrt{.75})/(.3 - .8\sqrt{.75})$
≈ -2.341 **B.8.** (a) $\sin 2x = \frac{120}{169}$, $\cos 2x = \frac{119}{169}$, $\tan 2x = \frac{120}{119}$; first quadrant
(c) $\sin 2x = \frac{24}{25}$, $\cos 2x = -\frac{7}{25}$, $\tan 2x = -\frac{24}{7}$; second quadrant **B.9.** (a) $\dfrac{1 + \cos 2x}{\sin 2x} =$
$\dfrac{1 + (\cos^2 x - \sin^2 x)}{2\sin x \cos x} = \dfrac{(1 - \sin^2 x) + \cos^2 x}{2\sin x \cos x} = \dfrac{\cos^2 x + \cos^2 x}{2\sin x \cos x} = \dfrac{2\cos^2 x}{2\sin x \cos x} = \dfrac{\cos x}{\sin x} =$
$\cot x$ (c) $\dfrac{2\cot x}{\csc^2 x} = \dfrac{(2\cos x/\sin x)}{(1/\sin^2 x)} = \dfrac{2\cos x}{\sin x} \cdot \dfrac{\sin^2 x}{1} = 2\cos x \sin x = \sin 2x$
B.11. (a) $.96$ (c) $.28$ (e) $\sqrt{.1} \approx .31622$ **B.12.** (a) $\sin (x/2) =$
$\sqrt{10}/10$, $\cos (x/2) = -3\sqrt{10}/10$ **B.13.** (a) $\cos x$ (c) $\sin 4y$ (e) 1

SECTION 9 PAGES 434–438

A.1. (a) $\sin \theta = 3/\sqrt{13}$, $\cos \theta = 2/\sqrt{13}$, $\tan \theta = \frac{3}{2}$, $\cot \theta = \frac{2}{3}$, $\sec \theta = \sqrt{13}/2$, $\csc \theta =$
$\sqrt{13}/3$ (c) $\sin \theta = 6/\sqrt{61}$, $\cos \theta = -5/\sqrt{61}$, $\tan \theta = -\frac{6}{5}$, $\cot \theta = -\frac{5}{6}$, $\sec \theta = -\sqrt{61}/5$,
$\csc \theta = \sqrt{61}/6$ (e) $\sin \theta = -\sqrt{2/11}$, $\cos \theta = -3/\sqrt{11}$, $\tan \theta = \sqrt{2}/3$, $\cot \theta = 3\sqrt{2}/2$,
$\sec \theta = -\sqrt{11}/3$, $\csc \theta = -\sqrt{11}/2$ **A.2.** (a) $\sin \theta =$
$2/\sqrt{13}$, $\cos \theta = 3/\sqrt{13}$, $\tan \theta = \frac{2}{3}$, $\cot \theta = \frac{3}{2}$, $\sec \theta = \sqrt{13}/3$, $\csc \theta = \sqrt{13}/2$
(c) $\sin \theta = \frac{4}{5}$, $\cos \theta = \frac{3}{5}$, $\tan \theta = \frac{4}{3}$, $\cot \theta = \frac{3}{4}$, $\sec \theta = \frac{5}{3}$, $\csc \theta = \frac{5}{4}$ (e) $\sin \theta = \sqrt{7}/4$,
$\cos \theta = \frac{3}{4}$, $\tan \theta = \sqrt{7}/3$, $\cot \theta = 3/\sqrt{7}$, $\sec \theta = \frac{4}{3}$, $\csc \theta = 4\sqrt{7}/7$ **A.3.** (a) $\sqrt{3}/2$
(c) $-\sqrt{3}$ (e) $-\sqrt{2}$ (g) $\frac{1}{2}$ **A.4.** (a) 36
(c) 36 (e) 8.4 **A.5.** (a) $.454$ (c) 1.15 (e) 2.366 **A.6.** (a) $.9848$
(c) -11.43 (e) 1.192 **B.1.** (a) $a = 6.43$, $c = 7.66$ (c) $b = 24.8$, $c = 24.06$
(e) $a = 10.72$, $c = 11.83$ (g) $a = 3.33$, $c = 1.08$ **B.2.** (a) $A = 33.69°$, $C = 56.31°$
(c) $A = 44.43°$, $C = 45.57°$ (e) $A = 48.19°$, $C = 41.81°$ (g) $A = 60.75°$,

$C = 29.25°$ **B.3.** (a) $A = 30°$, $C = 60°$, $c = 8.66$ (c) $A = 70.53°$, $C = 19.47°$, $a = 7.07$ (e) $C = 55°$, $a = 7$, $b = 12.21$ **B.5.** 171.01 ft   **B.7.** 263.44 ft   **B.9.** 13.63 ft   **B.11.** 555.31 ft   **B.13.** 173.21 mi   **B.15.** 25.39 m   **B.17.** 3.52 m   **C.1.** 449.12 ft

CHAPTER 7: TOPICS IN TRIGONOMETRY

SECTION 1 PAGES 446–448

A.1. (a) 2   (c) 0   (e) 1   **A.2.** (a) 2; third and fourth quadrants   (c) 2; second and fourth quadrants   (e) 2; first and fourth quadrants   **A.3.** (a) $x = k\pi$, where $k = 0, \pm1, \pm2, \ldots$ (c) $x = k\pi$, where $k = 0, \pm1, \pm2, \ldots$ (e) no solutions   **B.1.** (a) $x = \pi/3 + 2k\pi$ or $x = 2\pi/3 + 2k\pi$, where $k = 0, \pm1, \pm2, \ldots$ (c) $x = 2\pi/3 + k\pi$, where $k = 0, \pm1, \pm2, \ldots$ (e) $x = 5\pi/6 + 2k\pi$, or $x = 7\pi/6 + 2k\pi$, where $k = 0, \pm1, \pm2, \ldots$ (g) $x = 7\pi/6 + 2k\pi$ or $x = 11\pi/6 + 2k\pi$, where $k = 0, \pm1, \pm2, \ldots$ (i) $x = \pi/4 + k\pi$, where $k = 0, \pm1, \pm2, \ldots$ **B.2.** (a) $x = .1193$ or 3.0223 (c) $x = 2.9089$ or 6.0505 (e) $x = 1.9576$ or 4.3256 **B.3.** (a) $\theta = 22.95°$ or $157.04°$ (c) $\theta = 82.83°$ or $262.83°$ (e) $\theta = 114.83°$ or $245.17°$ **B.4.** (a) in $[0, 2\pi)$: $x = 2\pi/3$ or $5\pi/6$ or $5\pi/3$ or $11\pi/6$; all solutions: $x = 2\pi/3 + k\pi$ or $x = 5\pi/6 + k\pi$, where $k = 0, \pm1, \pm2, \ldots$ (c) in $[0, 2\pi)$: $x = 2\pi/9$ or $5\pi/9$ or $8\pi/9$ or $11\pi/9$ or $14\pi/9$ or $17\pi/9$; all solutions: $x = 2\pi/9 + 2k\pi/3$ or $x = 5\pi/9 + 2k\pi/3$, where $k = 0, \pm1, \pm2, \ldots$ (e) in $[0, 2\pi)$: $x = .7381$ or 1.3562 or 2.8325 or 3.4507 or 4.9269 or 5.5451; all solutions: $x = .7381 + 2k\pi/3$ or $x = 1.3562 + 2k\pi/3$, where $k = 0, \pm1, \pm2, \ldots$ **B.5.** (a) $\theta = 210°$ or $270°$ or $330°$ (c) $\theta = 60°$ or $120°$ or $240°$ or $300°$ (e) $\theta = 0°$ (g) $\theta = 30°$ or $150°$ or $210°$ or $330°$ **B.6.** (a) $x = 3.4814$ or $x = 5.9433$ (c) $x = 2.1588$ or $x = 5.3001$ or $x = 3\pi/4$ or $x = 7\pi/4$ (e) $x = 1.8235$ or $x = 4.4597$ or $x = 1.1864$ or $x = 5.0968$ **B.7.** (a) $x = \pi/2$ or $x = 3\pi/2$ or $x = \pi/4$ or $x = 5\pi/4$ (c) $x = \pi/2$ or $x = 3\pi/2$ or $x = \pi/6$ or $x = 5\pi/6$ (e) $x = 0$ or $x = \pi$ or $x = \pi/3$ or $x = 5\pi/3$ or $x = 2\pi/3$ or $x = 4\pi/3$ **B.8.** (a) $x = .8481$ or $x = 2.2935$ or $x = 1.7682$ or $x = 4.9098$ (c) $x = 1.8235$ or $x = 4.4597$ or $x = 1.2925$ or $x = 4.4341$ **B.9.** (a) $x = .8213$ or $x = 2.3203$ (c) $x = 1.2059$ or $x = 4.3475$ or $x = .3649$ or $x = 3.5065$ (e) $x = 1.0591$ or $x = 4.2007$ or $x = 2.8679$ or $x = 6.0095$ **B.10.** (a) no solutions   (c) $x = \pi/2$ or $x = 3\pi/2$ or $x = 7\pi/6$ or $x = 11\pi/6$ (e) $x = 0$ or $x = \pi/2$ or $x = \pi$ or $x = 3\pi/2$ **B.11.** (a) $x = \pi/6$ or $x = 5\pi/6$ (c) $x = \pi/2$ or $x = 7\pi/6$ or $x = 11\pi/6$ (e) $x = 0$ or $x = \pi$ or $x = 1.9552$ or $x = 4.3280$ **B.12.** (a) $x = 0$ or $x = \pi/3$ or $x = 5\pi/3$ (c) $x = \pi/6$ or $x = 5\pi/6$ (e) $x = 0$ or $x = \pi$ or $x = 2\pi/3$ or $x = 4\pi/3$ **C.1.** (a) $\sin x = c$ and $\cos x = c$ have no solutions when $|c| > 1$.   (c) $\tan x = c$ and $\cot x = c$ have solutions for every constant c **C.3.** $x = \pi/6n + 2k\pi/n$ or $x = 5\pi/6n + 2k\pi/n$, where $k = 0, \pm1, \pm2, \ldots$

SECTION 2 PAGES 456–458

A.1. (a) $C = 110°$, $b = 2.5$, $c = 6.3$ (c) $B = 14°$, $b = 2.2$, $c = 6.8$ **A.2.** (a) $a = 4.2$, $B = 125°$, $C = 35°$ (c) $c = 13.9$, $A = 22.5°$, $B = 39.5°$ **A.3.** (a) $A = 120°$, $B = 21.8°$, $C = 38.2°$ (c) $A = 24.1°$, $B = 30.8°$, $C = 125.1°$ **B.1.** (a) $B = 15.5°$, $C = 122.5°$, $c = 31.5$ (c) $B = 128°$, $C = 22°$, $b = 31.5$ (e) $A = 88.7°$, $B = 31.3°$, $a = 57.7$ **B.2.** (a) $C = 72°$, $b = 14.7$, $c = 15.2$ (c) $a = 9.8$, $B = 23.3°$, $C = 81.7°$ (e) no such triangle since $a + b = c$

B.3. 130.8 m **B.5.** 5.3° **B.7.** 334.9 km **B.9.** 231.9 ft **B.11.** 757.4 m
B.13. 762.9 mi

SECTION 3 PAGES 467–468

A.1. (a) $\pi/4$ (c) 0 (e) 0 (g) $\pi/6$ **A.2.** (a) .3576 (c) -1.2728
(e) .7168 (g) $-.8584$ (i) 2.5332 **B.1.** (a) $\pi/2$ (c) $\pi/2$ (e) $5\pi/6$
(g) $2\pi/3$ **B.2.** (a) $\pi/3$ (c) $-\pi/6$ (e) $\pi/6$ **B.3.** (a) $\frac{4}{5}$ (c) $\frac{4}{5}$
(e) $2\sqrt{5}/15$

B.4.

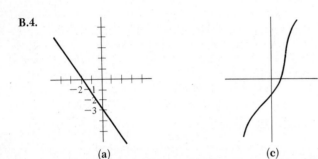

(a)   (c)

C.1. (a) Both sides are defined exactly when $-1 \le v \le 1$. Let $u = \sin^{-1} v$. Then $-\pi/2 \le u \le \pi/2$, so that $\cos u \ge 0$. Since $\cos^2 u = 1 - \sin^2 u$ and $\cos u \ge 0$, we have $\cos u = \sqrt{1 - \sin^2 u}$. But $u = \sin^{-1} v$ exactly when $\sin u = v$. Thus $\cos(\sin^{-1} v) = \cos u = \sqrt{1 - \sin^2 u} = \sqrt{1 - v^2}$. (c) Both sides are defined exactly when $-1 \le v \le 1$ *and* $v \ne 0$. Let $u = \cos^{-1} v$. Then $0 \le u \le \pi$ and $u \ne \pi/2$. Since $1 + \tan^2 u = \sec^2 u = 1/\cos^2 u$, we have $\tan^2 u = (1/\cos^2 u) - 1 = (1 - \cos^2 u)/\cos^2 u$. Since $u = \cos^{-1} v$ exactly when $\cos u = v$, we have $\tan^2 u = (1 - v^2)/v^2$. If $0 < v \le 1$, then $0 \le u < \pi/2$ so that $\tan u \ge 0$. Hence for $0 < v \le 1$, $\tan(\cos^{-1} v) = \tan u = +\sqrt{(1 - v^2)/v^2} = \sqrt{1 - v^2}/v$. If $-1 \le v < 0$, then $\pi/2 < u \le \pi$ so that $\tan u \le 0$. Hence for $-1 \le v < 0$, we have $\tan(\cos^{-1} v) = \tan u = -\sqrt{(1 - v^2)/v^2} = -\sqrt{1 - v^2}/\sqrt{v^2}$. But for $v < 0$, we have $\sqrt{v^2} = -v$ (why?). Thus for $-1 \le v < 0$, $\tan(\cos^{-1} v) = -\sqrt{1 - v^2}/\sqrt{v^2} = \sqrt{1 - v^2}/v$. (e) Both sides are defined for all v. Let $u = \tan^{-1} v$. Then $-\pi/2 < u < \pi/2$, so that $\cos u \ge 0$. Since $1 + \tan^2 u = \sec^2 u = 1/\cos^2 u$, we have $\cos^2 u = 1/(1 + \tan^2 u)$. Since $\cos u \ge 0$, we have $\cos u = \sqrt{1/(1 + \tan^2 u)}$. But $u = \tan^{-1} v$ exactly when $\tan u = v$. Thus $\cos(\tan^{-1} v) = \cos u = 1/\sqrt{1 + \tan^2 u} = 1/\sqrt{1 + v^2}$.

PAGE 473

B.1. (a) If there *were* an inverse function g, then we would have $x = (g \circ f)(x) = g(f(x)) = g(x^2)$ for *every* real number x. Thus for $x = 2$ we would have $g(4) = g(2^2) = 2$ and for $x = -2$ we would have $g(4) = g((-2)^2) = -2$. But *no* function can satisfy the condition $g(4) = 2$ *and* $g(4) = -2$. So there cannot be an inverse function for f.

B.1.

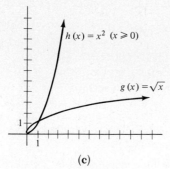

$h(x) = x^2 \ (x \geqslant 0)$

$g(x) = \sqrt{x}$

1

1

(c)

B.3.   The graph of f is just the graph of $h(x) = x^3$ shifted one unit to the right (see pp. 154 and 201), so the graph is always rising. Hence f is an increasing function and therefore *has* an inverse function g. The rule of g is $g(y) = \sqrt[3]{y} + 1$. [Verify that $(g \circ f)(x) = x$ and $(f \circ g)(y) = y$.]   **B.5.** $g(x) = (x/a) - (b/a)$.

CHAPTER 8: THE COMPLEX NUMBERS

SECTION 1   PAGES 479–482

A.1. (a) $8 + 2i$ (c) $-2 - 10i$ (e) $-\frac{1}{2} - 2i$ (g) $(\sqrt{2} - \sqrt{3})/2) + 2i$
A.2. (a) $1 + 13i$ (c) $-10 + 11i$ (e) $-21 - 20i$ (g) 4   **A.3.** (a) $-i$
(c) i (e) i **A.4.** (a) $\bar{z} = 1 - i$, $\bar{w} = -2i$, $\overline{zw} = -2 - 2i$
(c) $\bar{z} = 4 + 5i$, $\bar{w} = 2 + 3i$, $\overline{zw} = -7 + 22i$ **B.1.** (a) $\frac{5}{29} + \frac{2}{29}i$ (c) $\frac{-1}{3}i$
(e) $\frac{12}{41} - \frac{15}{41}i$ (g) $-2 - \frac{7}{2}i$ **B.2.** (a) $\frac{-5}{41} - \frac{4}{41}i$ (c) $\frac{10}{17} - \frac{11}{17}i$
(e) $\frac{31}{130} + \frac{27}{130}i$ **B.3.** (a) $\frac{7}{10} + \frac{11}{10}i$ (c) $\frac{-113}{170} + \frac{41}{170}i$ (e) $\frac{1}{25} - \frac{57}{25}i$ **B.4.**   Many
correct answers, including $z = 1$, $w = i$; in this case $|z + w| = |1 + i| = \sqrt{2}$, but $|z| + |w| = 1 + 1 = 2$ **B.5.** (a) 13   (c) $\sqrt{3}$ (e) 12   (g) $2\sqrt{10}$ (i) $1/\sqrt{17}$ **B.6.** -1
B.7. (a) $x = 2$, $y = -2$ (c) $x = -\frac{3}{4}$, $y = \frac{3}{2}$ **C.1.** (a) $(z + \bar{z})/2 =$
$[(a + bi) + (a - bi)]/2 = 2a/2 = a = \text{Re}(z)$ **C.2.** $\dfrac{1}{z} = \left(\dfrac{a}{a^2 + b^2}\right) - \left(\dfrac{b}{a^2 + b^2}\right)i$

C.3. (a) $\bar{z} = a - bi$; $z + \bar{z} = (a + bi) + (a - bi) = 2a$, which is a real number since a is one.   (c) $\bar{w} = c - di$; $\bar{z} + \bar{w} = (a - bi) + (c - di) = (a + c) - (b + d)i = \overline{(z + w)}$ **C.4.** (a) $\bar{z} = a - bi$; $|\bar{z}| = \sqrt{a^2 + (-b)^2} = \sqrt{a^2 + b^2} = |z|$
(c) $zw = (ac - bd) + (bc + ad)i$; $|zw| = \sqrt{(ac - bd)^2 + (bc + ad)^2} =$
$\sqrt{a^2c^2 - 2abcd + b^2d^2 + b^2c^2 + 2abcd + a^2d^2}$ $=$ $\sqrt{a^2c^2 + b^2d^2 + b^2c^2 + a^2d^2}$ $=$
$\sqrt{a^2(c^2 + d^2) + b^2(c^2 + d^2)}$ $=$ $\sqrt{(a^2 + b^2)(c^2 + d^2)}$ (e) $\dfrac{z}{w} = \dfrac{a + bi}{c + di} \cdot \dfrac{c - di}{c - di} =$
$\left(\dfrac{ac + bd}{c^2 + d^2}\right) + \left(\dfrac{bc - ad}{c^2 + d^2}\right)i$; $\left|\dfrac{z}{w}\right| = \sqrt{\left(\dfrac{ac + bd}{c^2 + d^2}\right)^2 + \left(\dfrac{bc - ad}{c^2 + d^2}\right)^2} =$
$\sqrt{\dfrac{(ac + bd)^2 + (bc - ad)^2}{(c^2 + d^2)^2}}$ $=$ $\sqrt{\dfrac{a^2c^2 + b^2d^2 + b^2c^2 + a^2d^2}{(c^2 + d^2)^2}}$ $=$ $\sqrt{\dfrac{(a^2 + b^2)(c^2 + d^2)}{(c^2 + d^2)^2}}$ $=$
$\sqrt{\dfrac{a^2 + b^2}{c^2 + d^2}}$

SECTION 2 PAGES 488–490

A.1. (a) $6i$ (c) $\sqrt{14}\,i$ (e) $-4i$ **A.2.** (a) $11i$ (c) $(\sqrt{15} - 3\sqrt{2})i$

(e) $-4\sqrt{3}$ (g) $\frac{2}{3}$ **A.3.** (a) $-41 - i$ (c) $(2 + 5\sqrt{2}) + (\sqrt{5} - 2\sqrt{10})i$

(e) $(15 + \sqrt{2}) + (5\sqrt{2} - 3)i$ (g) $\frac{1}{3} - \frac{\sqrt{2}}{3}i$ **B.1.** (a) $x = 1 + 2i$ or $1 - 2i$

(c) $x = \frac{1}{5} + \frac{2}{5}i$ or $\frac{1}{5} - \frac{2}{5}i$ (e) $x = 2$ or $-\frac{3}{2}$ **B.2.** (a) $x = 3$ or $\frac{-3}{2} + \frac{3\sqrt{3}}{2}i$ or

$\frac{-3}{2} - \frac{3\sqrt{3}}{2}i$ (c) $x = -2$ or $1 + \sqrt{3}i$ or $1 - \sqrt{3}i$ (e) $x = 1$ or i or -1 or $-i$

B.3. Many correct answers, including: (a) $x^2 - 4x + 5$ (c) $x^3 - 6x^2 + 13x - 10$

(e) $x^3 + x^2 + x + 1$ **B.4.** (a) $i, -i, 3, -2$ (c) $2 - i, 2 + i, i, -i$ **B.5.** Many

correct answers, including: (a) $x^2 - 2x + 5$ (c) $x^6 - 5x^5 + 8x^4 - 6x^3$

B.6. (a) $x = \frac{3}{5} + \frac{1}{5}i$ (c) $x = \frac{3}{7} + \frac{2}{7}i$ (e) $x = \frac{-11}{2} + \frac{3}{2}i$ (g) $x = -i$

B.7. (a) $x = \frac{-1}{2}i + (\sqrt{7 + 12i}/2)$ or $\frac{-1}{2}i - (\sqrt{7 + 12i}/2)$ (c) $x = ((1 + \sqrt{3})/2) + 2i$ or

$((1 - \sqrt{3})/2) + 2i$ (e) $x = (\frac{-3}{4} - \frac{3}{4}i) + (1 - i)\sqrt{7 - 8i}$ or $(\frac{-3}{4} - \frac{3}{4}i) - (1 - i)\sqrt{7 - 8i}$

B.8. Many correct answers, including: (a) $x^2 - (1 - i)x + (2 + i)$ (c) $x^2 - 2x + (1 + 2i)$

SECTION 3 PAGES 501–504

A.1.

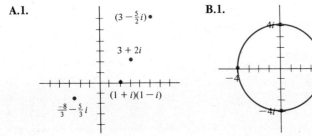

B.1.

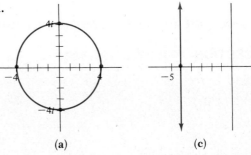

(a)   (c)

B.2. (a) $5(\cos(\pi/2) + i\sin(\pi/2))$ (c) $7(\cos\pi + i\sin\pi)$ (e) $\sqrt{5}(\cos(3\pi/2) + i\sin(3\pi/2))$ (g) $4(\cos(11\pi/6) + i\sin(11\pi/6))$ **B.3.** θ is approximated to four decimal places: (a) $5(\cos.9273 + i\sin.9273)$ (c) $13(\cos 5.1072 + i\sin 5.1072)$ (e) $\sqrt{5}(\cos 1.1071 + i\sin 1.1071)$ (g) $\sqrt{18.5}(\cos 2.191 + i\sin 2.191)$ **B.4.** (a) $6(\cos(2\pi/3) + i\sin(2\pi/3)) = -3 + 3\sqrt{3}i$ (c) $42(\cos(7\pi/6) + i\sin(7\pi/6)) = -21\sqrt{3} - 21i$ (e) $\frac{3}{2}(\cos(\pi/4) + i\sin(\pi/4)) = (3\sqrt{2}/4) + (3\sqrt{2}/4)i$ **B.5.** (a) $2\sqrt{2}(\cos(7\pi/12) + i\sin(7\pi/12))$ (c) $\cos(\pi/2) + i\sin(\pi/2)$ (e) $12(\cos(2\pi/3) + i\sin(2\pi/3))$ **B.6.** (a) i (c) $(-243\sqrt{3}/2) - (243/2)i$ **B.7.** (a) -64 (c) $(1/2) - (\sqrt{3}/2)i$ (e) i **B.8.** (a) $1, i, -1, -i$ (c) $1, i, -1, -i, (\sqrt{2}/2) + (\sqrt{2}/2)i, (-\sqrt{2}/2) + (\sqrt{2}/2)i, (-\sqrt{2}/2) - (\sqrt{2}/2)i, (\sqrt{2}/2) - (\sqrt{2}/2)i$ **B.9.** (a) $x = (1/2) + (\sqrt{3}/2)i$ or $(-1/2) + (\sqrt{3}/2)i$ or $(-1/2) - (\sqrt{3}/2)i$ or i or $-i$ (e) $x = 3i$ or $(-3\sqrt{3}/2) - (3/2)i$ or $(3\sqrt{3}/2) - (3/2)i$ (g) $x = \sqrt[4]{2}[(\sqrt{3}/2) + (1/2)i]$ or $\sqrt[4]{2}[(-1/2) + (\sqrt{3}/2)i]$ or $\sqrt[4]{2}[(-\sqrt{3}/2) - (1/2)i]$ or $\sqrt[4]{2}[(1/2) - (\sqrt{3}/2)i]$ **B.11.** (a) $x = -1$ or i or $-i$ (c) $x = \cos(2k\pi/n) + i\sin(2k\pi/n)$, where $k = 1, 2, \ldots, n - 1$ **B.13.** Since u is an nth root of unity, $u^n = 1$; hence $(1/u)^n = 1/u^n = 1/1 = 1$.   **C.2.** (a) b/a (c) $y - b = (d/c)(x - a)$

CHAPTER 9: TOPICS IN ALGEBRA

SECTION 1   PAGES 515–519

A.1. (b) and (c)   **A.2.** neither of them   **B.1.** (a) $x = \frac{11}{5}$, $y = -\frac{7}{5}$ (c) $x = \frac{6}{5}$, $y = \frac{13}{5}$ (e) $x = -\frac{7}{8}$, $y = \frac{5}{16}$ **B.2.** (a) $x = \frac{66}{5}$, $y = \frac{18}{5}$ (c) $x = -6$, $y = 54$
B.3. (a) $x = 4$, $y = 1$, $z = 2$ (c) $x = -\frac{17}{3}$, $y = \frac{68}{3}$, $z = \frac{22}{3}$ (e) $x = \frac{3}{2}$, $y = \frac{3}{2}$,
$z = -\frac{3}{2}$ **B.4.** (a) $z = \frac{3}{2}$, $y = t$, $x = -\frac{1}{2} - t$, where t is any real number   (c) $z = 0$,
$y = t$, $x = 1 - 2t$, where t is any real number   **B.5.** (a) $x = 1$, $y = 2$ (c) no solu-
tions   (e) no solutions   **B.6.** (a) dependent   (c) inconsistent   (e) inconsistent   **B.7.** (a) $x = 5$, $y = 6$, $z = 0$, $w = -1$ (c) no solutions   **B.8.** (a) $x = \frac{7}{11}$,
$y = -7$ (c) $x = \frac{1}{2}$, $y = \frac{1}{3}$, $z = -\frac{1}{4}$ (e) $x = -\frac{3}{4}$, $y = \frac{10}{3}$, $z = \frac{5}{2}$ **B.9.** (c) $w = 0$,
$z = s$, $y = 2 - 3s$, $x = -1 + 2s$, where s is any real number   (e) $v = r$, $w = -r$, $z = s$,
$y = t$, $x = -2t - 3s - 4r$, where r, s, t are any real numbers   **B.10.** (a) $x \approx .1845341968$,
$y \approx -.6239742478$ **B.11.** $a = -\frac{7}{20}$, $b = -\frac{19}{20}$, $c = \frac{23}{10}$ **B.12.** (a) $f(x) = (\frac{-7}{6})x^2 +$
$\frac{37}{6}x - 7$ (c) $f(x) = \frac{3}{2}x^2 - \frac{1}{2}x$ **B.13.** (a) $x = 3$, $y = 9$ or $x = -1$, $y = 1$
B.14.

    (a) $x = 1$, $y = -3$ or   (c) $x = 4$, $y = 3$ or   (e) $x = -3$, $y = 4$ or
        $x = -3$, $y = 5$ $x = -4$, $y = 3$ or $x = 1$, $y = 0$
                            $x = \sqrt{21}$, $y -2$ or
                            $x = -\sqrt{21}$, $y = -2$

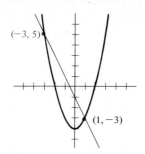

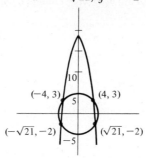

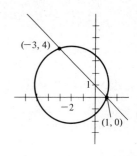

B.15. 13 and 19   **B.17.** no   **B.19.** 10 quarters, 28 dimes, 14 nickels   **B.21.** $x = 550$
mph, $y = 50$ mph

SECTION 2   PAGES 000–000

A.1. (a) -29 (c) $\frac{17}{2}$ (e) 13   (g) -13 (i) -26 **B.1.** (a) $x = 2$,
$y = 3$, $z = -1$ (c) $x = -2$, $y = -1$, $z = 4$ (e) $w = t$, $z = -4t$, $y = 6t$,
$x = 7t$, where t is any real number   (g) $x = -1$, $y = 1$, $z = -3$, $w = -2$
B.2. (a) $\begin{pmatrix} 10 & -3 \\ 3 & 7 \end{pmatrix}$ (c) $\begin{pmatrix} 2 & \frac{1}{2} \\ \frac{13}{2} & \frac{9}{2} \end{pmatrix}$ (e) $\begin{pmatrix} 3 & 5 \\ 7 & 9 \end{pmatrix}$ **B.3.** (a) $\begin{pmatrix} 14 & 9 \\ 8 & 7 \end{pmatrix}$ (c) $\begin{pmatrix} 7 & 9 \\ 3 & 5 \end{pmatrix}$

(e) $\begin{pmatrix} 1 & 2 \\ -6 & 10 \end{pmatrix}$ **B.5.** (a) $AB = \begin{pmatrix} 1 & 2 \\ 3 & 4 \end{pmatrix}\begin{pmatrix} -1 & 4 \\ 5 & 2 \end{pmatrix} = \begin{pmatrix} 9 & 8 \\ 17 & 20 \end{pmatrix}$, but $BA =$

$\begin{pmatrix} -1 & 4 \\ 5 & 2 \end{pmatrix}\begin{pmatrix} 1 & 2 \\ 3 & 4 \end{pmatrix} = \begin{pmatrix} 11 & 14 \\ 11 & 18 \end{pmatrix}$ (c) If $A = \begin{pmatrix} 0 & 2 \\ 0 & 0 \end{pmatrix}$, then $A^2 = \begin{pmatrix} 0 & 2 \\ 0 & 0 \end{pmatrix}\begin{pmatrix} 0 & 2 \\ 0 & 0 \end{pmatrix} = \begin{pmatrix} 0 & 0 \\ 0 & 0 \end{pmatrix}$.

(e) $A + B = \begin{pmatrix} 3 & 1 \\ 0 & 9 \end{pmatrix}$, so $(A + B)^2 = \begin{pmatrix} 3 & 1 \\ 0 & 9 \end{pmatrix}\begin{pmatrix} 3 & 1 \\ 0 & 9 \end{pmatrix} = \begin{pmatrix} 9 & 12 \\ 0 & 81 \end{pmatrix}$; but $A^2 + 2AB + B^2 = A^2$

$+ AB + AB + B^2 = \begin{pmatrix} 2 & -1 \\ 3 & 5 \end{pmatrix}\begin{pmatrix} 2 & -1 \\ 3 & 5 \end{pmatrix} + \begin{pmatrix} 2 & -1 \\ 3 & 5 \end{pmatrix}\begin{pmatrix} 1 & 2 \\ -3 & 4 \end{pmatrix} + \begin{pmatrix} 2 & -1 \\ 3 & 5 \end{pmatrix}\begin{pmatrix} 1 & 2 \\ -3 & 4 \end{pmatrix} +$

$\begin{pmatrix} 1 & 2 \\ -3 & 4 \end{pmatrix}\begin{pmatrix} 1 & 2 \\ -3 & 4 \end{pmatrix} = \begin{pmatrix} 1 & -7 \\ 21 & 22 \end{pmatrix} + \begin{pmatrix} 5 & 0 \\ -12 & 26 \end{pmatrix} + \begin{pmatrix} 5 & 0 \\ -12 & 26 \end{pmatrix} + \begin{pmatrix} -5 & 10 \\ -15 & 10 \end{pmatrix} = \begin{pmatrix} 6 & 3 \\ -18 & 84 \end{pmatrix}$

B.7.   (a) $ad - bc - dx + cy + bx - ay$ **B.10.**   (a) $x = -\frac{9}{31},\ y = \frac{7}{31}$ (c) $x = \frac{39}{41},$
$y = \frac{22}{41}$ (e) $x = 4,\ y = 2$ **B.11.**   (a) and (d)

SECTION 3   PAGES 537–539

A.1.   (a)   720   (c)   220   **A.2.**   (a)   0   (c)   64   (e)   3,921,225   **A.3.**   (a)
$x^5 + 5x^4y + 10x^3y^2 + 10x^2y^3 + 5xy^4 + y^5$ (c) $a^5 - 5a^4b + 10a^3b^2 - 10a^2b^3 +$
$5ab^4 - b^5$ (e) $32x^5 + 80x^4y^2 + 80x^3y^4 + 40x^2y^6 + 10xy^8 + y^{10}$ **B.2.**   (a) $x^3 +$
$6x^2\sqrt{x} + 15x^2 + 20x\sqrt{x} + 15x + 6\sqrt{x} + 1$ (c) $1 - 10c + 45c^2 - 120c^3 +$
$210c^4 - 252c^5 + 210c^6 - 120c^7 + 45c^8 - 10c^9 + c^{10}$ **B.3.**   (a)   56   (c) $-8i$
B.4.   (a) $10x^3y^2$ (c) $35c^3d^4$ (e) $\frac{35}{8}u^{-5}$ **B.5.**   (a)   4032   (c)   160
B.6. $-\frac{63}{8}$ **B.8.**   (b)   *Hint:* $0^n = (1-1)^n$ **B.9.**   (a) $\cos^4\theta +$
$4i\cos^3\theta\sin\theta - 6\cos^2\theta\sin^2\theta - 4i\cos\theta\sin^3\theta + \sin^4\theta$ (b) $\cos 4\theta + i\sin 4\theta$
(c) $\cos 4\theta = \cos^4\theta + \sin^4\theta - 6\cos^2\theta\sin^2\theta;$ $\sin 4\theta = 4\cos\theta\sin\theta(\cos^2\theta - \sin^2\theta)$
B.10.   (a) $f(x+h) - f(x) = \binom{5}{1}x^4h + \binom{5}{2}x^3h^2 + \binom{5}{3}x^2h^3 + \binom{5}{4}xh^4 + \binom{5}{5}h^5$ (c) $\binom{5}{1}x^4 =$
$5x^4$ since all the terms involving h are very, very close to 0 when h is small enough.   (e)   If h

is very close to 0, then $\dfrac{f(x+h) - f(x)}{h}$ is approximately equal to $\binom{12}{1}x^{11} = 12x^{11}$.   (f)   If h

is very close to 0, then $\dfrac{f(x+h) - f(x)}{h}$ is approximately equal to $\binom{n}{1}x^{n-1} = nx^{n-1}$.

SECTION 4   PAGES 547–549

B.1.   (a)   *Step 1:* for $n = 1$, the statement is $1 = 2^1 - 1$, which is true.   *Step 2:* Assume the
statement is true for $n = k$, that is,

$$1 + 2 + 2^2 + 2^3 + \cdots + 2^{k-1} = 2^k - 1.$$

Add 2^k to both sides, and rearrange terms:
$$1 + 2 + 2^2 + 2^3 + \cdots + 2^{k-1} + 2^k = 2^k - 1 + 2^k$$
$$1 + 2 + 2^2 + 2^3 + \cdots + 2^{k-1} + 2^{(k+1)-1} = 2^k + 2^k - 1 = 2(2^k) - 1$$
$$1 + 2 + 2^2 + 2^3 + \cdots + 2^{k-1} + 2^{(k+1)-1} = 2^{k+1} - 1$$

But this line says that the statement is true for $n = k + 1$. Therefore, by the Principle of
Mathematical Induction, the statement is true for every positive integer n.

Note: hereafter, in these answers, step 1 will be omitted if it is trivial [as in B.1.(a) above] and
only the essential parts of step 2 will be given.

B.1.   (c)   Assume that the statement is true for $n = k$:

$$1 + 3 + 5 + \cdots + (2k - 1) = k^2$$

Add $2(k + 1) - 1$ to both sides:

$1 + 3 + 5 + \cdots + (2k - 1) + (2(k + 1) - 1)$
$$= k^2 + 2(k + 1) - 1 = k^2 + 2k + 1 = (k + 1)^2$$

The first and last parts of this equation say that the statement is true for $n = k + 1$.

B.1. **(e)** Assume that the statement is true for $n = k$:

$$1^2 + 2^2 + 3^2 + \cdots + k^2 = \frac{k(k+1)(2k+1)}{6}$$

Add $(k+1)^2$ to both sides:

$$1^2 + 2^2 + 3^2 + \cdots + k^2 + (k+1)^2 = \frac{k(k+1)(2k+1)}{6} + (k+1)^2$$

$$= \frac{k(k+1)(2k+1) + 6(k+1)^2}{6} = \frac{(k+1)[k(2k+1) + 6(k+1)]}{6}$$

$$= \frac{(k+1)[2k^2 + 7k + 6]}{6} = \frac{(k+1)(k+2)(2k+3)}{6}$$

$$= \frac{(k+1)[(k+1)+1][2(k+1)+1]}{6}$$

The first and last part of this equation say that the statement is true for $n = k + 1$.

B.2. **(a)** Assume the statement is true for $n = k$:

$$\frac{1}{2} + \frac{1}{4} + \frac{1}{8} + \cdots + \frac{1}{2^k} = 1 - \frac{1}{2^k}$$

Add $\dfrac{1}{2^{k+1}}$ to both sides:

$$\frac{1}{2} + \frac{1}{4} + \frac{1}{8} + \cdots + \frac{1}{2^k} + \frac{1}{2^{k+1}} = 1 - \frac{1}{2^k} + \frac{1}{2^{k+1}} = 1 + \frac{-2+1}{2^{k+1}} = 1 - \frac{1}{2^{k+1}}$$

The first and last parts of this equation say the statement is true for $n = k + 1$.

B.3. Assume the statement is true for $n = k$:

$$\left(1 + \frac{1}{1}\right)\left(1 + \frac{1}{2}\right)\left(1 + \frac{1}{3}\right) \cdots \left(1 + \frac{1}{k}\right) = k + 1$$

Multiply both sides by $1 + (1/(k+1))$:

$$\left(1 + \frac{1}{1}\right)\left(1 + \frac{1}{2}\right)\left(1 + \frac{1}{3}\right) \cdots \left(1 + \frac{1}{k}\right)\left(1 + \frac{1}{k+1}\right) = (k+1)\left(1 + \frac{1}{k+1}\right) = (k+1) + 1$$

Therefore the statement is true for $n = k + 1$.

B.4. **(a)** Assume the statement is true for $n = k$: $k + 2 > k$. Adding 1 to both sides, we have: $k + 2 + 1 > k + 1$, or equivalently, $(k + 1) + 2 > (k + 1)$. Therefore the statement is true for $n = k + 1$. **(c)** Assume the statement is true for $n = k$: $3^k > 3k$. Multiplying both sides by 3 yields: $3 \cdot 3^k > 3 \cdot 3k$, or equivalently, $3^{k+1} > 3 \cdot 3k$. Now since $k \geq 1$ we know that $3k \geq 3$ and hence that $2 \cdot 3k \geq 3$. Therefore, $2 \cdot 3k + 3k \geq 3 + 3k$, or equivalently, $3 \cdot 3k \geq 3k + 3$. Combining this last inequality with the fact that $3^{k+1} > 3 \cdot 3k$ we see that $3^{k+1} > 3k + 3$, or equivalently, $3^{k+1} > 3(k + 1)$. Therefore the statement is true for $n = k + 1$. **(e)** Assume the statement is true for $n = k$: $3k > k + 1$. Adding 3 to both sides yields: $3k + 3 > k + 1 + 3$, or equivalently, $3(k + 1) > (k + 1) + 3$. Since $(k + 1) + 3$ is certainly greater than $(k + 1) + 1$, we conclude that $3(k + 1) > (k + 1) + 1$. Therefore the statement is true for $n = k + 1$.

B.5. Assume that the statement is true for $n = k$:

$$c + (c + d) + (c + 2d) + \cdots + (c + (k-1)d) = \frac{k(2c + (k-1)d)}{2}$$

Adding $c + kd$ to both sides, we have:

$$c + (c + d) + (c + 2d) + \cdots + (c + (k-1)d) + (c + kd) = \frac{k(2c + (k-1)d)}{2} + c + kd$$

$$= \frac{k(2c + (k-1)d) + 2(c + kd)}{2} = \frac{2ck + k(k-1)d + 2c + 2kd}{2}$$

$$= \frac{2ck + 2c + kd(k-1) + 2kd}{2} = \frac{(k+1)2c + kd(k-1+2)}{2}$$

$$= \frac{(k+1)2c + kd(k+1)}{2} = \frac{(k+1)(2c + kd)}{2}$$

$$= \frac{(k+1)(2c + ((k+1)-1)d)}{2}$$

Therefore the statement is true for $n = k + 1$.

B.7. **(a)** Assume the statement is true for $n = k$; then 3 is a factor of $2^{2k+1} + 1$; that is, $2^{2k+1} + 1 = 3M$ for some integer M. Thus $2^{2k+1} = 3M - 1$. Now

$$2^{2(k+1)+1} = 2^{2k+2+1} = 2^{2+2k+1} = 2^2 \cdot 2^{2k+1} = 4(3M - 1) = 12M - 4 = 3(4M) - 3 - 1$$
$$= 3(4M - 1) - 1$$

From the first and last terms of this equation, we see that $2^{2(k+1)+1} + 1 = 3(4M - 1)$. Hence 3 is a factor of $2^{2(k+1)+1} + 1$. Therefore the statement is true for $n = k + 1$.   **(c)** Assume the statement is true for $n = k$: 64 is a factor of $3^{2k+2} - 8k - 9$. Then $3^{2k+2} - 8k - 9 = 64N$ for some integer N, so that $3^{2k+2} = 8k + 9 + 64N$. Now

$$3^{2(k+1)+2} = 3^{2k+2+2} = 3^{2+(2k+2)} = 3^2 \cdot 3^{2k+2} = 9(8k + 9 + 64N)$$

Consequently,

$$3^{2(k+1)+2} - 8(k+1) - 9 = 3^{2(k+1)+2} - 8k - 8 - 9 = 3^{2(k+1)+2} - 8k - 17$$
$$= [9(8k + 9 + 64N)] - 8k - 17$$
$$= 72k + 81 + 9 \cdot 64N - 8k - 17 = 64k + 64 + 9 \cdot 64N$$
$$= 64(k + 1 + 9N)$$

From the first and last parts of this equation, we see that 64 is a factor of $3^{2(k+1)+2} - 8(k+1) - 9$. Therefore the statement is true for $n = k + 1$.   **B.9.** First observe that for every $n \geq 1$ we have $x_{n+1} = \sqrt{2 + x_n}$. Assume the statement is true for $n = k$, that is, assume $x_k < 2$. Then $(x_{k+1})^2 = (\sqrt{2 + x_k})^2 = 2 + x_k < 2 + 2 = 4$. Since $(x_{k+1})^2 < 4$ we must have $x_{k+1} < 2$. Therefore, the statement is true for $n = k + 1$.
B.10. **(a)** false; counterexample: $n = 9$ **(c)** true; proof: since $(1 + 1)^2 > 1^2 + 1$, the statement is true for $n = 1$. Assume the statement is true for $n = k$: $(k + 1)^2 > k^2 + 1$. Then $[(k+1) + 1]^2 = (k+1)^2 + 2(k+1) + 2 > k^2 + 1 + 2(k+1) + 1 = k^2 + 2k + 2 + 2 > k^2 + 2k + 2 = k^2 + 2k + 1 + 1 = (k+1)^2 + 1$. The first and last terms of this inequality say that the statement is true for $n = k + 1$. Therefore, by induction, the statement is true for every positive integer n.   **(e)** false; counterexample: $n = 3$ **B.11.** **(a)** The statement is true for $n = q$ by property (i). Consequently, by property (ii) (with q in place of k), the statement must also be true for $n = q + 1$. By property (ii) again (with $q + 1$ in place of k), the statement must also be true for $n = (q + 1) + 1 = q + 2$. Repeated use of property (ii) shows that the statement is true for $n = (q + 2) + 1 = q + 3$, for $n = (q + 3) + 1 = q + 4$, for $n = q + 5$, and so on.   **(c)** Since $2 \cdot 5 - 4 > 5$, the statement is true for $n = 5$. Assume the statement is true for $n = k$ (with $k \geq 5$): $2k - 4 > k$. Adding 2 to both sides shows

that $2k - 4 + 2 > k + 2$, or equivalently, $2(k + 1) - 4 > k + 2$. Since $k + 2 > k + 1$, we see that $2(k + 1) - 4 > k + 1$. So the statement is true for $n = k + 1$. Therefore, by the Extended Principle of Mathematical Induction, the statement is true for all $n \geq 5$. **(e)** Since $2^2 > 2$, the statement is true for $n = 2$. Assume that $k \geq 2$ and that the statement is true for $n = k$: $k^2 > k$. Then $(k + 1)^2 = k^2 + 2k + 1 > k^2 + 1 > k + 1$. The first and last terms of this inequality show that the statement is true for $n = k + 1$. Therefore, by induction, the statement is true for all $n \geq 2$.   **(g)** Since $3^4 = 81$ and $2^4 + 10 \cdot 4 = 16 + 40 = 56$, we see that $3^4 > 2^4 + 10 \cdot 4$. So the statement is true for $n = 4$. Assume that $k \geq 4$ and that the statement is true for $n = k$: $3^k > 2^k + 10k$. Multiplying both sides by 3 yields: $3 \cdot 3^k > 3(2^k + 10k)$,   or   equivalently, $3^{k+1} > 3 \cdot 2^k + 30k$.   But $3 \cdot 2^k + 30k > 2 \cdot 2^k + 30k = 2^{k+1} + 30k$. Therefore $3^{k+1} > 2^{k+1} + 30k$. Now we shall show that $30k > 10(k + 1)$. Since $k \geq 4$, we have $20k \geq 20 \cdot 4$, so that $20k \geq 80 > 10$. Adding $10k$ to both sides of $20k > 10$ yields: $30k > 10k + 10$, or equivalently, $30k > 10(k + 1)$. Consequently, $3^{k+1} > 2^{k+1} + 30k > 2^{k+1} + 10(k + 1)$. The first and last terms of this inequality show that the statement is true for $n = k + 1$. Therefore the statement is true for all $n \geq 4$ by induction.   **C.2.**   After 1 time period, the total amount $T(1)$ is the sum of the amount invested (namely, P dollars) and the interest earned on this amount in one time period (namely, rP); that is, $T(1) = P + rP = P(1 + r) = P(1 + r)^1$. So the statement $T(x) = P(1 + r)^x$ is true for $x = 1$. Assume that this statement is true for $x = k$. This means that the total amount after k time periods is $T(k) = P(1 + r)^k$. At the end of the next time period [the $(k + 1)$st] the new total $T(k + 1)$ will be the sum of the amount at the beginning of the period [namely, $T(k)$] and the interest earned by this amount in one time period [namely, $rT(k)$]; that is,

$$T(k + 1) = T(k) + rT(k) = P(1 + r)^k + rP(1 + r)^k$$
$$= (P + rP)(1 + r)^k = P(1 + r)(1 + r)^k$$
$$= P(1 + r)^{k+1}$$

The first and last terms of this equation show that the statement is true for $x = k + 1$. Therefore, by induction, it is true for every positive integer x.   **C.4.**   We shall prove that this statement is true for every positive integer n: "$\binom{n}{r}$ is an integer for every r with $0 \leq r \leq n$."

For $n = 1$, this statement reads "$\binom{1}{r}$ is an integer for every r with $0 \leq r \leq 1$"; in other words,

for $n = 1$, the statement says $\binom{1}{0}$ and $\binom{1}{1}$ are integers. Since $\binom{1}{0} = 1$ and $\binom{1}{1} = 1$, the statement is true for $n = 1$. Now assume that the statement is true for $n = k$, that is, assume that

(°) $\binom{k}{r}$ is an integer for every r with $0 \leq r \leq k$

We must show that $\binom{k + 1}{r}$ is an integer for every r with $0 \leq r \leq k + 1$. We know that $\binom{k + 1}{0} = 1$ and $\binom{k + 1}{k + 1} = 1$ (see p. 533). So we need only look at $\binom{k + 1}{r}$ when $1 \leq r \leq k$. It will be convenient to change notation slightly. Let $s = r - 1$, so that $s + 1 = r$. Then the numbers $\binom{k + 1}{r}$ with $r = 1, 2, \ldots, k$ are the same as the numbers $\binom{k + 1}{s + 1}$ with $s = 0, 1$, $2, \ldots, k - 1$. We must show that each of them is an integer. But by Exercise C.1.(c) on page

539 (with s in place of r and k in place of n) we know that:

$$\binom{k+1}{s+1} = \binom{k}{s+1} + \binom{k}{s} \qquad \text{for each } s \text{ with } s = 0, 1, 2, \ldots, k-1$$

But $\binom{k}{s+1} = \binom{k}{r}$ and $\binom{k}{s} = \binom{k}{r-1}$, both of which are integers by the induction hypothesis

(°). So their sum, namely, $\binom{k+1}{s+1}$, is also an integer. Thus, in the original notation, we have

shown that all of the numbers $\binom{k+1}{r}$ with $0 \leq r \leq k+1$ are integers. So the statement is

true for $n = k+1$ and hence, by induction, for every positive integer n.

ALGEBRA REVIEW   PAGES 572–580

A.1. (a) 7   (c) 26   (e) 70   (g) -50 **A.2.** (a) 0   (c) -1
A.3. (a) 5.7   (c) $-.7$ (e) $\frac{1}{10}$ **A.4.** (a) 36   (c) 73   (e) 3
A.5. (a) $-\frac{125}{64}$ (c) $-\frac{25}{4}$ (e) $\frac{29}{49}$ **A.6.** (a) x^8 (c) $24x^7$ (e) $2.6z^2y$
(g) $384w^6$ (i) x^7 **A.7.** (a) x^7 (c) u^{12}/v^2 (e) x^{10}/y^4 **A.8.** (a) $b^2c^2d^6$
(c) $9x^9y^6$ **A.9.** (a) $3x$ (c) $-2a^2b$ (e) $2u^4 + u^3 + 6$
A.10. (a) $6a^2b - 3ab - 6ab^2 + 3ac + abc$ (c) $-3x^3 + 15x + 8$ (e) $-x - 5xy$
A.11. (a) $2x^3 + 4x$ (c) $x^3y^2 - 6x^3y^3$ (e) $12z^4 + 30z^3$ (g) $12a^2b - 18ab^2 +$
$6a^3b^2$ **A.12.** (a) $4x^2 - 25$ (c) $9x^2 - y^2$ (e) $4x^4 - 81y^2$ (g) $16x^6 - 25y^4$
A.13. (a) $x^2 - x - 2$ (c) $y^2 - 9$ (e) $w^2 - 8w + 16$ (g) $2x^2 + 2x - 12$
(i) $xy + 2x + y + 2$ **A.14.** (a) $2c^3 - 7c^2 + 7c - 2$ (c) $2x^3 + 3x^2y - xy^2 + 2y^3$
(e) $r^2 + rs + ar + as + a + r$ (g) $-15w^3 + 2w^2 + 9w - 18$ **A.15.** (a) $24x^3 -$
$4x^2 - 4x$ (c) $x^3 - 6x^2 + 11x - 6$ (e) $6x^3 - 5x^2y - 17xy^2 + 6y^3$
A.16. (a) $(x + 2)(x - 2)$ (c) $(3y + 5)(3y - 5)$ (e) $(9x + 2)^2$
(g) $(\sqrt{5} + x)(\sqrt{5} - x)$ (i) $(2z + 7)^2$ (k) $(x^2 + y^2)(x + y)(x - y)$
A.17. (a) $(x + 3)(x - 2)$ (c) $(z + 3)(z + 1)$ (e) $(y + 9)(y - 4)$ (g) $(x - 3)^2$
(i) $(x + 2)(x + 5)$ (k) $(x + 2)(x + 9)$ **A.18.** (a) $(x + 1)(3x + 1)$
(c) $(z + 4)(2z + 3)$ (e) $9x(x - 8)$ (g) $2(5x + 1)(x - 1)$ (i) $(2u + 3)(4u - 3)$
(k) $(2x + 5y)^2$ **A.19.** (a) $(x - 5)(x^2 + 5x + 25)$ (c) $(x + 2)^3$
(e) $(2 + x)(4 - 2x + x^2)$ (g) $(-x + 5)^3$ (i) $(x + 1)(x^2 - x + 1)$
(k) $(2x - y)(4x^2 + 2xy + y^2)$ **A.20.** (a) $(x - 2)(x + 2)(x^2 + 2x + 4)(x^2 - 2x + 4)$
(c) $(y^2 + 2)(y^2 + 5)$ (e) $(9 + y^2)(3 + y)(3 - y)$
(g) $(z - 1)(z^2 + z + 1)(z + 1)(z^2 - z + 1)$ (i) $(x^2 + 3y)(x^2 - y)$
A.21. (a) $(x - y)(x + z)$ (c) $(a^2 - b)(a + 2b)$ (e) $(x + 4)(x + 2\sqrt{2})(x - 2\sqrt{2})$
A.22. (a) $x^2 + 4x + 4 = (x + 2)^2$ (c) $z^2 + 3z + \frac{9}{4} = (z + \frac{3}{2})^2$ (e) $x^2 +$
$12x + 36 = (x + 6)^2$ (g) $2(x^2 + 2x + 1) = 2(x + 1)^2$ (i) $2(z^2 + 7z + \frac{49}{4}) = 2(z + \frac{7}{2})^2$
A.23. (a) $\frac{9}{7}$ (c) $\frac{195}{8}$ (e) $(x - 2)/(x + 1)$ (g) $(a + b)/(a^2 + ab + b^2)$
(i) $1/x$ **A.24.** (a) $\frac{29}{35}$ (c) $\frac{121}{42}$ (e) $(ce + 3cd)/de$ (g) $(b^2 - c^2)/bc$
A.25. (a) $-1/(x(x + 1))$ (c) $(x + 3)/(x + 4)^2$ (e) $(2x - 4)/(x(3x - 4))$
(g) $\dfrac{x^2 - xy + y^2 + x + y}{x^3 + y^3}$ (i) $\dfrac{-6x^5 - 38x^4 - 84x^3 - 71x^2 - 14x + 1}{4x(x + 1)^3(x + 2)^3}$
A.26. (a) 2   (c) $2/3c$ (e) $3y/x^2$ **A.27.** (a) $12x/(x - 3)$ (c) $5y^2/(3(y + 5))$
(e) $(u + 1)/u$ (g) $\dfrac{(u + v)(4u - 3v)}{(2u - v)(2u - 3v)}$ **A.28.** (a) $\frac{35}{24}$ (c) u^3/vw^2

(e) $(x + 3)/2x$ (g) $x^2y^2/((x + y)(x + 2y))$ (i) $cd(c + d)/(c - d)$ **A.29.** (a) 12
(c) $\frac{8}{7}$ (e) $y/2$ (g) $2ab^4$ (i) $u^2\sqrt{u}$ (k) $4xy$ (m) $u + v$
(o) $3/(2(\sqrt[3]{6c^2}))$ **A.30.** (a) $2\sqrt{5}/5$ (c) $\sqrt{70}/10$ (e) $\sqrt{x}/x$
(g) $(r + 2\sqrt{rs} + s)/(r - s)$ (i) $(u^2 - v^2)(\sqrt{u + v} + \sqrt{u - v})/2v$ **A.31.** (a) $x = 8$
(c) $z = -\frac{5}{6}$ (e) $x = \frac{15}{11}$ **A.32.** (a) $x = 26$
(c) $x = \frac{5}{6}$ (e) no solution   (g) $x = 1$ (i) $x = -1$ **A.33.** (a) $b =$
$(2A/h) - b'$ (c) $h = (A/2\pi r) - r$ (e) $x = (3b^2 - 12ab - a)/(5a - b - 2)$
A.34. (a) $x = 8$ or -8 (c) $z = 2\sqrt{14}$ or $-2\sqrt{14}$ (e) $y =$
$-2 + \sqrt{5}$ or $-2 - \sqrt{5}$ (g) $y = 3$ **B.1.** (a) 3   (c) 1   (e) 6   (g) 1
B.2. (a) $3ax^2 + (3b + 2a)x + 2ab$ (c) $abx^2 + (a^2 + b^2)x + ab$ (e) $x^3 -$
$(a + b + c)x^2 + (ab + ac + bc)x - abc$ **B.3.** (a) 399   (c) 1599   (e) 2491
(g) 3536   **B.4.** (a) 3^{r+t+1} (c) $6y^{2k+2}$ (e) $y^{r+s} + y^s - 4y^r - 4$
B.5. (a) $(2x - y)(x + 3y + 3)$ (c) $(x - 3y)(x^2 + 3xy + 9y^2 + 1)$
B.6. (a) $(x - \frac{1}{8})(x + \frac{1}{8})$ (c) $(y - \frac{5}{6})(y + \frac{1}{6})$ (e) $(z + \frac{7}{4})(z + \frac{5}{4})$
B.7. (a) 1.23   (c) 7.77   **B.8.** (a) $3(y + 2) = 3y + 6$ (c) $(x + y)^2 =$
$x^2 + 2xy + y^2$ (e) $(y^5)^6 = y^{30}$ (g) $(\frac{1}{3}x)(\frac{1}{3}y) = \frac{1}{9}xy$ (i) $\sqrt{r + s} = \sqrt{r + s}$
(k) $x^2/(x^2 + x^6) = x^2/(x^2(1 + x^4)) = 1/(1 + x^4)$ (m) $(a - b)^2 = a^2 - 2ab + b^2$
B.9. (a) $x^2 - 5x - 6 = (x - 6)(x + 1)$ (c) $(-2)^5 = -32 = -2^5$ (e) $3 + (4/x) =$
$(3x + 4)/x$ (g) $a^3 + b^3 = (a + b)(a^2 - ab + b^2)$ (i) $\sqrt{2x + 1} +$
$\sqrt{3x + 2} = \sqrt{2x + 1} + \sqrt{3x + 2}$; no further simplification is possible   (k) $(u/v) +$
$(v/u) = (u^2 + v^2)/uv$ **C.1.** (a) If x is the chosen number, then adding one and squaring
the result gives $(x + 1)^2$. Subtracting one from the original number x and squaring the result
gives $(x - 1)^2$. Subtracting the second of these squares from the first yields $(x + 1)^2 - (x - 1)^2$
$= (x^2 + 2x + 1) - (x^2 - 2x + 1) = 4x$. Dividing by the original number x now gives
$4x/4 = 4$. So the answer is always 4, no matter what number x is chosen.

Index